Evolutionary
Medicine

Evolutionary
Medicine

Edited by

WENDA R. TREVATHAN

E. O. SMITH

JAMES J. MCKENNA

New York   Oxford

Oxford University Press

1999

Oxford University Press

Oxford   New York
Athens   Auckland   Bangkok   Bogotá   Buenos Aires   Calcutta
Cape Town   Chennai   Dar es Salaam   Delhi   Florence   Hong Kong   Istanbul
Karachi   Kuala Lumpur   Madrid   Melbourne   Mexico City   Mumbai
Nairobi   Paris   São Paulo   Singapore   Taipei   Tokyo   Toronto   Warsaw

and associated companies in
Berlin   Ibadan

Published by Oxford University Press, Inc.
198 Madison Avenue, New York, New York 10016

Oxford is a registered trademark of Oxford University Press

Library of Congress Cataloging-in-Publication Data
Evolutionary medicine / edited by Wenda R. Trevathan, Euclid O. Smith,
and James J. McKenna.
p.   cm.
Includes bibliographical references and index.
ISBN 0-19-510355-6; ISBN 0-19-510356-4 (pbk.)
1. Diseases—Causes and theories of causation.   2. Human
evolution.   3. Medicine—Philosophy.   I. Trevathan, Wenda.
II. Smith, Euclid O.   III. McKenna, James J. (James Joseph), 1948–   .
RB152.E96   1999
610—dc21   98-17323

9 8 7 6 5 4 3   2   1

Printed in the United States of America
on acid-free paper

H.L. Menken once said that "A teacher is one, who in his youth, admired teachers."

Indeed, if you are lucky, somewhere at sometime, you encounter those incredible teachers who inspire as they teach, and who leave you with an irrepressible desire to want to learn more. Such teachers give you an added reason to want to succeed, and they make it possible for you to do so.

In Jim McKenna's life those Professors were Larry Leach (San Diego State), Phyllis Dolhinow (University of California, Berkeley), and Sarah Mosko (UC Irvine School of Medicine).

Wenda Trevathan acknowledges Jack Kelso (University of Colorado), who encouraged his students to think outside the box.

Neal Smith remembers Lena M. Crain who made him understand that big ideas are important, but attention to detail is also essential. Richard M. Prenshaw introduced him to the importance of music in the tapestry of the human experience, and Irwin S. Bernstein (University of Georgia) patiently taught him about the rigor of scientific inquiry.

To these teachers, scholars and researchers, we owe a tremendous intellectual and personal debt. It is to them that we dedicate this book.

FOREWORD

Historical Overview

MICHAEL H. DAY

It is one of the minor mysteries of medicine that the publication by Darwin (1859) that identified evolution as a continuing biological process and that accounted for the diversity and successive change in living organisms was not immediately taken up by the physicians of the day. Their daily work then, as now, related to the whole animal biology of one species: one species that was better known in terms of its anatomy, physiology, and behaviour than any other species known to zoology at the time and indeed since. This failure to recognise the significance of the idea of evolution and the explanatory power of its mechanism, natural selection, may relate to the opposition of the established church, which focused all of its disbelief upon the perceived contradiction the theory posed to the biblical explanation of the universe as set out in the Book of Genesis. Be that as it may, it is still a matter of wonder that it has, until recently, been rare to hear an evolutionary reason put forward as an aetiological factor in any but a few disorders, and when they have been cited, it has often disclosed an ignorance of adaptation as a process—for example, backache is caused by *incomplete* evolution to upright posture.

Perhaps the most notable exception to the position of physicians and surgeons with regard to evolution was John Hunter (1728–1793), the father of scientific surgery, who had no doubt about the mutability of species. He wrote:

> To attempt to trace any natural production to its origin, or its first production, is ridiculous; for it goes back to that period, if ever such existed, of which we can form no idea. viz. the beginning of time. But, I think we have no reason to suppose that there was a period in time in which every species of natural production was the same; there being no variety in any species; but the variations taking place on the surface of the earth. ... (Owen, 1861)

It is clear that Hunter accepted the principle of the change of species through time. In other writings he accepted, in Darwin's words, the "theory of descent with slow and slight successive modifications," that fossils are the remains of extinct species, and that mankind and the monkeys bore a physical relationship, and that within species there are a great number of varieties. Even further, Hunter's evolutionary view is exemplified by his choice of the background of a portrait of him by Sir Joshua Reynolds that shows a folio of drawings that illustrate an evolutionary series of skulls and hands (Qvist, 1981).

Since Hunter's time there has been no shortage of medical men who have, through the study of comparative anatomy, accepted and contributed to an evolutionary view of the natural world. Perhaps much of the reason for this, at least in Europe, is that dissection of the human cadaver was almost entirely confined to those in medical practice and that opportunities for comparative study were available in the great schools of medicine in London and on the continent. This tradition is one that continued into the modern era with names such as Frederick Wood-Jones, Sir Arthur Keith, Raymond Dart, Sir Wilfrid LeGros Clark, Lord Zuckerman, John Napier, Philip Tobias, and many others. Interestingly, the contribution of these scholars has usually depended on their special knowledge of primates and human osteology coupled with soft tissue anatomy in making the interpretations of fossil materials with which they were presented, not on their training as doctors, other than an acceptance of the scientific method upon which modern orthodox medicine is based. It seems that it is in this sense that the application of Darwinian principles to the practice of medicine has slipped through the intellectual net of medical education, and it has been left to the evolutionary biologists to point out the need to take natural selection into account in an increasing number of fields of medical practice.

It is well known that Darwin formulated the theory of evolution without the benefit of any knowledge of genetics because his work preceded that of the founders of that science, but he was influenced by the writing of Malthus (1826), the country parson who feared that population increase would outstrip the means of the world to produce food for the people. While the worst fears of Malthus have not yet been realised, a link between genetics and population studies has been forged that is a powerful factor in our understanding of the way in which evolution works. It is an equally powerful factor in our understanding in our appreciation of the effects of evolutionary theory on the etiology and natural history of disease in the form of infection and epidemiology, resistance to infection, responses to allergens, cellular dynamics, and the development of tumours. A number of chapters in this volume address these problems directly.

It is equally well known that Darwin (1872) was interested in sexual selection and the expression of the emotions in humans and animals. Once again, the sciences of normal behaviour such as psychology and ethology as well as the medical science of abnormal behaviour, psychiatry, have been remarkably slow to take into consideration the consequences of the application of the idea of natural selection in these fields. An understanding of the origins and meaning of innate or genetically determined behaviour, as well as that of acquired or learned behaviour would seem to be central to any attempt to modify or "correct" what is diagnosed as an abnormality. The reasons for the range and the limits of what may, or may not, be acceptable should, at the very least, be understood before parents, doctors, or society outlaw behaviour for their own reasons or convenience. Several chapters address problems of this kind from fretful infants to substance abuse.

Finally, we all know that longevity brings its own problems, problems that most of us would rather face than the alternative. But it is clear that medical science has, over the centuries, increased potential life span and even more dramatically has increased the proportion of the population that reaches that potential, at least in affluent societies. This has resulted in changes of the proportion of the elderly in a given population and in turn a rise in the incidence of chronic and degenerative diseases and a rise in the number of post-menopausal women. (Menopause itself is a mystery, is it a "good" thing, a "bad" thing or a chance consequence of an exceptional biological longevity?) All of these topics—life span, chronic disease, reproductive death—are intensely bound with the Darwinian view of biology and the ways in which environmental change, nutritional change, and reproductive patterns are altering our perceptions of the normal and abnormal and of health and disease. The doctors of the future will find it necessary to take account of the biological imperatives that Darwin revealed if they are to have any real understanding of their trade as it will be practiced in the new millenium.

References

Darwin, C. (1859) *On the Origin of Species by Means of Natural Selection, or the Preservation of Favoured Races in the Struggle for Life*. London: Murray.

Owen, R. (1861) *Essays and Observations on Natural History, Anatomy, Physiology, Psychology and Geology by John Hunter*. Vol. 1, p. 4. Ed. R. Owen. London: John Van Voorst.

Qvist, George (1981) *John Hunter, 1728–1793*. London: Heinemann.

Malthus, T. R. (1826) *An Essay on the Principle of Population: or a View of Its Past and Present Effects on Human Happiness, with an Inquiry into Our Prospects respecting the Future Removal or Mitigation of the Evils Which it Occasions*, 6th Ed. London: Murray.

Darwin, C. (1872) *On the Expression of Emotions in Man and Animals*. London: Murray.

CONTENTS

CONTRIBUTORS

Robert Anderson, D.O., Ph.D.
Department of Anthropology and
  Sociology
Mills College
Oakland, CA 94613

George J. Armelagos, Ph.D.
Department of Anthropology
Emory University
Atlanta, GA 30322

Kathleen C. Barnes, Ph.D.
Division of Clinical Immunology
Department of Medicine
The Johns Hopkins University
  School of Medicine
Baltimore, MD 21224

Ronald G. Barr, MA, MDCM,
  FRCP (C)
Departments of Pediatrics and
  Psychiatry
McGill University and Montreal
  Children's Hospital
Montréal, Québec H3H 1P3 Canada

John Brett, Ph.D.
Department of Anthropology
University of Colorado at Denver
Denver, CO 80217-3364

Douglas E. Crews, Ph.D.
Department of Anthropology
The Ohio State University
Columbus, OH 43210

Hal J. Daniel III, Ph.D.
Departments of Biology and
  Anthropology
East Carolina University
Greenville, NC 27834

Michael H. Day, M.B.B.S., D.Sc.,
  Ph.D., M.R.C.S., L.R.C.P., F.C.S.P.
  (Hon)
Department of Palaeontology
The Natural History Museum
Cromwell Road
London, SW7 5BD, England

S. Boyd Eaton, M.D.
Departments of Anthropology
  and Radiology
Emory University
Atlanta, GA 30322

S. Boyd Eaton III
School of Education
Marshall University
Huntington, WV

Mark T. Erickson, M.D.
Alaska Psychiatric Institute
Anchorage, Alaska 99508-4677

Paul W. Ewald
Department of Biology
Amherst College
Amherst, MA 01002-5000

Linda M. Gerber, Ph.D.
Department of Public Health
Cornell University Medical College
New York, NY 10021

A. Magdalena Hurtado, Ph.D.
Department of Anthropology
University of New Mexico
Albuquerque, NM 87131

I. Arenas de Hurtado, M.D., Ph.D.
Centro de Microbiologia
Instituto Venezolano de
    Investigaciones Científicas
Caracas, Venezuela

Kim Hill, Ph.D.
Department of Anthropology
University of New Mexico
Albuquerque, NM 87131

Melvin J. Konner, M.D., Ph.D.
Departments of Anthropology
    and Psychiatry
Emory University
Atlanta, GA 30322

Lynnette E. Leidy, Ph.D.
Department of Anthropology
University of Massachusetts at
    Amherst
Amherst, MA 01003-4805

James McKenna, Ph.D.
Department of Anthropology
University of Notre Dame
Notre Dame, IN 46556

Steven C. Morreale
Department of Anthropology
Emory University
Atlanta, GA 30322

Sarah Mosko, Ph.D.
Department of Neurology
School of Medicine
University of California, Irvine
Orange, CA 92668

Randolph Nesse, M.D.
Institute for Social Research
University of Michigan
Ann Arbor, MI 48106

Susan Niermeyer, M.D.
Department of Pediatrics
Section of Neonatology
University of Colorado School of
    Medicine
Denver, CO 80262

Chris Richard, Ph.D.
Sleep Disorders Laboratory
School of Medicine
University of California, Irvine
Orange, CA 92668

Robert Sapien, M.D.
Department of Emergency Medicine
University of New Mexico Hospital
Albuquerque, NM 87131

E. O. Smith, Ph.D.
Department of Anthropology
Emory University
Atlanta, GA 30322

Wenda R. Trevathan, Ph.D.
Department of Sociology and
    Anthropology
New Mexico State University
Las Cruces, NM 88003

Carol M. Worthman, Ph.D.
Department of Anthropology
Emory University
Atlanta, GA 30322

Evolutionary
Medicine

INTRODUCTION

Evolutionary medicine takes the view that many contemporary social, psychological, and physical ills are related to incompatibility between the lifestyles and environments in which humans currently live and the conditions under which human biology evolved. Unfortunately, much of modern medical practice demonstrates a misunderstanding of the evolution of physical responses to stresses that were faced by our ancestors. The contributions in this book explore these topics, proposing ways in which questions about disease and disorders can be reframed to consider the evolutionary perspective and suggesting new approaches to treatment.

First and foremost, evolutionary medicine accords human beings their evolutionary past; but it also accords evolutionary histories and strategies to viruses and bacteria to which humans play host. This means that concepts of natural selection, life-history strategies, ecological effects of development, and distribution of disease and disorders are pertinent to the ideas discussed in this volume. The natural history of our species is a fundamental beginning point for new analyses of human disorders and degenerative disease, as well as the effects of culture on human environments. This sets the stage for exploring the notion of mismatches or discordances between more evolutionarily stable or "expected" human physical, social, and psychological environments and the "actual" environments encountered by modern humans.

Our biology, and many of our behaviors, are the result of millions of years of evolutionary history, beginning with the earliest forms of life. Characteristics that identify us as mammals (e.g., viviparity, mammary glands, homoiothermy, hair) may have appeared with the earliest mammals 225 million years ago. Characteristics that place us in the taxonomic order Primates (e.g., prehensile digits, structure of the middle ear, arboreal adaptations) trace their origins to the earliest primates 65 million years ago. Selection for visual acuity, grasping hands and feet, greater encephalization (larger brains relative to body size), and long periods of maturation may have begun 40 million years ago or more with the origin of the "higher primates" (monkeys, apes, and humans). Characteristics of our taxonomic family, Hominidae, began to be selected in our ancestors approximately 5 million years ago.

The major physical characteristic that identifies us as members of the family Hominidae ("hominids"), and one that is easily recognized in the fossil record, is bipedalism, the characteristic behavior of walking on two legs. Natural selection for this unique mode of locomotion began about 5 million years ago. It has been blamed for a number of disorders encountered in modern medical research and clinical practice. These include orthopedic disorders of the back, hip, spine, neck, and foot, complications of pregnancy and childbirth, and even otitis media, one of the most common diseases of childhood in Western nations.

Other characteristics of modern humans and our hominid ancestors that affect health include giving birth to relatively helpless young, even longer periods of maturation, and longer absolute life span than seen in most primates, and the addition of an apparently unique life-cycle phase, a long post-reproductive life in females. These topics are covered in chapters on women's cancers, bone loss associated with aging, infant sleep, sudden infant death syndrome, neonatal jaundice, high blood pressure, and obesity.

The earliest genus in the family Hominidae was *Australopithecus*,[1] the several species of which have been identified in geological deposits dating from 4.4 million years ago to 1 million years ago (Conroy, 1997). These were small-brained, bipedal creatures who, so far as is known, lived only on the continent of Africa. Between 2 and 2.5 million years ago, our genus, *Homo*, appeared in Africa. Further encephalization, greater dependency on material culture ("tools"), and expansion to Eurasia characterize the first 1.5 million years of its existence. Approximately 200,000 years ago, we find evidence of our own species, *Homo sapiens*. This is the large-brained, language-using species whose ever-expanding material culture enabled occupation, and even domination, of every known land area and habitat on earth. Although biological changes have been extensive since the origin of *Homo sapiens*, perhaps the most significant changes in terms of contemporary health problems have occurred in the environments in which humans live, beginning especially with the origins of agriculture, approximately 10,000 years ago.

Taking a conservative estimate of the origin of humanness as 200,000 years ago, we can safely say that more than 95% of our biology, and presumably many of our behaviors, were shaped during the time period in which our ancestors lived as gatherers of wild food resources. This lifestyle included low-fat diets, low technology, adequate but not overly abundant food resources, low birth rate, near-exclusive breastfeeding for the first two years, long (4 years) birth intervals, low total population, high mobility, and living in small (25–50 people), kin-based social groups. Natural selection thus favored characteristics and behaviors that were advantageous and suitable for that kind of lifestyle. The human environments changed with the origin of agriculture, however, and many of those behaviors are no longer advantageous. These changes included sedentism, a narrowing of the food base, increase in fat intake, larger communities, infectious diseases, decreased birth interval, increased birth rate, increased total population, and changes in interpersonal relations (e.g., male–female and intercommunity dynamics).

The changing human environments had profound effects on health. A better understanding of many modern health problems will emerge when we consider that most of human evolution took place when our ancestors were gatherer-hunters. This is often referred to as the "environment of adaptation" or the "environment of evolutionary adaptedness." Biologically, and perhaps behaviorally, we are gatherer-hunters, but clearly most of us do not live that kind of lifestyle. In fact, many of the characteristics and behaviors that evolved in that 95% of our evolutionary history had adaptive significance then, but may even be maladaptive today.

An example is our ability to quickly and easily store fat. In the past, that ability would have been favorably selected because overabundance of food was rare. People who could consume more than their daily needs during periods of abundance and could store calories as fat would have more resources on which to draw during the inevitable lean times. Obesity was rarely a problem because fat reserves were being drawn upon frequently. Today, however, many of us still have the ability to eat more than our daily needs and to store the excess as fat, but we rarely encounter environmentally imposed shortages, so those reserves are not used often enough. The result is obesity, diabetes, and other problems associated with high fat intake and excess body fat. What was selectively advantageous in the environment of adaptation, is now, for most people in industrialized nations at least, a maladaptive or even pathological condition.

As each chapter in this book will show, the methods of study used in evolutionary medicine and the interpretations of disease phenomena are eclectic and cross-disciplinary. The methods employed bridge the social, biological, and clinical sciences in ways few fields can. An organizing and explicit theme in evolutionary medicine, and hence, in this book, is the assumption that many contemporary people live in extraordinarily different social, physical, and psychological environments from the body's evolutionary "expected," or evolutionary stable, environment. The present environment places enormous strains on physiological and psychological systems designed by natural selection to develop under vastly different circumstances (see, for example, chapters on women's cancers, emotional disorders such as anxieties and phobias, substance abuse, and infant sleep). Evolutionary medicine considers seriously the implications and nature of these strains and the physiological system's response to them (illustrated in chapters on sudden infant death syndrome, high blood pressure, obesity, vitamin and mineral deficiencies, sensory disorders, and reproductive disorders including cancers).

Chapters in this volume also make clear how cultural ideologies, values, and the socialization experiences of medical researchers often prevent disease and human disorders from being conceptualized in evolutionary terms, even in the face of much relevant data that make it necessary and logical. A powerful argument to support this idea has occurred in the field of neonatology, specifically research on the value of skin-to-skin contact for premature babies (known as "kangaroo baby care"), which has virtually been ignored by the medical community. No fewer than 20 research papers published in respectable medical journals have demonstrated that premature infants benefit (i.e., gain weight faster and are released earlier from hospitals) when held by their caregivers (mothers and fathers). But the work has had little impact. It is clear that cultural context often influences what is "heard" by the medical community. This volume provides a new context for other perspectives to be heard and acted on.

In summary, evolutionary medicine encourages biomedical research and clinical practice to consider not only the immediate context of the individuals being treated and their own life histories but also the evolutionary history of the human species and of the organisms that afflict it (see figure I.1). The rest of this book provides evidence that this perspective is crucial for understanding and perhaps treating many contemporary medical problems.

BIOMEDICAL CONCERNS
exposure to a disease-causing organism
virulence of the pathogen
age, sex, state of the patient
previous health history of the patient
current state of the patient's health
treatments available for the illness or disorder

SOCIOMEDICAL CONCERNS
socioeconomic circumstances of the patient
medical resources available
medical beliefs held by the patient

EVOLUTIONARY MEDICINE CONCERNS
co-evolutionary history of pathogen and human hosts
evolutionary history of the human species

Figure I.1

Note

1. A recently described genus, Ardipithecus, dated at 4.4 million years old, may claim the distinction of being the earliest in the family Hominidae (Conroy, 1997).

Reference

Conroy, G. C. (1997) *Reconstructing Human Origins: A Modern Synthesis.* New York: W. W. Norton.

1

IS NEONATAL JAUNDICE A DISEASE OR AN ADAPTIVE PROCESS?

JOHN BRETT

SUSAN NIERMEYER

> Most people see bilirubin only if they get a bruise or liver disease. Neonatologists and newly birthed moms see it more often, and to them it is frequently a source of concern and confusion. Concern, because the pigment itself is poisonous and occasionally the herald of deep disease. Confusion, because it is often difficult to decide which jaundiced infants require treatment and because bilirubinology has become a morass of bizarre names and structures where even experts disagree (McDonagh and Lightner 1985: 443).

The Problem

Jaundice of the newborn is one of the most common, complex, well-studied, and misunderstood phenomena in modern pediatric medicine. Despite thousands of studies spanning nearly 50 years, neonatal jaundice, or hyperbilirubinemia, remains one of the most baffling and frustrating aspects of health in newborns.

We suggest that the primary reason for this ongoing confusion is the inappropriate conceptualization of the role of bilirubin in the newborn period. We argue that if placed in a broader, evolutionary perspective, bilirubin can be seen to serve an important adaptive role. Rather than the current view of bilirubin as toxin and jaundice as disease, we present a scenario in which bilirubin and hyperbilirubinemia of the newborn are beneficial components of postnatal adaptation in the majority of babies and indicators of underlying pathology in only a small minority (Brett and Niermeyer 1990; Brett and Niermeyer 1998). Emerging data on bilirubin as a potentially important antioxidant in the newborn (McDonagh 1990; Stocker et al. 1987a) better explain the presence and pattern of hyperbilirubinemia than does a model based on pathology.

In spite of decades of intensive research on hyperbilirubinemia, a clear understanding of its role in the newborn period has yet to emerge. It could be said that the more we know about this condition, the less we understand; more data and more detail fail to clarify the overall picture (e.g., see Gartner et al. 1994; Maisels 1987; Newman and Maisels 1992) We propose that an evolutionary view provides a strong analytical perspective which organizes and makes coherent many of the ambiguous research findings of the last 30–40 years (Brett and Niermeyer 1990). Bilirubin in the healthy newborn ceases to be a confusing and troubling aspect of the newborn period but rather an expected and valued part of the transition to extrauterine life if viewed from the stance of an evolutionary theory of medicine. We stress at this point that our argument addresses the healthy, term newborn without hemolysis.[1] Circumstances of prematurity and certain disease states alter the implications of hyperbilirubinemia. Immaturity of major organ systems compromises preterm infants in bilirubin management, yet the antioxidant properties of the molecule are perhaps even more pertinent than in the term infant. Term infants with hemolytic disease constitute another population that falls outside of this discussion.

The association between bilirubin, hyperbilirubinemia, and neurological damage or death from kernicterus is fundamental to an understanding of the research on bilirubin over the last several decades. The association between elevated levels of bilirubin and kernicterus, the staining and disruption of the basal ganglia of the brain by bilirubin, was first established in relation to hemolytic diseases, principally Rh isoimmunization (Hsia et al. 1952; Mollison and Cutbush 1949; Vaughan et al. 1950). Early researchers noted that as the level of serum bilirubin approached 20 mg/dl, the incidence and severity of neurological sequellae increased markedly.[2] Motor disability and/or hearing loss or death could occur in what might have been an otherwise normal child. Thus, the association of bilirubin with the devastating consequences of kernicterus meant that bilirubin was no longer merely a metabolic waste product but a powerful toxin of clinical importance. Entrance of bilirubin into the brain and subsequent damage are generally thought to be the cause of death or subsequent disability of the infant. The neurologically devastating consequences of kernicterus make bilirubin toxicity highly salient to clinicians because it occurs in children whose developmental trajectory likely would have been normal (Brett and Niermeyer 1998).

Arising from this early research was a wide range of studies concerned with epidemiology, chemistry, physiology, and toxicity of elevated bilirubin levels, not only in hemolytic but also in nonhemolytic jaundice. Hyperbilirubinemia became identified as a significant risk not just in one segment of sick newborns, but in all newborns (Brett and Niermeyer 1998). Jaundice of the newborn was perceived as a disease for which any infant was at risk, and tremendous resources were devoted to testing and/or treatment for this "ailment" (Newman et al. 1990). Research during the 1950s focused primarily on the pathogenesis of hyperbilirubinemia and kernicterus resulting from Rh isoimmunization (Hsia et al. 1952; Mollison and Cutbush 1949). Between 1959 and 1966 the National Collaborative Perinatal Project (NCPP) sought to establish norms for bilirubin in healthy white and black neonates in the United States (Hardy et al. 1979). Much of the research in the 1970s focused on bilirubin-binding determinations and efforts to identify and measure the particular molecular form of bilirubin responsible for brain injury (Broderson 1977, 1978; Cashore et al. 1978). Phototherapy,[3] already in wide use, was demonstrated to provide effective, apparently safe treatment for hyperbilirubinemia (Brown and McDonagh 1980; Tan 1975). In the 1980s research largely focused on the clinical observation that more breast-fed than bottle-fed infants developed hyperbilirubinemia (Kivlahan and James 1984; Maisels and Gifford 1986), and attempts increased to understand the way in which breast-feeding influenced jaundice (De Carvalho et al. 1982, 1985). Noninvasive methods of determining bilirubin level (Hanneman et al. 1982; Hegyi et al. 1982) and techniques for home phototherapy (Eggert et al. 1985; Slater and Brewer 1984) were technological efforts aimed at improving medical management of babies with elevated bilirubin levels. Most current research is attempting to elucidate biochemical mechanisms for measuring bilirubin production (Rogers et al. 1994) and preventing hyperbilirubinemia, primarily through heme oxygenase inhibitors (Kappas et al. 1995). Increasingly, however, researchers are questioning the status of hyperbilirubinemia as a disease state in the healthy newborn (Cashore 1991; McDonagh 1990; Newman and Maisels 1990, 1992; Newman et al. 1990).

Although most of the original research on neonatal jaundice was motivated by the relationship between bilirubin and hemolytic disease, a conjunction of social, cultural, and economic forces shifted the focus of bilirubin research from the relatively small segment of the newborn population affected by hemolytic diseases to include all newborns, most significantly those viewed as perfectly healthy *except* for their yellowness (Brett and Niermeyer 1998).

Evolutionary Bases of Neonatal Jaundice

The argument for an evolutionary basis of physiologic jaundice can be divided into two broad areas: (1) all normal babies will have bilirubin levels above those of the adult because there are physiological mechanisms that produce and maintain these higher levels; and (2) there is a distinct populational and genetic component to bilirubin production, expression, and metabolism. Evidence from phylogenetic relationships with nonhuman primates and populational variation among human and nonhuman primates strongly suggest that bilirubin production and expression are or have been subject to natural selection pressures. As detailed below, bilirubin apparently serves as an important antioxidant in newborns, prior to maturation of primary antioxidant enzyme systems, and thus elevated levels confer a selective advantage.

Common Occurrence of Neonatal Hyperbilirubinemia

All newborns have bilirubin levels well above the adult norm, and the majority (50–65%) develop visible jaundice within the first postnatal week (Gartner and Lee 1992; Maisels 1987). High bilirubin values arise from a number of normal metabolic processes which in concert elevate and maintain the level of serum bilirubin during the first postnatal weeks. Increased production, a slower rate of uptake, conjugation and excretion by the immature liver, and the process of enterohepatic circulation all result in an elevated bilirubin level that only gradually falls to the adult norm over 2–3 weeks.

Elevated bilirubin in the neonate arises because the rate of bilirubin production exceeds the rate of bilirubin clearance. As is the case in later life, the major source of bilirubin is the normal turnover of circulating red blood cells. Heme oxygenase catalyzes conversion of the iron-containing heme portion of red cells into bilirubin (Maisels 1987). Additionally, immediately after birth there is a substantial contribution to the bilirubin load from nonhemoglobin heme proteins and the normal breakdown of immature red cell precursors upon exposure to the oxygen-rich extrauterine environment. This "early peak" from bone marrow and hepatic sources (Robinson 1968), combined with a higher circulating red cell volume (on a per kilogram body-weight basis) and a shorter mean red cell life span, results in the high bilirubin load found in newborns. Thus, the normal daily bilirubin production (on a per kilogram basis) in the newborn is more than double that of the adult (Bartoletti et al. 1979).

The rate of bilirubin clearance, the other major variable determining bilirubin level, is affected by uptake, conjugation, and excretion by the liver and elimination via the intestine. Each of these metabolic systems (hepatic uptake, conjugation, and excretion) operates at a substantially lower rate relative to the mature systems found in the older infant (>2 months). The activity of these systems increases exponentially from birth to reach full maturity within several weeks after birth (Gartner and Lee 1992).[4] Elimination of bilirubin via the intestinal tract is conditioned by enterohepatic circulation, a process whereby the conjugated (excretable) form of bilirubin is converted back to the unconjugated form in the gut, reabsorbed into the bloodstream, and presented again to the liver for conjugation, effectively increasing the circulating bilirubin load on the liver (Brodersen and Hermann 1963). This cycling process results in part from differences in the neonatal intestinal tract (less reduction of conjugated bilirubin to urobilinogen by intestinal bacteria) and in part from the relatively low volume of milk intake characteristic of the immediate newborn period. A variety of studies in humans and animals suggest that enterohepatic circulation significantly contributes to physiologic jaundice in neonates (Gourley et al 1989; Poland and Odell 1971).

An important factor involved in neonatal jaundice in breast-fed infants is the limited maternal milk production characteristic of the immediate postpartum period. Milk production is stimulated and maintained by the infant's suckling and generally does not reach full volume for several days (De Carvalho, et al. 1983; Fleiss and Jeliffe 1987; Klaus 1987). Infrequent nursing or ineffective suckling fails to stimulate milk production and results in much smaller volumes of milk delivered to the intestine in the early days of life relative to formula-fed infants. During fasting, intestinal bilirubin absorption is increased, and the fasting state may further augment heme turnover in the liver and reduce hepatic bilirubin clearance (Whitmer and Gollan 1983), all of which maintain elevated levels of bilirubin. Frequent, effective nursing delivers a higher volume and more continuous supply of breast milk to the intestine, where it blocks absorption of unconjugated bilirubin, helps establish the intestinal flora that reduce conjugated bilirubin to the form excreted in the stool, and reverses the metabolic mechanisms responsible for fasting hyperbilirubinemia.

This metabolic scenario of a large bilirubin load and immature metabolic pathways is generally referred to as "physiologic jaundice" and is, strictly speaking, any elevation above the adult norm of 1–2 mg/dl not due to underlying pathology. "Pathologic jaundice" suggests abnormal processes beyond the usual course of neonatal development. Conditions for which bilirubin may serve as a sign of underlying pathology include isoimmunization, intrinsic red cell abnormalities, blood volume abnormalities, and assorted infectious, anatomic, metabolic, and drug-associated problems.

Even in the absence of pathologic processes, bilirubin may rise to levels above the statistically expected norm if bilirubin production increases or bilirubin clearance decreases further than usual. Physiologic jaundice is variable among populations because of different rates of bilirubin production. Using carbon monoxide as a marker for bilirubin production, several studies have demonstrated that in Navajo neonates (Johnson et al. 1986) and in some Asian populations (Bartoletti et al. 1979), the most significant cause of higher bilirubin levels is increased production (compared to white controls). At high altitude, increased bilirubin production, based on high carboxyhemoglobin levels and decreased clearance, indicated by a later peak and more prolonged elevation in serum bilirubin, are thought to account for observed differences (Leibson et al. 1989; Moore et al. 1984).

The fully breast-fed infant may exhibit exaggerated jaundice indicating decreased clearance. There has been no clear association between serum bilirubin in breast-fed infants and bilirubin production as measured by carbon monoxide generation (Meyers et al. 1984; Stevenson et al. 1980), suggesting that enterohepatic circulation (thus, decreased clearance) is the primary source of elevated plasma bilirubin levels in breast-fed infants. Breast milk (or formula) in the gut during the first several days minimizes this enterohepatic circulation (Gartner et al. 1983). Because the initiation of milk production and its increase is stimulated by infant suckling (Klause 1987), the volume of breast milk will be small until full flow is established, usually about the third day postpartum. The limited amount of liquid and calories in the neonatal intestinal tract fosters enterohepatic circulation during the first few days of breast-feeding, resulting in elevated bilirubin levels (Cashore and Stern 1982).[5]

Populational and Genetic Factors in Neonatal Jaundice

Several lines of evidence argue for genetic and populational components to the pattern of jaundice seen in human neonates: (1) the conversion of biliverdin, the first product of heme degradation, to bilirubin occurs through an energetically expensive metabolic step that, in effect, serves to maintain bilirubin in the mammalian system; (2) the pattern of neonatal bilirubin expression is similar in human and nonhuman primate species; and (3) bilirubin expression is partially under genetic control.

Biliverdin, the first product of heme degradation, is found in blue-green algae, many plant species, and virtually all animals (McDonagh 1990). Biliverdin is water soluble and is excreted directly by birds, amphibians, and reptiles (Colleran and O'Carra 1977). Bilirubin is found in one form or another in all mammals, though the metabolic pathways by which it is produced and managed vary. It is not found, with a few exceptions, in birds, reptiles, or amphibians. In mammals, an energy-dependent metabolic process converts biliverdin to non–water-soluble bilirubin. The bilirubin molecule is transported in the circulation bound to albumin and is conjugated with glucuronide to facilitate excretion in bile and ultimately in stool. The energetically expensive steps of reduction (biliverdin to bilirubin) and the presence of an enzyme system dedicated to that conversion, a specialized transport and conjugation system for excretion, and enterohepatic circulation (which can maintain levels above the norm) suggest a metabolic role for bilirubin (Colleran and O'Carra 1977; Maisels 1987). It is hypothesized that bilirubin represented an evolutionary adaptation of placental animals that facilitates elimination of heme degradation products from the fetus (Schmid 1976). This has been disputed by McDonagh (1990:361) who points out that this "was not the driving force for evolution of biliverdin reductase, however, because non-placental mammals and [some] non-mammalian vertebrates make bilirubin too, presumably by the same pathway."

The presence of bilirubin in mammals, similar metabolic processes for production and elimination in a wide range of species (McDonagh 1990), and a like pattern of expression in many primate species argue against bilirubin as a necessarily harmful waste product. The presence and similar pattern of expression of bilirubin in neonates of a wide range of nonhuman primates (Cornelius 1982; Gartner et al. 1977) suggest a shared evolutionary heritage. The metabolic pathway of heme degradation to bilirubin is thus a very old process that has been highly conserved during evolution, arguing for an adaptive rather than a deleterious role.

There is mounting evidence in human and nonhuman primates that bilirubin production, metabolism, and excretion are all under at least partial genetic control and exhibit populational differences. Asian and American Indian populations have higher mean bilirubin levels and an increased incidence of moderate and exaggerated physiologic jaundice than do populations of European and African ancestry, regardless of place of birth (Brown and Boon 1965; Fischer et al. 1988; Fisher et al. 1978; Fok et al. 1986; Horiguchi and Bauer

1975; Johnson et al. 1986; Postl et al. 1982; cf. Cohen et al. 1982). The increased levels of serum total bilirubin in Japanese and Navajo infants are attributable, in part, to an increased rate of bilirubin production (Fischer et al. 1988; Johnson et al. 1986). Lower glucuronide-conjugating capacity also may contribute to higher bilirubin levels in Asian infants (Chen and Lee 1980). Albumin, the primary carrier molecule for bilirubin in the circulation, also exhibits populational variation in both human and nonhuman primate species (Lorey et al. 1984, 1985). Although the majority of population-based studies have indicated that the differences in bilirubin level are independent of place of birth (e.g., Brown and Boon 1965; Fischer et al. 1988) and therefore reflect genetic differences, some have raised the question of a possible combination of environmental and genetic factors influencing neonatal jaundice (Cohen et al. 1982; Drew and Kitchen 1976), though they were unable to state possible mechanisms.

Studies attempting to define the risk of recurrence of neonatal hyperbilirubinemia in siblings have added evidence for a strong familial component. Nielsen et al. (1987) found a highly significant correlation in the peak bilirubin levels of siblings, independent of obstetrical and infant variables (e.g., oxytocin administration, cephalohematoma) that might account for the familial association. Khoury et al. (1988) found that newborns whose older siblings were moderately or severely jaundiced were more likely to experience hyperbilirubinemia than infants whose older siblings were not hyperbilirubinemic. Despite controlling for environmental factors known to influence the likelihood (risk) of elevated bilirubin levels, Khoury et al. concluded that the familial effect may be due to genetic or, perhaps, unidentified environmental factors.

Finally, the occurrence of clinically significant neonatal hyperbilirubinemia has been correlated with the presence of specific genetic polymorphisms (Lepore et al. 1989; Lucarini et al. 1991). These authors proposed that a complex of polymorphic enzymes that show quantitative variations of enzymatic activity among phenotypes may influence bilirubin levels in the first few days of life. Though the variables examined accounted for a relatively small part of the variance in bilirubin level, this represents direct evidence that genetic factors are involved in neonatal hyperbilirubinemia.

If there is a genetic basis (and thus, individual and populational variation) for bilirubin production and expression as the above evidence indicates, then natural selection pressures likely maintained bilirubin levels in newborns above those of the older infant or adult. There should be identifiable adaptive value to bilirubin even though it is a potent neurological toxin under certain circumstances.

Adaptive Significance

Although bilirubin and hyperbilirubinemia have been strongly linked with severe neurological sequellae in a number of disease states, the generalization of risk to healthy neonates seems unwarranted (Brett and Niermeyer 1998;

Newman and Maisels 1992). Recognition of this overgeneralization is appearing in the pediatric literature (American Academy of Pediatrics 1994; Newman and Maisels 1990, 1992), if only gradually in practice. The question remains: if bilirubin is generally harmful in the term, healthy neonate, as it is usually assumed to be in the pediatric literature, why then do levels so consistently rise above the adult norm, and why do so few normal children suffer any adverse consequences? Given that all babies (including nonhuman primates [Cornelius 1982]) will be jaundiced well above the adult level within the first postnatal week and that more than half will be visibly jaundiced (Maisels 1982), it is difficult to imagine that something is wrong with all of these infants. Two possibilities can be argued: first, bilirubin is simply a by-product of the transition from intra- to extrauterine life, resulting largely from the immaturity of several developmental processes. Selectively chosen, the medical literature on hyperbilirubinemia supports this argument, and it is the predominant view underlying the emerging consensus that elevated bilirubin in the healthy newborn is not, in and of itself, a pathologic condition (American Academy of Pediatrics 1994; Newman and Maisels 1992). We prefer a second alternative: that neonatal jaundice is not merely a passive consequence of, but plays an adaptive role in, the transition to extrauterine life.

In addition to the genetic and populational evidence cited earlier, there is increasing evidence that bilirubin acts as an antioxidant of physiological importance (McDonagh 1990; Stocker et al. 1987a). Using a variety of in vitro models, investigators have demonstrated bilirubin, at physiological concentrations, to be an efficient scavenger of several reactive oxygen species (oxygen free-radicals), which may initiate oxidative destruction of fatty acids (Stocker and Peterhans 1989; Stocker et al. 1987a, b). Because bilirubin is strongly associated with albumin, it is distributed widely throughout the circulatory system and extravascular spaces. Bilirubin has been demonstrated to scavenge peroxyl radicals in homogeneous solutions as well as in multilamellar liposome systems. Furthermore, this antioxidant activity has been shown to increase as ambient oxygen tensions are decreased to levels physiologically relevant at the intracellular level (Stocker et al. 1987a, b). Recent in vivo work in jaundiced Gunn rats (Dennery et al. 1992) confirmed inhibition of lipid peroxidation by bilirubin and consumption of the bilirubin molecule under conditions of oxidative stress. Bilirubin is also one of the most efficient deactivators of singlet oxygen (McDonagh 1990).

Heme oxygenase, the enzyme that initiates bilirubin formation, belongs to the family of heat-shock proteins (Stocker 1990). Heat-shock proteins are produced in increased amounts in response to a variety of stressors, including heat and oxidative stress. Although the process is yet to be worked out in detail, it appears that heme oxygenase is induced under circumstances of oxidative stress (McDonagh 1990). The result of heme oxygenase induction is twofold: (1) depletion of intracellular levels of heme proteins which can catalyze, or initiate, oxygen radical reactions, and (2) increased formation of the lipid-soluble, membrane-permeable antioxidant bilirubin. Thus, the effect of this enzyme is to help decrease the potential free-radical formation and to generate a molecule that protects against free-radical damage.

The newborn, even at term, is not yet mature in many functions, including antioxidant enzyme systems, yet the transition to extrauterine existence is one of significant stress. The placental oxygen transport system introduces a substantial gradient in oxygen tension resulting in low fetal pO_2. Relative to this carefully controlled system, the extrauterine environment is hyperoxic. After birth the neonate must adapt rapidly to a relatively oxygen-rich environment. As the lungs are exposed to ambient oxygen concentrations, the arterial pO_2 rises threefold and oxygen consumption increases. These circumstances are associated with accelerated production of oxygen free-radicals (Frank and Groseclose 1984). Antioxidant defenses become important to avoid damage to protein, lipid, and nucleic acids.

Even though the lungs of the normal infant are generally able to meet the demands of oxygenating the body at birth, the antioxidant enzyme defense systems remain immature. The important enzymes superoxide dismutase, catalase, and glutathione peroxidase have been documented at lower levels in newborns compared to adults (Frank and Groseclose 1984; McElroy et al. 1992; Tanswell and Freeman 1984). In animal models these enzyme systems begin at very low levels and rise rapidly in the first several weeks postnatally, reaching maturity in weeks to months (Autor et al. 1976; Gartner and Lee 1992). In addition to antioxidant enzymes, there are several nonprotein, nonenzymic molecules that protect against oxidative damage, especially in plasma and cell membranes. This group includes vitamin E, beta-carotene, ascorbic acid, and bilirubin. Importantly, when enzymic and other circulating antioxidant systems are at their lowest activity, bilirubin rises rapidly after birth and falls slowly over several weeks (Kivlahan and James 1984), suggesting the possibility that bilirubin serves an antioxidant role later carried out by the maturing antioxidant enzyme systems.

Consequences for Management of Neonatal Jaundice

If, as we propose, neonatal jaundice is at worst a benign consequence or, more likely, a necessary part of the adaptation to the extrauterine environment, what have been and are the consequences of considering bilirubin as a toxin and hyperbilirubinemia as a disease state? Ultimately, the primary purpose of monitoring bilirubin levels and treating hyperbilirubinemia in newborns is to prevent damage to the central nervous system. Kernicterus and related neurological deficits in survivors were first noted in erythroblastosis fetalis (Rh isoimmunization). Early research indicated the association between high levels of bilirubin in neonates and kernicterus or central nervous system impairment (Hsia 1952; Vaughn et al. 1950). Early studies of children with hemolytic and nonhemolytic jaundice suggested delays in cognitive and motor development (Boggs et al. 1967; Hardy et al. 1979; Scheidt et al. 1977). Follow-up (outcome) studies of older children argue convincingly that there are no serious sequellae following moderate bilirubin levels (Cully et al. 1970; Rubin et al. 1979). More recent research confirms there is minimal risk for adverse outcome or impairment to the normal child with moderate levels of hyperbilirubinemia (<15–18 mg/dl) (Newman and Klebanoff 1993).

There is no clear evidence that a certain level of bilirubin necessarily causes brain damage associated with kernicterus. Rather, concurrent insults or inadequacy of the usual physiologic processes for bilirubin metabolism may be more at issue. Compromises in the blood–brain barrier resulting from the underlying disease process may affect the permeability of the barrier, allowing passage of bilirubin into the brain. Thus, it is the status of the barrier, not the absolute level of bilirubin, that is of greatest concern (Levine 1979, 1988). Nevertheless, the devastating consequences of kernicterus and the apparent link between a certain level of serum bilirubin and negative neurological outcome, combined with a complex of factors involving cultural expectations of pathology and changes in neonatal care, resulted in the perception of hyperbilirubinemia as a pathological condition to which all infants were susceptible. This perception of bilirubin as potentially dangerous to all infants had a profound impact on neonatal care and affected in important ways the direction of research on bilirubin and the healthy neonate (Brett and Niermeyer 1998).

Both the perceived need for and the levels at which treatment would be instituted were largely determined by this association (expectation) of bilirubin as pathological agent and hyperbilirubinemia as a disease state. In Rhisoimmunized infants, the risk of kernicterus was found to increase markedly if serum bilirubin levels exceeded 20 mg/dl (Hsia et al. 1952). In standard medical practice this was the point at which the radical therapy of a double-volume exchange transfusion would be done to rapidly lower bilirubin content of blood and tissue. All other therapeutic approaches were designed to more easily manage hyperbilirubinemia and, ultimately, to prevent reaching what became "the magic number 20" (Watchko and Oski 1983). In the 1970s phototherapy gained acceptance as an excellent method to gradually lower bilirubin levels and prevent increase to the point of exchange transfusion. With a few simple precautions, phototherapy provides effective treatment with apparently minimal risk. Guidelines for the point at which to initiate phototherapy developed informally, largely based on preventing a rise of serum bilirubin to "exchange level." Thus, a value <15 mg/dl became the accepted standard at which to initiate phototherapy. In practice that level coincided with 12–13 mg/dl, representing the 95th percentile of peak neonatal bilirubin values in large population studies (Hardy et al. 1979). Above this point children were outside the statistically defined range of normal, and therefore presumably "abnormal" and at risk. The important point to note here is that guidelines for treatment of exaggerated physiologic jaundice were derived from an informal extrapolation of data from poorly controlled studies in a small population of sick newborns (erythroblastosis fetalis). They were not based on carefully controlled studies of the course and outcome of physiologic jaundice in the healthy newborn (Newman and Maisels 1990).

Recently, guidelines for treatment of jaundice in healthy, term infants have been formalized to reflect that "therapeutic interventions for hyperbilirubinemia in the healthy term infant may carry significant risk relative to the uncertain risk of hyperbilirubinemia in this population" (American Academy of Pediatrics 1994:559). Although the suggested levels for intervention with both phototherapy and exchange transfusion have been raised for infants with non-hemolytic jaundice, the emphasis remains on treatment rather than understanding and modification of causative factors.

The expectation of jaundice in the healthy newborn as a pathological state has had considerable effect on other aspects of newborn care. Probably the most troubling link was that made between jaundice and breast-feeding. It was noted informally that many fully breast-fed infants reached a higher mean bilirubin level than their bottle-fed compatriots. This observation was empirically verified in several carefully controlled studies (Maisels and Gifford 1986). But, we would argue that this link represents an example of what Williams and Nesse (1991) call an "abnormal environment" in evolutionary terms. In this case the environment is largely the result of rapid social and cultural change in the care environment surrounding birth and newborn care (Trevathan and McKenna 1994). It is the social milieu of the hospital routine and the cultural expectations of medicine that establish the abnormal environment (Brett and Niermeyer 1990, 1998). In the hospital, clinical routines dictate feedings on a schedule of every 3–4 hours rather than on demand. This contrasts with the normal human pattern in which mother–infant contact is virtually continuous and breast-feeding may occur several times per hour. Studies of ad libitum breast-feeding mother–infant pairs in Western hospitals indicated a feeding interval of every 1–2 hours (Cable and Rothenberger 1984; De Carvalho et al. 1983; Maisels et al. 1994; Rubatelli 1993).

Serum bilirubin levels correlate inversely with the frequency of breast-feeding (De Carvalho et al. 1982) and caloric intake of neonates (Wu et al. 1985). Frequent infant suckling directly stimulates milk production (De Carvalho et al. 1983). Yet, common practice in U.S. hospitals was to supplement breast-fed babies with water or glucose solution in the first day or two postpartum. These liquids have no effect on serum bilirubin levels (De Carvalho et al. 1981) and may decrease the intake of milk and consequently increase bilirubin levels. Routine early discharge from the hospital provides enough opportunity for parents to learn by example the hospital pattern of infrequent breast-feeding separated by several hours and perhaps punctuated by supplementation, but lack of follow-up support does not provide opportunities to assure that maternal milk supply is adequately established or that infants are receiving adequate milk volume. One of the earliest signs of insufficient milk intake can be exaggerated jaundice, as the small volume of milk present in the intestine does not block de-conjugation of bilirubin, thus permitting reabsorption.

Exaggerated jaundice is widely viewed as a potential disease state itself, rather than as a sign for underlying pathology (Brett and Niermeyer 1998). Determination of a serum bilirubin level triggers a cascade of laboratory tests—

usually one or more follow-up bilirubin values and perhaps other hematologic tests (Newman et al. 1990). While it is important to distinguish hemolytic from nonhemolytic jaundice, the majority of these tests provide little useful information in the management of bilirubin (Newman and Maisels 1992; Newman et al. 1990). However, the act of monitoring an elevated bilirubin level in an otherwise healthy infant, with or without treatment, can lead to adverse psychological and behavioral consequences. Kemper et al. (1989, 1990) compared mothers of jaundiced infants (bilirubin >12 mg/dl) and mothers of infants who were not jaundiced. Mothers of jaundiced infants were less likely to leave their child with anyone, sought more medical attention for their infants, and were more likely to have stopped breast-feeding; this situation is termed the "vulnerable child" syndrome by Kemper et al. (1989). Treatment of moderately elevated bilirubin levels may include temporary interruption of breast-feeding and/or initiation of phototherapy. Both interventions have been shown to result in shorter total duration of breast-feeding (Kemper et al. 1989).

Conclusion

There is an emerging consensus that moderate levels of hyperbilirubinemia (<15 mg/dl) are not necessarily harmful to the healthy newborn (Cashore 1991; Newman and Maisels 1992). This consensus is occurring not because of findings defining the role of bilirubin, but rather because of consistent inability to clearly identify pathological consequences in the normal newborn. For each study in which statistical significance was achieved on some variable, a similar study often failed to do so. In all too many cases, studies are methodologically incomparable, making generalizations difficult, if not impossible (American Academy of Pediatrics 1994; Newman and Maisels 1990, 1992). Several prominent researchers have stated repeatedly in leading journals that, based on their studies and reviews of the literature, there appears to be minimal risk of negative outcome in the healthy newborn (Cashore 1991; McDonagh 1990; Newman and Maisels 1990, 1992; cf. Johnson 1991; Stevenson and Brown 1991). We suggest that, while important and clinically likely to lead to better care of the newborn, this consensus opinion can be expanded by using an evolutionary perspective to examine neonatal jaundice.

An evolutionary perspective provides a framework for elucidating the reasons for elevated bilirubin in the majority of newborns—the possibility of beneficial effects measured against the potential risks. This perspective then calls into question the continued emphasis on treatment by phototherapy, discontinuation of breast-feeding, and exchange transfusion as well as the increasing emphasis on pharmacological management of hyperbilirubinemia in the normal newborn.

We have outlined elsewhere specific research needed to confirm the hypotheses we propose here (Brett and Niermeyer 1990). We summarize and augment the points below in light of recent research.

(1) If bilirubin is necessary/important in the newborn period, then there should be disease states secondary to its absence.

(2) Additional attention should be devoted to the genetic bases of bilirubin expression.

(3) Carefully designed studies to compare optimal breast-feeding (i.e., on demand) with the more usual 3 to 4 hour schedule common in many North American hospitals need to be done. If exaggerated physiological jaundice is the result of inadequate calories and liquid, then babies fed on a 4-hour schedule should exhibit higher bilirubin levels and a greater incidence of exaggerated hyperbilirubinemia than babies fed on demand.

(4) Greater attention should be devoted to factors that permit bilirubin to cross the blood–brain barrier. Identifying these conditions rather than continuing the unproductive focus on the bilirubin molecule itself will demonstrate the risk factors for kernicterus.

(5) Additional population-based studies should be designed to demonstrate that there are clear population-based differences in mean serum bilirubin in term, healthy babies and to identify the mechanisms that account for these differences (e.g., production, clearance, etc.).

In conclusion, we suggest that an evolutionary view of jaundice in the healthy newborn challenges the understanding of neonatal jaundice as a disease. This perspective allows us to unify conceptually much seemingly disparate and contradictory research that has defied explanation from the proximate, ahistorical perspective of biomedicine.

Notes

1. Hemolyis is the accelerated destruction of red blood cells due to antibody-mediated isoimmunization (Rh or ABO) or intrinsic defects of red cells. Rh isoimmunization results from an incompatibility between the Rh-negative mother and her Rh-positive baby. In essence the mother produces antibodies to the Rh+ blood group, resulting in a breakdown of the fetal red blood cells *in utero*. This process continues after birth and, if severe and untreated, is fatal. Bilirubin is a product of red blood cell breakdown (hemolysis). In the absence of significant preventive measures like Rhogam, Rh incompatibility affects approximately 0.6% of pregnancies. ABO incompatibility is similar in that if the mother is type O, and the fetus is type A or B, there will be cross-antigen reaction. ABO incompatibility is more common (3% of pregnancies), though generally clinically less severe than Rh isoimmunization (Oski 1981).

2. The adult norm for serum bilirubin is <2 mg/dl. This is only a marginally useful value because nearly all newborns will have levels above the adult norm (Maisels 1987), and more than half will have bilirubin levels above 5–7 mg/dl, making the bilirubin visible as jaundice in the newborn.

3. Phototherapy provides a simple, apparently safe method of reducing bilirubin levels. Placing the affected infant in light from the blue end of the

visible spectrum (or in sunlight) results in photoisomerization of the bilirubin molecule, making it water soluble and easily eliminated from the
body.

4. The consideration of these metabolic processes as *immature* is in
marked contrast to the usual medical
model that views this immaturity in
terms of *deficiency* and potential pathology. What is really at issue is the
normal immaturity of the human neonate resulting from an "evolutionary
compromise" where the size of the fetus (and therefore the degree of maturity) must be balanced against maternal
risks associated with delivering a very
large fetus (Gould 1977).

5. There is some evidence indicating that the apparent increased rate of
enterohepatic recirculation may be an
artifact of research design rather than a
real phenomenon. De Carvhalo et al.
(1982) have demonstrated that on-
demand breast-feeding results in a bilirubin level essentially equal to that of
bottle-fed infants. If, in the original
studies demonstrating enterohepatic
circulation (e.g., Poland and Odell
1971), some of the study population
included breast-fed infants on a
hospital-mediated four-hour feeding
schedule, then these "starved" infants
would bias the results, falsely indicating an increase in enterohepatic recirculation in the healthy newborn.

References

American Academy of Pediatrics Professional Committee for Quality Improvement and Subcommittee on
Hyperbilirubinemia Practice Parameter (1994) Management of hyperbilirubinemia in the healthy term neonate. *Pediatrics* 94:558–565.

Autor, A. P., Frank, L., and Roberts, R.
J. (1976) Developmental characteristics of pulmonary superoxide dismutase: Relationship to idiopathic respiratory distress syndrome. *Pediatric
Research* 10:154–158.

Bartoletti, A. L., Stevenson, D. K., Ostrander, C. R., and Johnson, J. D.
(1979) Pulmonary excretion of carbon monoxide in the human infant
as an index of bilirubin production:
I. Effects of gestational and postnatal
age and some common neonatal abnormalities. *Journal of Pediatrics* 94:
952–955.

Boggs, T. R., Hardy, J. B., and Frazier,
T.M. (1967) Correlation of neonatal
serum total bilirubin concentrations
and developmental status at age
eight months. *Journal of Pediatrics*
71:553–560.

Brett, J., and Niermeyer, S. (1990) Neonatal jaundice: A disorder of transition or an adaptive process? *Medical Anthropology Quarterly* 4:149–
161.

Brett, J., and Niermeyer, S. (1998) Neo-
natal jaundice: Medical culture and
the construction of a neonatal dis-
ease. In *Small Wars: The Cultural
Politics of Childhood* Nancy Scheper-
Hughes and Carolyn Sargent, eds.
Berkeley: University of California
Press.

Broderson, R. (1977) Prevention of kernicterus based on recent progress in
bilirubin chemistry. *Acta Paediatrica Scandinavica* 66:625–634.

Broderson, R. (1978) Free bilirubin in
blood plasma of the newborn: Effects of albumin, fatty acids, pH, dis-
placing drugs, and phototherapy. In
Intensive Care in the Newborn, II. L.
Stern, W. Oh, and B. Friis-Hansen,
eds. New York: Masson Publishing,
pp. 331–345.

Brodersen, R., and Sparre Hermann, L.
(1963) Intestinal reabsorption of unconjugated bilirubin: A possible con-

tributing factor in neonatal jaundice. *Lancet* 1:1242.

Brown, A. K., and McDonagh, A. F. (1980) Phototherapy for neonatal hyperbilirubinemia: Efficacy, mechanism and toxicity. In *Advances in Pediatrics*, vol. 27. L. A. Barness, ed. Chicago: Year Book Medical Publishers, pp. 341–349.

Brown, W. R., and Boon, W. H. (1965) Ethnic group differences in plasma bilirubin levels of full-term, healthy Singapore newborns. *Pediatrics* 36: 745–751.

Cable, T. A., and Rothenberger. L. A. (1984) Breast-feeding behavioral patterns among La Leche league mothers: A descriptive survey. *Pediatrics* 73:830–835.

Cashore, W. (1991) Neonatal hyperbilirubinemia: Comment. *New York State Journal of Medicine* 91:477–478.

Cashore W. J., Gartner, L. M., Oh, W., and Stern, L. (1978) Clinical application of neonatal bilirubin-binding determinations: Current status. *Journal of Pediatrics* 93:827–833.

Cashore, W. J., and Stern, L. 1982. Bilirubin metabolism and neonatal nutrition. *Journal of Pediatric Gastroenterology and Nutrition* 1:459–461.

Chen, S. H., and Lee, T. C. (1980) Carboxyhemoglobin and glucuronide formation in Chinese newborn infants with hyperbilirubinemia. *Journal of the Formosan Medical Association* 79:314–322.

Cohen, R. S., Hopper, A. O., Ostrander, D. R., and Stevenson, D. K. (1982) Total bilirubin production in infants of Chinese, Japanese, and Korean ancestry. *Journal of the Formosan Medical Association* 81:1524–1529.

Colleran, E., and O'Carra, P. (1977) Enzymology and comparative physiology of biliverdin reduction. In *Chemistry and Physiology of Bile Pigments*. P. D. Berk and N. I. Berlin, eds. Washington, DC: U.S. Government Printing Office, pp. 69–80.

Cornelius, C. (1982) The use of nonhuman primates in the study of bilirubin metabolism and bile secretions. *American Journal of Primatology* 2: 243–254.

Cully, P., Powell, J., Waterhouse, J., and Wood, B. (1970) Sequelae of neonatal jaundice. *British Medical Journal* 3:383–286.

De Carvalho, M., Hall, M., and Harvey, D. (1981) Effects of water supplementation on physiologic jaundice in breast-fed babies. *Archives of Disease in Childhood* 56:568–569.

De Carvalho, M., Klaus, M. H., and Merkatz, R. B. (1982) Frequency of breast-feeding and serum bilirubin concentrations. *American Journal of Diseases of Children* 136:737–738.

De Carvalho, M., Robertson, S., Friedman, A., and Klaus, M. (1983) Effect of frequent breast-feeding on early milk production and infant weight gain. *Pediatrics* 72:307–311.

De Carvalho, M., Roberston, S., and Klaus, M. (1985) Fecal bilirubin excretion and serum bilirubin concentrations in breast-fed and bottle-fed infants. *Journal of Pediatrics* 107:786–790.

Dennery, P. A., McDonagh, A. F., and Stevenson, D. K. (1992) In vivo role of bilirubin in neonatal antioxidant defenses. *Pediatric Research* 31: 200A.

Drew J. H., and Kitchen, W. H. (1976) Jaundice in infants of Greek parentage: The unknown factor may be environmental. *Journal of Pediatrics* 892:248–52.

Eggert, L. D., Pollary, R. A., Folland, D. S., and Jung, A. L. (1985) Home phototherapy treatment of neonatal jaundice. *Pediatrics* 76:579–584.

Fischer, A. F., Nakamura, H., Vetani, Y., Vreman, H. .J., and Stevenson, D. K. (1988) Comparison of bilirubin production in Japanese and Caucasian infants. *Journal of Pediatric Gastroenterology and Nutrition* 7:27–29.

Fisher, Q., Cohen, M. I., Curda, L., and McNamara, H. (1978) Jaundice and breast-feeding among Alaskan Eskimo newborns. *American Journal of Diseases of Children* 132:859–860.

Fleiss, P. M., and Jelliffe, D. B. (1987) Breast-feeding and jaundice. *Pediatrics* 80:306.

Fok, T. F., Lau, S. P., and Hui, C. W. (1986) Neonatal jaundice: Its prevalence in Chinese babies and associating factors. *Australian Paediatric Journal* 22:215–219.

Frank, L. and Groseclose, E. E. (1984) Preparation for birth into an O_2-rich environment: The antioxidant enzymes in the developing rabbit lung. *Pediatric Research* 18:240–244.

Gartner, L. M., Catz, C. S., and Yaffe, S. J. (1994) Neonatal bilirubin workshop. *Pediatrics* 94:537–540.

Gartner, L. M., and Lee, K. S. (1992) Unconjugated hyperbilirubinemia. In *Neonatal-Perinatal Medicine: Diseases of Fetus and Infant*, vol. 2. A. A. Fanaroff and R. J. Martin, eds. 5th ed. St. Louis: Mosby Year Book, 1075–1117.

Gartner, L. M., Lee, K.-S., and Moscioni, A. D. (1983) Effect of milk feeding on intestinal bilirubin absorption in the rat. *Journal of Pediatrics* 103:464–471.

Gartner, L. M., Lee, K. S., Vainman, S., Lane, D., and Zarafu, I. (1977) Development of bilirubin transport and metabolism in the newborn rhesus monkey. *Journal of Pediatrics* 90:513–531.

Gould, S. J. (1977) *Ever Since Darwin: Reflections in Natural History.* New York: Norton.

Gourley, G. R., Kreamer, B., and Arend, R. (1989) Neonatal serum bilirubin levels correlate with β-Glucuronidase activity in the first stool of life. *Hepatology* 10:728.

Hanneman, R. E., Schreiner, R. L., Dewitt, D. P., Norris, S. A., and Glick, M. R. (1982) Evaluation of the Minolta Bilirubin Meter as a screening device in white and black infants. *Pediatrics* 69:107–109.

Hardy, J. B., Drage, J. S., and Jackson, E. C. (1979) *The First Year of Life: The Collaborative Perinatal Project of the National Institute of Neurological and Communicative Disorders and Stroke.* Baltimore, Maryland: The Johns Hopkins University Press.

Hegyi, T. I., Hiatt, M., and Indyk, L. (1982) Transcutaneous bilirubinometry: I. Correlations in term infants. *Journal of Pediatrics* 98:454–457.

Horiguchi, T., and Bauer, C., (1975) Ethnic differences in neonatal jaundice: Comparisons of Japanese and Caucasian newborn infants. *American Journal of Obstetrics and Gynecology* 121:71–74.

Hsia, D. Y., Allen, F. H., Jr., Gellis, S. S., and Diamond, L. K. (1952) Erythroblastosis fetalis: VIII. Studies of serum bilirubin in relation to kernicterus. *New England Journal of Medicine* 247:668–671.

Johnson, J. D., Angelus, P., Aldrich, M., and Skipper, B. J. (1986) Exaggerated jaundice in Navajo neonates, the role of bilirubin production. *American Journal of Diseases of Children* 140:889–890.

Johnson, L. (1991) Hyperbilirubinemia in the term infant: When to worry, when to treat. *New York State Journal of Medicine* 91:483–489.

Kappas, A., Drummond, G. S., Henschke, C., and Valaes, T. (1995) Direct comparison of Sn-mesoporphyrin, an inhibitor of bilirubin production, and phototherapy in controlling hyperbilirubinemia in term and near-term newborns. *Pediatrics* 95:468–474.

Kemper, K., Forsyth, B., and McCarthy, P. (1989) Jaundice, terminating breast-feeding and the vulnerable child. *American Journal of Diseases of Children* 143:773–778.

Kemper, K., Forsyth, B., and McCarthy, P. (1990) Persistent perceptions of vulnerability following neonatal jaundice. *American Journal of Diseases of Children* 144:238–241.

Khoury, M. J., Calle, E. E., and Joesoef, R. M.. (1988) Recurrence risk of neonatal hyperbilirubinemia in siblings. *American Journal of Diseases of Children* 142:1065–1069.

Kivlahan, C., and James, E. J. (1984) The natural history of neonatal jaundice. *Pediatrics* 74:364–370.

Klaus, M. H. (1987) The frequency of suckling: A neglected but essential ingredient of breast-feeding. *Obstetric and Gynecologic Clinics of North America* 14:623–633.

Leibson, C., Brown, M., Thibodeau, S., Stevenson, D., Vreman, H., Cohen, R., Clemons, G., Callen, W., and Moore, L. G. (1989) Neonatal hyperbilirubinemia at high altitude. *American Journal of Diseases of Children* 143:983–987.

Lepore, A., Lucarini, N., Evangelista, M. A., Palombaro, G., Londrillo, A., Ballarini, P., Borgiani, P., Gloria-Bottini, F., and Bottini, E. (1989) Enzyme variability and neonatal jaundice: The role of adenosine deaminase and acid phosphatase. *Journal of Perinatal Medicine* 17:195–201.

Levine, R. L. (1979) Bilirubin: Worked out years ago? *Pediatrics* 64:380–385.

Levine, R. L. (1988) Neonatal jaundice. *Acta Paediatrica Scandinavica* 77:177–182.

Lorey, F., Ahlfors, C., Smith, D., and Neel, J. V. (1984) Bilirubin binding by variant albumins in Yanamama Indians. *American Journal of Human Genetics* 36:1112–1120.

Lorey, F., Smith, D., and Ahlfors, C. (1985) Bilirubin binding by the fractionated alternate allelic components of heterozygous monkey albumin. *American Journal of Physical Anthropology* 68:169–171.

Lucarini, N., Gloria-Bottini, F., Tucci-arone, L., Carapella, E., Maggioni, G., and Bottini, E. (1991) Role of genetic variability in neonatal jaundice. A prospective study on full-term, blood group-compatible infants. *Experientia* 47:1218–1221.

Maisels, M. J. (1982) Jaundice in the newborn. *Pediatrics in Review* 3:305–319.

Maisels, M. J. (1987) Neonatal jaundice. In *Neonatology, Pathophysiology and Management of the Newborn*. G. B. Avery, ed. Philadelphia: J. B. Lippincott Co, pp. 534–628.

Maisels, M. J., and Gifford, K. (1986) Normal serum bilirubin levels in the newborn and the effect of breast-feeding. *Pediatrics* 78:837–843.

Maisels, M. J., Vain, N., Acquavita, A. M., De Blanco, N. V., Cohen, A., and DeGregorio, J. (1994) The effect of breast-feeding frequency on serum bilirubin levels. *American Journal of Obstetrics and Gynecology* 170:880–883.

McDonagh, A. F. (1990) Is bilirubin good for you? *Clinics in Perinatology* 17:359–369.

McDonagh, A. F., and Lightner, D. A. (1985) 'Like a shrivelled blood orange'—Bilirubin, jaundice, and phototherapy. *Pediatrics* 75:443–455.

McElroy, M. C., Postle, A. D., and Kelly, F. J. (1992) Catalase, superoxide dismutase and glutathione peroxidase activities of lung and liver during human development. *Biochimica et Biophysica Acta* 1117:153–158.

Meyers, C. H., Kwong, L. K., Vreman, H. J., and Stevenson, D. K. (1984) The role of bilirubin production in breast-fed infants with elevated serum bilirubin concentrations at 2 weeks of life. *Clinical Pediatrics* 23: 480–482.

Mollison, P. L., and Cutbush, M. (1949) Haemolytic disease of the newborn: Criteria of severity. *British Medical Journal* 1:123–130.

Moore, L. G., Newbury, M. A., Freeby, G. M., and Crnic, L. S. (1984) Increased incidence of neonatal hyperbilirubinemia at 3,100m in Colorado. *American Journal of Diseases of Children* 138:157–161.

Newman, T., Easterling, M. J., Goldman, E. S., and Stevenson, D. K. (1990) Laboratory evaluation of jaundice in newborns: Frequency, cost and yield. *American Journal of Diseases of Children* 144:364–368.

Newman, T., and Klebanoff, M. A. (1993) Neonatal hyperbilirubinemia and long-term outcome: Another look at the Collaborative Perinatal Project. *Pediatrics* 92:651–657.

Newman, T., and Maisels, M. J. (1990) Does hyperbilirubinemia damage the brain of healthy, full-term infants? *Clinics in Perinatology* 17:331–358.

Newman, T., and Maisels, M. J. (1992) Evaluation and treatment of jaundice in the term newborn: A kindler, gentler approach. *Pediatrics* 89:809–818.

Nielsen, H. E., Haase, P., Blaabjerg, J., Stryhn, H., and Hilden, J. (1987) Risk factors and sib correlation in physiological neonatal jaundice. *Acta Paediatrica Scandinavica* 76: 504–511.

Oski, F. A. (1981). Hematological problems. In *Neonatology: Pathophysiology and Management of the Newborn*. G. B. Avery, ed. Philadelphia: J. B. Lippincott Co, pp. 545–582.

Poland, R. L., and Odell, G. B. (1971) Physiologic jaundice: The enterohepatic circulation of bilirubin. *New England Journal of Medicine* 284:1–6.

Postl, B. D., Nelson, N., and Carson, J. (1982) Hyperbilirubinemia in Inuit neonates. *Canadian Medical Association Journal* 126:811–813.

Robinson, S. H. (1968) The origins of bilirubin. *New England Journal of Medicine* 279:143–149.

Rogers, P. A., Vreman, H. J., Dennery, P. A., and Stevenson, D. K. (1994) Sources of carbon monoxide (CO) in biological systems and applications of CO detection technologies. *Seminars in Perinatology* 18:2–10.

Rubatelli, F. F. (1993) Unconjugated and conjugated bilirubin pigments during perinatal development, IV: The influence of breast-feeding on neonatal hyperbilirubinemia. *Biology of the Neonate* 64:104–109.

Rubin, R. A., Balow, B., and Fisch, R. O. (1979) Neonatal serum bilirubin levels related to cognitive development at ages 4 through 7 years. *Journal of Pediatrics* 94:601–604.

Scheidt, P. C., Mellits, E. D., Hardy, J. B., Drage, J. S., and Boggs, T. R. (1977) Toxicity to bilirubin in neonates: Infant development during first year in relation to maximum neonatal serum bilirubin concentration. *Journal of Pediatrics* 91:292–297.

Schmid, R. (1976) The distinguished lecture: Pyrrolic victories. *Transactions of the American Association of Physicians* 89:64–76.

Slater, L., and Brewer, M. F. (1984) Home versus hospital phototherapy for term infants with hyperbilirubinemia: A comparative study. *Pediatrics* 73:515–519.

Stevenson, D. K., Bartoletti, A. L., Ostrander, C. R., and Johnson, J. D. (1980) Pulmonary excretion of carbon monoxide in the human infant as an index of bilirubin production: IV. Effects of breast-feeding and caloric intake in the first postnatal week. *Pediatrics* 65:1170–1172.

Stevenson, D. K., and Brown, A. K. (1991) Observation on neonatal hyperbilirubinemia: comment. *New*

York State Journal of Medicine 91: 477–478.

Stocker, R. (1990) Induction of haem oxygenase as a defence against oxidative stress. *Free Radical Research Communications* 90:101–112.

Stocker, R., Glazer, A. N., and Ames, B. N. (1987a) Antioxidant activity of albumin-bound bilirubin. *Proceedings of the National Academy of Sciences (USA)* 84:5918–5922.

Stocker, R., and Peterhans, E. (1989) Antioxidant properties of conjugated bilirubin and biliverdin: biologically relevant scavenging of hypochlorous acid. *Free Radical Research Communications* 6:57–66.

Stocker, R., Yamamoto, Y., McDonagh, A. F., Glazer, A. N., and Ames, B. N. (1987b) Bilirubin is an antioxidant of possible physiological importance. *Science* 235:1043–1046.

Tan, K. L. (1975) Comparison of the effectiveness of phototherapy and exchange transfusion in the management of nonhemolytic neonatal hyperbilirubinemia. *Journal of Pediatrics* 87:609–612.

Tanswell, A. K., and Freeman, B. A. (1984) Pulmonary antioxidant enzyme maturation in the fetal and neonatal rat: I: Developmental profiles. *Pediatric Research* 18:584–587.

Trevathan, W., and McKenna, J. (1994) Evolutionary environments of human birth and infancy: Insights to apply to contemporary life. *Children's Environment* 11:88–104.

Vaughan, V. C. III, Allen., F. A. Jr., and Diamond, L. K. (1950) Erythroblastosis fetalis: IV. Further observations on kernicterus. *Pediatrics* 6:706–716.

Watchko, J., and Oski, F. (1983) Bilirubin 20 mg/dl = vigintiphobia. *Pediatrics* 71:660–663.

Whitmer, D., and Gollan, J. L. (1983) Mechanisms and significance of fasting and dietary hyperbilirubinemia. *Seminars in Liver Disease* 3:42–51.

Williams, G. C., and Nesse, R. M. (1991) The dawn of Darwinian medicine. *The Quarterly Review of Biology* 66:1–22.

Wu, P. Y. K. Hodgman, J. E., Kirkpatrick, B. V., White, N. B., Jr., Phil, M., and Bryla, D. A. (1985) Metabolic aspects of phototherapy. *Pediatrics* (suppl.) 75:427–433.

2

INFANT CRYING BEHAVIOR AND COLIC

An Interpretation in Evolutionary Perspective

RONALD G. BARR

On 15 December 1984, *Le journal de Montreal* reported that a 26-year-old anthropology student beat his 27-day-old infant daughter to death because he was unable to stop her persistent crying (*Le journal* 1984). This dramatic and tragic event is eloquent testimony to the power of crying to frustrate and antagonize parents and bring about maladaptive consequences for the child. While crying has been recognized increasingly as a proximate cause of child abuse and occasionally death (O'Neill, 1993; Singer & Rosenberg, 1992; Weston, 1968), the consequences are usually—and fortunately—less extreme. Nevertheless, early crying may signal considerable distress in both the infant and its caregivers. It is now well-documented that excessive crying is associated with clinically significant levels of maternal emotional stress (Miller et al., 1993). It is a major reason for weaning from breast-milk (and its health benefits) because mothers believe that crying indicates a hungry infant and an inadequate milk supply (Cable & Rothenberger, 1984; Cunningham et al., 1991; Forsyth et al., 1985a). It is also the most frequent complaint to physicians in the first 3 months after birth (Forsyth et al., 1985b).

In popular parlance and the clinical literature, excessive crying is referred to as "colic." In most medical formulations, colic is a condition that some infants *have* in the first 3 months of life. However, with the exception of cases in which an underlying organic disease is found [fewer than 10% (Sauls & Redfern, 1997; Miller & Barr, 1991; Treem, 1994)], the origin and the reasons for the crying are unknown. Here we consider different, and admittedly hypothetical, possibilities. (1) that colic is an activity infants *do* (rather than a condition they have), (2) that caretaking styles typical of industrialized Western societies exacerbate colic crying, and (3) that this early crying that produces such clinical concern in our society might even "make sense" in terms of the "environment of our evolutionary adaptedness" (Bowlby, 1969).

To address these possibilities, we first review our current understanding of the behavioral manifestations of colic syndrome and the medical conditions that might cause it. We then consider evidence suggesting that (1) colic crying is continuous with normal crying, rather than being a distinct syndrome; (2) increased crying is a "behavioral universal" of human infancy; (3) under caregiving conditions typical in Western industrialized countries, crying tends to occur in *prolonged* (rather than short) bouts typical of colic behavior; and (4) long crying bouts predispose to negative consequences, while short bouts have potentially positive consequences for infants. Finally, an argument is made that (5) a predisposition to increased crying might serve an adaptive function in the evolutionary sense of promoting survival and reproductive success. If so, the origin and reasons for early increased crying and colic in the absence of disease may not be as paradoxical as they seem from a medical point of view.

Colic as a Condition Infants "Have"

The Behavioral Manifestations of Colic Syndrome

Until recently, accounts of colic have been based on insightful but primarily anecdotal and uncontrolled clinical descriptions of the behaviors of infants brought to the physician with complaints of excessive crying (Barr, 1991). The recurring behavior patterns inclined parents and clinicians to regard such crying as a specific behavioral syndrome. Descriptions have focused on three dimensions. First, the syndrome has distinctive diurnal and age-dependent features. In the typical case, crying begins to increase about 2 weeks after birth, reaches a peak in the second month, and is over by 5 months of age. In addition, crying tends to cluster in the late afternoon and evening hours. Second, the infant is described as manifesting prolonged bouts of crying (sometimes called "colic" bouts) that are resistant to attempts to soothe, including feeding. The infant may clench its fists, flex its legs over its abdomen, and arch its back. Its face is usually active and grimacing, and may be flushed. The configuration leads caretakers to believe that the infant is in pain (a "pain facies"). In fact, clinical accounts resemble a particularly sensitive description in Darwin's *Expression of Emotion in Man and Animals* (Darwin, 1872/1979). He noted that "while thus screaming, [a baby's] eyes are firmly closed, so that the skin round them is wrinkled, and the forehead contracted in a frown. The mouth is widely opened with the lips retracted in a peculiar manner, which causes it to assume a squarish form; . . . An excellent observer, in describing a baby crying whilst being fed, says, 'it made its mouth like a square, and let the porridge run out at all four corners' " (pp. 148, 151). The abdomen is described as hard and distended. The crying bout is often associated with regurgitation and the passing of gas from the rectum. The third characteristic dimension is that the crying bouts are "paroxysmal," meaning that they begin and end apparently without warning, unrelated to other events in the environment. They appear to be spontaneous.

Clinicians differ in the criteria that must be met to decide that an infant has colic syndrome (Barr, 1991). The most frequently applied diagnostic criteria are Wessel et al.'s "rule of threes": crying occurs for more than 3 hours per day for more than 3 days per week for more than 3 weeks (Wessel et al., 1954). Lack of agreement on how to define colic syndrome has hampered attempts to accurately describe how commonly it occurs. Estimates of incidence range from 8% to more than 40% depending on the definition used (Hide & Guyer, 1982; Lehtonen & Korvenranta, 1995; Stahlberg, 1984). For similar reasons, it is as yet unresolved as to whether the prevalence of colic varies by geographical regions (Weissbluth & Weissbluth, 1992). However, there is little doubt that, in Western countries at least, excessive crying is a prevalent and distressing concern in the first postnatal months.

Medical Conditions as the Cause of Colic

The pediatric medical literature has focused on disease entities that may cause excessive crying, and for good reason. Crying unresponsive to soothing is a time-honored behavioral symptom of organic disease (Poole, 1991). Unfortunately, neither the consistency nor the consolability of crying generally correlates with the presence or the seriousness of the underlying etiology (Poole, 1991). Furthermore, when colic has been attributed to organic disease, the behavioral syndrome has been poorly defined, and the description often limited to the fact that crying was "excessive," "screaming," or "unresponsive." Of course, almost any disease entity may present with crying in infants; however, such diverse entities as otitis media, constipation, urinary tract infection, fructose intolerance, hiatal and inguinal hernias, Chiari malformation, gastroesophageal reflux/esophagitis, and corneal abrasions have specifically been reported to present as colic (Andersson & Nygren, 1978; Archibald, 1942; Berezin et al., 1995; Du, 1976; Gormally & Barr, 1997; Guttman, 1972; Harkness, 1989; Heine et al., 1995; Joos & Don Steepe, 1992; Listernick & Tomita, 1991; Poole, 1991). Furthermore, persistent crying can be due to maternal drug use, either from withdrawal to prenatal exposure (Pierog et al., 1977) or through passage in breast milk (Lester et al., 1993; Mortimer, 1977; Rogers, 1979). In a report of a series of infants coming to an emergency room, 1 of every 400 visits was for the primary complaint of crying in the absence of fever in infants up to 2 years of age (median 3.5 months; Poole, 1991). Of these, 61% were judged to have a serious organic condition as a cause for their crying. Two had suffered trauma (subdural hematoma or tibial fracture) subsequently determined to be "nonaccidental." About 1 in 10 had classic colic not otherwise related to organic disease.

In primary care settings, the incidence of crying (or colic) as a complaint in the first 3 months is higher (Forsyth et al., 1985b), and of organic diseases as a cause of that crying is lower, than seen in emergency room settings. Prevailing opinion and most reports of clinical experience in office practice support the conclusion that the majority of infants with colic are otherwise normal (Carey, 1984; Illingworth, 1954; Paradise, 1966; Treem, 1994; Wessel et al., 1954). Indeed, many clinicians limit the diagnosis of "colic" to persistent crying in an "otherwise well infant" (Barr, 1991). In office practice settings, both clinical practice and clinical research have focused on the role of nutrients (cow's milk protein and carbohydrate) in exacerbating crying. The interest in dietary factors follows from clinical observations that feeding is a sufficient stimulus to arrest crying in normal infants, while persistent crying after feeding often occurs in infants with colic. It is common clinical practice to switch formulas from cow's milk protein to hydrolyzed casein or noncasein-based formulas in infants with colic or to eliminate ingestion of cow's milk in mothers who are breast-feeding their infants. Both the term "colic" (Carey, 1984) and the apparent increase in abdominal distension and expired gas implicate a disturbed gastrointestinal system.

Clinical research has both supported and delimited the roles of cow's milk protein and carbohydrate in the pathogenesis of colic symptoms (Barr, 1996). Cow's milk protein is presumed to act as an antigenic stimulus for a gastrointestinal hypersensitivity reaction. There are about 20 potentially allergenic protein antigens in cow's milk. The most likely and possibly most antigenic (beta-lactoglobulin, casein, and bovine IgG) are also found in human breast milk (Clyne & Kulczycki, 1991; Jakobsson et al. 1985; Stuart et al., 1984). For a few weeks in the postnatal period, the small intestine is relatively permeable to dietary protein, and this permeability may be increased in infants with colic (Lothe et al., 1990b). Increased plasma cell content has been reported in the small intestinal mucosa in response to milk challenge in a few infants with colic, as might be expected with an IgE-mediated hypersensitivity reaction to milk protein (Harris et al., 1977). On the other hand, there is no difference in the prevalence, pattern, or amount of crying or colic in breast- versus formula-fed infants (Barr et al., 1989; Forsyth et al., 1985b; Illingworth, 1954; Paradise, 1966; Stahlberg, 1984). A difference might be expected because of the greater quantity and concentration of most potentially antigenic proteins in formula. There is also a lack of correlation between symptoms and indices of permeability of the small intestine (Lothe et al., 1990), and attempts to demonstrate small intestinal tissue damage report conflicting results (Lothe et al., 1990a, b; Thomas et al., 1987).

Controlled diet trials, the ultimate test of the clinical significance of dietary protein intolerance, have had mixed results (Evans et al., 1981; Jakobsson & Lindberg, 1983; Jakobsson et al., 1985; Lothe & Lindberg, 1989; Lothe et al., 1982). In the most carefully designed study comparing cow's milk and casein hydrolyzed formula, there was a statistically significant reduction in crying, fussing, and "colic" crying after the first, but not after the next two, changes to hydrolyzed formula. Furthermore, only 2 of the 17 infants showed the predicted pattern of increased crying when switching to cow's milk and decreased crying when switching to casein hydrolyzate. In addition, the level of crying was still high (2–3 hours/day) even on hydrolyzed formula (Forsyth, 1989).

Most studies are hampered by limitations in samples, measurement, and design (Miller & Barr, 1991; Sampson, 1989). Diet studies tend to have small, select patient samples from specialty referral services (such as gastroenterology clinics), and the infants are more likely to manifest additional symptoms (diarrhea, vomiting, weight loss) and more severe crying than is typical in samples of infants with colic from general pediatric practice (Miller & Barr, 1991). Consequently, although early studies suggested that up to 70% of infants with colic may have protein intolerance (Lothe et al., 1982), more recent estimates suggest that protein intolerance accounts for less than 10% (Sauls & Redfern, 1997; Miller & Barr, 1991; Treem, 1994).

The role of carbohydrate is equally intriguing but less clear. The most likely mechanism by which carbohydrate might produce crying is an increase in colonic gas production secondary to the utilization of incompletely absorbed carbohydrate (lactose) and endogenous glycoproteins as metabolic substrates by colonic bacteria (Barr, 1996; Perman & Modler, 1982). Incomplete carbohydrate absorption is consistent with many clinical features of colic: (1) there is large intraindividual variability in colonic gas production; (2) it is equally likely in breast- and formula-fed infants; (3) increased carbohydrate absorption occurs during the third month, at the same age when crying tends to resolve (Barr et al., 1984); (4) evening crying may be due to the accumulation of partial, incompletely absorbed carbohydrate from more frequent daytime feeds; and (5) persistent crying after feeding may be due to persistent colonic gas, which is unaffected by burping. However, as with protein nutrients, attempts to demonstrate an etiologic role for intestinally derived gas have been mixed and marked by methodological difficulties (Barr, 1996; Hyams et al., 1989; Miller et al., 1990). In normal infants, there was no reduced crying associated with level of lactose ingestion in a cross-over trial (Barr et al., 1991c). To date, there has not been a convincing demonstration that modifying incomplete lactose absorption or reducing intestinal gas by surface active agents (simethicone) reduces colic or crying (Metcalf et al., 1994; Miller et al. 1990; Sethi & Sethi, 1988; Stahlberg & Savilahti, 1986).

In sum, an excessively crying infant *may* have an underlying organic disease condition, and diagnostic efforts to detect these are indicated. If diagnostic tests for organic disease are negative, a further 5–10% may respond to a cow's milk protein-free diet, especially in the presence of vomiting, diarrhea, or weight loss. However, it is increasingly unlikely that most cases of colic will be accounted for by organic disease. From the medical perspective, colic presents the paradox of a symptom of disease in an otherwise well infant.

Colic as an Activity That Infants "Do"

For clinical purposes, it is appropriate to establish criteria to decide which infants have a condition. However, the crying of colic may simply represent the upper end of a continuum of crying in a single population, rather than distinctly different crying in two populations of infants who have, or do not have, colic. To say that they *do* it (rather than *have* it) is to say that the behaviors of the "syndrome" are continuous with normal behavior but that their frequency or duration is simply greater in infants with colic. A number of lines of evidence support the continuity of normal and colic crying.

Colic Crying as Continuous with Normal Crying

In their original proposal for a quantitative definition of colic, Wessel and colleagues (Wessel et al., 1954:428) observed that "the time distribution and frequency of diurnal regularity are similar for the mild fussy periods of the 'contented babies,' and for the more prolonged periods of the 'fussy infants.' " This continuity in *temporal patterning* has been confirmed repeatedly in studies of normal infants who do not come to the physican with a complaint of crying (Barr, 1990a). The crying pattern typical of infants without colic also shows more crying in the first 3 months, a peak pattern, and clustering in the late afternoon and evening (Barr, 1990a; Brazelton, 1962; Hunziker & Barr, 1986; St.James-Roberts, 1989; St.James-Roberts & Halil, 1991). Furthermore, there is no apparent "cut-off" in the quantity of crying. Most comparative studies document similar relationships: while there may be *a greater quantity of* a particular crying characteristic, it is also found in the crying of infants without colic. For example, during crying before a feed, infants with colic are more likely to manifest a "pain facies" (closed eyes, open mouth, furrowed brow, and deepened nasolabial folds), but pain facies are also manifest in infants without colic (Barr et al., 1992). Similarly, infants with colic are more likely to manifest unsoothable crying bouts, but such bouts also occur in infants without colic. Regardless of whether the infants have colic, these unsoothable bouts occur in proportion to the overall amount of crying and fussing the infants do (St.James-Roberts et al., 1995a). Finally, clinical studies show that there is considerable overlap in amounts of crying of infants who are or are not brought to health services with the complaint of colic (Barr et al., 1992; St.James-Roberts & Halil, 1991).

There have been fewer studies of cry *quality* in infants with colic. Parents are more likely to perceive the cries differently, and some acoustic features are more common, but none of these features are specific to infants with colic (Barr et al., 1992; Lester et al., 1992; Zeskind & Barr, 1997). Thus, despite the impression that there is something distinctly different about the behavior of infants with colic in the clinical setting, it is more likely that they do what otherwise normal infants do, but more so because they cry for longer periods of time.

Evolutionary Perspective

A central tenet of an evolutionary approach to understanding problems in clinical medicine is the notion that clinical conditions may emerge when there is a mismatch, or discordance, between the organism's predispositions to act in a certain way and the environment in which it finds itself (see Eaton et al., 1988). It is presumed that the organism's predispositions to act were matched (or "co-adapted") to expectable environments in which it evolved, but that this reciprocal organism–environment concordance was strained by relatively rapid changes in social, physical, and/or psychological settings. Applied to early crying and colic, the question is whether crying observed in the clinical setting might be a manifestation of such a discordance between an evolution-arily defined species predisposition to cry and contemporary Western care-giving practice. It was argued earlier that the increased duration of crying and the prolonged unsoothable crying bouts typical in the clinical phenomenology of colic syndrome are features of crying that all infants *do*, rather than a con-dition infants *have*. If so, the question from the point of view of evolutionary medicine is whether current caregiving practices exacerbate crying character-istics that would be otherwise "appropriate" in evolutionarily expectable en-vironments. If they do, then we would expect more infants to have longer, unsoothable crying bouts—that is, to manifest behavioral features of the "colic syndrome."

Crying in the Environment of Our Evolutionary Adaptedness

One way to approach this question is to ask what the characteristics of early crying might have been under caretaking conditions characteristic of those in which human beings evolved or, in Bowlby's felicitous phrase, "the environ-ment of our evolutionary adaptedness" (Bowlby, 1969). In this regard, a study of crying in infants of !Kung San hunter-gatherers of Botswana is of particular interest. Their caregiving is usually considered "indulgent" (DeVore & Kon-ner, 1974; Konner, 1976) and includes features that differ significantly from those typical of contemporary Western caregiving (Barr, 1990b). First, com-pared to Western norms, !Kung San infants are carried constantly in their mother's arms or in a sling (kaross). As a result, they are in direct body contact most of the time (rather than in a crib or a stroller). Second, both because of the sling and because of a cultural belief that the upright position enhances development, they are kept upright (instead of supine). Third, in a significant departure from "scheduled" and even "demand" feeding in Western societies, !Kung San infants are allowed to feed "continuously" (approximately three to four times an hour for 1–2 minutes/feed) [Konner & Worthman, 1980] rather than in a "pulse" pattern. And fourth, on classic measures of contingent res-ponsivity, !Kung San mothers are much more responsive than Western moth-ers. Specifically with regard to crying, virtually every fret is responded to immediately (92% of the time within 10 seconds [Barr et al. 1987]). Western mothers respond as quickly when they are in direct contact with their infants. However, they are more often separated from their infants and, overall, they deliberately refrain from responding as much as 40% of the time. In addition to the separation, this low rate of responding may be due in part to a cultural belief that always reponding to infant cries will "spoil" the child (Bell & Ains-worth, 1972). If differences in caregiving practices do modify crying behavior, crying patterns in the !Kung San infants should indicate what aspects of crying may have been similar and what aspects may have differed in these different caregiving contexts.

The "picture" of crying in !Kung San infants is telling (Barr et al., 1991a). First, !Kung San infants still manifest an early crying/fretting peak within the first 3 months of life. Second, relative to American and Dutch infants, the frequency of crying is essentially the same, but the duration of crying is only about half as long. Finally, prolonged crying bouts are rare. Crying that persists for more than 30 seconds accounts for less than 10% of the crying events in the first 9 months of life. Overall, these findings suggest that caregiving style can affect infant crying and that it specifically affects crying duration but not the frequency or the early "peak" pattern.

Whether !Kung San caregiving and crying is truly representative of that which occurred in the environment (more accurately, the environments) of our evolutionary adaptedness is an unanswerable question. However, there is at least some support for the commonality of these caregiving practices. Many of the caregiving features cluster together in other hunter-gatherer societies. Carrying, immediate nurturant response to crying, and prolonged breastfeed-ing were typical of 10 of the most clearly "foraging" societies in Murdoch and White's collection of reported observations from 186 nonindustrialized soci-eties (Lozoff & Brittenham, 1979; Murdock & White, 1969). Other nonindus-trialized societies tend to have intermediate levels of these caregiving features. Consequently, while it would be a mistake to assume that the !Kung San ex-perience represents *the* evolutionary caregiving context, variants of the !Kung San pattern are historically more venerable, very broadly distributed in human cultures, and apparently strongly associated with hunter-gatherer modes of existence. In contrast, Western caregiving is radically different from a cross-cultural point of view, and very recent from an evolutionary point of view, being typical of caregiving for less than 1% of human evolutionary experience (Kennel, 1980; Lozoff & Brittenham, 1979).

In addition, this cluster of "indulgent" caregiving behaviors appears to be a typical higher primate caregiving pattern. Similar high levels of mother–infant contact and contingent responsivity to infant signals occur in rhesus, squirrel, and pigtailed monkeys ((Hinde & Spencer-Booth, 1967, 1968; Jensen et al., 1968; Kaplan, 1972; Wolfheim et al., 1970). A "continuous" pattern of feeding is typical for most monkey species (Horwich, 1974). In addition, it is particularly evident for free-living chimpanzees, whose short (1–2 minutes), frequent (<20-minute interval) feeds closely resemble those of the !Kung San (Clark, 1977; Nicolson, 1977; Plooij, 1984; van Lawick-Goodall, 1968). Fur-thermore, Blurton-Jones (1972) has demonstrated cross-species regularities in mammals between breast milk composition and caregiving strategies. Species with low protein and fat content have low sucking rates, frequent feeding at short intervals, and continuous proximity with their young (so-called carrying species); species with high protein and fat content exhibit the opposite behaviors (so-called caching species [Ben Shaul, 1962; Blurton-Jones, 1972; Ewer, 1968]). Extended to humans, both sucking rates and breast milk composition would predict frequent, short feeds and continuous proximity—characteristics typical for the !Kung San but not for Western caregivers.

Analogously, there is considerable indirect support that the crying peak seen in !Kung San infants represents a stable, relatively "hard-wired" behavioral characteristic that both was and is a feature of early infant behavior. As already described, the crying peak is a feature of the clinical colic syndrome in contemporary society. In addition, Brazelton (1962) reported the increased amounts of crying and a crying peak in a diary study in the first 3 months in North American middle-class infants. Other studies have since confirmed the robustness of this pattern for normal and colicky infants in a variety of settings (Alvarez & St.James-Roberts, 1996; Barr, 1990a; St. James-Roberts & Halil, 1991). For example, a similar peak pattern was found in a Montreal study almost a quarter century after Brazelton's early report (Hunziker & Barr, 1986). Mothers in Manali, India, are much more responsive to infant cries, more likely to take their infants to bed with them and to breast-feed than mothers in London. Despite these caregiving differences, Manali and London infants both have similar patterns of crying with a distinct evening peak, an increase in crying from 2 to 6 weeks, and a tendency to decline between 6 and 12 weeks (St.James-Roberts et al., 1994). In Korean infants whose mothers are in direct contact with their infants more than 90% of the time, crying is also greater during the second month than later (Lee, 1994). Despite 8 additional weeks of extrauterine life and caregiving exposure, otherwise well preterm infants born at approximately 32 weeks gestation manifest their crying "peak" at about six weeks *corrected* age (i.e., 14 weeks postnatally), and peak timing is unrelated to a variety of postnatal medical complications (Barr et al., 1996).

This convergence of findings raises the likelihood that the early increase and peak crying pattern is a cross-culturally and evolutionarily stable developmental characteristic. Indeed, it is arguably the case that this pattern is a behavioral universal—that is, a propensity for increased crying characteristic of all groups of infants in the human species, if not of all infants (Konner, 1989). However, no empirical data directly address the !Kung San-Western difference in crying; namely, that crying duration is reduced while frequency remains the same. Interestingly, a randomized controlled trial of increased carrying of the infant does provide some convergent support for this observation (Hunziker & Barr, 1986). Parents in an experimental ("supplemental carrying") group increased the carrying of their infants to 4.4 hours/day between weeks 4 and 12 of age, while parents in the control condition carried their infants 2.7 hours/day. At the peak age of crying (6 weeks), the increased carrying produced a 43% reduction in the duration of daily crying and fussing, but no change in the frequency of crying/fussing bouts. These results provide rather striking support for the suggestion that the feature of crying most amenable to change by caregiving style is duration (or bout length) rather than frequency of crying and fussing (Barr, 1990b).

Taken together, these studies suggest that, in caregiving contexts typical of the environments of our evolutionary adaptedness, early crying might well have shared the early peak pattern seen in contemporary normal and colicky infants. Furthermore, crying may well have been equally frequent, but when crying occurred, it was shorter in duration. On the face of it, this makes considerable sense. All four of the prominent differences in caregiving (i.e., constant carrying and contact, upright posture, shorter interfeed intervals, and rapid and consistent response to infant distress signals) might reasonably be expected to be calming and to help the infant regulate its distress (Barr, 1990b).

The Function of Crying in the Environment of Our Evolutionary Adaptedness

If an evolutionary medicine approach is applicable to the phenomenology of early crying and colic, it follows that crying might be expected to have functioned for the benefit of the infant, at least in its historical evolutionary context. Crying may, of course, produce "positive" or "negative" short-term consequences for the infant. For a truly defensible evolutionary account, however, one would have to demonstrate that the behavior was adaptive in the long term—that is, maximized survival and reproductive success. Given the negative effects that have been attributed to colic in contemporary clinical contexts, the challenge to demonstrate that crying could be (or could have been) beneficial in the short or long term would seem to be considerable.

No such case for the adaptive benefits of crying has been demonstrated for infants. However, based on known relationships between crying and caretaker responses, a *prima facie* case for its adaptive value can be outlined. To the extent that the case is persuasive, it provides a framework for generating testable predictions concerning the function of crying. In the case proposed, it is argued not only that crying can have positive immediate consequences that could lead to long-term adaptive outcomes, but also that positive immediate and long-term outcomes are more likely when crying bouts are frequent and short (as in the !Kung San case) rather than prolonged (as in the Western case).

We begin with a model that relates feeding in a unidirectional causal relationship first to growth and then to survival (depicted in figure 2.1A). Understandably, feeding usually "stands for" the provision of calories (energy) essential for growth. Although simple, this model has benefit of summarizing one of the most basic of clinical principles. When feeding is compromised for any reason, we go to extraordinary efforts to compensate for this constraint. Furthermore, assuring adequate calories through feeding underlies virtually all health intervention projects in developing countries directed toward improving growth and survival (Dettwyler & Fishman, 1992).

In the nonclinical setting, crying plays a significant role in initiating a feed. In contemporary Western settings, maternal response to crying usually begins with a cluster of nonspecific soothing activities (talking to, picking up, moving

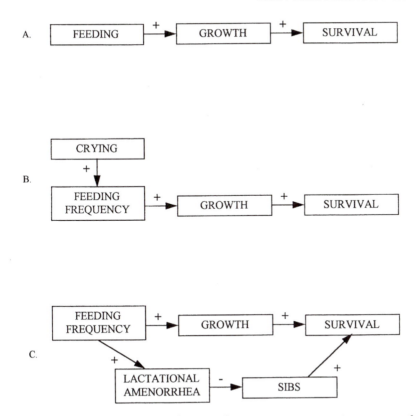

Figure 2.1. Proposed models to illustrate how crying may act to promote adaptation (increased survival). (+) indicates a positive influence of the precursor on the outcome that follows; (−) indicates a negative influence. (A) An assumed basic casual relationship between feeding, growth, and survival. (B) The same model modified by the addition of crying as a stimulus for feeding (and, especially, feeding frequency). Crying may work directly through this basic nutritional pathway to enhance survival. (C) Enlarged model to incorporate a second non-nutritional mechanism (feeding frequency determined lactational amenorrhea) by which feeding frequency may enhance survival. (See text for details.)

the infant), but feeding is the most likely next behavior if nonspecific soothing fails (Green et al., 1995; Gustafson & DeConti, 1990; Gustafson & Harris, 1990). The likelihood that parents will respond to crying with a feed is situationally determined, increasing as the time from the previous feed increases (Bernal, 1972; Murray, 1979). In the !Kung San, however, feeding (and feeding interval) is much more closely determined by crying. Following the smallest fret or cry, the probability that a !Kung San infant will be fed is 50% in the first 3 months and 40% for the next 6 (Barr et al., 1987). We can now expand and modify the basic model slightly (figure 2.1B) to recognize that this behavior—at least in the !Kung context—ensures a high feeding frequency and, by implication, the short interfeed interval characteristic of their "continuous" feeding pattern (Konner & Worthman, 1980).

It is a reasonable assumption that frequent feeding helps to assure adequate (perhaps even optimal) nutritional intake, growth, and survival. One mechanism by which this is accomplished is through the influence of frequent feeding on high-energy fat content. In those Gambian and Thai women who feed their infants more frequently, breast milk has a higher fat concentration in the next feed, and there is a smaller decrease in fat between the end of one feed and the beginning of the next (Jackson et al., 1988a, b; Prentice & Prentice, 1988; Prentice et al., 1981). This is probably mediated through suckling-induced pituitary prolactin release that suppresses lipoprotein lipase action in maternal adipose tissue but enhances it in breast tissue, so that serum lipids are diverted into breast milk production (Dettwyler & Fishman, 1992; Quandt, 1984).

Infant-driven breast-feeding frequency may recruit another mechanism contributing to infant survival—namely, delaying the appearance of sibling competitors for maternal caregiving resources by inducing interfeed interval-dependent lactational amenorrhea in the mother (figure 2.1C). Breast-feeding has received considerable attention as a major determinant of birth spacing (e.g., Howie & McNeilly, 1982). It is likely that lactational amenorrhea is mediated by elevated concentrations of pituitary prolactin maintained by frequent suckling (and/or suckling duration [Vitzthum, 1989]), suppressing ovarian cycling (Konner & Worthman, 1980; Wood et al., 1985). Indeed, suckling interval among the highland Papua New Guinea Gainj was shown to be the principal determinant of maternal prolactin concentration. The associated lactational anovulation accounted for 75% of the typical interval between a live birth and the next successful conception (Wood et al., 1985). The behavior of frequent crying, when responded to "indulgently" by the mother, may contribute to survival by both nutritional and non-nutritional pathways.

There is a third pathway by which frequent crying may facilitate infant growth directly and survival indirectly (figure 2.2). It is reasonable to ask whether the feeding frequency "feeds back" to change (increase or decrease) crying frequency. If so, one might expect crying frequency to be different in !Kung San and contemporary Western settings. As we have seen, frequency of crying appears to be relatively independent of the caregiving context (Barr et al., 1991a; Hunziker & Barr, 1986). However, frequent feeding is associated with shorter duration of crying. Because of the muscular activity associated with crying, the noncrying versus the crying state is associated with an energy savings, recently determined to be on the order of 13% in newborns (Rao et al., 1993). Reduced crying duration could therefore contribute energy to growth needs, a margin of safety which might be particularly important in the face of scarcity or illness.

This cry-survival "pathway" illustrates that growth and survival through feeding may not be determined solely by the caregiver. They may also be

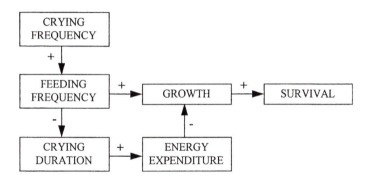

Figure 2.2. Modification of the basic model (figure 2-1) to illustrate how maternal feeding responses to crying produce a more proximal reduction in infant cry duration to save energy for growth and subsequent survival. (+) Indicates positive influences; (−) indicates negative influences.

determined by infant behavior when within a caregiving context (such as that provided by the !Kung San) that does not proscribe feeding again if the infant cries. Furthermore, this relationship is probably mediated by complementary and overlapping pathways, some nutritional and some not. Interestingly, highly analogous pathways can be described for two other short-term benefits (emotional attachment and decreased predation) with presumed contributions to long-term survival. In figure 2.3, the cry-feeding-survival pathway of figure 2.2 has been modified by substituting mother–infant attachment (figure 2.3A) and (decreased) predation (figure 2.3B) for growth.

In the attachment pathway, frequent, short cries function to elicit positive social-communicative responses (smiling, talking, face-to-face interaction). Repeated many times daily, these engender in the infant the confidence that the caregiver predictably responds to its needs. This caregiver–infant attachment relationship provides a secure base for meeting emotional and exploratory challenges in later infancy. This hypothesis, articulated most clearly in Bowlby's seminal work *Attachment and Loss* (1969), has been the stimulus for numerous productive investigations of infant emotional and social development (e.g., Bell & Ainsworth, 1972; Bowlby, 1969; Waters et al., 1995). Bowlby proposed that crying and other infant behaviors (smiling, suckling, clinging, calling, and following) functioned as "attachment behaviors" that contributed to proximity and subsequent secure caregiver–infant attachment relationships. Others have proposed that crying may be even more important than smiling in this role because a contingent caregiver response to crying is more predictable (Gekoski et al., 1983; Lamb, 1981; Lamb & Malkin, 1986). In support of this pathway, it is indeed the case that the most common first responses to crying and fretting are positive, social verbal responses in both !Kung San and contemporary Western caregivers (Barr et al., 1987; Green et al., 1995; Gustafson & DeConti, 1990; Gustafson & Harris, 1990; Moss, 1974).

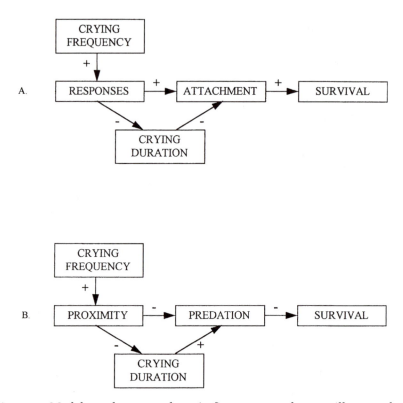

Figure 2.3. Models analogous to those in figures 2-1 and 2-2 to illustrate that infant crying may contribute to survival through parallel "attachment" and "reduced predation" effects, proximally mediated by caregiver responses that (A) are social positive and that (B) facilitate infant–caregiver proximity. (+) Indicates positive influence; (−) indicates negative influence. Not that the "logic" of all three models (figures 2-1–2-3) is identical: frequent crying is directly facilitatory and, in a responsive caregiving context, indirectly shortens crying duration that would otherwise increase the risk of predation.

Analogously, in the predation pathway, cries function to assure caregiver proximity, reducing the risk of predation if left alone.

It is important to note that feeding, social-communicative responses, and proximity also serve to soothe the infant, reducing the likelihood of prolonged crying bouts. Prolonged crying would be likely to disrupt secure caregiver–infant relationships (Murray, 1979) and increase the risk of predation. Thus, both the frequency and the short durational characteristics of crying contribute to its putative adaptive value. Frequent short crying bouts are sufficient, and better suited than long bouts, to promote all of the positive and presumably adaptive functions of infant crying.

Results from an Evolutionary Conceptualization

All three of the pathways discussed above may have been more important for survival in our evolutionary past than they are now. Significant environmental "selection pressures" such as starvation, being left without a caregiver, and predation have diminished significantly in contemporary Western society. Indeed, infant growth can clearly be achieved through sufficient provision of calories (even in the face of increased crying duration) using widely spaced "pulse" feedings of breast milk substitutes. Analogously, some Western societies are "natural" experiments for the hypothesis that normal emotional development and attachment may be possible through alternative caregiving settings in which parents and infants are separated for significant periods of time even during the first year of life.

This relative—and relatively recent—caregiver–infant separation, both with regard to physical proximity and temporal patterning of interaction, has stimulated basic questions about the importance of the caregiver–infant dyad. This separation might well affect the infant's abilities to meet its crucial developmental challenges in the first few weeks and months of life. Since the seminal work of Harlow (Harlow & Harlow, 1965), a substantial literature has validated the importance of a close mother–infant dyadic relationship for infant development.

In mammals at least, these challenges are faced first within the mother–infant pairing. Prototypical components of this relationship are the delivery of nutrients (such as breastmilk) and contact (broadly including suckling, touch, warmth, odor, and movement). It should not, therefore, be surprising if these caregiving components were important determinants of the infant's ability to cope with early challenges. Current research focuses not simply on whether separation has effects, but on identifying behavioral and biological mechanisms (or "hidden regulators" [Hofer, 1994]) by which these effects come about. Indeed, a substantial literature has eloquently described a large number of biological and behavioral regulators in animals (e.g., Blass, 1990; Blass & Ciaramitaro, 1994; Hofer, 1987, 1994, 1995; Krasnegor & Bridges, 1990; Rosenblum & Moltz, 1983; Smotherman & Robinson, 1992, 1994; Wilson & Sullivan, 1994).

Analogously, nutrient and contact caregiving components are likely to be important determinants of the infant's ability to regulate distress in the first few months of life. A number of studies in human infants have confirmed the potential importance of both contact and nutrients as regulators of infant behavioral state. As previously reported, increasing carrying from 3 to more than 4 hours a day reduces duration (but not frequency) of crying/fussing behavior by 43% at 6 weeks of age (Hunziker & Barr, 1986). Different nutrient combinations differentially affect behavioral states of the infant. A complete milk formula feed specifically increases post-feed sleep by 46% and 118% compared to water or carbohydrate-only feedings, but both milk formula and carbohydrate-only feeds reduce crying relative to water (Oberlander et al., 1992). The act of feeding (regardless of nutrient content) concentrates wakefulness in the first 10 minutes following the feed (Oberlander et al., 1992). In apparent confirmation of the !Kung San findings, studies with LaLeche League breast-feeding mothers showed that both shorter interfeed intervals and quicker response times to crying were independently associated with reduced crying activity (Barr & Elias, 1988). Interestingly, these relationships only held for crying at 2 months of age; no similar pattern was found at 4 months of age, after the crying peak was over.

In more recent work, evidence has accumulated for complementary but distinct orotactile and orogustatory (taste) pathways for infant calming. As little as two drops (250 µl) of sucrose applied to the tongue of a crying infant induces rapid and substantial calming, which persists for up to 5 minutes following the taste stimulus (Barr et al., 1994; Blass & Ciaramitaro, 1994; Blass et al., 1989, 1990; Smith et al., 1990). Similar calming effects seen with a pacifier do not persist after the pacifier is removed (Smith et al. 1990; Wolff, 1969). Studies in nonhuman animal infants suggest that the orogustatory (or taste) calming system is subsumed by an opioid-dependent central pathway, whereas the orotactile (or contact) pathway is not (Blass et al., 1990; Kehoe & Blass, 1986a, b, c; Panksepp et al., 1978, 1980). Furthermore, crying newborns of methadone-dependent mothers are not responsive to sucrose calming, but are responsive to pacifier calming (Blass & Ciaramitaro, 1994). These observations imply that, even at the level of oral stimulation within the feeding act, complementary but functionally separate nutrient and tactile pathways may contribute to distress regulation in the infant.

Clinical Application

The evolutionary perspective provides us with a picture of crying that stands in stark contrast to that encountered in the clinical setting. With the possible exception of the welcome sound of the first cry after birth, few physicians or parents take an increase in crying as a sign that the infant is developing well and thriving. Rather than being seen as an infant strategy to assure nutrition, crying is seen to reflect poor or inadequate feeding. Rather than facilitate attachment, crying may challenge mothers' parenting efficacy, self-worth, and emotional stability. Rather than protection from predators, crying may provoke anger, abuse, or even death from its caregivers. The previously described clinical perspective left us with the paradox of accounting for a symptom of a disease in an otherwise well infant. The evolutionary perspective provides a possible resolution of that paradox: the increasing spatial and temporal separation between caregivers and infants has reduced the caregiver contribution to distress regulation in infants, resulting (perhaps quite specifically) in longer bouts of otherwise normal crying and fretting behavior. With this general increase in crying, more infants are likely to manifest more prolonged, unsoothable, "colic" bouts. However, these manifestations of colic need not represent pathophysiology in the infant. Rather, the organism's natural predispositions to cry function the way they should from an evolutionary perspective, where crying encouraged feeding, attachment, and safety from predators and therefore was a fitness-maximizing behavior.

When faced with a complaint of increased crying, there are typically three general categories of intervention available to clinicians: patient education, direct intervention to modify the cause of the behavior, and prevention of secondary negative consequences. Most medical interventions are attempts at direct intervention. In the 10% or fewer cases in which a specific organic etiology is detected, direct intervention (e.g., elimination of cow's milk protein; antibiotics for urinary tract infection) is appropriate and desirable. In the remainder, however, interventions aimed at reducing crying behavior have been mixed, and the results have been far from impressive (for reviews, see Barr, 1996; Miller & Barr, 1991). If the evolutionary interpretation described here is correct, it offers potential alternatives for the clinical approach to crying complaints.

In regard to patient education, being able to inform parents that increases in daily crying are predicted and need not implicate disease, insufficient milk, nor bad caregiving can provide much needed reassurance. By itself, this advice need not be specific to an evolutionary account; it can be supported simply by current empirical accounts of clinical experience. What is specific to the evolutionary perspective is that this pattern of increased early infant crying "makes sense" (that is, the infant is engaging in fitness-maximizing behavior), and this may reasonably account for its presence in the absence of disease. For parents who are made anxious by its seeming inexplicability, such an understanding may suffice for dealing with the behavior.

With regard to direct intervention, a logical implication of this perspective is that caregiving more closely approximating that of the !Kung San may provide an alternative therapeutic strategy. This would be directed at changing normative caregiving, rather than treating intrinsic or extrinsically induced pathology in the infant. Whether changes in caregiving strategy will be effective, however, remains unresolved. Studies assessing the effects of simply increasing carrying have not been successful either as a treatment in already colicky infants (Barr et al., 1991b; Parkin et al., 1993) or as a prevention for crying complaints by parents (St.James-Roberts et al., 1995b). On the one hand, this relative ineffectiveness supports the clinical impression that infants with colic are relatively unsoothable compared to noncolicky infants. On the other hand, even the 6 hours of carrying reported in therapy trials does not approach the "constant" carrying characteristic of an "indulgent" caregiving strategy. Similarly, changing any one component (say, carrying) may be insufficient to recruit the complementary and overlapping calming pathways exploited in !Kung San caregiving.

Even if it is demonstrated that sufficiently large changes in caregiving could be effective in treating or preventing colic, the value of the evolutionary approach lies not in insisting that caregiving patterns be changed, but rather in providing a wider range of caregiving styles that are normative than is typically assumed to be true. In the face of competing ideological, cultural, or workplace demands of potential benefit to caregivers, "indulgent" caregiving may be neither advantageous nor necessary. On the other hand, in the face of an excessively crying infant, trading off the benefits of more parental freedom with a more typical Western strategy for the benefits of a quieter infant with a more indulgent caregiving strategy may be both appropriate and therapeutic. An indulgent caregiving strategy is a possible choice, not a prescription, for therapeutic intervention in colic.

Finally, in regard to prevention of secondary consequences, the evolutionary perspective provides alternative advice to current practice. Rather than advise deliberate nonresponse and proscription of frequent feeding, there is both precedent and a rationale for facilitating maternal responsiveness and shorter interfeed intervals. Being responsive is intuitive and probably "hardwired" for caregivers (Papousek & Papousek, 1990), and deliberately constraining this responsiveness is difficult for parents (Taubman, 1984). In the absence of significant weight loss, there is no medical rationale for discontinuing breast-feeding and switching to formula. In contrast, an evolutionary perspective implies that increased breast-feeding frequency may both reduce distress and enhance growth. Finally, rather than dismiss parental complaints about crying as "over-concern" on the part of the mother, it implies that such concern should be acknowledged and welcomed as a sign that the mother–infant system is working as it should.

Acknowledgments
This work was supported by the Sessenwein Academic Award to R.G.B. I acknowledge many helpful discussions with Melvin Konner, Carol Worthman, James McKenna, Marjorie Shostak, E.O. Smith, Robert Levine, Sara Harkness, and Nicole Calinoiu, and the helpful critical comments of an unknown reviewer.

References

Alvarez, M. & St. James-Roberts, I. (1996) Infant fussing and crying patterns in the first year in an urban community in Denmark. *Acta Paediatrica.* 85: 463–466.

Andersson, D.E. & Nygren, A. (1978) Four cases of long-standing diarrhoea and colic pains cured by fructose-free diet—A pathogenetic discussion. *Acta Med Scandinavica*: 230: 87–92.

Archibald, H.C. (1942) Colic or otitis media? *Archives of Pediatrics* 59 (12): 767–771.

Barr, R.G. (1990a) The normal crying curve: what do we really know? *Developmental Medicine and Child Neurology* 32: 356–362.

Barr, R.G. (1990b) The early crying paradox: a modest proposal. *Human Nature* 1: 355–389.

Barr, R.G. (1991) Colic and gas. In W.A. Walker, P.R. Durie, J.R. Hamilton, J.A. Walker-Smith & J.G. Watkins, eds., *Pediatric Gastrointestinal Disease: Pathophysiology, Diagnosis and Management*, vol. 1., Burlington, VT: B.C. Decker, pp. 55–61.

Barr, R.G. (1996) Colic. In W.A. Walker, P.R. Durie, J.R. Hamilton, J.A.

Walker-Smith & J.B. Watkins eds., *Pediatric Gastrointestinal Disease: Pathophysiology, Diagnosis, and Management*, 2nd ed., vol. 1. St. Louis: Mosby, pp. 241–250.

Barr, R.G., Bakeman, R., Konner, M. & Adamson, L. (1987) Crying in !Kung infants: distress signals in a responsive context. *American Journal of Diseases of Children* 141: 386.

Barr, R.G., Bakeman, R., Konner, M. & Adamson, L. (1991a) Crying in !Kung infants: a test of the cultural specificity hypothesis. *Developmental Medicine and Child Neurology* 33: 601–610.

Barr, R.G., Chen, S-J., Hopkins, B. & Westra, T. (1996) Crying patterns in preterm infants. *Developmental Medicine and Child Neurology* 38: 345–355.

Barr, R.G. & Elias, M.F. (1988) Nursing interval and maternal responsivity: effect on early infant crying. *Pediatrics* 81: 529–536.

Barr, R.G., Hanley, J., Patterson, D.K. & Wooldridge, J.A. (1984) Breath hydrogen excretion of normal newborn infants in response to usual feeding patterns: evidence for "functional lactase insufficience" beyond the first month of life. *Journal of Pediatrics* 104 (4): 527–533.

Barr, R.G., Kramer, M.S., Pless, I.B., Boisjoly, C. & Leduc, D. (1989) Feeding and temperament as determinants of early infant cry/fuss behaviour. *Pediatrics* 84: 514–521.

Barr, R.G., McMullan, S.J., Spiess, H., Leduc, D.J., Yaremko, J., Barfield, R., Francoeur, T.E. & Hunziker, U.A. (1991b) Carrying as colic "therapy": a randomized controlled trial. *Pediatrics* 87: 623–630.

Barr, R.G., Quek, V., Cousineau, D., Oberlander, T.F., Brian, J.A. & Young, S.N. (1994) Effects of intraoral sucrose on crying, mouthing and hand-mouth contact in newborn and six-week old infants. *Developmental Medicine and Child Neurology* 36: 608–618.

Barr, R.G., Rotman, A., Yaremko, J., Leduc, D. & Francoeur, T.E. (1992) The crying of infants with colic: a controlled empirical description. *Pediatrics* 90 (1): 14–21.

Barr, R.G., Wooldridge, J.A. & Hanley, J. (1991c) Effects of formula change on intestinal hydrogen production and crying and fussing behavior. *Journal of Development and Behavioural Pediatrics* 12: 248–253.

Bell, S.M. & Ainsworth, D.S. (1972) Infant crying and maternal responsiveness. *Child Development* 43: 1171–1190.

Ben Shaul, D.M. (1962) The composition of the milk of wild animals. *International Zoo Yearbook* 4: 333–342.

Berezin, S., Glassman, M.S., Bostwick, H. & Halata, M. (1995) Esophagitis as a cause of infant colic. *Clinical Pediatrics* 34: 158–159.

Bernal, J. (1972) Crying during the first ten days of life. *Developmental Medicine and Child Neurology* 14: 362–372.

Blass, E.M. (1990) Suckling: determinants, changes, mechanisms, and lasting impressions. *Developmental Psychobiology* 26: 520–533.

Blass, E.M. & Ciaramitaro, V. (1994) A new look at some old mechanisms in human newborns: taste and tactile determinants of state, affect and action. *Monographs of the Society for Research in Child Development* 59 (1): 1–80.

Blass, E.M., Fillion, T.J., Rochat, P., Hoffmeyer, L.B. & Metzher, M.A. (1989) Sensorimotor and motivational determinants of hand-mouth coordination in 1-3-day-old human infants. *Developmental Psychology* 25: 963–975.

Blass, E.M., Fillion, T.J., Weller, A. & Brunson, L. (1990) Separation of opioid from nonopioid mediation of affect in neonatal rats: nonopioid mechanisms mediate maternal contact influences. *Behavioral Neuroscience* 104: 625–636.

Blurton-Jones, N. (1972) Comparative aspects of mother-infant contact. In N. Blurton-Jones, ed., *Ethological Studies of Child Behaviour*. Cambridge: Cambridge University Press, 305–328.

Bowlby, J. (1969) *Attachment and Loss*. New York: Basic Books.

Brazelton, T.B. (1962) Crying in infancy. *Pediatrics* 29: 579–588.

Cable, T.A. & Rothenberger, L.A. (1984) Breast-feeding behavioral patterns among LaLeche League mothers: a descriptive survey. *Pediatrics* 73(6): 830–835.

Carey, W.B. (1984) "Colic"—Primary excessive crying as an infant-environment interaction. *Pediatric Clinics of North America* 31: 993–1005.

Clark, C.B. (1977) A preliminary report on weaning among Chimpanzees of the Gombe National Park, Tanzania. In S. Chevalier-Solnikoff & F.E. Poirier, eds., *Primate Bio-social Development: Biological, Social and Ecological Determinants*. New York: Garland, 235–260.

Clyne, P.S. & Kulczycki, A. (1991) Human breast milk contains bovine IgG. Relationship to infant colic? *Pediatrics* 87(4): 439–444.

Cunningham, A.S., Jelliffe, D.B. & Jelliffe, E.F.P. (1991) Breast-feeding and health in the 1980's: a global epidemiological review. *Journal of Pediatrics* 118: 659–666.

Darwin, C. (1872/1979) *The Expression of the Emotions in Man and Animals*. London: Julian Friedmann.

Dettwyler, K.A. & Fishman, C. (1992) Infant feeding practices and growth. *Annual Review of Anthropology* 21: 171–204.

DeVore, I. & Konner, M.J. (1974) Infancy in the hunter-gatherer life: an ethological perspective. In N.F. White, ed., *Ethology and Psychiatry*. Toronto: University of Toronto Press, 113–141.

Du, J.N.H. (1976) Colic as the sole symptom of urinary tract infection in infants. *Canadian Medical Association Journal* 115: 334–337.

Eaton, S. B., Shostak, M. & Konner, M.J. (1988) *The Paleolithic Prescription: A Guide to Diet and Exercise and a Design for Living*. New York: Harper and Row.

Evans, R.W., Fergusson, D.M., Allardyce, R.A. & Taylor, B. (1981) Maternal diet and infantile colic in breast-fed infants. *Lancet*; 1340–1342.

Ewer, R.F. (1968) *Ethology of Mammals*. London: Logos Press.

Forsyth, B.W.C. (1989) Colic and the effect of changing formulas: a double-blind, multiple-crossover study. *Journal of Pediatrics* 115: 521–526.

Forsyth, B.W.C., McCarthy, P.L. & Leventhal, J.M. (1985a) Problems of early infancy, formula changes, and mothers' beliefs about their infants. *Journal of Pediatrics* 106: 1012–1017.

Forsyth, B.W., Leventhal, J.M. & McCarthy, P.L. (1985b) Mothers' perceptions of problems of feeding and crying behaviours. *American Journal of Diseases of Children* 139: 269–272.

Gekoski, M.J., Rovee-Collier, C.K. & Carulli-Rabinowitz, V. (1983) A longitudinal analysis of inhibition of vocal distress: the origins of social expectations? *Infant Behavior and Development* 6: 339–351.

Gormally, S.M. & Barr, R.G. (1997) Of clinical pies and clinical clues: proposal for a clinical approach to complaints of early crying and colic. *Ambulatory Child Health* 3: 137–153.

Green, J.A., Gustafson, G.E. & Kalinowski, L.L. (1995) Surprising similarities in mothers' responses to infants' cries and vocalizations.

Abstracts for Society for Research in Child Development, p. 10.

Gustafson, G.E. & DeConti, K.A. (1990) Infants' cries in the process of normal development. *Early Child Development and Care* 65: 45–56.

Gustafson, G.E. & Harris, K.L. (1990) Women's responses to young infants' cries. *Developmental Psychology* 26: 144–152.

Guttman, F.M. (1972) On the incidence of hiatal hernia in infants. *Pediatrics* 50 (2): 325–328.

Harkness, M.J. (1989) Corneal abrasion in infancy as a cause of inconsolable crying. *Pediatric Emergency Care* 5: 242–244.

Harlow, H.F. & Harlow, M.K. (1965) The affectional systems. In A.M. Schrier, H.G. Harlow & F. Stollnitz, eds., *Behavior of nonhuman primates, vol. 2*. New York: Academic Press.

Harris, M.J., Petts, V. & Penny, R. (1977) Cow's milk allergy as a cause of infantile colic: immunofluorescent studies on jejunal mucosa. *Australian Paediatric Journal* 13: 276–281.

Heine, R.G., Jaquiery, A., Lubitz, L., Cameron, D.J.S. & Catto-Smith, A.G. (1995) Role of gastro-oesophageal reflux in infant irritability. *Archives of Disease in Childhood* 73: 121–125.

Hide, D.W. & Guyer, B.M. (1982) Prevalence of infantile colic. *Archives of Disease of Children* 57: 559–560.

Hinde, R. & Spencer-Booth, Y. (1967) The effect of social companions on mother-infant relations in Rhesus monkeys. In D. Morris, ed., *Primate Ethology*. Chicago: Aldine.

Hinde, R. & Spencer-Booth, Y. (1968) The study of mother-infant interaction in captive group-living Rhesus monkeys. *Proceedings of the Royal Society of London* 169: 177–201.

Hofer, M.A. (1987) Early social relationships: a psychobiologist's view. *Child Development* 58: 633–647.

Hofer, M.A. (1994) Hidden regulators in attachment, separation and loss. In N.A. Fox, ed., *The Development of Emotion Regulation: Biological and Behavioral Considerations*. Chicago: The University of Chicago Press, 192–207.

Hofer, M.A. (1995) Early symbiotic processes: hard evidence from a soft place. In R.A. Glick & S. Bone, eds., *Pleasure beyond the Pleasure Principle*. New Haven, CT: Yale University Press, 55–78.

Horwich, R.H. (1974) Regressive periods in primate behavioral development with reference to other mammals. *Primates* 15: 141–149.

Howie, P.W. & McNeilly, A.S. (1982) Effect of breast-feeding pattern on human birth intervals. *Journal of Reproductive Fertility* 65: 545–557.

Hunziker, U.A. & Barr, R.G. (1986) Increased carrying reduces infant crying: a randomized controlled trial. *Pediatrics* 77: 641–648.

Hyams, J.S., Geertsma, A., Etienne, N.L. & Treem, W.R. (1989) Colonic hydrogen production in infants with colic. *Journal of Pediatrics* 115: 592–594.

Illingworth, R.S. (1954) Three months' colic. *Archives of Disease of Children* 29: 165–174.

Jackson, D.A., Imong, S.M., Silprasert, A., Preunglumpoo, S. & Leelapat, P. et al. (1988a) Estimation of 24h breast-milk fat concentration and fat intake in rural northern Thailand. *British Journal of Nutrition* 59: 365–371.

Jackson, D.A., Imong, S.M., Silprasert, A., Ruckphaopunt, S. & Woolridge, M.W. et al. (1988b) Circadian variation in fat concentration of breast-milk in a rural northern Thai population. *British Journal of Nutrition* 59: 349–363.

Jakobsson, I. & Lindberg, T. (1983) Cow's milk proteins cause infantile colic in breast-fed infants: a double-blind crossover study. *Pediatrics* 71: 268–271.

Jakobsson, I., Lindberg, T., Benediktsson, B. & Hansson, B-G. (1985) Dietary bovine beta-lactoglobulin is transferred to human milk. *Acta Paediatrica Scandinavica* 74: 342–345.

Jensen, G.D., Bobbitt, R.A. & Gordon, B.N. (1968) Sex differences in the development of independence of infant monkeys. *Behaviour* 30: 1–14.

Joos, T.H. & Don Steepe, C.A. (1992) Inguinal hernia as a cause of colic in infancy. *Journal of the Michigan State Medical Society* 61: 1108–1109.

Kaplan, J. (1972) Differences in the mother-infant relations of Squirrel monkeys housed in social and restricted environments. *Developmental Psychology* 5: 54–62.

Kehoe, P. & Blass, E.M. (1986a) Central nervous system mediation of positive and negative reinforcement in neonatal albino rats. *Developmental Brain Research* 27: 69–75.

Kehoe, P. & Blass, E.M. (1986b) Opioid-mediation of separation distress in 10-day-old rats: reversal of stress with maternal stimuli. *Developmental Psychobiology* 19: 385–398.

Kehoe, P. & Blass, E.M. (1986c) Behaviorally functional opioid systems in infant rats: II. Evidence for pharmacological, physiological, and psychological mediation of pain and stress. *Behavioral Neuroscience* 100: 624–630.

Kennel, J.H. (1980) Are we in the midst of a revolution? *American Journal of Diseases of Children* 134: 303–310.

Konner, M.J. (1976) Maternal care, infant behavior and development among the !Kung. In R.B. Lee & I. DeVore, eds., *Kalahari Hunter-Gatherers, Studies of the !Kung San and Their Neighbors*. Cambridge, MA: Harvard University Press, 218–245.

Konner, M. (1989) Spheres and modes of inquiry: integrative challenges in child development research. In P.H. Zelazo & R.G. Barr, eds., *Challenges to Developmental Paradigms: Implications for Theory, Assessment and Treatment*. Hillsdale, NJ: Lawrence Erlbaum Associates, 227–258.

Konner, M.J. & Worthman, C. (1980) Nursing frequency, gonadal function, and birth spacing among !Kung hunter-gatherers. *Science* 207: 788–791.

Krasnegor, N.A. & Bridges, R.S. (1990) *Mammalian Parenting: Biochemical, Neurobiological, and Behavioral Determinants*. Oxford: Oxford University Press.

Lamb, M.E. (1981) Developing trust and perceived effectance in infancy. In L.P. Lipsitt, ed., *Advances in Infancy Research*. Norwood, NJ: Ablex.

Lamb, M.E. & Malkin, C.M. (1986) The development of social expectations in distress-relief sequences: a longitudinal study. *International Journal of Behavioral Development* 9: 235–249.

Lee, K. (1994) The crying pattern of Korean infants and related factors. *Developmental Medicine and Child Neurology* 36: 601–607.

Lehtonen, L. & Korvenranta, H. (1995) Infantile colic: seasonal incidence and crying profiles. *Archives of Pediatrics and Adolescent Medicine* 149: 533–536.

Lester, B.M., Boukydis, C.F.Z., Garcia-Coll, C.T., Hole, W. & Peucker, M. (1992) Infantile colic: Acoustic cry characteristics, maternal perception of cry, and temperament. *Infant Behavior and Development* 15: 15–26.

Lester, B.M., Cucca, J., Andreozzi, L., Flanagan, P. & Oh, W. (1993) Possible association between fluoxetine hydrochloride and colic in an infant. *Journal of the American Academy of Child and Adolescent Psychiatry* 32: 1253–1255.

Listernick, R. & Tomita, T. (1991) Persistent crying in infancy as a presentation of Chiari type I malformation. *Journal of Pediatrics* 118: 567–569.

Lothe, L., Ivarsson, S-A., Ekman, R. & Lindberg, T. (1990a) Motilin and infantile colic. *Acta Paediatrica Scandinavica* 79: 410–416.

Lothe, L. & Lindberg, T. (1989) Cow's milk whey protein elicits symptoms of infantile colic in colicky formula-fed infants: A double-blind crossover study. *Pediatrics* 83: 262–266.

Lothe, L., Lindberg, T. & Jakobsson, I. (1982) Cow's milk formula as a cause of infantile colic: a double-blind study. *Pediatrics* 70: 7–10.

Lothe, L., Lindberg, T. & Jakobsson, I. (1990b) Macromolecular absorption in infants with infantile colic. Acta *Paediatrica Scandinavica* 79: 417–421.

Lozoff, B. & Brittenham, G. (1979) Infant care: cache or carry. *Journal of Pediatrics* 95: 478–483.

Metcalf, T.J., Irons, T.G., Sher, L.D. & Young, P.C. (1994) Simiethicone in the treatment of infant colic: a randomized placebo-controlled, multicenter trial. *Pediatrics* 94 (1): 29–34.

Miller, A.R. & Barr, R.G. (1991) Infantile colic: Is it a gut issue? *Pediatric Clinics of North America* 38 (6): 1407–1423.

Miller, A.R., Barr, R.G. & Eaton, W.O. (1993) Crying and motor behavior of six-week-old infants and postpartum maternal mood. *Pediatrics* 92: 551–558.

Miller, J.J., McVeagh, P., Fleet, G.H., Petocz, P. & Brand, J.C. (1990) Effect of yeast lactase enzyme on "colic" in infants fed human milk. *The Journal of Pediatrics* 117: 261–263.

Mortimer, E.A. (1977) Drug toxicity from breast milk? *Pediatrics* 60: 780–781.

Moss, H.A. (1974) Communication in mother-infant interaction. In L. Krames, P. Pliner & T. Alloway, eds., *Nonverbal Communication.* New York: Plenum, 171–191.

Murdock, G.P. & White, D.R. (1969) Standard cross-cultural sample. *Ethnology* 8: 329–369.

Murray, A.D. (1979) Infant crying as an elicitor of parental behavior: an examination of two models. *Psychological Bulletin* 86: 191–215.

Nicolson, N.A. (1977) A comparison of early behavioral development in wild and captive Chimpanzees. In S. Chevalier-Solnikoff & F.E. Poirier, eds., *Primate Bio-Social Development: Biological, Social and Ecological Determinants.* New York: Garland, 529–560.

Oberlander, T.F., Barr, R.G., Young, S.N. & Brian, J.A. (1992) Short-term effects of feed composition on sleeping and crying in newborn infants. *Pediatrics* 90 (5): 733–740.

O'Neill, T. (1993) Hard shaking suspected in infant's death. *St. Louis Post-Dispatch,* 5A.

Panksepp, J., Meeker, R. & Bean, N.J. (1980) The neurochemical control of crying. *Pharmacology Biochemistry and Behavior* 12: 437–443.

Panksepp, J., Vilberg, T., Bean, N.J., Coy, D.H. & Kastin, A.J. (1978) Reduction of distress vocalization in chicks by opiate-like peptides. *Brain Research Bulletin* 3: 663–667.

Papousek, M. & Papousek, H. (1990) Excessive infant crying and intuitive parental care: buffering support and its failures in parent-infant interaction. *Early Child Development and Care* 65: 117–125.

Paradise, J.L. (1966) Maternal and other factors in the etiology of infantile colic. *Journal of the American Medical Association* 197: 123–131.

Parkin, P.C., Schwartz, C.J. & Manuel, B.A. (1993) Randomized controlled trial of three interventions in the management of persistent crying in infancy. *Pediatrics* 92: 197–201.

Perman, J.A. & Modler, S. (1982) Glycoproteins as substrates for production of hydrogen and methane by colonic bacterial flora. *Gastroenterology* 83: 388–393.

Pierog, S., Chandavasu, O. & Wexler, I. (1977) Withdrawal symptoms in infants with the fetal alcohol syndrome. *Journal of Pediatrics* 90 (4): 630–633.

Plooij, FX. (1984) *The Behavioural Development of Free-living Chimpanzee Babies and Infants*. Norwood: Ablex.

Poole, S.R. (1991) The infant with acute, unexplained, excessive crying. *Pediatrics* 88 (3): 450–455.

Prentice, A.M. & Prentice, A. (1988) Energy costs of lactation. *Annual Review of Nutrition* 8: 63–79.

Prentice, A., Prentice, A.M. & Whitehead, R.G. (1981) Breast-milk fat concentration of rural African women. 1. Short-term variations within individuals. *British Journal of Nutrition* 45: 483–494.

Quandt, S.A. (1984) Nutritional thriftiness and human reproduction: beyond the critical body composition hypothesis. *Social Science and Medicine* 19: 117–182.

Rao, M., Blass, E.M., Brignol, M.J., Marino, L. & Glass, L. (1993) Effect of crying on energy metabolism in human neonates. *Pediatric Research* 33: 309.

Rogers, W.B. (1979) Fussy baby: A new cause? *Pediatrics* 63: 347–348.

Rosenblum, L.A. & Moltz, H., eds. (1983) *Symbiosis in Parent-Offspring Interaction*. New York:Plenum.

Sampson, H.A. (1989) Infantile colic and food allergy: fact or fiction? *The Journal of Pediatrics* 115(4): 583–584.

Sauls H.S. & Redfern D.E. (eds.) (1997) *Colic and Excessive Crying*. Columbus, Oh: Ross Products Division Abbott Laboratories.

Sethi, K. & Sethi, J.K. (1988) Simethicone in the management of infant colic. *The Practitioner* 232: 508.

Singer, J.I. & Rosenberg, N.M. (1992) A fatal case of colic. *Pediatric Emergency Care* 8(3): 171–172.

Smith, B.A., Fillion, T.J. & Blass, E.M. (1990) Orally mediated sources of calming in 1 to 3-day old human infants. *Developmental Psychology* 26: 731–737.

Smotherman, W.P. & Robinson, S.R. (1992) Prenatal experience with milk: fetal behavior and endogenous opioid systems. *Neuroscience and Biobehavioral Reviews* 16: 351–364.

Smotherman, W.P. & Robinson, S.R. (1994) Milk as the proximal mechanism for behavioral change in the newborn. *Acta Paediatrica Supplement* 397: 64–70.

St. James-Roberts, I. (1989) Persistent crying in infancy. *Journal of Child Psychology and Psychiatry* 30: 189–195.

St.James-Roberts, I., Bowyer, J., Varghese, S. & Sawdon, J. (1994) Infant crying patterns in Manali and London. *Child: Care, Health and Development* 20: 323–337.

St.James-Roberts, I., Conroy, S. & Wilsher, K. (1995a) Clinical, developmental and social aspects of infant crying and colic. *Early Development and Parenting* 4: 177–189.

St.James-Roberts, I. & Halil, T. (1991) Infant crying patterns in the first year: normal community and clinical findings. *Journal of Child Psychology and Psychiatry* 32: 951–968.

St.James-Roberts, I., Hurry, J., Bowyer, J. & Barr, R.G. (1995b) Supplementary carrying compared with advice to increase responsive parenting as interventions to prevent persistent crying. *Pediatrics* 95: 381–388.

Stahlberg, M-R. (1984) Infantile colic: occurrence and risk factors. *European Journal of Pediatrics* 143: 108–111.

Stahlberg, M.-R. & Savilahti, E. (1986) Infantile colic and feeding. *Archives of Disease in Childhood* 61: 1232–1233.

Stuart, C.A., Twiselton, R., Nicholas, M.K. & Hide, D.W. (1984) Passage of cow's milk protein in breast milk. *Clinical Allergy* 14: 533–535.

Taubman, B. (1984) Clinical trial of the treament of colic by modification of parent-infant interaction. *Pediatrics* 74: 998–1003.

Thomas, D.W., McGilligan, K., Eisenberg, L.D., Lieberman, H.M. & Rissman, E.M. (1987) Infantile colic and type of milk feeding. *American Journal of Diseases of Childhood* 141: 451–453.

Treem, W.R. (1994) Infant colic: a pediatric gastroenterologist's perspective. *Pediatric Clinics of North America* 41: 1121–1138.

Un pere de 26 ans soupconne d'avoir battu a mort sa fille de 27 jours. (1984) *Le Journal de Montreal* 15 December, p. 3.

van Lawick-Goodall, J. (1968) The behaviour of free-living chimpanzees in the Gombe Stream Reserve. *Animal Behaviour Monographs* 1: 161–311.

Vitzthum, V.J. (1989) Nursing behaviour and its relation to duration of post-partum amenorrhoea in an Andean community. *Journal of Biosocial Science* 21: 145–160.

Waters, E., Vaughn, B.E., Posada, G. & Kondo-Ikemura, K., eds. (1995) Caregiving, cultural, and cognitive perspectives on secure-base behavior

and working models: new growing points of attachment theory and research. *Monographs of the Society for Research in Child Development* 60(2,3).

Weissbluth, M. & Weissbluth, L. (1992) Colic, sleep inertia, melatonin and circannual rhythms. *Medical Hypotheses* 38 (3): 224–228.

Wessel, M.A., Cobb, J.C., Jackson, E.B., Harris, G.S. & Detwiler, A.C. (1954) Paroxysmal fussing in infancy, sometimes called "colic." *Pediatrics* 14: 421–434.

Weston, J. (1968) The pathology of child abuse. In R. Helfer & C. Kempe, eds., *The Battered Child.* Chicago: University of Chicago Press.

Wilson, D.A. & Sullivan, R.M. (1994) Neurobiology of associative learning in the neonate: early olfactory learning. *Behavioral and Neural Biology* 61: 1–18.

Wolff, P.H. (1969) The natural history of crying and other vocalizations in early infancy. In B.M. Foss, ed., *Determinants of Infant Behavior.* London: Methuen, 81–108.

Wolfheim, J.J., Jensen, G.D. & Bobbitt, R.A. (1970) Effects of group environment on the mother-infant relationship in pig-tailed monkeys (*Macaca nemestrina*). *Primates* 11: 119–124.

Wood, J.W., Lai, D., Johnson, P.L., Campbell, K.L. & Maslar, I.A. (1985) Lactation and birth spacing in Highland New Guinea. *Journal of Biosocial Science (supplement)* 9: 159–173.

Zeskind, P.S. & Barr, R.G. (1997) Acoustic characteristics of naturally occurring cries of infants with "colic." *Child Development* 68 (3): 393–403.

3

BREAST-FEEDING AND MOTHER-INFANT COSLEEPING IN RELATION TO SIDS PREVENTION

JAMES MCKENNA

SARAH MOSKO

CHRIS RICHARD

> Though medical therapies (in most cases) are constructed from the data of biology, medicine in general pays little attention to what is probably the single most important concept in biology: the theory of evolution (Brown 1993: A3)

> Sudden Infant Death Syndrome: The sudden death of any infant or young child which is unexpected by history and in which a thorough post-mortem examination fails to demonstrate an adequate cause of death. (Beckwith 1971).

Since being formally recognized as a medical entity in 1970, sudden infant death syndrome (SIDS, cot death, or crib death), a presumed sleep-related disorder primarily affecting infants in the first year of life (Fitzgerald 1995), has contributed to the deaths of well more than 290,000 babies in the Western world alone. Described as a "diagnosis by exclusion," the causes of SIDS still remain elusive; and, yet, for the first time in the history of SIDS research, death rates per 1000 live births are declining to unprecedented levels worldwide (Rognum 1995).

The breakthrough in reducing SIDS risks came from an unexpected direction—child care practices—specifically, the avoidance of the prone infant sleep position in favor of the side, supine, or back position. A warning about the dangers of prone sleeping was first published in 1944 by Abramson, who found that 68% of infants found dead in their cribs in New York City (55% of whom were under 5 months of age) were found lying face down. He recommended that: "the routine practice of placing infants in the prone position be avoided except during such times as the babies are constantly attended. The practice should, furthermore, be entirely done away with at night" (1944:413).

Three years later Werne and Garrow (1947) echoed Abramson's research, arguing for the elimination of prone infant sleeping. It was not until 40 years later, however, following observations by Beal (1988), Beal and Bundell (1988) Davies (1985), de Jonge et al. (1989), Nelson (1989), and Tonkin (1986), that any attention was paid to the relatively high number of SIDS deaths associated with infants sleeping prone. National state-of-the art epidemiological studies in Australia, New Zealand, and the United Kingdom were initiated, all of which confirmed that the prone infant sleeping position was, indeed, a significant risk factor for SIDS (see Fleming 1994; Guntheroth and Spiers 1992 for reviews). Between 1990 and 1992, following well-organized national campaigns recommending a change away from the prone infant sleep position to the supine (back) or side position, dramatic declines in SIDS were experienced in at least eight nations: Tasmania, New Zealand, the United Kingdom, Australia, Denmark, Norway, Sweden, and the Netherlands. Since 1992 in the United States, when the National Institutes of Health and the American Academy of Pediatrics initiated the "Back to Sleep" campaign, SIDS rates have fallen 30% nationally (Schaefer et al. 1996).

SIDS researchers are astonished that a behavioral child care practice over which parents assert control, rather than a surgical, pharmacological, or otherwise biological intervention, reduces SIDS rates so significantly. In fact, this discovery marked the beginning of an important paradigm shift toward associating suspected SIDS deficits with the infant sleep environment within which they occur (cf. Gantly et al. 1993; Einspieler et al. 1988; Johnson 1995). This discovery raised the possibility that other child care practices might also prove important.

Our research team studies child care practices in relation to SIDS prevention from the standpoint of the human infant's evolutionary legacy. For example, compared with other mammalian young, human infants are born neurologically much less mature, with only 25% of their adult brain volume (Konner 1981), and they exhibit a remarkably slow rate of biological and social maturation, during which time proximity and contact with the caregiver asserts a significant amount of external regulation of the infant's physiology, both day and night. Natural selection, we suggest, probably favored postnatal infant responsivity to parental sensory stimuli similar to the way it favored fetal regulation by the mother prenatally (McKenna and Mosko 1993), particularly during the 2- to 5-month period when infants are most susceptible to SIDS (Fleming 1994). Infants and mothers sleeping within arms reach (cosleeping) with nighttime breast-feeding represents the evolutionary stable sleeping arrangement and, indeed, the two practices (mother–infant cosleeping in tandem with breast-feeding) remain both inevitable and inseparable for most contemporary people (Konner 1981; Whiting 1981). We suggest, therefore, that not only are these two child care practices biologically appropriate, but that for the human infant, who develops extremely slowly and is born in such a helpless and dependent state, cosleeping may facilitate a unique sensory bridge within the mother–infant dyad that maximizes the chances of optimal development and, in some circumstances, compensates for the infant's delayed maturity (McKenna 1986).

Historically, studies of pediatric sleep disorders and SIDS have universally failed to consider the potential benefits of breast-feeding and parental nighttime contact (cosleeping), probably because cultural values favor both early weaning and solitary sleep environments as a presumed strategy to produce independent children as early in life as possible (see Ferber 1985; Pinilla and Birch 1993; Spock and Rothenberg 1985). Moreover, Western scientists and SIDS researchers in particular are exposed to bed sharing (as representative of only one form of infant–parent cosleeping) mostly as it occurs under unsafe conditions as, for example, in urban and impoverished environments where infants are injured (or die from SIDS) while bed sharing sometimes at numbers approximately equal to (or higher than) the numbers of infants dying alone in cribs (Hauck 1997). Unsafe sleeping conditions such as an infant sleeping on the same surface next to a mother who smokes, or sleeping with an adult on a soft mattress (which is risky) come to be understood as an inevitable feature of all bed sharing and, thus, all bed sharing outcomes must be negative (Mitchell 1995). But just as solitary sleep environments for infants can vary from safe to risky, so can bed-sharing environments. That dangerous bed-sharing environments exist is no more of an argument against all bed sharing than the existence of unsafe cribs constitutes an argument against the safety of all solitary infant sleep.

The inference we draw from an evolutionary perspective is that, although mothers sleeping on a bed with an infant did not evolve per se, infants and mothers lying next to each other in some form or another for nighttime breast-feeding and nurturing did. We argue that unless infant sleep is recorded and measured within diverse cosleeping/breast-feeding contexts, "normal" species-wide infant sleep development in the first year of life will never be completely understood. More importantly, unless we know what constitutes biologically and socially "normal" infant sleep, we will be less likely to identify pathophysiologies that can conspire with environmental risk factors to increase SIDS risks (McKenna et al. 1990; Mosko et al. 1993). This is especially true where explanations for sleep-related disorders such as SIDS appear to be multifactorial and have multiple etiologies (Konner and Super 1987). It is possible, for example, that some of the babies that die from SIDS do so because solitary infant sleep may act to increase exogenous stressors, pushing some infants beyond their adaptive limits (McKenna et al. 1993).

Hypothesis

An evolutionary perspective on infancy predicts that infant–parent cosleeping with breast-feeding should be fitness enhancing for both the infant and parent; however, because there is no single cause of SIDS, no one prevention strategy, including cosleeping with breast-feeding or even the supine infant sleep position, will necessarily eliminate SIDS, nor necessarily have a significant positive effect. In fact, because SIDS is relatively rare, there is no reason to suppose that cosleeping evolved specifically to prevent SIDS. Indeed, when practiced in combination with maternal smoking, soft-bedding, medicated or desensitized parents, or the prone infant sleep position, bed sharing is known to increase SIDS risks (Bass et al. 1986; Byard and Cohle 1994; Fleming 1996; Scragg et al. 1995; Wilson et al., 1994). It is, therefore, essential to acknowledge that dangerous sleeping conditions can transform an otherwise beneficial sleeping arrangement into a risky one (McKenna 1995).

With this caveat in mind, we hypothesize that, on the other hand, there may exist a subclass of SIDS-prone infants for whom the chances of succumbing to SIDS could be reduced in situations where mothers and infants remain close enough during sleep to monitor, exchange and/or respond to each other's sensory stimuli (e.g., movements, sounds, touches, smells, and exhaled gases). These sensory exchanges, we suggest, can change the infant's sleep and arousal patterns in potentially beneficial ways, while simultaneously increasing the chances of successful interventions by the caregiver should the infant experience a crisis during sleep. We do not suggest that solitary sleep causes SIDS per se, only that separation from the mother (or from some other primary caregiver) for extended sleep periods potentially enhances the deleterious effects of SIDS-related congenital deficits. This may be especially true for deficits that interfere with an infant's ability to arouse to reinitiate breathing following apnea, which can occur during deep stages of sleep when arousal threshold is high.

The background for our hypothesis emerges from four lines of evidence: (1) cross-cultural (ethnographic) data documenting the universal prevalence of infant–parent cosleeping and considerably reduced SIDS rates in cultures where, in the absence of maternal smoking, high maternal contact with cosleeping occurs (Barry and Paxton 1971; Davies 1985; Lee et al., 1989; Takeda 1987; Tasaki 1988; Whiting 1964, 1981); (2) cross-species laboratory studies of monkeys and apes showing the deleterious physiological consequences on infant monkeys following separation from their mothers for periods as short as 3–9 hours (see McKenna 1986 for a review); (3) human clinical studies showing the positive physiological effects of skin-to-skin contact ("kangaroo care") on preterm infants and neonates (Anderson 1991; Acolet et al. 1989; Field et al. 1986); and (4) SIDS research findings suggesting that increased breast-feeding frequency and duration (Fredrickson et al. 1993) and avoidance of the prone sleep (Guntheroth and Spiers 1992) can reduce SIDS risks and that some SIDS victims may have problems arousing from sleep before succumbing to SIDS (see Mosko et al. 1997a).

It should be noted that no controlled epidemiological studies have yet demonstrated a protective effect of cosleeping and that at least three major epidemiological studies show increased SIDS risk when mothers smoke, infants sleep prone, and/or mothers fail to breast-feed in the context of bed sharing. But where rates of maternal smoking are probably low and breast-feeding is high, SIDS rates are among the lowest in the world (Lee et al. 1989; Davies

1985; Takeda 1987; Tasaki 1988). That close contact with a parent during sleep contributes to these low SIDS rates is suggested by the recent increase in SIDS rates in Japan that has been paralleled by a shift from a tradition of mothers and infants sleeping together on futons to solitary sleeping in separate beds or futons (Wantanabe et al. 1994).

Observing Mother–Infant Cosleeping and Potential Adaptations in a Sleep Laboratory

Methods

Our studies of mothers and infants focus on bed sharing, which is only one of many different forms of mother–infant cosleeping. (Bed sharing occurs when a mother and infant sleep together on a raised mattress surface, often with a box spring underneath, all of which is ordinarily framed by both a head and foot board constituting a piece of furniture.) The research took place at the Sleep Disorders Laboratory, University of California, Irvine (UCI), under the direction of Sarah Mosko, a certified sleep specialist. We completed two small preliminary studies (McKenna et al. 1990; Mosko et al. 1993) and a larger study in which 35 healthy Latino mother–infant pairs (20 routine bed sharers and 15 routine solitary sleepers) spent 3 consecutive nights in our sleep laboratory (Mosko et al. 1997a, b, 1996; Richard et al. 1996). The methods and data described below are abbreviated and intended to provide an overview.

Subjects

The mothers in our study were Latina (mean 24.5 years of age, range 18–37) nonsmokers, had prenatal care, and delivered at UCI Medical Center, had no history of drug or alcohol use, and experienced uncomplicated pregnancies, labors, and deliveries. They were in good health, free of sleep disorders, and were taking no medications which could affect sleep patterns. Mothers had to meet our criteria for routinely cosleeping or routinely sleeping solitarily (see below) and have chosen their sleeping practice for reasons other than infant temperament. Mothers were all breast-feeding nearly exclusively, with little formula or food supplements, as determined from sleep and feeding logs kept daily by each potential recruit (see next section).

At the time of the study, infants averaged 12.6 weeks of age (range 11–16 weeks, the peak age for SIDS) and were in good health with normal growth and development as assessed by a pediatrician. At delivery all infants were >2500 in weight and >37 weeks gestation age with 5-minute Apgar scores of 8 or higher. All infants came from families with no history of SIDS in first-degree relatives, and none had experienced prolonged apneas or life-threatening events.

Identification of Routine Solitary and Routine Cosleeping Groups

Routine bed sharers (RB) were identified by cosleeping for at least 4 hours per night, 5 days per week; routine solitary sleepers (RS) shared beds no more than once per week for any part of the night. Mother–infant pairs were categorized into one of these groups (or excluded) based on daily sleep logs kept for 2 weeks that were completed at home before the sleep recordings to confirm maternal reports of the infant's usual home sleep environment. For the 33 pairs who completed 14 nights of the log, the mean number of bed-sharing nights was 13.7 for the RB group versus 0.6 for the RS group. All mothers were nearly exclusively breast-feeding (no more than 4 ounces of milk per any given day) as assessed from these same logs.

Procedures

Beginning with an "adaptation night" in the laboratory, mother–infant pairs slept as they routinely did at home. On one of the next 2 nights (the order being randomly chosen), the mother and infant slept again in the routine condition and the other night in the nonroutine condition. Continuous all-night polysomnographic and video recordings using infrared lamps were performed nightly. A single polygraph recorded all standard physiological parameters (EEG, EOG, EKG, air flow and respiratory effort, and oxygen saturation for the infants) simultaneously in the mother and infant. Axillary temperature and oxygen saturation were also recorded in the infants. All signals were written out simultaneously each night on a single 22-channel polygraph (Grass 8 plus).

Electrodes were attached first to the infant because they retire before mothers at home. Mothers positioned their infants for sleeping as usual each night with no instructions. Recordings begin when infants retired at their usual time. Mothers retired later, also at their normal bedtimes. They slept in rooms adjacent to their infants' rooms on solitary sleeping nights with the bedroom doors open so that they remained in auditory contact. Mothers responded to infant crying, etc., on an ad libitum basis, performing all caregiver interventions themselves each night.

Polygraphic recordings were scored for sleep stages in 30-second epochs according to accepted criteria. A widely used sleep scoring system for mothers was used which differentiates between brain, eye, and chin electrical signals needed to identify the sleep stages or awake periods, whereas the scoring system for 3-month olds, developed by Guilleminault and Souquet (1979), was used for the infants. Identification of sleep–wake stages in both scoring systems depends on three simultaneous parameters: EEG, EOG, and chin EMG. Five sleep stages are identified in adults: rapid eye movement (REM) stage plus four stages of non-REM sleep delineated as stages 1, 2, 3, and 4. In the infant, only three stages are defined: REM stage, stage 1–2, and stage 3–4. In the process of data reduction, stages 1 and 2 in the adult are combined to obtain a combination stage 1–2, and likewise stages 3 and 4 are combined for comparability to infant sleep stages (because stages 1 and 2 and 3 and 4 are not easily distinguished for infants).

Sleep stage scoring was based on assigning to each 30-second epoch either wakefulness or one stage of sleep based on the predominant (greater than 50%) sleep or wakefulness pattern occupying the epoch. Although awakenings of 15 seconds or longer that meet these criteria (i.e., epochal awakenings) are automatically identified by this epochal system, shorter duration arousals occupying less than 50% of an epoch are not. Because of our interest in all arousal phenomena in sleep, however, we quantify these shorter arousals, calling them transient arousals, which were 3 seconds or more and were also identified by multiple signals, including EEG, following standard accepted criteria.

Infrared video recordings on all individuals across all nights were hand scored in real time in their entirety. Based on an ethogram of sleep-related behaviors, every observable behavior was recorded as it occurred, sequentially, and where appropriate its duration measured.

Selected Findings

Breast-feeding Patterns

Breast-feeding was defined as oral infant breast attachment and scored through observation of the video recordings combined with the sucking artifact (chin and tongue movement) on the polysomnograph (McKenna et al. 1997). Breast-feeding episodes began and ended with breast attachment and detachment, respectively, but also included very short interruptions of less than a minute during which the mother switched from one breast to the other. All breast-feeding episodes were elicited by the behavior of the infant but initiated by the mother placing her nipple in the infant's mouth and defined to capture a single intentional and continuous act of breast-feeding on her part. Thus, if breast-feeding was interrupted by maternal behaviors that indicated an apparent attempt to terminate feeding (e.g., by closing her bra or nightgown) but the infant's subsequent refusal to settle prompted the mother to reinitiate feeding, a new breast-feeding episode was scored. Three breast-feeding variables were computed each night: number, mean duration, and total duration, (sum) of breast-feeding episodes.

For the RB group, both the number and total duration of breast-feeding episodes were significantly higher on the bed-sharing night (BN) than on the solitary night (SN) (see McKenna et al. 1997 for details). The mean duration of breast-feeding episodes was also higher on the BN, but the difference just failed to reach statistical significance. Since no significant differences were found in either the infants' total time in bed or total sleep time on the two nights, the differences in breast-feeding could not be attributed to different-length sampling periods. The net result was that RB infants spent more than twice as long breast-feeding on the BN than on the SN.

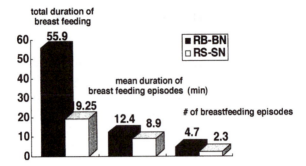

Figure 3.1. Number, mean duration (minutes per episode), and total nightly duration of breast-feeding episodes (minutes) among routinely bed-sharing (RB) and routinely solitarily sleeping (RS) mother–infant pairs on their bed-sharing night (BN) and solitary sleeping (SN) in a sleep laboratory (McKenna et al. 1997).

All three breast-feeding variables were also larger in the RS group on the BN than on the SN, but none of the differences reached significance. Similarly, comparing the two groups on the BN, the three breast-feeding variables were all larger in the RB group than in the RS group, but only the difference in total duration of breast-feeding episodes was significant (McKenna et al., 1997).

The largest differences found in breast-feeding were consistently between the two groups in their routine conditions (figure 3.1). Comparing the RB group on the BN to the RS group on the SN, the number of breast-feeding episodes was doubled, the total duration of breast-feeding episodes was nearly three times as great, and the mean duration of breast-feeding episodes was 39% larger in the RB group. A modestly longer total bed time in RB infants on the BN (544.2 ± 40.0 minutes) than for RS infants on the SN (512.7 ± 49.4 minutes) was insufficient to explain the large differences seen in breast-feeding.

Significance   Breast-feeding is known to reduce infant morbidity and mortality worldwide (Cunningham 1995; Cunningham et al. 1991; Jeliffe and Jeliffe 1978). These are the first data collected comparing breast-feeding in bed-sharing and solitary sleeping conditions, and the first data on nocturnal breast-feeding based on direct observation rather than on maternal recall, which has proven to be inaccurate (Quandt 1987; Vitzhum 1994a, b). The results are important because, although not all epidemiological studies suggest that breast-feeding lowers the risks of SIDS (see Gilbert et al. 1995), others do (Fredrickson et al. 1993; Hoffman et al. 1988; Mitchell et al. 1992), and many international SIDS prevention campaigns recommend breast-feeding as a way to help reduce SIDS risks. How breast-feeding protects some infants is not known, although the degree of protection offered by breast-feeding may be "dose dependent" (see Blackwell et al. 1993; Fredrickson et al. 1993; Jura et al. 1994); i.e., the more frequently infants are fed, the greater the protection.

The reduced average interval between feeds that accompanied the increased breast-feeding episodes on the bed-sharing night is extremely important in understanding how infants potentially regulate their mothers' ability to ovulate and, hence, to conceive. Mother–infant cosleeping, thus, reflects mutual regulatory processes. For example, reduced intervals between feeds are known to keep maternal prolactin levels high enough in the bloodstream to block ovulation (i.e., induce lactational amenorrhea), which potentially lengthens the birth interval. It is argued that the potential contraceptive effect of breast-feeding depends more on the structure of breast-feeding, including the daily intervals between feeds, rather than on simply breast-feeding or not breast-feeding (Ellison 1995; Knauer, 1985; Konner and Worthman 1980; McNeilly et al. 1994; Vitzhum 1989). The structure of nocturnal feeds we report here appears to maximize the chances of increasing the birth interval, thereby reducing the chances of exploiting (or exhausting) maternal resources, which theoretically can benefit both mother and infant.

Mother–Infant Interactions

Infrared videotape analysis revealed that, irrespective of the routine sleeping arrangement, during bed sharing all mothers were highly responsive to the emotional and physiological status of their infants. For example, all mothers exhibited behaviors during bed sharing that we have termed non-nursing maternal interventions that they rarely exhibited on the solitary night. These behaviors appear sometimes to be the result of something the infant needs or does; but often they involve affectionate kisses, hugs, or whispers that appear to be simply expressions of affection. We divide this behavioral category into two subdivisions: (1) protective behaviors involving the physical management and safety of the infant (e.g., blanket adjustments for thermal control, protecting the infants air passages, or repositioning the infant, visually inspecting the infant; and (2) affectional behaviors such as patting, kissing, whispering, speaking, singing, hugging, and rocking. Any of these behaviors can occur spontaneously without infant participation, but most occur in response to any audible or movement-related infant behaviors. In fact, in a subset of bed-sharing pairs that have been analyzed, of the 308 non-nursing maternal interventions observed on the BN, 87% of them were provoked by the infant, whereas 13% appeared to be spontaneously initiated by the mother. On average, and irrespective of routine sleeping behavior, mothers exhibited 6.8 (SD ±5.03) protective maternal interventions and an average of 8.1 (SD ±7.4) affectional interventions during the BN. There was little spontaneous checking up on the baby in the laboratory on solitary nights, except for breast-feeding episodes, which were elicited by the infant crying.

Significance   Maternal proximity during bed sharing appears to provide the infant with a range of both protective and affectional maternal behaviors simply not available in the solitary environment. Some of the advantages of maternal proximity are obvious, such as responding to an infant in physical distress or preventing it from sleeping prone. Other benefits may be less obvious and may be in evidence at some point later in the child's life, given that some of the infant's nighttime emotional needs fulfilled by a bed-sharing mother may be quite subtle. Children from military families who cosleep appear to be under-represented in psychiatric populations and receive higher prosocial scores from their teachers than do children who do not cosleep (Forbes et al. 1992). Certainly, the functional significance of mother–infant nighttime behavioral interactions are complex, and they are likely to have both immediate and long-terms effects on the infant's social and psychological development, including the quality and strength of attachment to its caregiver and, possibly, the in-fant's reaction to stress later in life (Lewis and Haviland 1993). Although our study does not directly measure these outcomes, affectionate maternal inter-ventions are known to significantly influence the direction of early emotional development to promote, for example, more empathy and social cooperation among children (see Lewis and Haviland 1993).

Bed-sharing Effects on Sleep Architecture and Arousals

An analysis of polysomnographic recordings scored for sleep stages and arous-als revealed significant differences between the bed-sharing and solitary con-ditions in both infants and mothers (Mosko et al. 1996 1997a, b). All data were restricted to the time the mother was in bed each night. Repeated measures analysis of variance was used for the statistical analysis.

Infants   On the BN, infants had longer total sleep time, more minutes and a greater percentage of stage 1–2, and fewer minutes and a lower percentage of stage 3–4 than on the SN, but REM stage was not affected. For the two stage 3–4 variables, there was a significant interaction effect, reflecting a larger night-effect in the RB group than in the RS group (figure 3.2). The BN was also associated with shorter mean duration episodes of uninterrupted stage 3–4 sleep but longer episodes of both stage 1–2 and REM stage irrespec-tive of the infant's routine sleeping condition. In contrast, no significant dif-ferences were obtained for the longest duration or number of occurrences of any sleep stage. In addition, none of the sleep architecture variables were af-fected by the sex of the infant, as determined by covariant analysis (Mosko et al. 1996).

Bed sharing was also found to promote infant arousals (Mosko et al., 1997b). Stage comparisons revealed that stage 3–4 sleep is associated with a striking paucity of rousals, compared to the other sleep stages. Bed sharing facilitated both transient arousals and the more sustained epochal awakenings *selectively* in stage 3–4 sleep. Epochal awakenings from stage 3–4 were more frequent on the BN than on the SN in both infant groups. Furthermore, RB infants exhib-ited more frequent transient arousals in stage 3–4 than RS infants in both sleeping conditions. Arousals from sleep are potentially very important in that observations in victims of SIDS support a role of arousal deficiency in the etiology of SIDS (see Mosko et al., 1997 a, b). The greatest effect of bed

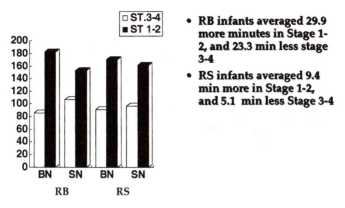

Figure 3.2. Comparison on infant sleep architecture in routinely bedsharing and routinely solitary sleeping infants on the bedsharing (BN) and solitary (SN) nights (Mosko et al., 1996).

sharing on infant arousals concerned temporal overlap with maternal arousals: the number of overlapping arousals doubled on the BN in both RB and RS infants (figure 3.3). The greatest differences were in those where the infant aroused first. All within- and between-group comparisons of the BN and SN resulted in significant differences for all three types of arousals: those where the mother aroused first, those where the infant aroused first, and those where arousals appeared simultaneous (Mosko et al. 1997b).

Mothers   On the BN, mothers of both groups had slightly less stage 3–4 sleep (minutes and percentage) and more stage 1–2, with no change in REM (rapid eye movement) or total sleep time. As in infants, the decrease in stage 3–4 on the BN was due to a significant decrease in the mean duration of stage 3–4 episodes rather than in the number of occurrences of stage 3–4. Contrary to popular expectations, bed sharing did not result in shorter maternal sleep, despite the fact that there was a uniform effect on arousals in that the frequencies of total arousals were more in both groups of mothers while bed sharing (Mosko et al. 1996 Mosko et. al 1997b). Moreover, all significant interaction effects in mothers reflected enhancement of the effects on sleep and arousal variables in the RB compared to the RS group (Mosko et al. 1997a, b). This indicates that, rather than habituation to the baby's presence, there was evidence of increased sensitization by the RB mothers to the effects of the baby's presence.

The increase on the BN in overlapping arousals was even greater in the mothers than in the infants, with an approximate threefold increase. By far the largest magnitude increase was in the number of arousals where the infant aroused first, although all group comparisons (within or between groups) of the BN and SN revealed significant differences for all three types of arousals (Mosko et al. 1997a).

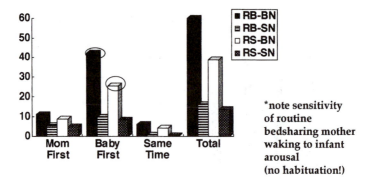

Figure 3.3 Mean number of infant arousals overlapping maternal arousals when the mother aroused first, or the baby aroused first, or they aroused simultaneously (from Mosko et al. 1997b).

Significance   The findings of an increase in infant arousals and a decrease in stage 3–4 sleep in the bed-sharing versus solitary condition are important when considered in the context of the hypothesis proposed by several researchers that SIDS is attributable at least in part to a defect in arousal mechanisms. If the arousal-deficiency hypothesis is correct, then factors that facilitate arousals or arousability might act to minimize the risk for SIDS in vulnerable infants. Stage 3–4 sleep is associated with a high arousal threshold; limiting the total amount and duration of episodes of this stage by bed sharing might increase the likelihood that an infant would be able to arouse if faced with a prolonged apnea or other potentially life-threatening event in sleep (Mosko et al., 1996). The increase in infant arousals that occurs with bed sharing might also promote arousability by ensuring a basal level of practice in arousing, practice which might be critical to infants deficient in intrinsic arousal mechanisms (Mosko et al. 1996, 1997a). The temporal overlap with maternal arousals is important to the arousal hypothesis because the overlap suggests that many arousals are occurring at times they otherwise would not have; the occurrence of these apparently nonspontaneous arousals is evidenced by those instances where the mother arouses first or the arousals appear simultaneous (Mosko et al. 1996, 1997a & b). The presence of all these effects of bedsharing in the RB infant group suggests that the effects are not due to the novelty of bedsharing and, furthermore, that the effects do not habituate (Mosko et al. 1996).

That the percentage of stage 3–4 is only mildly reduced in mothers while bed sharing, plus the facts that the mothers' total sleep time was not reduced by bed sharing and that both RB and RS mothers rated their sleep on the BN as satisfactory and representative of their usual sleep at home, act to dispel a popular notion that bed sharing is unacceptably disruptive to a parent's sleep (Mosko et al. 1997b in press). Finally, bed-sharing factors that enhance a parent's arousability (less stage 3–4 sleep and more frequent arousals) would be expected to be protective to infants because the parent would be more likely to detect and, hence, reverse any potentially dangerous condition developing in the infant (such as a prolonged breathing pause, apnea, or choking). That all of the effects of bed sharing on maternal sleep and arousals were measured in both RB and RS mothers suggests they are not novelty effects and do not habituate. Furthermore, that the increase in total arousals on the BN was significantly larger in the RB mothers suggests that this effect of bed sharing is actually enhanced by routine bed sharing (Mosko et al. 1996, 1997a, b).

Body Positions in Bed

Prone infant sleeping position is the most recently identified risk factor for SIDS, carrying a relative risk of 8.8 (Fleming 1994). Although the mechanisms for this increased susceptibility remain unknown, some investigators have proposed that hyperthermia (Fleming et al. 1990; Nelson 1989) or hypercapnia (excessive CO_2) (Chiodini and Thach, 1993) are involved. Our analysis of infant sleeping position arose from an observation that bed-sharing infants are almost always placed on their backs, in the safer supine position (which appears to facilitate easier breast-feeding), while solitary sleeping infants were more often placed prone for sleep. Mothers in our study were always allowed to place their infants in bed without instruction to best replicate their behavior at home. In addition, our preliminary experiments on infant exposure to maternal CO_2 indicated that the orientation and proximity of an infant to the mother is a critical factor in the level of CO_2 exposure. This led us to measure the amount of time bed-sharing mother–infant pairs spent facing each other (Mosko et al. 1995).

Data presented here were derived from videotape recordings of the first six RB pairs and six RS pairs on both their solitary and bed-sharing nights. A clock in the field of view was used to partition the entire night into four position categories (prone, supine, left side, and right side) on a minute-by-minute basis. The same data were then recategorized to measure the amount of time that each member of the pair faced toward or away from the other member. Movement time (e.g., body position changes or feedings) was excluded from the analyses. The proximity of the mother's face to the baby's face was computed by measuring the distance on the video screen and converting it to actual distance by comparison to an object of known length placed in the field of view (see also Mosko et al. 1997b).

The most striking result was that both RB and RS infants were never placed prone while bed sharing (figure 3.4). Some infants (three RB and one RS) were placed prone on the SN and one additional RS infant was placed prone on the BN, but only until the mother got into bed. In most cases, the infant spent most of his or her time in just one of the position categories; i.e., most infants were repositioned once or not at all after going to bed and all but one were unable to reposition themselves. Due to this infrequent nature of infant position changes, many of the values in the analysis were zero. Therefore, variability was very high and statistical tests indicated no significant differences between positions. The most common infant sleeping position on the SN was supine, but on the BN infants showed a wider range of positions (excluding prone). No clear differences exist between RB and RS infants. Note that in our continuing analyses of the remaining pairs, we have documented two cases where infants were placed prone for bed sharing. However, the data continue to support a strong preference for the nonprone positioning (Richard et al. 1996).

Analyses of the number of minutes spent facing the sleeping partner during bed sharing clearly show that, irrespective of routine condition, both members of a pair orient toward the other member (figure 3.5). This behavior is most evident for the infants; our analysis demonstrated that the infants spent significantly more time facing their mothers than facing away, regardless of routine sleeping condition. Mothers also spent significantly more time facing the infant than not. More than half of the infants (7/12) faced the mothers 100% of the night. Only two infants spent less than 40% of the night oriented away from the mother (range 23–100%). Almost all mothers were oriented toward their infants for most of the night (range 37–98%). The difference between the time that either member spent facing the other and the time facing away was more pronounced for the RB pairs than the RS pairs. The ratio between the time mothers faced toward the infants and the time they faced away was about 3:1 for the RB mothers versus 1.25:1 for the RS mothers; similarly, that ratio for the infants was about 7:1 for the RB infants versus 4:1 for the RS infants. The average RB pairs spent about twice as much time facing each other as did RS pairs (Richard et al. 1996).

Significance   As reported elsewhere (Richard et al. 1996), these data suggest that bed sharing fosters face-to-face orientation in both RB and RS pairs, an effect which is further enhanced in mothers that routinely bed share. This orientation may be the reason for the minimization of the prone position in bed-

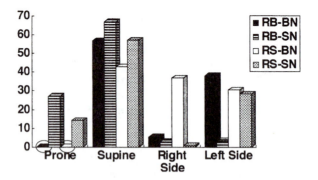

Figure 3.4 Percentage of nights routinely bed-sharing (RB) and routinely solitarily sleeping (RS) infants slept prone, supine, or on the right or left side (as positioned by their mothers) on the bed-sharing night (BN) and solitary nights (SN) in the laboratory (from Richard et al. 1996).

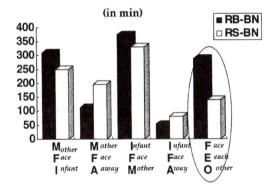

Figure 3.5 Number of minutes in which routinely bed-sharing (RB) mother–infant pairs and routinely solitarily sleeping (RS) mother–infant pairs faced each other and/or away (either partner) on the bed-sharing night (BN) in the sleep laboratory (from Richard et al. 1996). Note: routinely bedsharing mother/infant dyads, on their routine bedsharing night, face each for almost twice the amount of time per night than do routinely solitary sleeping mother/baby pairs on the bedsharing night; this suggests a kind of sensory bridge develops while bedsharing, which is not apparent when bedsharing is not routinely practiced.

sharing infants, but, whatever the reason, the bed-sharing environment seems to greatly diminish prone sleeping, which is important in terms of SIDS prevention. All of the infants in this study were placed in bed by the mother, which might preclude one from calling this infant "behavior"; however, infants that slept supine spent most of the night facing toward the mother, indicating that they, too, orient toward the mother. These behaviors during bed sharing have several implications; for example, they refute the idea that bed sharing imposes a substantial risk of overlying because the mother's preferred orientation toward the infant suggests she is highly cognizant of the infant's presence (Richard et al. 1996; Mosko et al. 1997a, b).

Summary

Our findings reveal that routinely bed-sharing infants breast-feed twice as frequently, which reduces the average interval between breast-feeding episodes. Moreover, infants breast-feed for almost three times the total nightly duration as do solitarily sleeping infants. Bed-sharing mothers almost always place their infants in the safer supine infant sleep position rather than in the more risky prone position, probably to facilitate side-by-side breast-feeding. Infants and mothers face in the direction of each other for much of the night. Mothers and infants exhibit more arousals and arousal overlaps while bed sharing and, by reducing the amount of time both mothers and infants spend in deeper stages of sleep (stage 3–4) and increasing the amount of time spent in stage 1–2, cosleepers assert an important regulatory effect on each other's sleep architecture. Contrary to the popular stereotype about how bed sharing disturbs or shortens the one's sleep, routinely bed-sharing mothers and infants slept as much on their routine bed-sharing night as did routinely solitary-sleeping mothers and infants under their customary condition. Behavioral analysis also reveals that mothers are highly sensitive to their infants' behaviors while bed sharing, exhibiting a high number of protective and affectionate responses toward them throughout the night, responses that were not exhibited when they slept in different rooms.

The data we present cannot provide direct support of our hypothesis that cosleeping might protect some infants from SIDS, but they do provide important insights into the ways infant and maternal behavior and physiology are entwined in adaptive or potentially adaptive ways. For example, if breast-feeding offers "dose-dependent" protection against SIDS as Fredrickson et al's (1993) data suggest, then by increasing both the frequency and total nightly duration of breast-feeding, bed sharing may enhance the protective effect. If the supine infant sleep position is safer than the prone position, then by placing their infants supine for sleep, bed-sharing mothers reduce risks incurred by prone sleeping. If some SIDS-prone infants exhibit arousal deficiencies, then bed-sharing might provide the infant with valuable opportunities to practice arousing (in response to the mothers' arousals), thereby developing abilities potentially applicable during life-threatening apneas. If some SIDS-prone infants have difficulties arousing to reinitiate breathing from deep stages of sleep (following apneas), when arousal thresholds are high, then bed sharing might alleviate this potential vulnerability by reducing the total time and average episode length of stage 3–4 sleep. And, finally, if maternal interventions such as touch or movement can induce arousals or stimulate infant breathing, then a mother's proximity to her infant in the bed-sharing context, and her increased time spent in lighter stages of sleep, increase the chances of her successfully detecting and responding quickly to acute infant distress.

Beyond SIDS: The Evolutionary Ecology of Infant Sleep

Sleep medicine informed by models of human infant evolution may potentially produce new and exciting interpretations and potential solutions to several contemporary pediatric sleep disorders. For example, it may be a small but nonetheless important consolation for the estimated 20–40% of parents struggling with questions concerning why their infants or children will not sleep alone to know that, from a biological perspective, such struggles are predicable (if not appropriate). These behaviors and the infant behaviors that underlie them evolved in attempts to ameliorate what surely amounted to a life-threatening situation (i.e., separation from mother). Indeed, we can speculate that natural selection probably favored infants and children who, upon finding themselves separated from their caregivers, protested (by crying). Infants crying during the night to elicit parental contact, perhaps had better chances of survival than did offspring who failed to protest, if indeed parents and infants were ever separated at night before recent times. What typically in Western culture is considered to be an infant "sleep problem" or "disorder" may instead represent an ancient emotional response (however out of synch with present cultural habits) reflecting a highly vigorous, well-adapted infant or child who needs parental contact to feel satisfied or protected. From this evolutionary perspective, it would be the infant or child who passively accepts the absence of the parent on whom its survival depends, and fails to protest that behaves biologically inappropriately, not the other way around. Which perspective one adopts (an evolutionary or a cultural one) determines how the behavior is to be interpreted: from a biological perspective, infants protesting sleep isolation from their parents (by crying) must be regarded as adaptive, whereas from the the perspective of a Western, urban parent, the behavior could be interpreted as a deficiency in the parent, the infant, or both.

The developmental significance placed on the ability of infants to "sleep through the night" at 2–4 months of age (see Ferber 1985) is, yet, another Western cultural construct having little to do with what an infant's biology is designed to experience (see Elias et al. 1986, 1987, Konner 1981; McKenna 1986). Breast-feeding and cosleeping practically guarantees, as our data show, that infants do not sleep without awakening briefly or arousing frequently throughout the night, specifically in relationship to mother's presence and/or activities, but also in response to its own internal needs. Yet, the expectation and perception of infant sleep consolidation and self-soothing back to sleep without parental intervention remains a widely accepted benchmark against which infant's developmental "progress" or maturity in Western cultures is measured and evaluated. This reflects the fact that an infant's ability to sleep alone and continuously without either feedings or parental contact is a *cultural* goal based on Western cultural judgments (see Pinilla and Birch 1993). It does not necessarily reflect the biological goals or needs of the human infant who, by awakening, regulates its oxygen supply, hunger, or communicates its need to be handled and reassured by the caregiver (Elias et al. 1987; McKenna et al. 1997a & b). It may be encouraging for parents to know that the real causes of infant– or child–parent sleep struggles are not necessarily deficient parenting skills or spoiled infants or children. Historically recent cultural ideologies and expectations about how infants *should* sleep, rather than how they were designed by evolution to want to sleep could be undermining physical and psychological health.

Pediatric Sleep Medicine: "Back to the Future"?

Child care practices change much faster than infant biology, and the majority of research on human infant development suggests strongly that it is far more likely to find negative socioemotional and physiological consequences when contact between parents and children is reduced rather than increased, irrespective of time of day (see McKenna and Mosko 1991; Lewis and Haviland 1993). Indeed, except where benefits are defined in terms of parental interests or other cultural values, or where bed sharing occurs under unsafe conditions, there is not one scientific study documenting the presumed socioemotional, psychological, or physiological benefits of solitary infant sleep or the presumed deleterious consequences of infant–parent cosleeping.

Isolated infant sleep is a recent cultural innovation, no more than a few 100 years old, affecting only a relatively small number of infants worldwide (Barry and Paxson 1971; McKenna 1986; Whiting 1964). Such a biologically novel sleeping arrangement, we speculate, places infants at odds with their innate need to feel protected and close to their caregivers for feeding and nurturing. It also deprives the infant of an "expected" source of constant sensory exchange with its mother's body, with which it is designed to be in constant physiological transaction.

The concept of evolution and human evolutionary data offer physicians, biomedical researchers, and parents a powerful, unbiased point to begin to understand and reconceptualize infant sleep and sleep-related disorders, only one of which is SIDS. While pediatric medical problems framed by evolutionary theory cannot guarantee solutions, it is clear that infant sleep research is enriched and made more accurate by recognizing at the outset the essential and irrefutable biological complimentarity between *evolved* infant needs (nutritional, emotional, social, and transportational) and intense parental contact both day and night.

Acknowledgments
This research was supported by National Institutes of Health grant RO1 HD27482. We thank Sean Drummond and Naz Kajani for their excellent technical contributions, James Ashurst for statistical advice, and Ron Harper and Marian Willinger for encouragement and support.

References

Abramson, H. (1944) Accidental mechanical suffocation in infants. *Journal of Pediatrics* 25:404–413.

Acolet, D., Sleath, K., and Whitelaw, A. (1989) Oxygen, heart rate, and temperature in very low birthweight infants during skin-to-skin contact with their mothers. *Acta Pediatrica Scandinavia* 78:189–189.

Anderson, G.C. (1991) Current knowledge about skin-to-skin (kangaroo care) for preterm infants. *Journal of Perinatology* 11:216–228.

Barry, H. III, and Paxson, L.M. (1971) Infancy and childhood: cross-cultural codes 2. *Ethology* 10:466–508.

Bass, M., Kravath, R.E., and Glass, L. (1986) Sudden infant death: death scene investigation. *New England Journal of Medicine* 315:100–106.

Beal, S. (1988) Sleeping position and SIDS. *Lancet ii*: 512.

Beal, S., and Blundell H. (1988) Sudden infant death syndrome related to position in the cot. *Medical Journal of Australia* 2:217–218

Blackwell C.C., Saadi, A.T., Raza, M.W. et al. (1993) The potential of bacterial toxins in sudden infant death syndrome (SIDS). *International Journal of Legal Medicine* 105: 333–338.

Beckwith, B. (1970) Observations on the pathological anatomy of the sudden infant death syndrome. In: Bergman, A.B., Beckwith, J.B., and Ray, C.G., eds., *Sudden Infant Death Syndrome: Proceedings of the Second International Conference On Causes of Sudden Death in Infants.* Seattle: University of Washington Press, pp. 83–107.

Brown, David. Darwinian Medicine (1993) *Washington Post*, April 3.

Byard R.W., and Cohle, S.D. (1994) *Sudden Death In Infancy and Adolescence.* Cambridge: Cambridge University Press.

Chiodini, B.A., and Thach, B.T. (1993) Impaired ventilation in infants sleeping facedown; potential significance for the sudden infant death syndrome. *Journal of Pediatrics* 5: 686–692.

Cunningham, A.S. (1995) Breastfeeding: adaptive behavior for child health and longevity. In: Stuart-Macadem, P., and Dettwyler, K., eds., *Breastfeeding: Biocultural Perspectives.* New York, Aldine de Gruyer, pp. 243–254.

Cunningham, A.S., Jelliffe, D.B., and Jelliffe, E.F. (1991) Breast-feeding and health in the 1980s: a global epidemiological review. *Journal of Pediatrics* 118(5):1–8.

Daves, D.P. (1985) Cot death in Hong Kong: a rare problem? *Lancet* ii: 1346–1347.

de Jonge, G.A., Enleberts, A.C., Koomen-Liefting, A.JM., and Kostense, P.J. (1989) Cot death and prone sleeping position in the Neth-

erlands. *British Medical Journal* l298: 722–725.

Einspieler, C., Widder, J. Holzer, A., and Kenner, T. (1988) The predictive value of behavioral risk factors for sudden infant death. *Early Human Development* 18:101–109.

Elias, M.F., Nicholson, N.A., Bora, C. and Johnston, J. (1986) Sleep/wake patterns of breast fed infants in the first two years of life. *Pediatrics* 77: 322–329.

Elias, M.F., Nicholson, N., and Konner, M. (1987) Two subcultures of maternal care in the United States. In: Taub, D., and King, F., eds., *Current Perspectives in Primate Social Dynamics.* New York: Von Nostrand Reinhold, pp. 37–45.

Ellison, P. (1995) Breastfeeding, fertility and maternal condition. In: Stuart-Macadam, P., Dettwyler, K., eds., *Breastfeeding: Biocultural Perspectives.* New York: Aldine de Gruyter, pp. 305–324.

Field, T., Schanberg, S.M., Scafidi, F., Bauer, C.R., Vega-Lahr, N., Garcia, R., Nystrom, J., and Kuhn, C. (1986) Tactile/kinesthetic stimulation effects on preterm neonates. *Pediatrics* 70:381–384.

Ferber R. (1985) *Solve Your Child's Sleep Problem.* New York: Simon and Schuster.

Fleming, P. (1994) Understanding and preventing SIDS, 1994. *Current Opinion in Pediatrics* 6:158–162.

Fleming, P. (1996) CESDI epidemiological study. *British Medical Journal* 340:233–242.

Fleming, P., Gilbert, R., Azaz, Y., Berry, P.J., Rudd, P.T., and Stewart, A.J. (1990) Interactions between bedding and sleeping position in the sudden infant death syndrome: a population based control study. *British Medical Journal* 301:85–89.

Forbes, J.F., Folen, R.A., Weiss, D.S. (1992) Cosleeping habits of Military

Children *Military Medicine.* 157 (4) 196–200.

Fredrickson, D.D., Sorenson, J.R., Biddle, A. K., and M. Kotelchack (1993) Relationship of sudden infant death syndrome of breast-feeding duration and intensity (abstract no. 191). *American Journal of Diseases in Children* 147:460.

Gantly, M., Davies, D.P., and Murcott, A. (1993) Sudden infant death syndrome: links with child care practices. *British Medical Journal* 306:16–20.

Gilbert, R.E., Wigfield, R.E., and Fleming, P.J. (1995) Bottle feeding and the sudden infant death syndrome. *British Medical Journal* 310:88–90.

Guilleminault, C., and Souquet, M. (1979) Sleep states and related pathology. In: Korobkin, R., and Guilleminault, C., eds., *Advances in Perinatal Neurology, vol. 1,* C. New York: S.P. Medical and Scientific Books, pp. 225–247.

Guntheroth, W. G., and Spiers, P.S. (1992) Sleeping prone and the risk of sudden infant death syndrome. *Journal of the American Medical Association* 267:2359–2362.

Hauck, F. (1997) Bedsharing: review of epidemiological data examining links to SIDS. (Abstract) Infant Sleep Environment and SIDS Risk Conference. Bethesda, Md. Jan. 9–10, 1997.

Hofer, M. (1981) *The Roots of Human Behavior.* San Francisco: W.H. Freeman.

Hoffman, H., Damus, K., Hillman, L., and Krongrad, E., (1988) Risk factors for SIDS: results of the Institutes of Child Health and Human Development SIDS cooperative epidemiological study, *Annals of the New York Academy of Sciences* 533:13–30.

Jeliffe, D.B., and Jeliffe, E.F.P. (1978) *Human Milk In the Modern World.* Oxford: Oxford University Press.

Johnson, P. (1995) Why didn't my baby wake up? In: Rognum,T.O., ed., *Sudden Infant Death Syndrome. New Trends in the Nineties.* Oslo: Scandinavian University Press, pp. 218–225.

Jura, J., Olejar, V., Dluholucky, S. (1994) Epidemiological risk factors of SIDS in Slovakia, 1993, 1994 (abstract). In: *Program and Abstracts, Third SIDS International Conference, Stavanger, Norway,* July 31–August 4. Oslo: Holstad Grafisk98.

Knauer, M. (1985) Breastfeeding and the return of ovulation in urban Canadian mothers practicing "natural" mothering. In: Hull, V., and Simpson, M., eds., *Breastfeeding, Child Health and Child Spacing: Cross-Cultural Perspectives.* London: Croom Helm, pp. 187–198.

Konner, M. (1981) Evolution of human behavior development. In: Monroe, R. H., Monroe, R. L., and Whiting, J. M., eds., *Handbook of Cross-Cultural Human Development.* New York: Garland STPM Press, pp 3–52.

Konner, M., and Super, C. (1987) Sudden infant death syndrome: an anthropological hypothesis. In: Harkness, S., and Super, C., eds., *The Role of Culture In Developmental Disorder.* New York: Academic Press, pp. 95–108.

Konner, M.J., and Worthman C. (1980) Nursing frequency, gonadal function and birth spacing among !Kung hunter-gathers, *Science* 207:788–791.

Lee, N.P., Chann, Y.F., Davis, D.P., Lau, E., and Yip, D.C.P. (1989) Sudden infant death syndrome in Hong Kong: confirmation of low incidence. *British Medical Journal* 298:721.

Lewis, M., and Havilland, J. (1993) *The Handbook of Emotion.* New York: Gulford Press.

McKenna, J.J. (1986) An anthropological perspective on the sudden infant death syndrome (SIDS): the role of parental breathing cues and speech breathing adaptations. *Medical Anthropology* 10:9–53.

McKenna, J.J. (1995) The potential benefits of infant-parent cosleeping in relation to SIDS prevention: overview and critique of epidemiological bed sharing studies. In: Rognum, T. O., ed., *Sudden Infant Death-Syndrome: New Trends in the Nineties.* Oslo:Scandinavian University Press, pp. 256–265.

McKenna, J. J. (1996) SIDS in cross-cultural perspective: is infant-parent cosleeping protective? *Annual Reviews in Anthropology* 25:201–216.

McKenna, J.J., and Mosko, S. (1993) Evolution and infant sleep: an experimental study of infant-parent cosleeping and its implications for SIDS: an experiment in evolutionary medicine. *Acta Paediatrica* 389:31–36.

McKenna, J.J., Mosko, S., Dungy, C.,and McAninch, J. (1990) Sleep and arousal patterns of co-sleeping human mother-infant pairs: a preliminary physiological study with implications for the study of sudden infant death syndrome (SIDS). *American Journal of Physical Anthropology* 83:331–347.

McKenna, J., Mosko, S., and Richard, C. (1997) Bedsharing promotes breast feeding. *Pediatrics* 100 (2):214–219).

McKenna, J.J., Thoman, E.B., Anders, T.F., Schechtman, V., and Glatzbach, S. (1993) Infant-parent cosleeping in an evolutionary perspective: implications for understanding infant sleep development and the sudden infant death syndrome. *Sleep* 16 (3):263–282.

McNeilly, A.S., Tay, C.C.K., and Glasier, A. (1994) Physiological mechanisms underlying lactational amenorrhea. *Annals of the New York Academy of Sciences* 709:145–155.

Mitchell, E.A. (1995) Smoking: the next modifiable risk factor. In: Rognum, T. O., ed., *Sudden Infant Death Syndrome: New Trends in the Nineties.* Oslo: Scandinavian University Press, pp. 225–235.

Mitchell, E.A., Taylor, B.J., and Ford, R.P.K. (1992) Four modifiable and other major risk factors for cot death: The New Zealand study. *Journal of Paediatrics and Childhood Health* (suppl. 1):S3–8.

Mosko, S., McKenna, J., Dickel, M., and Hunt, L. (1993) Parent-infant cosleeping: the appropriate context for the study of infant sleep and implications for SIDS. *Journal of Behavioral Medicine* 16:589–610.

Mosko, S., Richard, C., and McKenna, J. (1997a) Maternal sleep and arousals in the bedsharing environment. *Sleep* 20:142–150.

Mosko, S., Richard, C., and McKenna, J. (1997b) Infant arousals during mother-infant bedsharing: implications for infant sleep and SIDS research. *Pediatrics* 100 (5)841–849.

Mosko, S., Richard, C., McKenna, J., and Drummond. (1996) Infant sleep architecture during bedsharing and possible implications for SIDS. *Sleep* 19 (9):677–684

Nelson, E.A. (1989) *Sudden Infant Death Syndrome and Child Care Practices.* Hong Kong: Chinese University of Hong Kong.

Pinilla, T., and Birch, L.L. (1993) Help me make it through the night: behavioral entrainment of breast-fed infants' sleep patterns. *Pediatrics* 91(2):436–444.

Quandt, S.A. (1987) Maternal recall accuracy for dates of infant feeding transitions. *Human Organization* 46: 152.

Richard, C., Mosko, S., McKenna, J.J., Drummond, S. (1996) Sleeping position, Orientation, and Proximity in Bedsharing Mothers and Infants (1996) *Sleep* 19:685–690.

Rognum, T.O. Sudden Infant Death Syndrome: New Trends in the '90s. (1995) Scandinavia University Press: Oslo Norway.

Schaefer, S.J.M., Willinger, M., and Hoffman, H. (1996) SIDS trends in the USA 1990–1994. Paper delivered at the Fourth SIDS International Conference, Washington, D.C., June 23–26.

Scragg, R., Stewart, A.W., Mitchell, E.A., Ford, R.P.K., and Thompson, J.D. (1995) Public health policy on bed sharing and smoking in the sudden infant death syndrome. *New Zealand Medical Journal* 108:218–229.

Spock, B., and Rothenberg, M. (1985) *Dr. Spock's Baby and Child Care.* New York: Pocket.

Tonkin, S. (1986) Epidemiology of cot deaths in Auckland. *New Zealand Medical Journal* 99:324–3326.

Vitzthum, V.J. (1989) Nursing behaviour and its relation to duration of post-partum amenorrhoea in an Andean community. *Journal of Biosocial Science* 21:145–160.

Vitzhum, V. (1994a) Sucking patterns: lack of concordance between maternal recall and observational data. *American Journal of Human Biology* 6:551–669.

Vitzthum, V.J. (1994b) Comparative study of breastfeeding structure and its relation to human reproductive ecology. *Yearbook of Physical Anthropology* 37:143–163.

Wantanabe, N., Yotsukura, M., Kadoi, and Yashiro, N. (1994) Epidemiology of sudden infant death syndrome in Japan. *Acta Paediatric of Japan* 36: 329–332.

Werne, J., and Garrow, I. (1947) Sudden deaths of infants allegedly due to mechanical suffocation. *American Journal of Public Health* 37:675–688.

Whiting, J. (1964) Effect of climate on certain cultural practices. In: Goodenough, W., ed., *Explorations in Cultural Anthropology: Essays in Honor of George Peter Murdoch.* New York: McGraw-Hill, pp. 117–141.

Whiting, J.W.M. (1981) Environmental constraints on infant care practices. In: Munroe, R. H., Munroe, R. L., and Whiting, J. M., eds., *Handbook of Cross-Cultural Human Development.* New York: Garland STPM Press, pp. 155–189.

Willinger, M., and Rognum, T.O. (1995) SIDS Definition. In: Rognum, T. O. (ed.), *Sudden Infant Death Syndrome: Trends in the Nineties.* Oslo: Scandinavian Press, pp. 13–26.

Wilson, C.A., Taylor, B.J., Laing, R.M., Williams, S.M., and Mitchell, E.A. (1994) Clothing and bedding and its relevance to sudden infant death syndrome: further results from the New Zealand Cot Death Study. *Journal of Paediatrics and Childhood Health* 30: 506–510.

4

OTITIS MEDIA

An Evolutionary Perspective

HAL J. DANIEL, III

Primum non nocere
           —Hippocrates

In this chapter I provide a review and update of otitis media (OM) and suggest that the prevalence and severity of OM could be reduced by adopting an evolutionary perspective. Many of the ideas that follow are oriented toward a reduction in the expensive, contemporary medical treatment of OM in favor of conservative, less medicalized "complementary" interventions. I present information that demonstrates a need to seriously consider these nonconventional intervention strategies for the prevention and control of OM.

Acute otitis media (AOM), the presence of fluid in the middle ear, may be accompanied by signs and symptoms including pain, hearing loss, irritability, fever, and malaise, whereas otitis media with effusion (OME) may be accompanied by pain and hearing loss only (Daniel et al., 1988). The major difference between OME and AOM is the predominance of *Haemophilus influenzae* and the minor role of *Streptococcus pneumoniae* in the former (Ruuskanen and Heikkinen, 1994). The infection duration of either may vary from days to months and may result in alteration of middle ear structures, mastoiditis, hearing loss, language development deficiency, and increased susceptibility to recurrence (Daniel et al., 1988). Conversely, the infection may subside with no residual effects. If the disease becomes chronic (COM), the sequelae may include labyrinthitis, mastoiditis, tympanic membrane rupture, and cholesteatoma, as well as other intratemporal and intracranial complications (e.g., sepsis, meningitis, brain abscess, subdural empyema, and lateral sinus vein thrombosis) (Fliss et al., 1994; Berman, 1995a). Chronic suppurative otitis media (CSOM), in which the middle ear frequently drains, may differ in its bacterial etiology from both AOM and OME. In CSOM *Pseudomonas aeruginosa* is a common etiologic agent and is insensitive to antibiotics frequently used to treat other types of OM (Ruuskanen and Heikkinen, 1994). (More on the sequence and bacteriology of OM follows in the section "Microbiology and immunology"). The younger the age at which OM first occurs, the greater the risk of recurrences (Paradise, 1980). By age 5 or 6, the morbidity associated with infancy and early childhood decreases considerably as cranio-facial structures mature, specifically the eustachian tubes and mastoid processes (Daniel et al., 1988).

Diagnosis

OM is frequently misdiagnosed. Recent studies have shown that many of the previously mentioned physical and behavioral symptoms are of little diagnostic value. For example, Heikkinen and Ruuskanen (1995) found that fever and restless sleep are of no value in differentiating upper respiratory infection (URI) from OM. In the absence of URI, ear pulling, a frequently considered behavior cue, has also been shown to be nonpredictive (Baker, 1992). Fatigue, fever, anorexia, nausea, and vomiting are no more predictive of OM than other infectious diseases (Niemelä et al., 1994; Uhari et al., 1995). In a Swedish population, about 50% of the children were found to have otalgia in the absence of OM (Ingvarsson, 1982), which may be caused by other problems such as myringitis, laryngitis, tonsillitis, and teething. Apparently, the only reliable predictor of OM is otalgia in the presence of rhinitis and a cough (Heikkinen and Ruuskanen, 1995; Uhari et al., 1995).

Similar problems exist in diagnosing OM when using otoscopy. In AOM, the tympanic membrane may be red, bulging, or ruptured with an obscured light reflex, but the membrane is usually retracted in OME (Ruuskanen and Heikkinen, 1994). Although erythema of the membrane is often considered a major sign of AOM, a red tympanic membrane may be caused by crying, a viral infection, and cerumen removal (Berman, 1995b). According to Ruuskanen and Heikkinen (1994), a cloudy-colored membrane may be a better predictor of inflammation if otalgia is present. Froom et al. (1990) found the accuracy of otoscopic diagnosis for AOM to be only 58% for children up to 1 year of age. This percentage increased to 66% in those aged 13–30 months and 73% in those children older than 31 months of age. For these reasons, all types of OM are frequently misdiagnosed and are often mistreated.

Prevalence and Costs

During the OM-prone years, the human eustachian tube is relatively horizontal, short, and wide, and the mastoid process is incompletely developed. These morphological factors are related to the middle ear's aeration, clearance, and consequent vulnerability to URI. The most common finding in children with persistent or recurrent OM is tubal dysfunction, usually documented as the ineffective dilation of the eustachian tube lumen (Cantekin, 1985). A significant positive correlation exists between small mastoid volume and impaired eustachian tube function (Andreasson, 1976). A detailed discussion of the ontogeny, anatomy, and physiology of the human eustachian tube and mastoid process as they relate to OM is provided by Daniel et al. (1988).

OM occurs with great frequency and virulence among those populations with Asian origins (e.g., Eskimos, American Indians, Mongolians) (Reed et al., 1967; Jaffe, 1969; Kaplan et al., 1973, Wiet, 1979; Prellner et al., 1993), and it has been listed as the leading disease among these groups since 1969, with 65% of new cases occurring before the age of 5 (Wiet et al., 1980). Relative to American Indians, susceptibility to OM is about 15 times lower in Americans of Northern European origin (Zonis, 1970; Wiet, 1979) and is lowest in Americans of African origin (Griffith, 1979). The Greater Boston Study (Teele et al., 1989) showed that by 1 year of age, more than 60% of the children investigated had more than one episode of AOM, and by 3 years of age, more than 80% had experienced bouts of AOM, with 40% having more than three episodes. Within this group of Boston children, Hispanic children had the highest frequency of the disease, followed by non-Hispanics of Northern European origin, with children of African origin having the lowest occurrence.

In an epidemiological investigation of a mixed population in the West Indies (Karmody and Trinidade, 1988), an "overwhelming higher incidence" of OM was found among a population of individuals from India (88%) as opposed to a "comparatively low incidence" in those of African origin (10%). Although the nature and extent of anatomic and genetic factors involved in predisposition to OM are undetermined, current research continues to confirm a relationship between these factors and observed population differences in the incidence of middle ear infection. One could theoretically plot a frequency of occurrence against presumed geographical variation. Genetic traits may increase susceptibility to infectious organisms, perhaps through inherited immunoglobulin differences (Prellner et al., 1993). But most research has focused on inherited anomalies (e.g., eustachian tube morphology, mastoid size and pneumatization, cranial base variation) that facilitate OM (Daniel et al., 1988). In addition to population variation, there are various groups more prone to the disease than others: those with craniofacial anomaly, Down's syndrome, congenital velopharyngeal insufficiency, achondroplastic dwarfs, those with family histories of OM, and those with middle ear tumors and congenital cilia abnormalities. Moreover, in most populations males are more susceptible than females (Todd, 1986).

One of the most common childhood diseases, OM ranks second only to the common cold as the most frequent illness among children in the United States. It is the problem most frequently seen by pediatricians, and it has been estimated that up to 30% of all pediatric patients are those with OM. In addition, if pediatricians dealt with children during their first year of life only, the estimate would be as high as 50% (van Cauwenberge, 1986; Brooks, 1994). Nearly 70% of children born in the United States develop OM by age 2, representing health care costs between $3.5 and $5 billion dollars annually (Brooks, 1994; Gates, 1995). Office visits to physicians for OM increased 175% from 1975 through 1990 (McCaig and Hughes, 1995), and it has been estimated that in 1990 there were 24.5 million office visits of OM, with the most significant rise among children under 2 (Stool et al., 1994). It has also been estimated that there are about 1 million children surgically treated for OM in the United States annually and that each treatment averages $3000 per child (Gates, 1995). This surgical procedure, commonly called tympanostomy, involves implanting tubes to drain the middle ear of fluid; it is the most common surgery requiring general anesthesia performed on children under age 2 (Gates, 1995).

Susceptibility to OM varies as a function of age and sex, as well as other genetic and extrinsic predisposing factors (Daniel et al., 1988). The treatment and its costs vary from country to country and also from locale to locale within different countries. For example, there are areas in the United States where OM occurs with high frequency and virulence (areas having large populations of people of Asian origins), yet little is spent on its treatment; other areas exhibit comparatively lower prevalence (areas inhabited by populations of people mostly of Northern European descent), but considerably more is spent on treating the disease. A review of the epidemiology of OM (Daniel et al., 1988) suggests that expenditures are higher among those groups who have the financial resources to pay for services and lower among those who do not have such resources (Stoodard et al., 1994). This economic inequity forms a major foundation for this chapter: OM may or may not be a "civilized disease" (see Berman, 1995a; Nesse and Williams, 1995), but its contemporary intervention using western medical techniques as well as the exorbitant associated costs designate it as such. OM patients often receive modern "civilized" treatments, albeit not necessarily the most effective treatments, if these patients have the dollars, pounds, krona, or yen available. Alternatively, it is probable that OM's "civilized" treatment exacerbates its frequency and virulence and therefore designates the disease as one associated with affluence in certain countries and locales.

Ontogenetic and Phylogenetic Considerations

"Bipedalism, more than anything else, separates early hominids from apes" (Falk, 1992, p. 88). Habitual upright walking resulted in cranial and postcranial alterations (Leaky and Lewin, 1977), many of which decreased the predisposition to OM in adult human populations. Specifically, the developmental changes of the skull, mastoid process, eustachian tube, palatal musculature, and breathing apparatus produced structures that decreased OM in adulthood (Daniel et al., 1988). With advancing age, the human eustachian tube becomes more vertical and the mastoid process becomes larger due to the development and pull of the sterno-cleido-mastoideus muscle. The concomitant verticality of the palatal sling musculature and lack of postnatal velar–epiglottic contact significantly alters both respiratory and deglutitory structure and function.

These ontogenetic alterations relate to the child's ability to eventually "outgrow" predispositions to the disease. As stated earlier, the eustachian tube and mastoid process are intimately related to infections of the middle ear. The eustachian tube, extending from the nasopharynx to the anterior wall of the middle ear cavity, is an outgrowth of the dorsal end of the first pharyngeal pouch. The mastoid process, derived from the second branchial arch, develops into the posterior wall of the middle ear cavity. Both structures are involved in middle ear aeration and gas exchange, the major physiological requirements for middle ear function (Daniel et al., 1988).

In human infants the eustachian tube is shorter, wider, and more horizontal than that of adults. It neither drains nor aerates the middle ear cavity as well as an adult tube (see figure 4.1). In addition, it is more susceptible to retrograde influx from the nasopharynx than is the adult tube. Interestingly, significant facial maturation occurring after a child becomes ambulatory (e.g., increased verticality of the splanchnocranium and the increased basicranial kyphosis or flexion) increases the overall physiological effectiveness of tubal dilation and subsequently reduces the chance of OM (see figure 4.2).

As bipedalism has produced significant ontogenetic changes in the eustachian tube and its function, it has had similar effects on the mastoid process. Van der Klaauw (1931) was the first to relate the relatively large adult human mastoid to the evolution of bipedality. He believed that habitual bipedalism caused stronger development of the sterno-cleido-mastoideus muscle and thus increased the size of the adult human mastoid. Schultz (1950a) concluded that formation of the mastoid was the result of an ontogenetic innovation that led to its late development in chimps and gorillas and relatively early but constant development in humans. For this reason Schultz (1950b) considered the mastoid process, like bipedality, a distinction of humans. Human infants have small, undeveloped mastoids which enlarge following the cranio-facial alterations associated with bipedalism (see figure 4.2). As the mastoids enlarge and become pneumatized, the middle ear cavity is better aerated and less prone to infection (Daniel et al., 1988). The mastoid process, acting as an air reservoir, allows for various pressure changes, with small mastoids producing a larger percentage of carbon dioxide within the middle ear cavity than large mastoids (Holmquist, 1980). Carbon dioxide within the middle ear may originate within the mucosal capillaries or may be ambient and diffused across the tympanic membrane and/or vented by the eustachian tube (Ranade et al., 1980). High carbon dioxide and low oxygen content is a characteristic finding in middle ears prone to OM, as well as in other poorly aerated cavities of the skull (Aust and Drettner, 1974) and diseased ears are characteristically associated with smaller mastoids and lower oxygen volumes (Holmquist, 1980).

Also related to mastoid development and the possibility of decreased susceptibility to OM is an observation (Falk 1986) that during hominid evolution, increased numbers of mastoid emissary foramina (associated with blood supply and brain cooling) correlated with decreasing thickness of the mastoid process. Falk (1992) also noted that the veins within the mastoid help cool the brain—these diploic veins dilate when the human brain is overheated. This

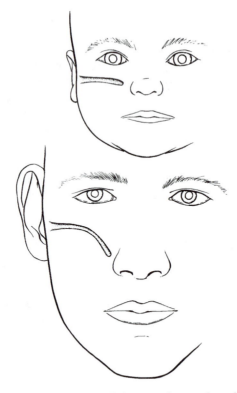

Figure 4.1. Schematic of child and adult eustachian tubes showing increased verticality of the tube with age.

"selective brain cooling" has relevance when considering middle ear vascularity and the reduction of pain associated with attacks of AOM. Equally relevant is Silver's (1995) observation that African populations differed from Asian populations in mastoid/cranial vasculature and placement of the emissary foramina. African populations had a greater overall frequency and a wider network of anterior emissary foramina, which may be related to differences in ambient temperature (Silver, 1995). These findings suggest that we should reevaluate the earlier finding of population differences of the mastoid that have been described (Howells, 1957; Giles and Eliot, 1962; Schulter, 1976; Schulter-Ellis, 1979) and reconsider how these findings relate to OM's occurrence and differential selection to various ambient habitats. Otitis media-prone children may have fewer and smaller mastoid foramina and therefore less blood supply to the middle ear system.

Other significant alterations in human respiratory structures and their functions have been noted in relation to bipedalism. Postnatal laryngeal descent (Lieberman et al., 1972), early velar–epiglottic contact (Magriples and Laitman, 1987), nasal aperture shape (Weiner, 1954), and longer exhalation periods (Bouhuys, 1974) are all by-products of habitual bipedal locomotion.

These evolutionary vocal and upper respiratory tract alterations relate to breathing mechanics, tubal dilation, and upper respiratory ventilation and are indirectly related to the onset and duration of OM (Fulghum et al., 1977; Daniel et al., 1988). Interestingly, children who acquire the habit of mouth breathing due to immature respiratory structure and function are more prone to OM (van Bon et al., 1989). The human nasal index, an indication of the shape of the nasal openings, varies as a function of humidity and habitat. For example, those populations inhabiting the tropics usually have flatter, more open noses than those inhabiting drier areas (Weiner, 1954) and in general, a lower frequency of OM.

The human middle ear appears to be selected to function optimally in an upright body position. Daniel et al. (1985, p. 79), in an investigation involving various body positions and acoustic admittance, concluded, "upright posture is optimal to the transmission of acoustic stimuli through the middle ear mechanism."

In addition, the human auricle, the external ear canal, and the tympanic membrane undergo considerable ontogenetic changes in humans (Peck, 1994).

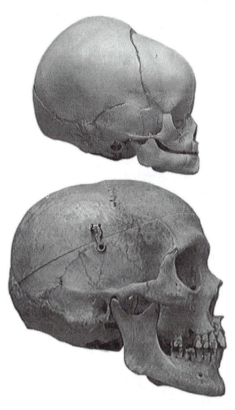

Figure 4.2. Child and adult skulls showing increased basicranial flexion, mastoid, tympanic, and splanchnocranial development in the adult.

While these changes have not been directly related to bipedality, it is likely that these external ear structures are by-products of other cranial alterations associated with habitual upright walking. The auricle continues to grow substantially until around 9 years of age and continues slight growth through adulthood (Peck, 1994). Heathcote (1996) reported that human auricles grow about 0.0087 of an inch a year, or about 0.5 inch over 50 years, due to the fact that auricles are primarily composed of cartilage that does not stop growing at puberty. The *external* ear canal, like the eustachian tube, also increases its verticality with continued mandibular and facial development (Kenna, 1990). Both the adult auricle and the external ear canal appear differentially selected to facilitate drainage of middle ear effusions. The intertragal notch of the human auricle, or pinna, could be viewed as a trough, like the ridges of the philtrum of the upper lip, selected to pass exudate (draining ears) from the middle ear and external canal. Just as the nose produces an effusion (sniffle) with URI, it is possible the intertragal notch may have been selected to enable draining ears to continue to drain effectively.

A similar suggestion is made for the human external auditory meatus. Its design as a "conveyor belt" to remove epidermal, sebaceous, and cerumenous debris and various foreign objects from the cul-de-sac ear canal may also facilitate drainage of the middle ear after tympanic membrane rupture. Rupturing of the tympanic membrane, which aerates and helps restore normal middle-ear microflora, might be viewed as something similar to a "sneeze" of the middle ear. Perhaps, it could be speculated that URI is to a sneeze as OM is to tympanic membrane rupture.

In human infants the tympanic membrane is nearly horizontal but becomes more vertical as the tympanic ring extends laterally to form the bony external ear canal (von Unge, 1994). In adults it becomes almost vertical and, if ruptured, drains more effectively than that of the human infant. This increasing verticality of the ear drum also increases the overall size and volume of the middle ear cavity by extending its lateral border, thus increasing the aeration potential of the middle ear (Daniel et al., 1988). The tympanic membrane can also be viewed as an air pressure regulator. The tympanic membrane is less displaced in normal ears than in those with poor tubal function, and the displacement is greater with small mastoids (Holmquist, 1978). A relationship exists between small mastoids and hypofunction of the tube; an inefficient tube produces more variation in pressure over time and subsequently more infection (Holmquist, 1978, 1980).

For these anatomical and physiological reasons, human OM is primarily a disease of childhood, an epigenetic consequence of the developmental process. This is not necessarily the case in other species. For example, OM is rare in young laboratory rats, but its incidence increases significantly with age (Daniel et al., 1971, 1973), just opposite from the human condition. More important, OM is rare in most other feral species. One could make the argument that OM in most nonhuman mammalian species is primarily a disease resulting from human intervention. Mongolian gerbils rarely get OM in the wild or in the laboratory (Means et al., 1975); in humans with Mongolian or Northern

Asian origins, however, the condition is nearly ubiquitous (Daniel et al., 1988). Even though there are definite ontogenetic predisposing factors in humans, the same factors do not necessarily increase the risk in other species. While inherent dangers exist in making anatomical and physiological comparisons across species, the distinctive cranial biology of certain species (e.g., the laboratory rat's small, underventilated middle ear cavity and horizontal eustachian tube) indeed lends itself to a higher disease incidence, especially when the rats are housed in crowded laboratory conditions. Conversely, the hypertrophied, well-ventilated and drained middle ear cavities of the gerbil and chinchilla are related to their respective nonsusceptibility in the wild and in the laboratory (Daniel et al., 1982, 1988).

Little is known about OM in higher nonhuman primates. The use of rhesus monkeys as models in OM has been limited thus far to descriptions of structure and function rather than to pathology (Doyle and Rood, 1979, 1980), thus no conclusive data on their natural and laboratory susceptibility has been established. OM is rarely observed in chimpanzees, orangutans, or gorillas, except occasionally when housed in laboratory conditions and is most often a by-product of URI. Of interest, however, are the recent observations that wild chimpanzees ingest certain plants possibly to rid themselves of parasites as well as to prevent URI (Huffman and Seifu, 1989; Sears, 1990), which may or may not relate to the lack of OM in chimps in the wild. For an interesting discussion of zoopharmcognosy, see Sapolsky (1994).

Environmental and Biocultural Influences

A number of possible environmental and biocultural influences on the predisposition to OM have been reported. Geographic prevalence was first noted by Hippocrates (460–375 B.C.), who described inhabitants of places exposed to wet southerly winds as having "moist heads full of phlegm" (Pahor, 1978). The incidence of the disease is highest in winter and lowest in summer (Lous and Fiellau-Nikolajsen, 1981; Casselbrant et al., 1985; Pukander and Karma, 1988), paralleling an increased incidence in episodes of URI. Children living in areas of low altitude and high humidity appear to be at greater risk than those living at higher altitudes and in areas of low humidity (Suehs, 1952; Anderson, 1965). Smoking in the household increases the risk (Kramer et al., 1983), and children of mothers who smoked during pregnancy also have an increased risk (van Cauwenberge, 1984), most likely a function of prematurity and/or low birth weight. There are fewer episodes of OM among breast-fed infants than among bottle-fed infants (Schaeffer, 1971; Duncan et al., 1993), as breast milk apparently inhibits the upper respiratory tract attachment of *Streptococcus pneumoniae* and *Haemophilus influenzae* (Anderson et al., 1986). The protective effect of prolonged breast-feeding against URI, which usually precedes bouts of OM, is well documented (Cunningham, 1979; Welsh and May, 1979; Klein, 1982). Breast-feeding protects a baby against infections, especially during the transient hypoimmunoglobulin period around the first birthday. Prolonged breast-feeding as prophylaxis for recurrent OM is also well documented (Saarinen, 1982; Pukander et al., 1986, 1993). In addition, "exclusive" breast-feeding has more protective value than does breast-feeding while simultaneously supplementing the child's diet (Duncan et al., 1993). Although the exact protective mechanism in breast milk has not been identified, a number of theories (e.g., mechanical, nutritional, immunological, or possible combined effects) have been recently reported (Newman, 1995). Nonetheless, exclusive, long-term breast-feeding is probably the most cost-effective method of preventing both URI and OM.

Other recently observed predisposing factors to OM include day-care attendance (Hardy and Fowler, 1993), allergies of the child and relatives (inhalant: Pukander et al., 1993; household pets and food: Hurst, 1990; Nsouli et al., 1994; Pizzorno, 1996), pacifier usage (Niemela et al., 1995), a history of OM within the child as well as exposure to relatives with OM (Pukander et al., 1993), previous antibiotic usage (Pukander et al., 1993), malnutrition and child abuse (Downs, 1995), trauma and malposition at birth (Fallon, 1995), caesarian section at birth (Mansfield et al., 1993), sleeping position (Gannon et al., 1995) and availability of health insurance or medicaid for payment (Stoodard et al., 1994; Hoffman et al., 1995). It is important to consider these predisposing factors in light of an evolutionary and cross-cultural perspective. Most of these factors did not exist a century ago, nor do they presently exist in all cultures. It becomes apparent that many of the modern technologies and interventions indeed exacerbate the prevalence of OM.

Microbiology and Immunology

Several types of OM are classified in the literature. Among these, AOM, OME, COM, and CSOM are thought to have different microbial or immunological etiologies. Lim and DeMaria (1982) reviewed the pathogenesis of the various types of OM due to bacterial infection or to immunological phenomena. They maintained that the various bacterial etiologies of OM exist as a continuum in which each type exhibits as a clinical entity seen at a particular time in the course of the disease.

Acute otitis media of microbial etiology is due to an active infection with viral (Henderson et al., 1982; Sarkkinen et al., 1985) or bacterial (Giebink and Quie, 1977) agents. Although OME may be caused by many factors, bacterial products remaining in the middle ear after an infection play a role in many cases (Lim et al., 1979; Nonomura et al., 1987). Sequential ascendency or change of the pathogenicity to members of polymicrobial (mixed culture) infections such a anaerobic infections (Fulghum et al., 1977; Beamer et al., 1986) or secondary infections may shift the disease from one type to another of these clinical entities.

It appears that OM follows four recognizable patterns in children: (1) occasional episodes that resolve with or without treatment and without sequela; (2) AOM or OME that resolves with or without treatment but recurs repeatedly;

(3) AOM that persists without subsiding for a long period of time; and (4) OM that in treated cases appears to respond but relapses about 11 days after cessation of treatment (Donaldson, 1987). It should be noted that these conditions follow medical and pharmaceutical antibiotic treatment. This type of disease continuum may not exist when other types of treatment are employed (e.g., herbal medicine, homeopathy; see the section "A few suggestions").

S. pneumoniae, H. influenzae, and *Branhamella catarrhalis* are the most common etiologic agents of OM, whereas other bacteria such as *Staphylococcus aureus, Streptococcus pyogenes,* coagulase-negative *Staphylococcus* species, mixed pathogens, and anaerobic bacteria are detected less often (Giebink and Quie, 1978; Klein, 1980; Fulghum et al., 1985; Brook, 1986). Polymicrobial infection flora (many including anaerobic bacteria) are found in a proportion of all OM, but especially in recurrent and COM (Fulghum et al., 1977; Brook and Finegold, 1979).

Knowledge of the etiology and pathology of OME have not greatly progressed despite numerous studies of the disease. Lim et al. (1979) proposed a microbial role in the etiology of OME, citing the presence of phagocytic cell types found in effusions from OME. Otitis media with effusion has also been explained by allergic and immunologic factors, as all major immunoglobulin classes, inflammatory cells, immune complexes, and several different chemical mediators of inflammation have been found in the effusions (see Daniel et al., 1988, for a more detailed discussion of the pathogenesis of OME).

The shifting patterns of antibiotic resistance now found in the etiologic agents of OM are of increasing importance and concern. The basis for the increase of antibiotic resistance lies in the ability of many bacterial species to share genetic material through the transfer of extrachromosomal elements of DNA (plasmids, etc.). Organisms exist in nature that have the genetic capability for resistance to antibiotics. Thus, the widespread misuse of antibiotics in medical settings has provided an environment in which antibiotic-resistant bacteria are selected and the transfer of plasmids and other non-nuclear DNA is an advantage. Antibiotic resistance and the mechanisms that prevent antibiotics from killing bacteria have been recently reviewed (Amabile-Cuevas et al., 1995).

The Politics of OM and Other Observations Regarding Its Prevalence

OM has been a problem for certain populations for a long time. The paleopathology literature (Daniel et al., 1988) indicates that the age and population variation in the prevalence of OM is the same today as it was in the past, but the overall prevalence of the disease probably never exhibited the epidemic proportions seen in certain contemporary populations. Before the availability of antibiotics, an unknown percentage of people developed the serious sequelae of OM (mastoiditis, labyrinthitis, meningitis), some dying as a result of these complications. Equally probable, however, is that many people who had

OM never developed life-threatening sequelae. It has been suggested that the paleopathologic literature needs to be reappraised. Daniel et al. (1988) warned of the need to distinguish temporal bone trauma and erosion caused by external or postmortem influences from that produced by the osteologic changes resulting from OM, as trauma and erosion have frequently been incorrectly assumed to indicate the presence of past OM and/or mastoiditis.

For the past 50 years or so, most medical treatment for OM has involved pharmaceutical antibiotics and/or tympanostomy tube surgery. It would be informative to know, although ethically difficult to ascertain, how many people who have had AOM, OME, COM, or CSOM actually developed the life-threatening sequelae. The risk of complications for each type of OM will probably never be known because most medical practitioners adopt the "better safe than sorry" philosophy in OM treatment. For all types of OM, the incidence of serious intracranial complications has been estimated to be only about 0.15% (Feigin and Cherry, 1992). However, Berman (1995a), reporting on the disease burden of OM in developing countries, mentioned that perforation, otorrhea, mastoiditis, and hearing loss were frequent complications of OM, with the latter being the most frequent effect of OM. In contrast, however, are the recent findings of Rahko et al. (1995) and Noordzij et al. (1995) that AOM and COM, respectively, do not seem to have significant long-term effects on hearing. It should also be pointed out that Berman's observations of OM's "disease burden" were from developing countries, and his findings could be related to recent observations (Ewald, 1993) of mutating virulence due to malnutrition and a lifestyle burdened by poverty and squalor. Many would also say that it is better to treat a child with COM than to have a child at risk for mastoiditis, which could lead to the more serious complication of meningitis. Of interest, however, is the observation that mastoiditis only developed in 2 out of 5000 (0.04%) Dutch children when treated with analgesics and local decongestants, the preferred treatment regime in Holland (van Buchem et al., 1995). The authors, relating this decline in mastoiditis to overall health improvement and variation in pathogenicity of causative bacteria, concluded by stating "the decrease in mastoiditis over the past 40 years cannot be an argument for the routine antibiotic treatment of all cases of acute otitis media" (van Buchem et al., 1995 p. 1151). Perhaps a "middle ground" could be reached between the fear of serious complication and the overuse of antibiotics.

McCaig and Hughes (1995) recently reported in the *Journal of the American Medical Association* that in 1992 the annual antibiotic use among children (younger than 15 years of age) was 928 per 1000 population. Antibiotic use for OM significantly increased between 1980 and 1992, from 12 to 24 million prescriptions. Still, the incidence of OM increased an estimated 64% between 1982 and 1991 despite the increasing prevalence of antibiotic and surgical treatments. Unfortunately, there does not appear to be any demonstrable scientific evidence to support the claim that antibiotics are effective against OME, according to various meta-analysis investigations (Cantekin et al., 1991; Cantekin, 1994; Bailar, 1995). Similarly, there is no scientific evidence supporting the effectiveness of tympanostomy tubes for children with OME or for the prevention of recurrent OM (Bodner et al., 1991; Cantekin, 1994; Kleinman et al., 1994). Furthermore, tympanostomy tubes have been reported to cause tympanosclerosis and tympanic atrophy (Pichichero et al., 1989; Tos and Stangerup, 1989). The ear–nose–throat specialists or otolaryngologists performing most of the tympanostomies in the United States, however, continue to support the operation's clinical effectiveness (Bluestone and Klein, 1995). Earlier surgical intervention, as compared to modern-day surgery in the United States, was usually performed without general anesthesia and involved lancing the tympanic membrane; allowing the middle ear effusion to drain. This operation, tympanocentesis, was frequently performed by pediatricians in their offices or clinics using topical anesthesia. Effective and inexpensive, it frequently had to be repeated and subsequently became perceived by medical practitioners to be inefficient and not medically cost effective.

At a recent international conference on OM, Rosenfeld (1996, p. 17) suggested that "host and epidemiologic factors may be more important as outcome determinants than the frequency and variety of antibiotics consumed." It was also noted that meta-analyses show no significant difference among the various antibiotics used for OM in terms of their ineffectiveness. More important, it now appears that children who take antibiotics for OM are at greater risk for reoccurrence (Cantekin, 1994). Even the recent guidelines on *Managing Otitis Media with Effusion in Young Children* (Stool et al., 1994), published for medical clinicians treating OME, suggest that "watchful waiting—allowing the problem to clear up by itself" is effective for most children. This "watchful waiting," no doubt, is the proper advice as the mechanisms preventing antibiotics from killing bacteria are appearing faster than are the means to control the rampant resistance (Amabile-Cuevas et al., 1995). Amabile-Cuevas et al. report that this problem is primarily the result of an experiment in evolution and natural selection as the indiscriminate, reckless use of antibiotics has selected for increasingly resistant bacteria, most of which are found in medical settings (iatrogenic and nosocomial origins). Schrag and Perrot (1996) summarized recent proposals suggesting reduction of antibiotic usage as a measure to reverse the growing public health problem of antibiotic resistance. Neu (1992, p. 1072) stated that

> the responsibility of producing resistance lies with the physician who uses antimicrobial agents and with patients who demand antibiotics when the illness is viral and when antibiotics are not indicated. It is also critical for the pharmaceutical industry not to promote inappropriate use of antibiotics for humans or for animals because this selective pressure has been what has brought us to this crisis.

As pointed out by Zajicek (1994 p. 552), "academic medicine is the main source for iatrogenesis" as well as the "main cause of rising medical expenditures." For a complete discussion of antibiotic resistance, see the volume 257, no. 5073, (1992) and volume 264, no. 5157, (1994) of *Science*. The postantibiotic era is quickly emerging, and public health measures may be the most effective way to reduce the incidence and prevalence of OM in the future.

Although it is tempting to blame the alarming increase in antibiotic use on mutually beneficial relationships between many clinician-investigators and the pharmaceutical companies, the drug companies and their respective economic foci are not totally to blame. Living in a "fast-pace, quick-fix" society causes many a parent to look for any intervention that will stop a crying child with OM and allow the parent to get some sleep and return to work. The medical clinician simply becomes the pacifier, many knowing full well that the child and parent will return to doctor's office before the month is up with the parent asking, Why did I just spend $200 last month on Johnny's ear infection? Can't you do anything?

The answer is yes, something can be done. In the next section I intend to show that the beginning of the solution to OM's epidemic problem in the Western world is for both parent and clinician to embrace a little knowledge of simplicity, wellness behaviors, and "low-tech" solutions to minor health problems as well as an appreciation of our evolutionary origins and history. In doing so, we might return the treatment of OM to the home and family, where it most probably belongs.

A Few Suggestions

My suggestions will not eliminate the morbidity of OM and its various types, but their implementation might possibly reduce its epidemic proportions. The first suggestion is that society embrace prevention rather than control of OM. As OM usually follows bouts of URI, one of the most important solutions would be more emphasis on the prevention of URI. Families having rigorous, healthy lifestyles with good nutrition and exercise along will the proper amount of rest will no doubt have fewer bouts of URI and subsequently fewer cases of OM. Pizzorno (1996) recently stated that "one-third of all OM kids would be free of otitis media if they stopped consuming dairy products," a frequent cause of allergies (Oski, 1985). Pizzorno also stated that children with OM are deficient in zinc and vitamin A (see Schmidt et al., 1994). Parents should definitely take control of their children's health and well-being, not leaving it exclusively to the clinician. If contemporary families regained the "locus of control" once found within the home, considerable improvement in health and well-being would be realized. This reemphasis on family control should begin even before the children are born. Family-planning philosophies should promote information stressing the importance of prenatal care, as compromised pregnancy and childbirth increase both the risk of URI and OM. Prolonged breast-feeding should be encouraged. Physicians in the United States need more information on the methods and procedures for encouraging, supporting, and counseling parents-to-be on the importance of breast-feeding and maternal/child health (Freed et al., 1995). As in various Scandinavian countries, parents should be encouraged to avoid cigarette smoking, and mothers should plan to breast-feed their infant as long as possible. The workplace should be "family friendly," encouraging and facilitating long-term breast-feeding as well as enabling maternity/paternity leave with continuation of wage and advancement opportunity. In addition, parents should strive to avoid placing their children in crowded day-care centers. Children younger than 2 years of age placed in large day-care centers are 36 times more likely than children who stay at home to acquire pneumococcal infections, a leading cause of OM (Takala et al., 1995). Marx et al. (1995) found that 15.4% of children cared for at home, as compared to 37% of those cared for in a large day-care center, experienced frequent ear infections. In a study of Finnish children, 50% of the cases of OM among 1 year olds were attributed to day-care attendance (Louhiala et al., 1995). For a review of the day-care and OM problem, see Hardy and Fowler (1993).

Another suggestion is that parents "nurse" their children more in a professional sense. A Dutch otolaryngologist recently related that when he sees a child with "hot red ears," he simply asks the parents to return home and "nurse" the child back to health (in any way they find that works—hot water bottles, heat packs, sweet oils in the ear canals, etc.) over the next 48 hours. He then asks them to bring the child back to his office if the child is not better. Ninety percent do not return to this conservative Dutch physician as their symptoms subside; equally important, the children do not have altered flora and/or compromised immune systems from antibiotic intervention (Jan Wind, personal communication, 1985).

In the United States clinicians have discussed the role of antibiotic therapy in the relief of symptoms of OM (Burke et al., 1994; Moser, 1994). Jacobson (1992, p. 3029), in a recent commentary in the *Journal of the American Medical Association,* illustrated the problem with antibiotic therapy by stating, "the principle outcome that parents and clinicians seek in treating otitis media is the relief of pain." However, ear pain does not justify the use of antibiotics when analgesics are indicated. Also relevant is Jacobsen's observation that 85% of children with OM will be free of pain within 24 hours, with or without antibiotic therapy (van Buchem et al., 1981). Less contact with the physician and the clinical setting is not completely out of the question. A suggestion worth considering is for parents and caregivers to patronize only those sites of care (the physician's office, the well-child centers, clinics, hospitals) optimally designed to avoid harboring and thus communicating antibiotic-resistant pathogens.

Another suggestion is for parents to elevate the child's head when he or she is sleeping. Elevating the head of infected children (about 20°) while they sleep reduces fluid pressure and eustachian tube edema, thus facilitating middle ear drainage (Rundcrantz, 1969; Loesche, 1975). Virtanen (1983), however, did not find head elevation to have any significant effect in the prevention and treatment of AOM. Avoiding bottle feeding infants in a supine position is also suggested. Beauregard (1971) discussed "positional otitis media" after observing that formula spurted jetlike into the eustachian tube and even into the middle ear cavity of children bottle-feeding in a supine position. Breast-fed infants are usually fed while being supported upright, and breast-feeding does not require as strong negative pressure as does bottle sucking (Beauregard, 1971).

Parents should reduce pharmaceutical antibiotic usage by the child and by the entire family. In some countries (e.g., The Netherlands, Sweden, etc.), use of antibiotics is limited to life-threatening problems. These countries most likely have fewer problems with OM as a "civilized disease." When children develop OM, parents in the United States frequently seek medical treatment. However, research has shown that many parents have the desire and ability to manage the disease at home (Wuest and Stern, 1990a,b). Evolutionary biologists believe that some symptoms of infection (e.g., fever, cough, mucus production) should not be routinely treated, as they are natural defense mechanisms that help rid the body of pathogens (see Nesse and Williams, 1995).

If a child does get a cold, there are many effective regimes frequently used in diverse societies that keep URI from developing into OM. For example, mothers in the U.K., France, and Germany frequently use homeopathic medicines, available without prescription at the local chemist, to keep URI from developing into OM (Hazard and Daniel, 1995). Many often labeled "alternative" or "complementary" practices often work with both URI and OM in terms of prevention and control, as they have natural antibiotic and vascular-enhancing properties. These include medicinal plants (Swarts and Doyle, 1993), herbs (Singh, 1986; Duke, 1987), the insertion of oils with garlic and ginger into the ear canal (Gupta, 1986), and eardrops with various ingredients, such as mother's milk, train oil, lard from reindeer hooves, and human and animal urine (Henriksen and Stenfors, 1993). In addition, heat application from stones and knife blades, and salt and smoke blown into the ear canal (Henriksen and Stenfors, 1993) have been reported to be effective in OM symptom amelioration. Schmidt et al. (1994) recommended 1000 mg of vitamin C three times a day and 15 mg of zinc once a day during episodes of OM. South American Indians treat OM with eardrops made from various plants. Those from Venezuela use *Chrysanthemum parthenium*; those from Chile use drops made from the sap of *Peumus boldus* (Duke, 1987). North American Indians have used tocacco (*Nicotiana tabacum*), persimmon (*Diospyros virginiana*), blue flag (*Iris versicolor*), purple sage (*Salvia carnosa*), wild ginger (*Asarum canadense*), and white milk wort (*Polygala alba*) for treatment (Train et al., 1957; Vogel, 1970). In Gloucestershire, England, an early practice for OM treatment involved pricking a snail and allowing the "snail juice" to fill the ear canal (Black, 1970). Elsewhere in England, cockroaches were boiled in oil and placed in the external meatus (Black, 1970). In Scotland, ground earthworms were boiled in goose grease and the strained "liquor" then placed in the ear canal (Hamilton, 1981). Sicilians placed the wool of the ram's testicles in the ear canal (Pitre, 1971), and the Egyptians inserted mouse heads in the canal for OM (Estes, 1989).

In 1886, R.V. Pierce wrote in *The People's Common Sense Medical Advisor* that a spirit vapor bath is advised in acute OM, and "the perspiration should be kept up with the Compound Extract of Smart Weed and small doses of aconite." In chronic OM, Pierce suggested purgative pellets and/or goldenseal and dogwood tonics. If the ear canal has discharge, Pierce suggested a douche with Dr. Sage's Catarrh Remedy for cleansing. In addition, chiropractic (Hendricks and Larkin-Thier, 1989), homeopathic (Daniel and Hazard, 1995), craniosacral therapy (Weil, 1995), acupressure (Gardner, 1989), massage and yoga (Mars, 1992), Chinese breathing exercises in Qigong (Koh, 1992), and Ayurvedic lymphatic massage (Goldberg, 1994) have been anecdotally reported as being effective, although studies on their physiologic mechanisms of action have not been reported. The *Handbook of Medicinal Herbs* (Duke, 1987) lists 27 specific herbs reported to be effective with earaches, none of which, however, have been subjected to rigorous clinical trial protocols. It should be noted that many folk and nonmedical treatment regimes do not lend themselves to single clinical trial investigations. They frequently are used in combination with others regimes as the symptomatology of OM changes from onset to resolution (Daniel and Hazard, 1995). Because a remedy's treatment effect has not been subjected to the scientific empiricism of modern western medicine does not mean it is necessarily effective or ineffective.

As stated earlier, OM is likely a "civilized disease," as its overtreatment exacerbates the problem in many Western cultures. There are places, however, where the medical and/or health delivery Zeitgeist is laissez-faire. A colleague giving a seminar on Tibetan medicine was asked by a student member of the audience, "Is it true that all the Tibetan children have runny noses?" To which the professor responded, "Yes, and they all have runny ears as well." He continued by stating the local Tibetan physicians viewed both effusions as part of the disease process, its normal course of action and no reason for significant alarm. He or she treats the disease with various herbs and teas and accepts the running ears and nose as "nature's way" (James L. Mathis, personal communication, 1977).

C. Everett Koop (1996) recently explained that the effects of time spent with patients relates to the healing process. Pointing to the fact that 33% of Americans, 60% of Europeans, and 80% of the world's population use complementary therapies, Koop stated that the healing process begins with the communicative interaction between patient and healer. He further discussed a study of physician–patient communication which showed that male physicians interrupt their patients on the average of 14 seconds after the interview begins, whereas with female physicians the average time before interruption is around 45 seconds. Koop felt that practitioners of conventional medicine could benefit from the approach used by the non-conventional practitioners who spend considerably more time listening and paying attention to their patients during the interview and treatment process. "Fast-paced, quick fix" conventional medicine does not provide much in the way of quality communication. Simply by taking the time and effort to effectively communicate and listen to the parents who present a child with OM, future health care providers could reduce the prevalence, morbidity, cost, and anxiety associated with OM. Ancient healers, according to Koop, knew the importance of active listening and therapeutic communication.

In conclusion, if those concerned with reducing the prevalence and morbidity of this costly disease considered an evolutionary perspective, we would see less of this major health care problem in the twenty-first century. A suggestion from Hippocrates prefaced this chapter: "First, do no harm" specifically relates to the apparent overmedicalization of OM. In closing, I offer another bit of Hippocratic advice: "Honor the healing power of Nature" (*vis medicatrix naturae*).

Acknowledgments
I thank C. Bland, M. Barnes, E. Cantekin, R. Connery, M. Bernstein, M. Curry, C. Dunham, R. Fulghum, A. Hazard, J. Hazard, C. Jolls, T. Lamb, J. Loesche, K. Morgan, E. Trevathan, E. Szathmary, J. Wind, and three anonymous reviewers for their contributions to this chapter.

References

Amábile-Cuevas, C. F., Cárdenas-García, M., and Ludgar, M. (1995) Antibiotic resistance. *American Scientist* 83:320–329.

Anderson, B., Porras, D., Hanson, L. A., Jorgensen, F., Leffler, H., Nylen, O., and Svanborg-Eden, C. (1986) Breast milk inhibits the attachment of *S. pneumoniae* and *H. influenzae*. In J. Sade, ed., *Acute and secretory otitis media*, pp. 239–342. Amsterdam: Kluger.

Anderson, D. (1965) The effect of climate on the incidence of hearing loss. *Journal of Speech and Hearing Disorders* 30:66–70.

Andreasson, L. (1976) Correlation of tubal function and volume of mastoid and middle ear space as related to otitis media. *Annals of Otology, Rhinology and Laryngology* 85:198–203.

Aust, R., and Drettner, B. (1974) The functional size of the maxillary ostium in vivo. *Acta Otolaryngology* 78:432–435.

Bailar, III, J. C. (1995) The practice of meta-analysis. *Journal of Clinical Epidemiology* 48:149–157.

Baker, R. B. (1992) Is ear pulling associated with ear infection? *Pediatrics* 90:1006–1007.

Beamer, M. E., Fulghum, R. S., Marrow, H. G., Allen, W. E., Brinn, J. E., and Daniel, III, H. J. (1986) The Mongolian gerbil as a model of otitis media caused by anaerobic bacteria. In J. Sadé ed., *Acute and secretory otitis media*, pp. 191–199. Amsterdam: Kluger.

Beauregard, W.G. (1971) Positioned otitis media. *The Journal of Pediatrics* 79:294–296.

Berman, S. (1995a) Otitis media in developing countries. *Pediatrics* 96:126–131.

Berman, S. (1995b) Otitis media in children. *The New England Journal of Medicine* 332:1560–1565.

Black, W.G. (1970) *Folk medicine: A chapter in the history of culture.* New York: Burt Franklin Inc.

Bluestone, C. D., and Klein, J. O. (1995) Clinical practice guideline on otitis media with effusion in young children: strengths and weaknesses. *Otolaryngology- Head and Neck Surgery* 112:507–511.

Bodner, E. E., Browning, G. G., Chalmers, F. T., and Chalmers, T. C.

(1991) Can meta-analysis help uncertainty in surgery for otitis media in children? *The Journal of Laryngology and Otology* 105:812–819.

Bouhuys, A. (1974) *Breathing.* New York: Grune and Stratton.

Brook, I. (1986) Direct and indirect pathogenicity of *Branhamella catarrhalis. Drugs* 31(suppl. 3):97–102.

Brook, I., and Finegold, S. M. (1979) Bacteriology of chronic otitis media. *The Journal of the American Medical Association* 241:487–488.

Brooks, A. (1994) Middle ear infections in children. *Science News* 146: 332–333.

Burke, P., Conroy, M., and Crossley, N. (1994) A night call for earache. *The Practitioner* 283:337–342.

Cantekin, E.I. (1985) Eustachian tube function in children. In J. Sadé, ed., *Abstracts from the International Symposium on Acute and Secretory Otitis Media*, p. 183). Tel-Aviv: Tel-Aviv University.

Cantekin, E. I. (1994) *The case against aggressive, expensive and ineffective treatment of a benign disease: Comments on the clinical practice guidelines on otitis media.* Report submitted to Public Health Service, Washington, DC.

Cantekin, E. I., McGuire, T. W., and Griffith, T. L. (1991) Antimicrobial therapy for otitis media with effusion ('secretory' otitis media). *The Journal of the American Medical Association* 266:3309–3317.

Casselbrant, M. L., Brostoff, L. M., Cantekin, E. I., Flaherty, M. R., Bluestone, C. D., Doyle, W. J., and Fria, T. J. (1985) Otitis media with effusion in preschool children. *The Laryngoscope* 95:428–436.

Cunningham, A. S. (1979) Morbidity in breast-fed and artificially fed infants II. *Pediatrics* 95:685–689.

Daniel, H. J., Carmine, F. H., and Cook, R. A. (1971) Otitis media in two strains of laboratory rats. *Journal of Auditory Research* 11:276–278.

Daniel, H. J., Fulghum, R. S., Barrett, K. A., and Brinn, J. E. (1982) Comparative anatomy of eustachian tube and middle ear cavity in animal models for otitis media. *Annals of Otolology, Rhinology and Larynogology* 91:82–89.

Daniel, III, H. J., and Hazard, A. A. (1995) Otitis media and homeopathy. In *Abstracts of the Sixth International Symposium on Recent Advances in Otitis Media*, p. 127. Pittsburgh, PA: University of Pittsburgh Medical Center.

Daniel, H. J., Hume, W. G., Givens, G. D., and Jordan, J. H. (1985) Body position and acoustic admittance. *Ear and Hearing* 6:76–79.

Daniel, H. J., Means, L. W., Dressel, M. E., and Loesche, P. J. (1973) Otitis media in laboratory rats. *Physiological Psychology* 1:7–8.

Daniel, III, H. J., Schmidt, R. T., Fulghum, R. S., and Ruckriegal, L. (1988) Otitis media: A problem for the physical anthropologist. *Yearbook of Physical Anthropology* 31: 143–167.

Donaldson, J. D. (1987) Otitis media 1987. *Journal of Otolaryngology* 16: 221–224.

Downs, M. P. (1995) The high risk concept extended. In D.J. Lim, C.D. Bluestone, M. Casselbrant, J.O. Klein, and P.L. Ogra (eds.), *Recent Advances in Otitis Media*, pp. 352–353. Hamilton, Ontario: B.C. Decker. University of Pittsburgh Medical Center.

Doyle, W. J., and Rood, S. R. (1979) Anatomy of the auditory tube and related structures in the rhesus monkey (*Macaca mulatta*). *Acta Anatomica* 105:209–225.

Doyle, W. J., and Rood, S. R. (1980) Comparisons of the anatomy of the eustachian tube in the rhesus monkey (*Macaca mulatta*) and man. *An-*

nals of Otology, Rhinology and Laryngology 89:49–57.

Duke, J. A. (1987) Handbook of medicinal herbs. Boca Raton, FL: CRC
Press.

Duncan, B., Ey, J., Holberg, C. J.,
Wright, A. L., Martinez, F. D., and
Taussig, L. M. (1993) Exclusive
breast-feeding for at least 4 months
protects against otitis media. Pediatrics 91:867–872.

Estes, J.W. (1989) The medical skills of
ancient Egypt. Canton, OH: Science
History Publications.

Ewald, P.W. (1993) The evolution of
virulence. Scientific American
268(4):86–93.

Falk, D. (1986) Evolution of cranial
blood drainage in hominids: Enlarged occipital/marginal sinuses
and emissary foramina. American
Journal of Physical Anthropology 70:
311–324.

Falk, D. (1992) Braindance. New York:
Henry Holt and Company.

Fallon, J. (1995) The role of birth
trauma and malposition in the de-
velopment of otitis media in chil-
dren under the age of 6 months. In
Abstracts of the Sixth International
Symposium on Recent Advances in
Otitis Media, p. 284. Pittsburgh, PA:
University of Pittsburgh Medical
Center.

Feigin, R. D., and Cherry, J. D. (1992)
Otitis media. In R. D. Feigin, M. W.
Kline, S. R. Hyatt, and K. L. Ford,
eds., Textbook of pediatric infectious diseases, vol. 2, 3rd. ed., pp.
174–189. Philadelphia: WB Saunders.

Fliss, D.H. Leiberman, A., and Dagan,
R. (1994) Medical sequelae and complications of acute otitis media. The
Pediatric Infectious Disease Journal
13 (Suppl.): 34–40.

Freed, G. L., Clark, S. J., Sorenson, J.,
Lohr, J. A., Cefalo, R., and Curtis, P.
(1995) National assessment of physicians' breast-feeding knowledge, attitudes, training, and experience.
The Journal of the American Medical Association 273:472–476.

Froom, J., Culpepper, L., Grob, P., Bar-
telds, A., Bowers, P., Bridges-Webb,
C., Grava-Gubins, I., Green, L., Lion,
J., Somaini, B., Stroobant, A., West,
R., and Yodfat, Y. (1990) Diagnosis
and antibiotic treatment of acute otitis media: Report from International
Primary Care Network. British Medical Journal 300:582–586.

Fulghum, R. S., Daniel, H. J., and Yar-
borough, J. G. (1977) Anaerobic bacteria in otitis media. Annals of Otology, Rhinology and Laryngology 85:
198–202.

Fulghum, R. S., Hoogmoed, R. P.,
Brinn, J. E., and Smith, A. M. (1985)
Experimental pneumococcal otitis
media: Longitudinal studies in the
gerbil model. International Journal
of Pediatric Otorhinolaryngology 10:
9–20.

Gannon, M. M., Haggard, M. P., Golding, J., and Fleming, P. (1995) Sleeping position—a new environmental
risk factor for otitis media. In Abstracts of the Sixth International
Symposium on Recent Advances in
Otitis Media, p. 24. Pittsburgh, PA:
University of Pittsburgh Medical
Center.

Gardner, J. (1989) The new healing
yourself. Freedom, CA: Crossing
Press.

Gates, G. A. (1996) Cost-effectiveness
considerations in otitis media treatment. In D.J. Lim, C.D. Bluestone, M.
Casselbrant, J.O. Klein, and P.L.
Ogra (eds.) Recent advances in otitis
media, pp. 1–4, Hamilton, Ontario:
B.C. Decker.

Giebink, G. S., and Quie, P. G. (1977)
Comparison of otitis media due to
types 3 and 23 Streptococcus pneumoniae in the chinchilla model.
Journal of Infectious Disease (suppl.)
136:191–195.

Giebink, G. S., and Quie, P. G. (1978)

Otitis media: The spectrum of middle ear inflammation. *Annual Review of Medicine* 29:285–306.

Giles, E., and Eliot, O. (1962) Race identification from cranial measurements. *Journal of Forensic Science* 7: 147–157.

Goldberg, B. (1994) *Alternative medicine: The definitive guide.* Washington, DC: Future Medicine Publications.

Griffith, T. E. (1979) Epidemiology of otitis media—An interracial study. *The Laryngoscope* 89:22–30.

Gupta, K. R. L. (1989) *Hindu practice of medicine.* Delhi: S. R. I. Satgaru.

Hamilton, D. (1981) *The healers: A history of medicine in Scotland*, p. 75. Edinburgh: Canongate Publishing.

Hardy, A. M., and Fowler, M. G. (1993) Child care arrangements and repeated ear infections in young children. *American Journal of Public Health* 83:1321–1325.

Heathcote, J. (1996) British ears. *Discover* 17(6):26.

Heikkinen, T., and Ruuskanen, O. (1995) Signs and symptoms predicting acute otitis media. *Archives of Pediatric and Adolescent Medicine* 149:26–29.

Henderson, F.M., Collier, A.M., Sanyal, M.A., Watkins, J.M., Fairclough, D.L., Clyde, W.A., and Denny, F.W. (1982) A longitudinal study of respiratory viruses and bacteria in the etiology of acute otitis media with effusion. *The New England Journal of Medicine* 306:1377–1383.

Hendricks, D.C., and Larkin-Thier, S.M. (1989) Otitis media in young children. *Chiropractic* 2:9–12.

Henriksen, A. Ö., and Stenfors, L.-E. (1993) The Lappish way of treating otitis media over the centuries. In D. J. Lim, C. D. Bluestone, J. O. Klein, J. D. Nelson, and P. L. Ogra, eds., *Recent advances in otitis media*, pp. 262–264. Philadelphia: B. C. Decker.

Hoffman, H.J., Overpeck, M. D., and Hildesheim, M. E. (1996) Factors in the U. S. affecting risk of frequent ear infections, deafness or trouble hearing and related conditions. In D.J. Lim, C.D. Bluestone, M. Casselbrant, J.O. Klein, and P.L. Ogra (eds.) *Recent advances in otitis media*, pp. 71–75. Hamilton, Ontario: B.C. Decker.

Holmquist, J. (1978) Aeration in chronic otitis media. *Clinical Otolaryngology* 3:279–284.

Holmquist, J. (1980) Middle ear ventilation in chronic otitis media. In G. Munker and W. Arnold, eds., *Physiology and Pathophysiology of Eustachian Tube and Middle Ear*, pp. 213–217. New York: Thieme-Stratton.

Howells, W. W. (1957) The cranial vault: Factors of size and shape. *American Journal of Physical Anthropology* 15:19–46.

Huffman, M. A., and Seifu, M. (1989) Observations on the illness and consumption of a possible medicinal plant *Vernonia amygdalina* by a wild chimpanzee in the Mahale Mountains National Park, Tanzania. *Primates* 30:51–63.

Hurst, D.S. (1990) Allergy management of refractory serous otitis media. *Otolaryngology-Head and Neck Surgery* 102:664–669.

Ingvarsson, L. (1982) Acute otalgia in children—Findings and diagnosis. *Acta Pædiatrica Scandinavica* 71: 705–710.

Jacobson, R. M. (1992) Secretory otitis media: The Cantekin affair [Letter to the editor]. *The Journal of the American Medical Association* 267: 3029.

Jaffe, B. F. (1969) The incidence of ear disease in the Navajo Indians. *The Laryngoscope* 79:2126–2134.

Kaplan, G.J., Fleshman, J. K., and Bender, T. R. (1973) Long-term effects of otitis media: A ten year cohort study of Alaskan Eskimo children. *Pediatrics* 52:557–585.

Karmody, C. S., and Trinidade, A. (1988) Epidemiology of otitis media in an ethnically mixed island population. In D.J. Lim et. al., eds., *Recent Advances in Otitis Media*, pp. 24–26. Philadelphia: BC Decker.

Kenna, M. A. (1990) Embryology and developmental anatomy of the ear. In C.D. Bluestone and S.E. Stool, eds., *Pediatric Otolaryngology*, pp 77–87. Philadelphia: WB Saunders.

Klein, J.O. (1980) Microbiology of otitis media. *Annals of Otology, Rhinology and Laryngology* 89(suppl. 68):98–101.

Klein, J.O. (1982) Prevention of persistent middle ear effusion: Immunoprophylaxis. *Pediatric Infectious Disease Journal* 1(suppl. 1):99–106.

Kleinman, L. C., Kosecoff, J., Dubois, R. W., and Brook, R. H. (1994) The medical appropriateness of tympanostomy tubes proposed for children younger than 16 years in the United States. *The Journal of the American Medical Association* 271:1250–1255.

Koh, T. C. (1992) Qigong: Chinese breathing exercise. *American Journal of Chinese Medicine* 10:86–91.

Koop, C.E. (1996) Keynote address: The World Congress on Complementary Therapies in Medicine, 24 May 1996. Washington, DC.

Kramer, M. J., Richardson, M. A., and Weiss, N. S. (1983) Risk factors for persistent middle ear effusions: Otitis media, catarrh, cigarette smoking exposure and atopy. *The Journal of the American Medical Association* 249:1022–1024.

Leakey, R.E., and Lewin, R. (1977) *Origins.* New York: E. P. Dutton.

Lieberman, P., Crelin, E. S., and Klatt, D. H. (1972) Phonetic ability and related anatomy of the newborn, adult human, Neanderthal man, and the chimpanzee. *American Anthropologist* 74:287–307.

Lim, D.J., and DeMaria, T.F. (1982) Panel discussion: Pathogenesis of otitis media, bacteriology and immunology. *Laryngoscope* 92:278–286.

Lim, D.J., Lewis, D.M., Schram, J.L., and Birck, H.G. (1979) Otitis media with effusion: Cytological and microbiological correlates. *Archives of Otolaryngology* 105:404–412.

Loesche, P.J. (1975) Posture and otitis media in laboratory rats and gerbils. Master's thesis. Greenville, NC: East Carolina University.

Louhiala, P.J., Jaakkola, N., Ruotsalainen, R., and Jaakkola, J.J.K. (1995) Form of day care and respiratory infections among Finnish children. *American Journal of Public Health* 85:1109–1112.

Lous, J., and Fiellau-Nikolajsen, M. (1981) Epidemiology of middle ear effusion and tubal dysfunction: A one year prospective study comprising monthly tympanometry in 387 non-selected seven year old children. *International Journal of Pediatric Otorhinolaryngology* 3:303–317.

Magriples, U., and Laitman, J. T. (1987) Developmental change in the position of the fetal human larynx. *American Journal of Physical Anthropology* 72:463–472.

Mansfield, C. J., Daniel, H. J., Sumpter, E. A., Barnes, J., Coggins, D., and Young, M. (1993) Mode de naissance et otite muqueuse [Manner of birth and otitis media]. *Archives of French Pediatrics* 50:97–100.

Mars, B. (1992) Natural remedies for ear ailments. In *American herbalism: Essays on herbs and herbalism by members of the American Herbalist Guild.* Freedom, CA: Grossing Press.

Marx, J., Osguthorpe, J. D., and Parsons, G. (1995) Day care and the incidence of otitis media in young children. *Otolaryngology- Head and Neck Surgery* 112:695–699.

McCaig, L. F., and Hughes, J. M. (1995) Trends in antimicrobial drug prescribing among office-based physicians in the United States. *The Journal of the American Medical Association* 273:214–219.

Means, L.W., Daniel, H.J., Jordan, L.H., and Loesche, P.J. (1975) Nonsusceptibility to otitis media of the laboratory gerbil, *Meriones unquiculatus*. *Physiological Psychology* 3:229–330.

Moser, R. (1994) Otitis media in children: Overdiagnosed and overtreated. *Advanced PA* 2(10):21–36.

Nesse, R. M., and Williams, G. C. (1995) *Why we get sick: The new science of Darwinian medicine*. New York: Times Books.

Neu, H. C. (1992) The crisis in antibiotic resistance. *Science* 257:1064–1073.

Newman, J. (1995) How breast milk protects newborns. *Scientific American* 273(6):76–79.

Niemelä, M., Uhari, M., Jounio-Ervasti, K., Luotonen, J., Alho, O.-P., and Vierimaa, E. (1994) Lack of specific symptomatology in children with acute otitis media. *Pediatric Infectious Disease Journal* 13:765–768.

Niemelä, M., Uhari, M., and Möttönen, M. (1995) A pacifier increases the risk of recurrent acute otitis media in children in day care centers. *Pediatrics* 96:884–888.

Nonomura, N., Nakano, Y., Fujioka, O., Niijima, H., Kawana, M., and Fujita, M. (1987) Experimentally induced otitis media with effusion following inoculation with the outer cell wall of nontypeable *Haemophilus influenzae*. *Archives of Oto-rhino-laryngology* 244:253–257.

Noordzij, J. P., Dodson, E. E., Ruth, R. A., Arts, H. A., and Lambert, P. R. (1995) Chronic otitis media and sensorineural hearing loss: Is there a clinically significant relation? *American Journal of Otology* 16:420–423.

Nsouli, T. M., Nsouli, S. M., Linde, R. E., O'Mara, F., Scanlon, R. T., and Bellanti, J. A. (1994) Role of food allergy in serous otitis media. *Annals of Allergy* 73:215–219.

Oski, F. A. (1985) Is bovine milk a health hazard? *Pediatrics* 75(suppl.): 182–186.

Pahor, A. L. (1978) An early history of secretory otitis media. *Journal of Laryngology and Otology* 92:543–560.

Paradise, J. L. (1980) Otitis media in infants and children. *Pediatrics* 65: 917–943.

Peck, J. E. (1994) Development of hearing. Part II. embryology. *Journal of the American Academy of Audiology* 5:359–365.

Pichichero, M. E., Berghash, L. R., and Hengerer, A. S. (1989) Anatomic and audiologic sequelae after tympanostomy tube insertion or prolonged antibiotic therapy for otitis media. *Pediatric Infectious Disease Journal* 8: 780–787.

Pierce, R. V. (1886) *The people's common sense medical adviser*. Buffalo, NY: World's Dispensary Medical Association Publications.

Pitre, G. (1971) *Sicilian folk medicine*. Lawrence, KS: Coronado Press.

Pizzorno, J. E. (1996) Medicine for the 21st Century: Examining Our Options. Plenary Session. The World Congress on Complementary Therapies in Medicine, 24 May 1996. Washington, DC.

Prellner, K., Bohm, J., Bretlau, P., and Hallberg, T. (1993) Genetic factors as related to chronic otitis media in polar Eskimos. In D. J. Lim, C. D. Bluestone, J. O. Klein, J. D. Nelson, and P. L. Ogra, eds., *Recent advances in otitis media*, pp. 36–38. Philadelphia: B.C. Decker.

Pukander, J.S., and Karma, P.H. (1988) Persistence of middle ear effusion and its risk factors after an acute attack of otitis media with effusion.

In D. J. Lim, C. D. Bluestone, J. O. Klein, and J. D. Nelson, eds., *Recent advances in otitis media*, pp 8–11. Philadelphia: B.C. Decker.

Pukander, J. S., Sipilä, M. M., Kataja, M. J., and Karma, P. H. (1993) The Bayesian approach to the evaluation of risk factors affecting the prolongation of middle ear effusion after acute otitis media. In D. J. Lim, C. D. Bluestone, J. O. Klein, J. D. Nelson, and P. L. Ogra, eds., *Recent advances in otitis media*, pp. 33–35. Philadelphia: B. C. Decker.

Pukander, J. S., Sipilä, M. M., Kataja, M. J., and Karma, P. H. (1986) The risk of an urban child to contract acute otitis media during the first two years of life. In J. Sadé, ed., *Acute and secretory otitis media*, pp. 119–124. Amsterdam: Kluger.

Rahko, T., Laitila, P., Sipilä, M., Manninen, M., and Karma, P. (1995) Hearing and acute otitis media in 13-year-old-children. *Acta Otolaryngologica* 115:90–192.

Ranade, A., Lambertsen, C. J., and Noordergraf, A. (1980) Inert gas exchange in the middle ear. *Acta Otolaryngology* 371(suppl.)

Reed, D., Struve, S., and Maynard, J. E. (1967) Otitis media and hearing deficiency among Eskimo children: A cohort study. *American Journal of Public Health* 57:1657–1662.

Rosenfeld, R. M. (1996) How can meta-analysis help in treatment of otitis media? In D.J. Lim, C.D. Bluestone, M. Casselbrant, J.O. Klein, and P.L. Ogra (eds.) *Recent advances in otitis media*, pp. 15–18. Hamilton, Ontario: B.C. Decker.

Rundcrantz, H. (1969) Posture and eustachian tube function. *Acta Otolaryngologica* 68:279–292.

Ruuskanen, O., and Heikkinen, T. (1994) Otitis media: Etiology and diagnosis. *Pediatric Infectious Disease Journal* 13(suppl. 1):23–26.

Saarinen, U. M. (1982) Prolonged breast feeding as prophylaxis for recurrent otitis media. *Acta Pædiatrica Scandinavica* 71:567–571.

Sapolsky, R. M. (1994) Fallible instinct. *The Sciences* 34:13–15.

Sarkkinen, H., Ruuskanen, O., Meurman, O., Puhakka, H., Virolainen, E., and Eskola, J. (1985) Identification of respiratory virus antigens in middle ear fluids of children with acute otitis media. *Journal of Infectious Disease* 151:444–448.

Schaeffer, D. H. (1971) Otitis media and bottle feeding. *Canadian Public Health* 62:478–489.

Schmidt, M. A., Smith, L. H., and Sehnert, K. W. (1994) *Beyond* antibiotics: 50 (or so) ways to boost immunity and avoid antibiotics. Berkeley, CA: North Atlantic Books.

Schrag, S. J., and Perrot, V. (1996) Reducing antibiotic resistance. *Nature* 381:120–121.

Schulter, F. P. (1976) A comparative study of the temporal bone in three populations of man. *American Journal of Physical Anthropology* 44:453–468.

Schulter-Ellis, F. P. (1979) Population differences in cellularity of the mastoid process. *Acta Otolaryngology* 87:461–465.

Schultz, A. H. (1950a) The physical distinction of man. *Proceedings of the American Philosophical Society* 94:428.

Schultz, A. H. (1950b) Morphological observations on gorillas. In *The Henry Cushier Raven Memorial Volume*, pp. 227–251. New York: Columbia University.

Sears, C. (1990) The chimpanzee's medicine chest. *New Scientist* 4:42–44.

Silver, W. G. (1995) Cranial emissary foramina and their relationship to climate. *American Journal of Physical Anthropology* 20 (suppl.):195.

Singh, Y. N. (1986) Traditional medicine in Fiji: Some herbal folk cures used by the Fiji Indians. *Journal of Ethnopharmacology* 15:57–88.

Stoodard, J. J., St. Peter, R.F., and Newacheck, P.W. (1994) Health insurance status and ambulatory care for children. *New England Journal of Medicine* 330:1421–1425.

Stool, S. E., Berg, A. O., Berman, S., Carney, C. J., Cooley, J. R., Culpepper, L., Eavey, R. D., Feagans, L. V., Finitzo, T., Friedman, E., et al. (1994) *Managing otitis media with effusion in young children. Quick reference guide for clinicians.* AHCPR Publication No. 94–0623. Rockville, MD: Agency for Health Care Policy and Research, Public Health Services, U. S. Department of Health and Human Services.

Suehs, O. W. (1952) Secretory otitis media. *Laryngoscope* 62:998–1027.

Swarts, M. C., and Doyle, W. J. (1993) Medicinal plants used for the treatment of otitis media. In D. J. Lim, C. D. Bluestone, J. O. Klein, J. D. Nelson, and P. L. Ogra, eds., *Recent advances in otitis media*, pp. 257–260. Philadelphia: B. C. Decker.

Takala, A.K., Jero, J., Kela, E., Rönnberg, P.-R., Koskenniemi, E., and Eskola, J. (1995) Risk factors for invasive pneumococcal disease among children in Finland. *The Journal of the American Medical Association*, 273:859–864.

Teele, D.W., Klein, J.O., Rosner, B.A., and the Greater Boston Otitis Media Study Group (1989) Epidemiology of otitis media during the first seven years of life in children in greater Boston: a prospective, cohort study. *The Journal of Infectious Diseases* 160:83–94.

Todd, N. W. (1986) High risk populations for otitis media. In J. F. Kavanaugh, ed., *Otitis media and child development*, pp. 32–59. Parkton, MD: York Press.

Tos, M., and Stangerup, S.-E. (1989)

Hearing loss in tympanosclerosis caused by grommets. *Archives of Otolaryngology- Head and Neck Surgery* 115:931–935.

Train, P., Henrichs, J. R., and Archer, W. A. (1957) *Medicinal uses of plants by Indian tribes of Nevada.* Lawrence, KS: Quarterman Publishing.

Uhari, M., Niemelä, M., and Hietala, J. (1995) Prediction of acute otitis media with symptoms and signs. *Acta Pædiatrica* 84:90–92.

van Bon, M. J. H., Zielhuis, G. A., Rach, G. H., and Van den Broek, P. (1989) Otitis media with effusion and habitual mouth breathing in Dutch pre-school children. *International Journal of Pediatric Otorhinolaryngology* 17:119–125.

van Buchem, F. L., Dunk, J. H. M., and van't Hof, M. A. (1981) Therapy of acute otitis media: Myringotomy, antibiotics, or neither? *The Lancet* 2: 883–887.

van Buchem, F. L., Knottnerus, J. A., and Peeters, M. F. (1995) Otitis media in children [Correspondence]. *The New England Journal of Medicine* 333:1151.

van Cauwenberge, P. B. (1984) Relevant and irrelevant predisposing factors in secretory otitis media. *Acta Oto-laryngologica* (suppl. 414): 147–153.

van Cauwenberge, P. B. (1986) The character of acute and secretory otitis media. In J. Sadé, ed., *Acute and secretory otitis media*, pp. 3–11. Amsterdam: Kluger.

van der Klaauw, C.J. (1931) The auditory bulla in some fossil mammals. *Bulletin of the American Museum of Natural History* 62:1–352.

Virtanen, H. (1983) Relation of body posture to eustachian tube function. *Acta Oto-laryngologica* 95:63–67.

Vogel, V.J. (1970) *American Indian medicines.* Norman: University of Oklahoma Press.

Von Unge, M. (1994) *Mechanical properties of the tympanic membrane: An experimental study.* Stockholm: Department of Otorhinolaryngology, Karolinska Institute.

Weil, A. (1995) *Spontaneous healing.* New York: Andrew A. Knopf.

Weiner, J.S. (1954) Nose shape and climate. *American Journal of Physical Anthropology* 12:615.

Welsh, J. K., and May, J. T. (1979) Anti-infective properties of breast milk. *Pediatrics* 94:1–9.

Wiet, R. J. (1979) Patterns of ear disease in the Southwestern American Indian. *Archives of Otolaryngology* 105:381–385.

Wiet, R., Stewart, J., DeBlanc, G., and Weider, D. (1980) Natural history of otitis media in the American native. *Annals of Otology, Rhinology and Laryngology* 89:14–19.

Wuest, J., and Stern, P. N. (1990a) Childhood otitis media: The family's endless quest for relief. *Issues in Comprehensive Pediatric Nursing* 13: 25–39.

Wuest, J., and Stern, P. N. (1990b) The impact of fluctuating relationships with the Canadian health care system on family management of otitis media with effusion. *Journal of Advanced Nursing* 15:556–563.

Zajicek, G. (1994) Benefits of primary care. *Nature* 371:552.

Zonis, R. D. (1970) Chronic otitis media in the Arizona Indian. *Arizona Medicine* 27:1–6.

5

THE EVOLUTIONARY ECOLOGY
OF CHILDHOOD ASTHMA

A. MAGDALENA HURTADO

I. ARENAS DE HURTADO

ROBERT SAPIEN

KIM HILL

National practice guidelines provide the following definition of childhood asthma:

a lung disease with the following characteristics: 1. airway obstruction that is reversible (but not completely so in some patients) either spontaneously or with treatment; 2. airway inflammation; and 3. increased airway hyperresponsiveness to a variety of stimuli (U. S. Department of Health and Human Services 1991: 1)

A disproportionate increase in childhood asthma morbidity and mortality over the past 2 decades in spite of pharmacotherapeutic improvements is known in the epidemiological literature as the "asthma paradox" (Page 1991). This, and other questions in epidemiology (Magana and Clark 1995; Markides and Coreil 1986; Tambyraja 1991) may not seem paradoxical when viewed from an evolutionary perspective (Ewald 1994; Hill 1993; Nesse and Williams 1995; Profet 1991). In contrast to epidemiology, modern evolutionary theory seeks to explain and predict the timing of biological events. The objective of evolutionary explanations of disease is to understand why throughout history human diseases have caused considerable morbidity and mortality under some conditions and little if any harm under others (Ewald 1994). In epidemiology, on the other hand, there is no consensus on what constitutes a unifying body of principles for explaining the distribution of disease in human populations (Koopman and Weed 1990), even though one of its most important tenets is the significance of theory for assigning causality to statistical associations (Rothman 1986). "Theory" usually refers to hypotheses informed by published statistical trends as opposed to deductively derived propositions, and methodological rigor informs research design (Kleinbaum et al. 1982). In most instances, explanations are devoid of evolutionary principles, a

paradoxical oversight for a field that seeks to explain biological phenomena.

Here we propose a theoretical framework for the study of the natural history
of childhood asthma informed by life history theory, the overarching body of
principles in evolutionary ecology (Roth 1992; Stearns 1992). The application
of life history principles is new to epidemiology, biological and medical anthropology, and evolutionary medicine (Nesse and Williams 1995). This chapter starts with an overview of fundamental information about the natural history of childhood asthma: a definition of this disease phenotype followed by
a chronological discussion of the contexts in which asthma occurs, its phylogenetic history, its emergence in human evolution, and its distribution in
contemporary populations. Recent advances in immunology are key to understanding the asthma phenotype and its expression in contemporary human
populations. It has been known for some time that immunological pathways
responsible for asthma are ancient features of the class Mammalia (Moqbel
1992). In human history these pathways served to contain and expel macro-
parasites such as helminths, whereas in contemporary populations they cause
atopy, a condition in which people with genetic predispositions to allergy
upon exposure to ubiquitous biological or chemical substances that are inhaled or ingested produce significantly greater quantities of the immunoglobulin E (IgE) than the average individual (Pope et al. 1993). IgE in turn sets off
a complex network of physiological events (Sutton and Gould 1993), leading
to allergy symptoms such as inflammation of the skin (eczema), lung (bronchoconstriction, wheeze, and asthma), upper respiratory tract (rhinitis), and
intestinal tissues (diarrhea and vomiting) (Barnes et al. 1992) and their more
severe anaphylactic end products. Anaphylaxis refers to severe allergic reactions that involve respiratory compromise and cardiovascular collapse or
shock.

Although the bulk of the evidence suggests that IgE-mediated defenses
against parasites are adaptive (Moqbel 1992 and references therein), research
in contemporary human populations have yet to demonstrate that asthma renders a fitness advantage to those who suffer from the condition (see discussion
below). While we wait for the evidence, it would perhaps be wise for clinical
practice guidelines (U. S. Department of Health and Human Services 1991) to
consider that the IgE network has evolved for good reasons. As long as we do
not understand its full range of functions and their impact on fitness, primary
emphasis should be given to patient care strategies that effectively alter the
environmental conditions under which the IgE network produces asthma
among patients at risk of the condition and/or at risk of its sequelae. The more
frequently used and invasive pharmacotherapeutic treatment (U. S. Department of Health and Human Services 1991), with as yet unknown long-term
side effects, should then be subsumed to this preventive scheme. A preventive
environmental approach would provide all children at risk of asthma morbidity and their parents an opportunity to assess the effectiveness of environmental changes and to adopt those changes when appropriate. Such oppor-

tunities are only available to the fortunate few who are in the care of health care providers knowledgeable of environmental control solutions and who believe in their effectiveness.

Although clinical practice guidelines for asthma care provide considerable direction to providers, the guidelines' primary focus is on medical algorithms to guide decisions about drug prescriptions, in what quantities, and what to do when they fail to work (U.S. Department of Health and Human Services 1991). No equivalent algorithms are included for screening children at risk of asthma onset (mainly, those born to families with a history of asthma), for determining which environmental changes might work for the patient, for prescribing to parents suitable changes, and for monitoring their effectiveness through time. In lieu of algorithms, general descriptions of environmental controls in asthma care are published in the guidelines (U.S. Department of Health and Human Services 1991). These descriptions make the erroneous assumptions that physicians have the time and know how to translate general information about indoor allergens, where they are found, and how to reduce their densities into protocols tailored to each child (there is considerable variation in the substances to which children are allergic and that trigger asthma attacks). Even if physicians were to devise such protocols, general descriptions also erroneously assume that patients will comply to their prescriptions or recommendations. Like the asthma phenotype, parental behavior is under selection pressure favoring decisions which optimally allocate resources among behavioral alternatives and in response to environmental cues; in childhood asthma these cues include perceived or actual net benefits incurred from spending effort and time on drug and environmental therapies. Given the haphazard nature of clinical practice in childhood asthma (Asthma Zero Mortality Coalition 1995), parents are probably prescribed ineffective treatments more often than not and are thus more likely to experience costs rather than benefits associated with prescribed treatment for their asthmatic children. If we expect parents to adopt environmental controls in combination with drug therapy, algorithms in the guidelines should help ensure that parents adhere to corresponding behavioral changes in the home.

Sound management of asthma in the home is unlikely to occur as long as existing clinical practice guidelines do not translate environmental and pharmacotherapeutic treatments into algorithms that tailor prescribed treatment to the patient's risk profile and to the daily contexts in which care takes place. National health-policy decisions are partially responsible: only a tiny fraction of funds allocated to asthma research by the National Institutes of Health is spent on areas most akin to this translation process. In addition, U.S. health insurance companies and health maintenance organizations do not pay for indoor allergen assessment costs. If we continue to ignore the natural history of childhood asthma as well as social and political obstacles to sensible treatment, natural selection processes rather than judicious clinical practice will determine the levels of morbidity and mortality experienced by children with asthma in present and future generations.

Background

The incidence (i.e., instantaneous increase or decrease in the number of new cases as measured longitudinally) and prevalence (i.e., proportion of cases as measured cross-sectionally) of asthma-related morbidity and mortality among children has increased at alarming rates in many parts of the world over the past 2 decades. At the same time, drug clinical trials have demonstrated considerable improvement in the efficacy of combined use of bronchodilator and anti-inflammatory metered-dose inhalers or pharmacotherapeutic management of asthma symptoms, although for the most part effectiveness in natural settings has not been determined. These improvements in pharmacotherapeutics are not compensating for what appears to be a worldwide childhood asthma epidemic. In this section we discuss epidemiological trends, the asthma paradox, and national practice guidelines designed by experts in response to this paradox.

The Epidemiological Entity

Rates of childhood asthma increased dramatically over the past decade throughout the world. In the United States, prevalence per 1000 persons between birth and 21 years of age increased from 38 to 51 cases from 1980 to 1990 (4.3% prevalence in children from birth to 17 years of age; *1988 National Health Interview Survey*; Halfon and Newacheck 1993). Throughout the decade children and adolescents experienced higher prevalence rates than individuals over 20 years of age (U.S. Department of Health and Human Services 1991). Moreover, hospitalizations rates due to asthma increased substantially between 1979 and 1987, and most of this increase was accounted for by pediatric patients under 15 years of age (U.S. Department of Health and Human Services 1991). Analysis of National Health Interview Survey data indicate that childhood asthma afflicted 2.7 million children under 18 years of age in 1988 and that the costs associated with the condition were substantial. Of these children with asthma, 30% experienced some limitation of activity, accounting for 10.1 million days missed from school and 200,000 hospitalizations (Taylor and Newacheck 1992). In 1990, national health–care related asthma expenditures (including adult asthma) were estimated at $6.2 billion, or nearly 1% of all U.S. health care costs (Mortality and Morbidity Weekly Report 1992; Sullivan and Weiss 1993).

In keeping with national trends, a rise in the use of asthma-related health care was observed among Native Americans a population showing negligible rates of asthma when compared to whites up to the early 1970s (Gerrard et al. 1976). Between 1979 and 1989, asthma-related hospital admissions on U.S. Indian reservations increased an average of 2.6% per year, with the highest increase (3.7%) in children between birth and 4 years of age. In contrast, hospitalization rates for all causes declined by 1% per year over the same time period for the same age group (Hishanick et al. 1994). In addition, the lifetime prevalence of childhood asthma was estimated at 12.3% as measured by parental reports of physician diagnoses for children between 3 and 13 years of age in one Pueblo Indian group (Clark et al. 1995).

Similar trends have been documented in many countries. In various regions of the United Kingdom, lifetime prevalence rates of asthma among children and adolescents increased from 3.1% to 12% between 1964 and 1988. Hospital admissions registered for the entire country increased from 34 to 76.3 per 10,000 among 0–4 year olds between 1980 and 1985 (Wells 1994). The annual percent increase in childhood asthma prevalence from 1956 to 1989 in England, New Zealand, Switzerland, France, Taiwan, Sweden, and Finland ranged between 1.3% and 12.4% (Burney 1992). Finally, a study conducted in Costa Rica in the early 1990s found a 23.4% lifetime prevalence rate of asthma among children between 5 and 17 years of age in a sample of 2682 children (Soto-Quiros et al. 1994).

This universal increase in childhood asthma prevalence is paralleled by several other consistent findings across continents. Prevalence rates are invariably higher in urban than in rural areas (Gambia, Venezuela, Finland; Godfrey 1975; Lynch et al. 1984, Poysa et al. 1991). A most striking finding is the almost complete absence of asthma among isolated Amerindian groups such as the Waorani of Ecuador, the Yanomamo and Hiwi of Venezuela (see below; Hurtado et al. 1996a; Kaplan et al. 1980; Lynch et al. 1983). In addition, higher prevalence rates of asthma are reported for low socioeconomic status populations than for middle or upper classes (Evans 1992; Gottlieb et al. 1995; Weiss et al. 1992; Willies-Jacob et al. 1993).

The Asthma Paradox and National Practice Guidelines

Demonstrations of efficacy in drug clinical trials (Hatoum et al. 1994) has led to the widespread assumption in the asthma literature that most patients whose doctors follow national practice guidelines and who in turn abide by their physicians' prescribed treatments should be able to lead active lives and to avoid hospitalizations and near deaths (Taylor and Newacheck 1992; U.S. Department of Health and Human Services 1991). However, a majority of patients with asthma report more disability, missed school and/or work days than clinical trials would lead us to expect (LeSon and Gershwin 1995; Page 1991). In response to this problem, the National Institutes of Health recently supported an initiative by the National Asthma Education Program (NAEP) to develop and disseminate national practice guidelines; that is, preformed recommendations issued for the purpose of influencing decisions affecting patients in clinical practice (Eddy 1990). In 1991 the NAEP published the "Guidelines for the Diagnosis and Management of Asthma" (GDMA; U.S. Department of Health and Human Services 1991) to guide medical decisions in clinics and hospitals attending children and adults with asthma throughout the United States. Given their salience in asthma treatment at the national level, we discuss here the therapeutic strategies supported by the guidelines.

Algorithms are important tools in ensuring that human error is minimized in clinical decisions (Lawler 1995). Here we use the term "algorithm" to refer to questionnaires, charts, tables, or other information-gathering aids that prompt health care providers or patients to record one or more criteria (e.g., oxygen saturation) at critical or ongoing junctures in care (e.g., to decide whether or not to discharge a patient, seek emergency care, or discontinue use of bronchodilators) to help ensure that the best decision is made given the circumstances of the particular patient and health condition. Only 4 out of the 10 chapters in the GDMA provide detailed algorithms. The entire sample of 13 algorithms in turn help inform pharmacotherapeutic decisions. In contrast, the rationale for environmental control and patient education measures is described in prose that is of little use to the general practitioner and nurses for whom the guidelines are intended (U.S. Department of Health and Human Services 1991: xi).

Some specific examples of algorithms published in the guidelines are the "algorithm for diagnosing asthma" (U.S. Department of Health and Human Services 1991: 9), which describes the interpretation of health outcomes in patients with asthma after use of bronchodilators and anti-inflammatory agents as the basis for establishing a diagnosis of asthma. A diagnosis is ruled out if symptoms do not subside after presumed compliant use of these agents. Other important criteria such as assessment of allergies and the allergens to which a child is exposed are *not* included. Similarly, "classification of asthma by severity of disease" (U.S. Department of Health and Human Services 1991:10) lists symptomatological criteria for establishing a diagnosis of severity. Symptoms are incorrectly assumed to have high sensitivity (i.e., proportion of those who truly have asthma who are accurately classified as asthmatic) and high specificity (i.e., proportion of those who truly do not have asthma who are accurately classified as not having the condition) for establishing severity.

To the contrary, use of symptoms to assign severity puts patients at risk of considerable misclassification because asthma symptoms are as much a function of true disease severity (differences in the extent of bronchial hypersensitivity upon exposure to an allergen to which a patient is sensitized) as they are a function of patient noncompliance (Alessandro et al. 1994), access to health care (Halfon and Newacheck 1993), and total exposure to allergenic substances (total number of allergens multiplied by allergen density). Consequently, in the worst-case scenario, a well-managed child (i.e., a patient with a good physician who is compliant to treatment, whose home is allergen-free, and who reports few exacerbations and few if any school days lost) whose true disease status is severe is likely to be classified as a mild case, whereas a poorly managed patient (i.e., a patient inadequately cared for and/or who is noncompliant) whose true disease status is mild will be classified as moderate or severe case according to the algorithm. In reality, a large proportion of asthmatics who are admitted to emergency rooms and who die are reported as mild cases. In one study, researchers reported that 22% of all patients who died of asthma at a hospital had been assigned mild severity, while in another study mild cases accounted for 29% of all patients with asthma attending an emergency room clinic, followed by moderate (52%) and severe cases (19%) (Matsuse et al. 1995). Interestingly, the severe cases were the least represented. Misclassification of severity probably occurs often because general practitioners and emergency room doctors who treat most children with asthma (Mortality and Morbidity Weekly Report 1992) do not routinely conduct bronchoprovocation tests, a more objective assessment of severity, in which patients inhale methacholine or other substances at incrementally higher doses causing bronchoconstriction only in individuals with asthma.

Algorithms for home and hospital management of chronic asthma and acute exacerbations also emphasize pharmacotherapeutics in lieu of a more well-rounded preventive approach. The "weekly asthma symptom and peak flow diary" (U.S. Department of Health and Human Services 1991:53) includes a chart on which patients record their peak flows and symptoms in the mornings and afternoons, but no room is provided for recording the timing of respiratory infections (viral or bacterial), environmental exposures, including the number of hours spent indoors, and at what locations, and whether the patient has followed the prescribed regimen. This is unfortunate because neither patient or physician are encouraged to identify on an ongoing basis the conditions under which children are at highest risk of respiratory problems. Instead, the weekly diary is used to inform decisions about altering drug dosage and types of drugs prescribed to the patient. The "family guide for managing a child's asthma episode" (U.S. Department of Health and Human Services 1991:55) includes a detailed series of pharmacotherapeutic decisions to follow when a child's condition begins to deteriorate; no such detail is given to inform decisions about what to do when a child's environment and family functioning begins to deteriorate. The "correct use of a metered-dose inhaler" (United States Department of Health and Human Services 1991:57) is also a detailed set of steps that helps children and parents learn the most efficacious pharmacotherapeutic inhaling techniques. This contrasts with one list with vague descriptions of allergen reduction recommendations such as "encase the mattress in an airtight cover" (What is an airtight cover? Do physicians provide them? How often should the cover be changed?), "use chemical agents to kill mites" (Should all parents do this, including those of children who are not allergic to mites? How often should these be used? How should the effects of the chemical agents be measured to determine whether or not they work?) (U.S. Department of Health and Human Services 1991:66).

Finally, charts for the management of pediatric asthma (chronic, moderate, and severe asthma), and "acute exacerbations of asthma in children: home management and emergency department management" describe detailed lung function indicators (mainly, peak expiratory flow) that should be used to guide decision making about pharmacologic therapy; which medications to prescribe for daily use and during asthma attacks, how often they should be administered, what beneficial outcomes to expect from drug therapy and what to do when these fail (U.S. Department of Health and Human Services 1991: 80, 82, 105, 107, 108). No equivalent algorithms are provided with precise measures of environmental (allergen exposure) and behavioral (poor access to medical care, noncompliance) risk factors for patients and physicians to make decisions about what environmental controls to implement in the home, what behavioral changes to teach or adopt, and how to proceed when these fail to produce the expected outcomes. Thus, the GDMA algorithms fare poorly in a preventive scheme because primary prevention would require the removal of environmental causes and changes in the pertinent parental and patient behaviors before onset or recurrence of adverse health events.

The GDMA reflect and sanction a universal emphasis on pharmacotherapy in the clinical care of children with asthma at several institutional levels. Government-endorsed patient education programs, although well studied and widely disseminated, are considered to be secondary rather than central to clinical care in pediatric asthma (U.S. Department of Health and Human Services 1984). These programs provide fundamental background information on asthma physiology, symptoms, and their control. However, once again, they do not provide algorithms that guide decisions about which environmental and behavioral instructions are most appropriate for each family. Furthermore, the National Institutes of Health spends a small portion of its asthma-related funds on research to enhance preventive, family-centered approaches to asthma management. In 1995 half of these funds were allocated to research immunological mechanisms (figure 5.1). The remainder was distributed among epidemiology, educational interventions, drug clinical trials, genetic studies, and research center activities. A small percentage of the total funds (6%) were in turn allocated to apparently less important areas such as assessments of environmental control interventions, development of medical curricula, patient noncompliance, methodological improvement, and family functioning (figure 5.1). Interestingly, it appears that none of these include research on physician noncompliance to national practice guidelines: in a sample of 990 physicians surveyed in 1994, only 22% of physicians reported use of therapeutic protocols for moderate and severe asthma that are consistent with the GDMA (Asthma Zero Mortality Coalition 1995). In keeping with these trends most health maintenance organizations and insurance companies in the United States today ignore the significance of environmental controls for patients with asthma; these agencies do not pay for indoor allergen assessments, purchases of mattress and pillow covers, carpet and bedding chemical treatments, vacuum cleaner filters, and other allergen reduction costs.

In summary, pharmacotherapeutic improvements may be failing to contain childhood asthma epidemics because they relieve symptoms but do not eliminate the causes of such epidemics. Simply put, better drugs may only help solve the asthma paradox when subsumed to a preventive plan of action that includes effective environmental control and behavioral management strategies. Although national practice guidelines acknowledge the significance of these causes, they de facto endorse exclusive use of pharmacotherapeutic agents by only providing easy-to-use algorithms for this form of therapy and by omitting environmental control and behavioral equivalents. Whether these omissions are warranted depends on an as yet unanswered biostatistical question: How much of the variance in childhood asthma morbidity and mortality is explained by pharmacotherapeutic therapies relative to environmental and

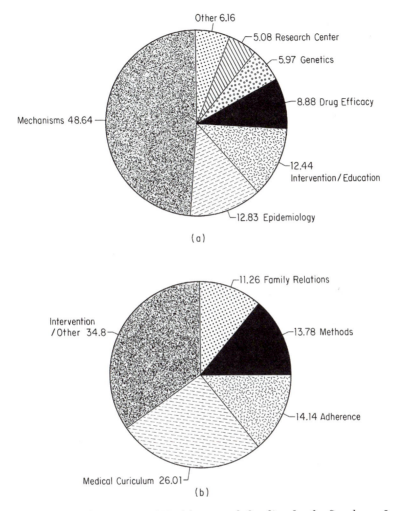

Figure 5.1. National Institutes of Health research funding for the fiscal year January 1, 1995 to December 30, 1995. (a) percentage of funds by category; (b) percentage of funds accounting for the "other" category. (Source: National Institutes of Health 1997.)

behavioral causes? The guidelines reviewed here make the implicit assumption that inadequate or minimal adoption of pharmacotherapeutic therapies account for most of the variance. In the absence of data, we can and should reformulate and approach the question from a deductive rather than from a biostatistical standpoint only: whether the omissions of environmental and behavioral factors in the guidelines are justified in light of epidemiological evidence and immunological insights into the etiology of childhood asthma and its place in human evolutionary history.

An Evolutionary View of Childhood Asthma

Informed by the theory of natural selection, life history principles explain the diversity of the life course across populations. Particularly relevant to the life course are the timing and probability of births and deaths, which are in turn determined in large part by disease, a physiological state leading to deviations from a culturally expected state of good health for a given age or sex. Within this framework, diseases are viewed as phenotypes whose manifestations are determined in part by ecology (i.e., environmental factors). The genotypes coding for these phenotypes are in turn subject to selective pressures depending upon the extent to which timing of deaths or disability (early as opposed to later in the life course) affect frequency of conceptions, live births, and viability of offspring.

Here we define asthma as a disease phenotype—the expression of heritable asthma genotypes (*Clinical and Experimental Allergy* 1995) in interaction with the environment during the course of growth and development (Stearns 1992). Two environmental exposures are key to understanding the distribution of childhood asthma in human populations: parasites and indoor allergens (Moqbel 1992). The timing of exposure to parasites and allergens across the life course are also significant; the probability that the asthma phenotype will be expressed in an individual is negatively associated with the age at which a child is exposed to inhalant allergens (Sporik et al. 1990). Furthermore, experimental studies with rats show that exposure to inhalant allergens before exposure to helminths is positively associated with the probability of developing allergic disease (Orr and Blair 1969; Turner et al. 1982).

The high prevalence of asthma phenotypes at the present time indicates that associated genotypes have been favored by natural selection throughout human evolution. By this we mean that asthma genotypes and their phenotypic expression have for millennia increased the probability of survival or births or both among its bearers over people without it. To examine this proposition we need to determine whether these genotypes are new only to humans or ancient features that we share with other orders in Mammalia. We then examine the relevant "reaction norms," or the expression of the phenotype as a function of environmental exposures such as parasites and inhalant allergens in past and present human environments (Stearns 1992). This chronological discussion of the phenotype helps us understand why asthma is a common condition in contemporary human populations and how likely it is that it will be prevalent in future generations. In particular, we present a trade-offs model of parental care behavior and the ways in which these behaviors might interact with levels of inhalant allergen exposure on morbidity and mortality of children with asthma. The extent to which parents can and do minimize the effect of childhood asthma on fertility and mortality before and/or during their offsprings' reproductive years has implications for the maintenance of asthma genotypes and thus the persistence of childhood asthma epidemics in future generations.

Immunological Phenotypes

Asthma is referred to as an atopic disease, a condition in which people with genetic predispositions to allergy upon exposure to ubiquitous biological or chemical substances that are inhaled or ingested produce significantly greater quantities of IgE than the average individual (Pope et al. 1993 see page 190). Interestingly, infestation of human hosts by helminths (parasitic metazoan parasites) is also associated with elevated levels of circulating IgE, and this response in turn predisposes individuals to asthma and allergy, although parasitic infection under other conditions seems to prevent asthma and allergies (Pritchard 1992). It is therefore possible that the genotypes involved in parasitic defense and asthma are similar, but that, whereas the former provides a clear fitness advantage, it is unclear whether asthma is adaptive. For the purposes of our discussion, the asthma phenotype refers to morbidity and mortality proximately caused by the mast cell IgE-mediated reaction to allergens in the lung (Pope et al.1993).

The immunological principle linking parasite load to asthma is the concept of *mast cell saturation*. IgE circulates in serum (total serum IgE) at varying levels (range: 0.05 IU/ml–46850 IU/ml) (Haus et al. 1988; Hochreutener et al. 1991). A portion of the IgE reacts specifically with identifiable environmental antigens (allergen-specific IgE) (Bazaral et al. 1973). IgE binds to mast cells in tissue and to basophilic leukocytes and sensitizes these cells to discharge pharmacologically active mediators when they are subsequently exposed to antigens to which the IgE is specifically reactive (Bazaral al. 1973). These IgE complexes are responsible for the development and maintenance of the inflammatory response in asthma and its concomitant spasms of bronchial smooth muscle fibers and nonspecific bronchial hyperactivity (Ishizaka 1973). However, under some conditions, animal and human mast cells become saturated and fail to express the allergen-specific response—as, for example, when all or most available receptors are occupied by IgE directed to nonallergen antigens, as may be the case in parasite infestation (Godfrey 1975 and references therein). Among individuals heavily parasitized by helminths, who also tend to be free of asthma and show high total IgE levels, it has been demonstrated that allergic reactions of the skin are not expressed. This has been interpreted to mean that although specific antiallergen IgE may be present in those individuals, it does not react functionally in the presence of high loads of unrelated IgE (Lynch et al. 1993). However, occasional parasitic infestation may activate allergen-specific production via a nonspecific stimulation of IgE synthesis. In this case total IgE synthesis could be large enough to substantially stimulate rather than inhibit allergen-specific IgE production (Hagel et al. 1993).

Genotypes

Although our understanding of the immunological mechanisms associated with the asthma phenotype has improved considerably over the past 2 decades, the genetics of asthma continue to remain elusive (Doull et al. 1996; Sanford et al. 1995). Total serum IgE production appears to be controlled by a dominant atopy gene unlinked to HLA class II, whereas allergen-specific IgE production is unequivocally linked to the HLA class II molecules. The releasability of inflammation mediators mediated by IgE production is controlled by poorly understood genetic coding.

In spite of our poor understanding of the genetics of asthma, it is nevertheless clear that childhood asthma is heritable (Pope et al. 1993). After controlling for other factors, children whose parents have allergies experience the highest incidence of asthma (*Clinical and Experimental Allergy* 1995; Luoma 1984). One study found that children whose mothers or fathers had a history of asthma were almost three times as likely to develop asthma up to the age of 7 (Jenkins et al. 1993). Two studies of histories of asthma in twins found that monozygotic pairs have a higher concordance of asthma onset (59%, 43%) than do dizygotic pairs (24%, 25%) (Nieminen et al. 1991; Sarafino and Goldfedder 1995; see also Akasawa et al. 1991), regardless of whether twins are reared together or apart (Hanson et al. 1991).

Differences in childhood asthma prevalence across ethnic groups have been attributed to genetics as well. Puerto Rican children with asthma have significantly larger numbers of the variant S or Z in α_1-antitrypsin than do nonasthmatic controls (Hurtado 1995 and refs therein). In 1984, this ethnic group, with considerably more African admixture than other Hispanics in the United States, was reported in a national survey as having higher lifetime prevalence rates of asthma than Mexican Americans, the group with substantial Amerindian admixture (21% versus 4%, respectively) (Hurtado 1995). Similarly, in a national sample of self-reports of asthma collected from 1976 to 1990, African Americans reported higher lifetime prevalence rates of asthma than did non-Hispanic whites (9.2% versus 6.2%) (Hurtado 1995; Turkeltaub and Gergen 1991; see also Gaddy et al. 1993; Grundbacher and Massie 1985). African Americans have been found to produce IgE at significantly higher levels than non-Hispanic whites in the United States (Grundbacher and Massie 1985), but the extent to which these differences are genetically determined has yet to be determined.

Is the Immunological Phenotype Fixed at Higher Taxonomic Levels?

The immunological mechanisms associated with asthma are ancient features of Mammalia. Immunoglobulin E is fixed at this taxonomic level, sharing a common ancestor, IgY, with other nonmammalian vertebrates (Warr et al. 1995). IgE and IgG evolved from IgY in mammals as two separate antibodies that had fulfilled two distinct immunological functions in their IgY ancestor. In mammals IgE is a major low molecular weight serum antibody that mediates allergic reactions; IgG defends tissues from systemic infections and neutralizes viruses. Upon divergence of immunoglobulin functions sometime in the phylogenetic history of mammals, IgE became a quantitatively minor isotype

(0.002% of total serum immunoglobulin), probably downregulated due to its life-threatening anaphylactic (i.e., allergic reactions resulting in respiratory compromise and/or cardiovascular collapse) side effects, and IgG assumed the role of the major serum antibody in mammals (80% of total serum immuno-globulin). IgG downregulates IgE production by mechanisms yet to be fully identified and is often referred to as a *blocking antibody*. Immunotherapy (allergy shots) stimulates the production of IgG to ensure this blocking action; IgG interferes with anaphylactic reactions through direct competition for antigens (i.e., foreign molecules bearing sites to which antibodies bind) (Durham 1995; Hussain et al. 1992). It has been proposed that IgE and IgG became mutually distinct in mammals and not in other animals in order to prevent maternal anaphylaxis from damaging the embryo, and thus IgE must behave independently in protecting the fetus from the onslaught of the maternal immune system in mammals (Ionov 1985). More definitive insights into the evolutionary reasons for bifurcation may someday shed light on the evolution of asthma.

How Does the Distribution of Asthma in Humans Compare to Other Mammals?

Although IgE is ubiquitous in Mammalia, chronic wheezing is rarely seen in noncaptive animals. However, data on zoo animals, pets, and experimental breeds suggest that asthma can be easily induced under some conditions. Halpern et al. (1989) reported a 12-year history of respiratory tract allergic disease in a 25-year-old, 70-kg female chimpanzee. The symptoms were controlled with treatment. This animal had strong positive skin reactions to grass, weed, and tree allergens but not to mold or mite antigens (Halpern et al. 1989). At least one case of atopic dermatitis in gorillas living in captivity has been reported (Hog and Schindera 1989). Feline asthma syndrome was identified in 29 cats, and treated effectively with prednisone, a commonly used steroid in human asthma treatment (Corcoran et al. 1995). Finally, the asthma phenotype or its immunological correlates have been experimentally induced in several species of mice, rats, guinea pigs, rabbits, sheep, dogs, horses, baboons, and rhesus monkeys (Breen et al. 1987; Karol 1994; Patterson and Harris 1993; Wegner et al. 1993). To our knowledge, asthma-induced deaths in these species have not been reported, although IgE mediated reactions have been induced in laboratories (Karol 1994).

What Might Be the Adaptive Function of the Asthma Trait?

The term "adaptive" has many definitions in evolutionary biology (Stearns 1992). Here we use a less mathematically rigorous meaning: a presumed change in a phenotype that occurs in response to a specific environmental signal that results in improvement in growth, survival, and/or fertility. More mathematically rigorous definitions include precise and valid measures of age-specific fertility and mortality for individuals bearing alternative phenotypes (e.g., "has asthma," "does not have asthma") under specified environmental conditions (e.g., "parasitized," "not parasitized") to measure the effects of the phenotypic expression on fitness in different contexts of interest (Stearns 1992). It would be advantageous for us to know whether individuals bearing the asthma phenotype have higher survivorship rates and/or fertility than do others in populations with endemic parasite infestation. For example, some studies show a strong negative correlation between asthma and worm load (Venge and Shaw 1989). However, these short-term observations tell us little about the fitness profiles of individuals (i.e., age-specific fertility and survival rates) necessary to determine whether asthma is truly adaptive.

Total serum IgE levels have been shown to increase and provide a survival advantage in the face of multiple and massive endo- and ectoparasitic infestation (Abdel Fattah et al. 1994; Hagan et al. 1991; Kigoni et al. 1986; Matsuda et al. 1990). Because parasitic infestations are naturally hyperendemic in all species including humans, it is possible that asthma is a problem arising in recent times and one that is limited to mammals who have drastically reduced their exposure to parasites over the past century (Bazaral et al. 1973). Thus, it appears that the asthma phenotype is in part a cost that humans have incurred for getting ahead in the host–pathogen arms race in human history, if only momentarily. Similar inferences apply to diseases such as diabetes, one of the costs of having superabundant nutrient supplies, particularly carbohydrate and fats, in the modern world (Eaton et al. 1988, Sapolsky 1994 and references therein).

Total serum IgE also appears to play a role in other debilitating or life-threatening conditions: risk of sudden cardiac death (IgE depresses clot formation) (Szczeklik et al. 1993); development of cancerous tumors (IgE decreases tumor growth and incidence); and risk of death during traumatic injury (IgE is inversely associated with organ failure and mortality in trauma patients) (DiPiro et al. 1994); onset of *Staphylococcus aureus* related infections in persons unable to produce other immunoglobulins (Nesse and Williams 1995); and containment of inhalant and ingested toxins of all sorts (Profet 1991), which in turn may reduce the risk of developing cancer (Profet 1991 and references therein).

What Are Some Examples of Reaction Norms for the Phenotypic Expression of Childhood Asthma?

The mast cell saturation principle provides the foundation for understanding reaction norms between parasite load, exposure to allergens, and the expression of asthma. One way to learn about these reaction norms is to examine the relationship between total serum IgE levels (as determined by parasite load) and allergen-specific IgE (a strong correlate of asthma prevalence) (see figure 5.2). The mast cell saturation principle would lead us to predict an inverse relationship between parasite-induced total IgE and allergen-specific IgE; as total IgE serum levels increase, mast cells become saturated, impeding

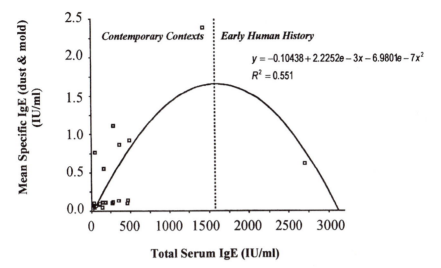

Figure 5.2. The relationship between total blood serum IgE levels and the mean house dust and mold-specific IgE levels for 27 patients, ages 13–67 (data from Lynch et al. 1985: table 1).

allergen-specific IgE from accessing receptor sites. The relationship between IgEs, however, is more complex: effective inhibition of allergen-specific IgE may only occur at high levels of parasite load exposure and not at intermediate levels. At intermediate levels, serum IgE synthesis could be large enough to substantially increase total serum IgE levels, but not large enough to block mast cell activation of *allergen-specific* IgE levels (Hagel et al. 1993). Figure 5. 2 illustrates this complexity: allergen-specific IgE is positively associated with total IgE up to a point (~1500 IU/ml), but beyond that point it appears to be negatively correlated.

The reaction norm between parasite load (as measured by total serum IgE) and the expression of the allergic response (as measured by allergen-specific IgE) (figure 5.2) provides insight into the epidemiology of childhood asthma in present and past environments. Research on isolated Amerindian groups support the proposition that thresholds >1500 IU/ml total IgE may be typical of our evolutionary past. Loads of helminth infestation in remote South American Indian populations are extremely high, ranging between 60% and 98% prevalence for individuals of all ages (population prevalence rates are positively associated with worm count or load per individual; see Bundy and Medley 1992: figure 2.7). These estimates may be our best guess of what parasite prevalence was throughout human history (Salzano 1990). Not surprisingly, total IgE levels in these populations are the highest ever measured in extant populations (11,975 and 13,088 IU/ml among the Waorani and various Venezuelan Amazon groups, respectively, see table 5.1), and asthma rates are extremely low or nonexistent (Hurtado, Arenas and Hill 1996; Kaplan et al. 1980; Lynch et al. 1983). Other extant populations residing in rural areas also

Table 5.1 Total serum IgE and childhood asthma in diverse populations

Population	Age group	Mean total serum IgE (IU/ml)	% Positive skin test	Asthma prevalence	Allergen[a]	Sample size	Reference
South American Indians							
Waorani	Adults	11,975	5.67	n.d.	V	227	Kaplan et al. (1980)
Amazonas groups, Venezuela	Adults	13,088	6.70	1.02[b]	V	49	Lynch et al. (1983)
Rhodesia							
Rural	Adults	1,613	6.00	n.d.[c]	DF	100	Merrett et al. (1976)
Gambia							
Rural	Children	962	0.01	0	V	131	Godfrey (1975)
Urban	All ages	368	0.01	0.008[d]	V	191	Godfrey (1975)
United States, Mexicans							
Parasite infested	Adults	2,219	n.d.	n.d.	n.d.	20	Bazaral et al. (1973)
No parasite infestation	Adults	458	n.d.	n.d.	n.d.	12	Bazaral et al. (1973)
Java laborers							
Low parasite load	Adults	2,193	n.d.	n.d.	n.d.	15	Noerjasin (1973)
High parasite load	Adults	5,460	n.d.	n.d.	n.d.	21	Noerjasin (1973)
Mestizos, Venezuela							
Rural	All ages	975	6.42	6.00	D(2)	327	Lynch et al. (1984)
Urban	All ages	745	14.53	25.00	D(2)	475	Lynch et al. (1984)
Urban, Venezuela							
Pre-parasite treatment	Children	2,543	17	n.d.	HD	86	Lynch et al. (1993)
Post-parasite treatment	Children	1,124	68	n.d.	HD	86	Lynch et al. (1993)
No treatment, beginning of study	Children	1,649	26	n.d.	HD	75	Lynch et al. (1993)
No treatment, end of study	Children	3,697	12	n.d.	HD	46	Lynch et al. (1993)
Medium-high SES	Children	557	42	n.d.	HD	391	Hagel et al. (1993)
Low SES	Children	1,870	23	n.d.	HD	455	Hagel et al. (1993)
Rural	Children	3,140	8	n.d.	HD	319	Hagel et al. (1993)

[a]DF, dust mite species *Dermatophagoides farinae*; V, various; D(2), two species of *Dermatophagoides*; HD, house dust.
[b]Estimate based on 75,013 medical consultations among Venezuelan Amazonian Indians over a 5-year period.
[c]n.d., no data.
[d]No systematic data were collected: 44 individuals with asthma were referred to the investigators. In contrast, only 1 person with asthma was reported in a population of 1200 rural residents.

show IgE levels above the 1500 IU level, high parasite load, and low rates of asthma and/or low percentage of positive skin tests (rural Rhodesia, rural Gambia, rural mestizos, urban poor; table 5.1). Finally, Venezuelan children of higher socioeconomic status are less parasitized and show lower total IgE levels than do their lower socioeconomic status counterparts (table 5.1).

Parasite load per se may not only play a role in inhibiting allergen-specific IgE production among South American Indian groups but also the *timing* of exposure in growth and development may be critical. Extensive exposure to helminths before exposure to inhalant allergens early in life may greatly dampen the expression of allergies in humans as has been shown in the rat (Orr and Blair 1969; Turner et al. 1982). Among the less acculturated South American Indian groups, children become infested with parasites at a very young age (Santos et al. 1995). However, they are exposed to few if any indoor allergens because they live in well-ventilated huts and use traditional bedding that is a poor breeding site for dust mites and cockroaches (Hurtado et al. 1996a). As groups adopt less ventilated housing (concrete block shelters and homes with machine-made wooden boards), Western bedding such as mattresses and blankets, and spend more time indoors (National Research Council 1991; Spengler and Sexton 1983), the risk of exposure to high levels of indoor allergens soon after birth increases. Moreover, in contemporary urban centers with a high prevalence of childhood asthma, children are not only chronically exposed to indoor allergens throughout infancy and the school-age years but may also be intermittently, although minimally, exposed to helminth infestation (Bosman et al. 1991; Kappus et al. 1994). Lastly, acculturation and urban residence leads to an increase in the probability of helminth infestation, however slight, *after exposure* to indoor allergens.

Hence, decreases in parasite load and adoption of insulated housing and Western bedding in recent human history help explain why total IgE levels <1500 IU/ml are characteristic of populations residing in metropolitan areas throughout the world today (Hetman et al. 1988; Yadav et al. 1994). Within this lower range (figure 5.2), elevated total IgE is a important and consistent predictor of asthma in children. Total serum IgE levels are strongly associated with a history of asthma, whereas no asthma is present in subjects with the lowest IgE levels (Burrows et al. 1989; Criqui et al. 1990). Among children diagnosed with asthma, those with elevated IgE levels (>500 IU/ml in older children and 100 IU/ml in infants) experience higher asthma-related morbidity than children with lower total IgE serum levels (Maruo et al. 1990). Finally, total IgE predicts the onset of asthma but not the onset of other atopy-related phenotypes such as allergic rhinitis (DiPiro et al. 1994).

Throughout the world today, children living in poverty are more likely than their affluent counterparts to develop asthma (Evans 1992; Gottlieb et al. 1995; Halfon and Newacheck 1993; Weiss et al. 1992) and to have elevated allergen-specific IgE and total serum IgE (but <1500 IU/ml) (Berciano et al. 1987; Pollart et al. 1989; Willies-Jacob et al. 1993). Although both high and low socioeconomic status individuals find themselves at the contemporary lower end of the distribution (see figure 5.2), socioeconomic status is associated negatively with total IgE production. This is possibly because exposures or factors that elevate total IgE occur more frequently and at higher levels in low than in high socioeconomic status families. Factors of primary interest are intermittent exposure to parasites, respiratory viruses and/or chronic exposure to indoor and outdoor allergens (Bardin et al. 1992; Sporik et al. 1990), and the number of hours that children spend indoors (Silvers et al. 1994).

Public health programs have eliminated exposure to helminths in high socioeconomic status communities but failed to do so entirely, if at all, in low socioeconomic status areas (Stephenson 1987). Low socioeconomic status families are therefore faced with the synergistic effects of parasite and indoor allergen exposure on childhood asthma (Hagel et al. 1993). The co-occurrence of parasite infestation and allergens and the incidence of asthma has been studied in depth by Lynch and co-workers in Venezuela (but see Hagel et al. 1993; Moqbel 1992). The implications of this work may be far reaching and extremely relevant to U.S. populations: Puerto Ricans show both higher rates of childhood asthma (Hurtado 1995) and hookworm infestation (Maldonado 1993) than any other ethnic group in the United States. This raises the possibility that parasitism may also play a role in asthma among African Americans residing in low socioeconomic status neighborhoods.

In addition to parasite exposure, low socioeconomic status children experience respiratory viral infections at significantly higher rates than do their more affluent counterparts (Margolis et al. 1992; McConnochie et al. 1995; Porro et al. 1992). Viruses are major culprits of asthma exacerbations in children and adults (Bardin et al. 1992), and they do so through stimulation of IgE production (Weiss and Stein 1993). Prospective studies have shown that 28% of subjects with undetectable IgE to respiratory syncytial virus had subsequent wheezing (an important asthma symptom), whereas 70% with elevated IgE to respiratory syncytial virus wheezed (Welliver et al. 1986). Another study found a strong relationship between infection with parainfluenza 3 virus and respiratory syncytial virus and the subsequent development of allergy symptoms coupled with significant elevations in total IgE within 3 months of infection (Frick 1986).

Children of low socioeconomic status families are also at higher risk of indoor and outdoor allergen exposure than are children of high socioeconomic status. This is true of all the major allergens currently under extensive investigation throughout the world: dust mite, cockroach, animal dander and cigarette smoke (Barnes and Brenner 1996; Malveaux and Fletcher-Vincent 1995; Pattermore et al. 1989). Exposure to dust mites is of particular concern; Sporik et al. (1990) found a positive dose–response relationship between the age at which children were exposed to dust mites above a critical threshold and the probability of developing asthma before adolescence. The relationship between exposure to the other major indoor allergens and the incidence of asthma in children is less well understood; nevertheless, correlational (i.e., cross-sectional) studies clearly indicate that levels of cockroach, cigarette smoke, and animal dander are strongly associated with the prevalence of asthma in children (Etzel 1995). Synergistic effects between indoor allergens and inhalant pollutants (e.g., nitrogen dioxide: Infante-Rivard 1993; ozone: Bethel 1984; Biagini et al. 1986; Peden et al. 1995), known to be more densely distributed in low socioeconomic status areas (Bryant and Mohai 1992), may further increase the risk of developing asthma among children residing in these areas. Cellular responses to allergens such as dust mites among children sensitized to this allergen are significantly more pronounced in the presence of ozone than in its absence (Biagini et al. 1986; Peden et al. 1995).

The impact of indoor allergens on the onset of asthma among children living in poverty is further exacerbated by the number of hours that children spend indoors (National Research Council 1991). Children of low socioeconomic status tend to spend more hours per day watching television and are less active than children of high socioeconomic status families (Evers and Hooper 1995; Rosenfeld 1992; Wolf et al. 1993). Under such conditions, the percentage of time spent indoors by children between 5 and 12 years of age may exceed the national average of 60% (n = 1000 households; Silvers et al. 1994).

In summary, part of the solution to the asthma paradox involves an appreciation for the relationship between levels of parasite infestation and the timing and extent of allergen exposure during early human development. It appears that in more recent human history, ample opportunities for IgE stimulation occur below levels necessary to saturate and impede the expression of allergen-specific IgE. These opportunities in turn are more ubiquitous in disadvantaged populations and they tend to occur together; e.g., children exposed to high levels of cockroach allergen are also more likely to be exposed to ozone and nitrogen oxygen, cigarette smoke, and to spend more time indoors.

How Does the Evolutionary Biology of Asthma Interact with Human Parental Care Strategies?

Environmental factors associated with an increase in the frequency of childhood asthma cases are mediated by parental care behaviors. Table 5.2 illustrates how little we know about these behaviors. In contrast to biology and anthropology (Clutton-Brock 1991; Hrdy 1992), pediatrics curricula do not include the study of parental care behaviors, even though the evolutionary ecology of parental decisions is key to understanding the distribution of childhood diseases in human populations. This relevance is evident because most forms of investment necessary to ensure a productive life in children is primarily under the control of parents. The role of ecology and curvilinear relationships are as important for understanding the etiology of parental care as is true for understanding the IgE network. Parents face difficult trade-offs between allocating time and effort to activities that are beneficial to the child with asthma but are also mutually exclusive (e.g., ensuring allergen-free environments versus directly monitoring pulmonary function; administering medication versus providing other direct forms of care) (table 5.2). These

Table 5.2 Behavioral changes associated with medication and environmental control protocols

Prescribed treatment[a]	Behavioral changes associated with the prescribed treatment	Actual treatment or actual behavioral changes
Pharmacological agents		
Purchase, administer medication, and report adverse effects	Income expenditures to purchase medication, transportation (gas, bus, taxi) to go to pharmacy	?
	Time spent traveling to pharmacy, waiting for the prescription to be filled, etc.	?
	Time spent calling the provider to talk about adverse effects, waiting for the return call, etc.	?
	Activities foregone in order to purchase and administer the medication (leisure, work, etc.)	?
Monitor signs and symptoms and peak expiratory flow rate	Time spent convincing children to use the peak flow meter for measurement on a regular basis	? ?
Environmental control		
Encase mattress in an airtight cover	Income expenditure to purchase of cover	?
	Time expenditure to time to find a store that sells the cover, time to actually purchase the right cover (e.g., waiting in line, exchange)	
Wash the bedding in water of 130° weekly	Time expenditure	?
Remove carpets from the bedroom	Income expenditure to pay for cleaning or adding a new floor	?
	Time expenditure to find professional to remove the carpet, time to monitor workers	

[a]Source of prescribed treatment examples: Guidelines for the diagnoses and treatment of asthma (U.S. Department of Health and Human Services 1991).

trade-offs can be hypothetically described to shed light on the reasons that, even though parents may agree with physicians about ideal health goals for their children with asthma, they are not in a good position to attain those goals. Parents seek those goals in a constrained environment; health care officials, on the other hand may naively assume that parents can attain unconstrained optimal levels by simply taking home drug prescriptions and asthma management recommendations.

A Hypothetical Model of Parental Care Decisions

A hypothetical optimization graph of the relationship between maternal time and child health outcomes helps us identify the behavioral causal pathway that links environmental constraints such as allergen exposure and levels of education with health outcomes among children with asthma (see also Hurtado et al. 1996c). In this thought experiment, only families with one child who has asthma are considered. We then consider the options available to mothers over a 12-hour period; these options are stratified by differences in educational levels (figure 5.3). We plot on the x axes of figures 5.3a and the goods obtained from working outside the home and from providing care in terms of income and health insults avoided (e.g., accidents, infections, malnutrition). Income yields food, shelter, clothing, and medical care. Direct care protects children from experiencing health insults of a diverse nature. In asthma care, health insults of relevance include morbidity (e.g., school days lost) and near deaths that can be avoided by monitoring environmental triggers, respiratory function and planning daily and weekly environmental control and medication use schedules. Good care depends entirely on parental ability to predict symptoms and attacks based on information about changes in the child's exposure to triggers and lung function (peak expiratory flow). Parents must be able to identify and avoid triggers such as respiratory viruses and indoor allergens and implement preventive measures to avoid school absences and severe asthma attacks and assure participation in leisure activities (table 5.2). Parental inability to behave effectively and in a timely manner is one of the most important predictors of asthma mortality in children (Robertson et al. 1990; Sears et al. 1986).

At this point we describe differences in hypothetical budget constraints (figure 5.3c, d) that might characterize the ecologies of less and more educated mothers. In our example, if less and more educated women were to spend all their time in child care, they would avoid a maximum of 65 and 35 health insults, respectively, over a 12-hour period (y axis). We chose these numbers to reflect published findings showing that women who are less educated are also more likely to be of low socioeconomic status (Grogger and Bronars 1993) and more likely to reside in homes and areas with high allergen density. In contrast, if less educated mothers were to spend all their daylight hours working outside the home, they would obtain a maximum of $48 over a 12-hour period ($4 per hour), while more educated mothers would obtain a maximum of $240 dollars over the same time period ($20 dollars per hour) (x-axis). These

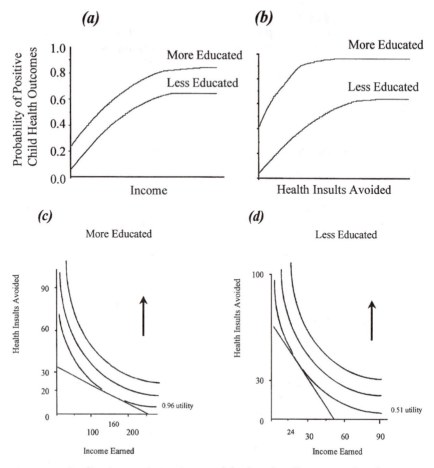

Figure 5.3. Qualitative optimization model of trade-offs women face between working outside the home and providing child care for more and less educated mothers. (a) Relationship between the number of health insults avoided by providing child care and the probability of positive child health outcomes. (b) Relationship between income earned and the probability of positive child health outcomes. (c,d) Optimal combination of time expenditures for more and less educated mothers, respectively.

numbers also reflect a common wage difference in the United States today. The budget constraint lines connect the maximum amount of goods that would be obtained if women spent all their time in one activity or another; these show that even though mothers may want to both avoid the maximum number of health insults per day and to earn the maximum income per day possible, they can only choose from a limited set of combinations.

Which combination should a mother choose? The answer depends on the effect of income earned and health insults avoided on the probability of positive child health outcomes. Monotonically decreasing returns curves are used to describe these relationships (figure 5.3a, b). For both types of goods (income and health insults avoided), the baseline probability of positive health outcomes among children with asthma is higher for more educated than for less educated mothers. This is because on average infants and young children of less educated mothers experience higher co-morbidity unrelated to asthma than their more educated counterparts (Scholl et al. 1987). Moreover, more educated women are likely to hold jobs that include health benefits. Furthermore, it is also hypothesized that the marginal child health returns are higher and reach a point of diminishing returns sooner among more educated than among less educated women because skills and knowledge greatly increase parental ability to implement medical information (targeted to educated individuals in the first place). Consequently, we propose that more educated mothers are able to attain higher levels of child health utility (defined in our example as the probability of positive child health outcome) than are less educated mothers at all levels of income earned and health insults avoided through child care.

For both sets of women, the public health, medical, and parental ideal of 100% probability of positive child health outcome is not an option; this is because they cannot both earn the maximum income and avoid the maximum number of health insults on a given day (figs 5.3c, d). Instead mothers must choose among realistic possibilities, and preferably choose the one that yields the highest utility among alternatives (i.e., the optimal solution). In our example, less educated women can attain a maximum of 56% probability of positive child health outcomes, and they can only realize this maximum if they choose to earn $24 per day and to avoid 30 health insults (6 hours working outside the home, 6 hours in child care). More educated women, on the other hand, can attain a maximum of 96% probability of positive child health outcome, and they can only realize this maximum if they choose to earn $160 per day and avoid 20 health insults per day (8 hours working outside the home, 4 hours in child care). This thought experiment raises questions relevant to the asthma paradox: are ecological constraints on behavioral options the reason rates of parental and patient compliance with prescribed asthma medication are so dismally low (Range: 14.5–66%; Alessandro et al. 1994; Coutts et al. 1992; Gibson et al. 1995; Pachter and Weller 1993; Rand and Wise 1994; Schoni 1993; Warner 1995; Wood et al. 1985)?

What Are the Life History and Parental Care Implications for Fitness of the Asthma Phenotype?

The behavioral trade-offs that parents of diverse socioeconomic status face in the care of children with asthma may have fitness implications: if low socioeconomic status parents provide less care than wealthier parents, then their children will be at risk of morbidity and mortality at higher rates than the rest of the population. Research findings are consistent with this proposition. Children of low socioeconomic status who have asthma experience more days lost from school, more hospitalizations, more deaths due to poorly managed asthma attacks at home, more psychological problems, and lower grades in school than their more well-to-do counterparts (Fowler et al. 1992; MacLean et al. 1992). From the published data, it is not clear whether these short-term effects result in lower fitness, for which sex, and at what levels of asthma severity. We know that asthma tends to delay puberty, marriage (Kokkonen 1995), and employment (Sibbald et al. 1992), but effects on age-specific mortality and fertility have not been measured. Predictions about increases in the prevalence of asthma in future generations depend on knowing something about life expectancy at reproductive age and subsequent fertility among asthmatics. Is this highly heritable phenotype (Neiminen et al. 1991) under directional, neutral, or negative selection in contemporary populations?

Conclusions

An appreciation for the evolutionary etiology of childhood asthma leads us to conclude that there is little incongruity to the asthma paradox. There are probably good evolutionary reasons that childhood asthma is prevalent in contemporary populations: it may be a consequence of an active IgE network favored by natural selection to provide defense against ubiquitous and endemic parasites throughout human history. But this is only part of the story; IgE-mediated mechanisms may also provide a survival advantage in heart disease, bacterial respiratory infection, cancer, and injury. However, whether childhood asthma is an *adaptive* consequence is an important empirical question: does asthma provide a survival or fertility advantage to individuals growing up in contemporary allergen-infested environments? We need more than qualitative arguments to test this proposition and show that children with asthma experience increased survivorship and fertility as adults when compared to children free of chronic conditions.

For as long as we do not fully understand the range of functions of the IgE system, it may be prudent for clinical practice guidelines to focus on environmental controls with pharmacotherapy as a complementary, albeit fundamental, aspect and the parental care behaviors that are key to their effective implementation. At the present time, national practice guidelines do not provide such practical algorithms. Instead, practitioners are given ample instruction on how to tailor medication prescriptions to individual patients and on ways to evaluate the effectiveness of their decisions. Excessive dependence on pharmacotherapeutics for asthma management de facto assumes that the benefits from drug intake can in fact counteract incremental increases in allergen exposure. As is true in most biological systems, there is probably a threshold point at which the morbidity caused by additional exposure to allergens cannot be relieved by further increments in drug therapy. Are some populations of children living at that threshold or beyond it? If so, a call for guidelines with a focus on environmental and behavioral changes would be easy to justify. In the meantime, we should perhaps consider the possibility that many children in contemporary populations may be at that threshold.

To assess the usefulness of an evolutionary approach to medical problems in childhood asthma, we propose several areas of research that focus on the development, implementation, and evaluation of detailed algorithms and their use in clinical practice. These algorithms are information-gathering aids that longitudinally assess (1) allergic sensitization (including total IgE levels) in children at risk of asthma (as determined by family history) or that have been diagnosed as having asthma using bronchoprovocation tests; (2) allergen exposure levels relevant to the child's sensitization profile (e.g., if the child is allergic to dust mites, measure levels of dust mite allergen in the home) with corresponding environmental control measures; (3) parental ability to provide the care necessary to attain effectiveness and parental noncompliance with specific courses of action to follow when parents are unable to comply or choose not to comply; (4) the effectiveness of these algorithms (i.e., computer-based data-tracking procedures that are user friendly for general practitioners and nurses); and (5) the cost effectiveness and quality of life outcomes of interventions outlined in items 1–4 as compared to existing pharmacotherapeutic-based interventions. Research findings on these clinically based algorithms with proven effectiveness should then be translated into home-based algorithms. This transfer could be made possible by developing technologies that allow parents to monitor allergen exposure in the home, at school, and other places where children spend most of their time (i.e., cheap and easy-to-use kits) and by instituting health insurance mechanisms for reimbursement of environmental control-related costs (e.g., installation of adequate heating, ventilation, and air conditioning systems; purchase of high-efficiency particulate-arresting [HEPA] air filters; elimination of carpeting; purchase of nonallergenic bedding, mattresses, and pillow covers).

An evolutionary approach to public health problems in childhood asthma also helps inform a basic science research agenda in epidemiology and immunology. Three areas of research largely neglected to date seem especially promising to undertake in developing nations, among U.S. ethnic groups with the highest prevalence rates of asthma (African Americans and Puerto Ricans), and among groups showing a recent rise in asthma prevalence (American Indians): (1) the relationship between parasite exposure, allergen exposure, and the onset of allergic sensitization between birth and adolescence; (2) changes in the timing, sequence, and levels of allergen and parasite exposure early in life among populations in the process of adopting more insulated housing and Western bedding; (3) the relationship between levels and forms of parental investment and the survivorship and fertility of adults who as children had asthma as compared to those free of chronic childhood conditions; and (4) the effect of allergen exposure levels on the efficacy of pharmacotherapy.

In addition, basic research on more immunologically precise definitions of "asthma phenotypes" is equally timely. Given the critical role of IgG in blocking IgE-mediated inflammation, it may be the case that a rise in total IgE levels among children with asthma could be entirely a function of IgG responses. If so, definitions of asthma phenotypes should be based on IgE- and IgG-informed criteria. Among individuals exposed to the same level of allergens and having the same levels of total serum IgE, IgG may successfully keep in check over-activation of IgE in some (no asthma), in others this may be accomplished intermittently (true mild asthma), while still in others this may not be accomplished at all (true moderate to severe asthma). The reasons IgG fails to provide its blocking function in some individuals and under some conditions needs to be investigated. One possible reason might be that total IgE production increases under conditions where IgG is being used to fight other concurrent problems such as bacterial infections, and the differences in the expression of asthma that we observe reflect these physiological trade-offs. Future research projects guided by evolutionary principles should help resolve some of these uncertainties.

Finally, current pediatric asthma management guidelines would be considerably strengthened by a prevention addendum that includes health care delivery decision-making algorithms for providers and home care algorithms for parents of children who have asthma. These algorithms will help providers recommend lifestyle changes that are within reach to families and ensure that parents and patients make the necessary medication use and environmental control-related behavioral adjustments at home and at school as allergen exposures change and as the severity of the condition waxes and wanes.

References

Abdel Fattah, S. M., el Sedfy, H. H., el Sayed, H. L., Abdel Aziz, H., Morsy, T. A., Salama, M. M., Hamdi, K. N. (1994) Seropositivity against pediculosis in children with cervical lymphadenopathy. *Journal of Egyptian Social Parasitology* 24(1):59–67.

Akasawa, A., Koya, N., Iikura, Y. (1991) Investigation of the genetics in allergic children: the first report—Twin study on HLA in allergic disease. *Arerugi* 40(4):428–34.

Alessandro, F., Vincenzo, Z. G., Marco, S., Marcello, G., Enrica, R. (1994) Compliance with pharmacologic prophylaxis and therapy in bronchial asthma. *Annals of Allergy* 73(2):135–40.

Asthma Zero Mortality Coalition (1995) The Asthma in America Report: What primary care doctors and consumers say about asthma and its treatment [on-line]. Available: http://www.

Bardin, P. G., Johnston, S. L., Pattemore, P. K. (1992) Viruses as precipitants of asthma symptoms. II. Physiology and mechanisms. *Clinical Experimental Allergy* 22(9):809–22.

Barnes, K. C., Brenner, R. J. (1996) Quality of housing and allergy to cockroaches in the Dominican Republic. *International Archives of Allergy and Immunology* 109(1):68–72.

Barnes, P. J., Rodger, I. W., Thomson, N. C., eds. (1992) *Asthma: Basic Mechanisms and Clinical Management.* 2nd ed. New York: Academic Press.

Bazaral, M., Orgel, H. A., Hamburger, R. N. (1973) The influence of serum IgE levels of selected recipients, including patients with allergy, helminthiasis and tuberculosis, on the apparent P-K titre of a reaginic serum. *Clinical Experimental Immunology* 14:117–125.

Berciano, F. A., Crespo, M., Bao, C. G., Alvarez, F. V. (1987) Serum levels of total IgE in non-allergic children. Influence of genetic and environmental factors. *Allergy* 42(4):276–83.

Bethel, R. A. et al. (1984) Interaction of sulfur dioxide and dry cold air in causing bronchoconstriction in asthmatic subjects. *Journal of Applied Physiology* 57:419.

Biagini, R. E., Moorman, W. J., Lewis, T. R., Bernstein, I. L. (1986) Ozone enhancement of platinum asthma in a primate model. *American Review of Respiratory Disease* 134(4):719–25.

Bosman, A., De Giorgi, F., Kandia Diallo, I., Pizzi, L., Bartoloni, P., Cancrini, G. (1991) Prevalence and intensity of infection with intestinal parasites in areas of the Futa Djalon, Republic of Guinea. *Parasitologia* 33(2–3): 203–8.

Breen, P. H., Becker, L. J., Ruygrok, P., Mayers, I., Long, G. R., Leff, A., Wood, L. D. (1987) Canine bronchoconstriction, gas trapping, and hypoxia with methacoline. *Journal of Applied Physiology* 63(1):262–69.

Bryant, B., Mohai, P. (1992) *Race and the incidence of environmental hazards.* Boulder, CO: Westview Press.

Bundy, D. A. P., Medley, G. F. (1992) Epidemiology and population dynamics of helminth infection. In Moqbel, R., ed., *Allergy and Immunity to Helminths: Common Mechanisms or Divergent Pathways?* London: Taylor and Francis; pp. 38–62.

Burney, P. G. (1992) Epidemiology. In Clark, T. J., Godfrey, S., Lee, T. H., eds., *Asthma.* 3rd ed. New York: Chapman, pp. 150–152.

Burrows, B., Martinez, F. D., Halonen, M., Barbee, R. A., Cline, M. G. (1989) Association of asthma with serum IgE levels and skin-test reactivity to allergens. *The New England Journal of Medicine* 320(5):271–77.

Chapman, M. D., Heymann, P. W., Sporik, R. B., Platts-Mills, T. A.

(1995) Monitoring allergen exposure in asthma: new treatment strategies. *Allergy* 50(suppl. 25):29–33.

Clark, D., Gollub, R., Green, W. F., Harvey, N., Murphy, S. J., Samet, J. M. (1995) Asthma in Jemez Pueblo schoolchildren. *American Journal of Respiratory Critical Care Medicine* 151(5):1625–27.

Clinical and Experimental Allergy (1995) Recent advances in the genetics of asthma and allergy symposium. Vol. 25. New York: Blackwell Science.

Clutton-Brock, T. H. (1991) *The Evolution of Parental Care.* Princeton, NJ: Princeton University Press.

Corcoran, B. M., Foster, D. J., Fuentes, V. L. (1995) Feline asthma syndrome: a retrospective study of the clinical presentation of 29 cats. *Journal of Small Animal Practice* 36(11):481–88.

Coutts, J. A., Gibson, N. A., Payton, J. Y. (1992) Measuring compliance with inhaled medication in asthma. *Archives of Diseases in Children* 67(3):332–33.

Criqui, M. H., Seibles, J. A., Hamburger, R. N., Coughlin, S. S., Gabriel, S. (1990) Epidemiology of immunoglobulin E levels in a defined population. *Annals of Allergy* 64(3): 308–13.

DiPiro, J. T., Hamilton, R. G., Howdieshell, R., Adkinson, N. F., Mansberger, A. R. (1994) Lipopolysaccharide-reactive immunoglobulin E is associated with lower mortality and organ failure in traumatically injured patients. *Clinical Diagnosis Laboratory Immunology* 1(3):295–98.

Doull, I.J., Lawrence, S., Watson, M., Begishvili, T., Beasley, R.W., Lampe, F., Holgate, T., Morton, N.E. (1996) Allelic association of gene markers on chromosomes 5q and 11q with atopy and bronchial hyperresponsiveness. *American Journal of Respi-*

ratory and Critical Care Medicine 153 (4):20–32

Durham, S. (1995) New insights into the mechanisms of immunotherapy. *European Archives of Otorhinolaryngology (Supplement)* S64–67.

Eaton, S. B., Konner, M., Shostak, M. (1988) *The Paleolithic Prescription: A Program of Diet and Exercise and a Design for Living*, 1st ed. New York: Harper & Row.

Eddy, D. M. (1990) Practice policies—what are they? *Journal of the American Medical Association* 263(6):877–80.

Etzel, R.A. (1995). Indoor air pollution and childhood asthma: effective environmental interventions. *Environmental Health Perspectives* 103 (suppl. 6): 55–58.

Evans, R. (1992) Asthma among minority children. A growing problem. *Chest* 101(6 suppl.):368S–371S.

Evers, S.E., Hooper, M.D. (1995) Dietary intake and anthropometric status of 7 to 9 year olds. *Journal of the American College of Nutrition* 14 (6): 595–603.

Ewald, P. W. (1994) *Evolution of Infectious Disease*. New York: Oxford University Press.

Fowler, M. G., Davenport, M. G., Garg, R. (1992) School functioning of US children with asthma. *Pediatrics* 90(6):939–44.

Frick, O. L. (1986) Effect of respiratory and other virus infections on IgE immunoregulation. *Journal of Allergy and Clinical Immunology* 78: 1013–18.

Gaddy, J. N., Busse, W. W., Sheffer, A. L. (1993) Fatal asthma. In Weiss, E. B., Stein, M., eds., *Bronchial Asthma*. 3rd ed. Boston: Little, Brown and Company; pp. 1154–66.

Gerrard, J. W. et al. (1976) Serum IgE levels in white and metis communities in Saskatchewan. *Annals of Allergy* 37:91.

Gibson, N. A., Ferguson, A. E., Aitchison, T. C., Paton, J. Y. (1995) Compliance with inhaled asthma medication in preschool children. *Thorax* 50(12):1274–79.

Godfrey, R. C. (1975) Asthma and IgE levels in rural and urban communities of the Gambia. *Clinical Allergy* 5: 201–207.

Gottlieb, D. J., Beiser, A. S., O'Connor, G. T. (1995) Poverty, race and medication use are correlates of asthma hospitalization rates. A small area analysis in Boston. *Chest* 108(1):28–35.

Grogger, J., Bronars, S. (1993) The socioeconomic consequences of teenage childbearing: findings from a natural experiment. *Family Planning Perspectives* 25(4): 156–61.

Grundbacher, F. J., Massie, F. S. (1985) Levels of immunoglobulin G, M, A and E at various ages in allergic and nonallergic black and white individuals. *Journal of Allergy and Clinical Immunology* 75(6):651–58.

Hagan, P., Blumenthal, U. J., Dunn, D., Simpson, A .J., Wilkins, H. A. (1991) Human IgE, IgG4 and resistance to reinfection with Schistosoma haematobium. *Nature* 349(6306):243–45.

Hagel, I., Lynch, N. R., DiPrisco, M. C., Lopez, R. I., Garcia, N. M. (1993) Allergic reactivity of children of different socioeoconomic levels in tropical populations. *International Archives Allergy Immunology* 101: 209–214.

Halfon, N., Newacheck, P. W. (1993) Childhood asthma and poverty: differential impacts and utilization of health services. *Pediatrics* 91(1):56–61.

Halpern, G. M., Gershwin, L. J., Gonzales, G., Fowler, M. E. (1989) Diagnosis of inhalant allergy in a chimpanzee using in vivo and in vitro tests. *Allergology and Immunopathology* 17(5):271–76.

Hanson, B., McGue, M., Roitman-Johnson, B., Segal, N.L., Bouchard, T.J. Jr, Blumenthal, M.N. (1991) Atopic disease and immunoglobulin E in twins reared apart and together. *American Journal of Human Genetics* 48(5): 873–79.

Hatoum, H.T., Schumock, G.T., Kendzierski, D.L. (1994) Meta-analysis of controlled trials of drug therapy in mild chronic asthma: the role of inhaled corticosteroids. *Annals of Pharmacotherapy* 28(11): 1285–89.

Haus, M., Heese, H. D., Weinberg, E. G., Potter, P. C., Hall, J. M., Malherbe, D. (1988) The influence of ethnicity, an atopic family history, and maternal ascariasis on cord blood serum IgE concentrations. *Journal of Allergy and Clinical Immunology* 82(2):179–89.

Hetman, S., Kivity, S., Greif, J., Fireman, E. M., Topilsky, M. (1988) IgE values in the allergic and healthy Israeli population. *Annals of Allergy* 61(2):123–8.

Hill, K. (1993) Life history theory and evolutionary anthropology. *Evolutionary Anthropology* 2(3): 78–88.

Hishanick, J. J., Coddington, D. A., Gergen, P. J. (1994) Trends in asthma-related admissions among American Indian and Alaskan native children from 1979 to 1980. Universal health care in the face of poverty. *Archives of Pediatric and Adolescent Medicine* 148(4):357–63.

Hochreutener, H., Wuthrich, B., Huwyler, T., Schopfer, K., Seger, R., Baerlocher, K. (1991) Variant of hyper IgE syndrome: the differentiation from atopic dermatitis is important because of teatment and prognosis. *Dermatologica* 182(1):7–11.

Hog, M., Schindera, I. (1989) Neurodermatitis in primates: a case description of a female gorilla. *Hautarzt* 40(3):150–52.

Hrdy, S. B. (1992) Fitness tradeoffs in the history and evolution of dele-gated mothering with special reference to wet-nursing, abandonment and infanticide. *Ethology and Sociobiology* 13: 409–442.

Hurtado, A. M. (1995) Childhood asthma prevalence among Puerto Ricans and Mexican Americans: Implications for behavioral intervention research. *Hispanic Journal of Behavioral Sciences* 17(3):362–370.

Hurtado, A. M., Arenas, I., Hill, K. (1996a) Evolutionary contexts of chronic allergic disease: the Hiwi of Venezuela. *Human Nature* 8(1):1–20.

Hurtado, A. M., Hill, K., James, S. (1996c) Maternal care, child health epidemiology and optimization modeling. Submitted to the *Journal of Behavioral Medicine*.

Hussain, R., Poindexter, R. W., Ottesen, E. A. (1992) Control of allergic reactivity in human filariasis. Predominant localization of blocking antibody to the IgG4 subclass. *Journal of Immunology* 148(9):2731–37.

Infante-Rivard, C. (1993) Childhood asthma and indoor environmental risk factors. *American Journal of Epidemiology* 137(8):834–44.

Ionov, I. D. (1985) Anaphylaxis as a physiological mechanism of masking of embryonic antigens. *Medical Hypotheses* 17(4):313–20.

Ishizaka, T., Soto, C. S., Ishizaka, K. (1973) Mechanisms of passive sensitization: number of IgE molecules and their receptor sites on human basophil granulocytes. *Journal of Immunology* 111:500.

Jenkins, M. A., Hopper, J. L., Flander, L. B., Carlin, J. B., Giles, G. G. (1993) The associations between childhood asthma and atopy, and parental asthma, hay fever and smoking. *Paediatric Perinatal Epidemiology* 7(1): 67–76.

Kaplan, J. E., Larrick, J. W., Yost, J. A. (1980) Hyperimmunoglobulinemia E in the Waorani, and isolated Amerindian population. *American Journal of Tropical Medicine and Hygiene* 29(5):1012–17.

Kappus, K. D., Lundgren, R. G. Jr, Juranek, D. D., Roberts, J. M., Spencer, H. C. (1994) Intestinal parasitism in the United States: update on a continuing problem. *American Journal of Tropical Medicine and Hygiene* 50(6):705–13.

Karol, M. H. (1994) Animal models of occupational asthma. *European Respiration* 7(3):555–68.

Kigoni, E. P., Elsas, P. P., Lenzi, H. L., Dessein, A. J. (1986) IgE antibody and resistance to infection. II. Effect of IgE suppression on the early and late skin reaction and resistance of rats to *Schistosoma mansoni* infection. *European Journal of Immunology* 16(6):589–95.

Kleinbaum, D. G., Kupper, L. L. and Morgenstern, H. (1982) *Epidemiological Research: Principles and Research Methods*. New York: Van Rostrand.

Kokkonen, J. (1995) The social effects in adult life of chronic physical illness since childhood. *European Journal Pediatrics* 154(8):676–81.

Koopman, J. S., Weed, D. L. (1990) Epigenesis theory: a mathematical model relating causal concepts of pathogenesis in individuals to disease patterns in populations. *American Journal of Epidemiology* 132(2):366–90.

Lawler, F. H. (1995) Clinical use of decision analysis. *Primary Care* 22 (2):281–93.

LeSon, S., Gershwin, M. E. (1995) Risk factors for asthmatic patients requiring intubation. I. Observations in children *Journal of Asthma* 32(4):285–94.

Luoma, R. (1984) Environmental allergens and morbidity in atopic and nonatopic families. *Acta Paediatrica Scandinavia* 73:448–453.

Lynch, N. R., Hagel, I., Perez, M., Di-

Prisco, M., Lopez, R., Alvarez, N. (1993) Effect of antihelminthic treatment on the allergic reactivity of children in a tropical slum. *Journal of Allergy and Clinical Immunology* 92:404–11.

Lynch, N. R., Hurtado, J. C., Verde, O., Lopez, R. I., Soyano, A. (1985) Interrelationships between immediate, intermediate and delayed cutaneous reactions and their in vitro correlates. *Annals of Allergy* 55:848–853.

Lynch, N. R., Lopez, R., Isturiz, G., Tenias-Salazar, E. (1983) Allergic reactivity and helminthic infection in Amerindians of the Amazon Basin. *International Archives Allergy Applied Immunology* 72:369–372.

Lynch, N. R., Medouze, L., DiPrisco-Fuenmayor, M. C., Verde, O., Lopez, R. I. and Malave, C. (1984) Incidence of atopic disease in a tropical environment: partial independence from intestinal helminthiasis. *Jounal of Allergy and Clinical Immunology* 73:229–233.

MacLean, W. E., Perrin, J. M., Gortmaker, S., Pierre, C. B. (1992) Psychological adjustment of children with asthma: effects of illness severity and recent stressful life events. *Journal of Pediatric Psychology* 17(2):159–71.

Magana, A., Clark, N. M. (1995) Examining a paradox: does religiosity contribute to positive birth outcomes in Mexican American populations? *Health Education Quarterly* 22(1):96–109.

Maldonado, A. E. (1993) Hookworm disease: Puerto Rico's secret killer. *Puerto Rican Health Sciences Journal* 12(3):191–6.

Malveaux, F. J., Fletcher-Vincent, S. A. (1995) Environmental risk factors of childhood asthma in urban centers. *Environmental Health Perspectives* 103:(suppl. 6):59–62).

Margolis, P.A., Greenberg, R. A., Keyes, L. L., LaVange, L. M., Chapman, R. S., Denny, F. W., Bauman, K. E., Boat , B. W. (1992) Lower respiratory illness in infants and low socioeconomic status. *American Journal of Public Health* 82(8):1119–26.

Markides, K. S., Coreil, J. (1986) The health of Hispanics in the southwestern United States: an epidemiologic paradox. *Public Health Reports* 101(3):253–65.

Matsuda, H., Watanabe, N., Kiso, Y., Hirota, S., Ushio, H., Kannan, Y., Azuma, M., Koyama, H., Kitamura, Y. (1990) Necessity of IgE antibodies and mast cells for manifestation of resistance against larval *Haemaphysalis longicornis* ticks in mice. *Journal of Immunology* 144(1):259–62.

Matsuse, H., Shimoda, T., Kohno, S., Fujiwara, C., Sakai, H., Takao, A., Asai, S., Hara, K. (1995) A clinical study of mortality due to asthma. *Annals of Allergy and Asthma Immunology* 75(3):267–72.

Mauro, H., Hashimoto, K., Shimoda, K., Shimanuki, K. et al (1990) Long term follow up studies of bronchial asthma in children. II. Prognosis and complications, treatment and allergic evaluations. *Arerugi* 39(8):662–69.

McConnochie, K. M., Roghmann, K. J., Liptak, G. S. (1995) Hospitalization for lower respiratory tract illness in infants; variation in rates among counties in New York State and areas within Monroe County. *Journal of Pediatrics* 126(2):220–29.

Merrett, T. G., Merrett, J., Cookson, J. B. (1976) Allergy and parasites: the measurement of total and specific IgE levels in urban and rural communities in Rhodesia. *Clinical Allergy* 6:124–131.

Moqbel, R. (1992) *Allergy and Immunity to Helminths: Common Mechanisms or Divergent Pathways?* London: Taylor and Francis.

Mortality and Morbidity Weekly Report (1992) Asthma—United States 1980–1990. 41(39):733–35.

National Research Council (1991) *Indoor Pollutants.* Washington, DC: National Academy Press.

Nesse, R., and Williams, G. C. (1995) *Why We Get Sick: The New Science of Darwinian Medicine.* New York: Vintage Books.

Nieminen, M.M., Kapriom, J., Koskenvuo, M. (1991) A population-based study of bronchial asthma in adult twin pairs *Chest* 100(1): 70–75.

Noerjasin, B. (1973) Serum IgE concentrations in relation to antihelminthic treatment in a Javanese population with hookworm. *Clinical Experimental Immunology* 13:545–551.

Orr, T. S., Blair, A. M. (1969) Potentiated reagin response to egg albumin and conalbumin in Nippostrongylus brasiliensis-infected rats. *Life Sciences* 8:1073–77.

Pachter, L. M., Weller, S. C. (1993) Acculturation and compliance with medical therapy. *Journal of Developmental Behavioral Pediatrics* 14(3): 163–68.

Page, C. P. (1991) One explanation of the asthma paradox: inhibition of natural inflammatory mechanism by 2-agonists. *Lancet* 337(8743):717–20.

Pattermore, P. K., Asher, M. I., Harrison, A. C., Mitchell, E. A., Rea, H. H., Stewart, A. W. (1989) Ethnic differences in prevalence of asthma symptoms and bronchial hyperresponsiveness in New Zealand schoolchildren. *Thorax* 44(3):168–76.

Patterson, R., Harris, K. E. (1993) Substance P and IgE-mediated allergy. II. Reduction of rhesus IgE antibody after aerosol exposure to substance P and allergen. *Allergy Proceedings* 14(1):53–59.

Peden, D. B., Setzer, R. W., Devlin, R. B. (1995) Ozone exposure has both a priming effect on allergen-induced responses and an intrinsic inflammatory action in the nasal airways of perenially allergic asthmatics. *American Journal of Respiratory Critical Care* 151(5):1336–45.

Pollart, S. M., Chapman, M. D., Fiocco, G. P., Rose, G., Platts-Mills, T. A. (1989) Epidemiology of acute asthma: IgE antibodies to common inhalant allergens as a risk factor for emergency room visits. *Journal of Allergy and Clinical Immunology* 83(5):875–82.

Pope, A. M., Patterson, R., Burge, H. (1993) *Indoor Allergens: Assessing and Controlling Adverse Health Effects.* Washington: National Academic Press.

Porro, E., Calamita, P., Rana, I., Montini, L., Criscione, S. (1992) Atopy and environmental factors in upper respiratory infections: an epidemiological survey on 2304 school children. *International Journal of Pedatric Otorhinolaryngology* 24(2):111–20.

Poysa, L., Korppi, M., Pietikainen, M., Remes, K., Juntunene-Backman K. (1991) Asthma, allergic rhinitis and atopic eczema in Finnish children and adolescents. *Allergy* 46(3):151–61.

Pritchard, D. I. (1992) Parasites and allergic disease: a review of the field and experimental evidence for a 'cause-and-effect' relationship. In Moqbel, R., ed., *Allergy and Immunity to Helminths: Common Mechanisms or Divergent Pathways?* London: Taylor and Francis; pp. 38–62.

Profet, M. (1991) The function of allergy: immunological defense against toxins. *Quarterly Review of Biology* 66(1):23–62.

Rand, C. S., Wise, R. A. (1994) Measuring adherence to asthma medication regimens. *American Journal Respiratory Critical Medical Care* 149(2):S69–76.

Robertson, C. F., Rubinfeld, A. R., Bowes, G. (1990) Deaths from asthma in Victoria: a 12-month survey. *Medical Journal Australia* 152(10):511–17.

Rodger, P. J., I. W. and Thomson, N. C., eds. (1992) *Asthma: Basic Mechanisms and Clinical Management,* 2nd ed. New York: Academic Press.

Rosenfeld, J. A. (1992) Maternal work outside the home and its effect on women and their families. *Journal of American Medical Women's Association* 47(2):47–53.

Roth, D. A. (1992) *The Evolution of Life Histories: Theory and Analysis.* New York: Routledge.

Rothman, K. J. (1986) *Modern Epidemiology.* Boston, MA: Little, Brown and Company.

Salzano, F. M. (1990) Parasitic load in South American tribal populations. In Swedlund, A. C., Armelagos, G. J., eds., *Disease in Populations in Transition: Anthropological and Epidemiological Perspectives.* New York: Bergin and Garvey; pp. 201–21.

Sanford, A.J., Moffatt, M.F., Daniels, S.E., Nakamura, Y., Lathrop, G.M., Hopkin, J.M., Cookson, W.O. (1995) A genetic map of chromosome 11q, including the atopy locus. *European Journal of Human Genetics* 3(3): 188–94.

Santos, R.V., Coimbra, Junior C.E., Flowers, N.M., Silva, J.P. (1995) Intestinal parasitism in the Xavante Indians, central Brazil. *Review Institute of Tropical Medicine at Sao Paulo* 37(2): 145–8.

Sapolsky, R. M. (1994) *Why Zebras Don't Get Ulcers: A Guide to Stress, Stress-related Diseases, and Coping.* New York: W.H. Freeman.

Sarafino, E.P., Goldfedder, J. (1995) Genetic factors in the presence, severity, and triggers of asthma. *Archives of Diseases in Childhood* 73(2): 112–16.

THE EVOLUTIONARY ECOLOGY OF CHILDHOOD ASTHMA 133

Scholl, T., Karp, R., Theophano, J., Decker, E. (1987) Ethnic differences in growth and nutritional status: a study of poor schoolchildren in southern New Jersey. *Public Health Report* 102:278–83.

Schoni, M. H. (1993) Compliance with inhalation therapy in children with respiratory diseases. *Schweiz Rundsch Prax* 82(44):1218–21.

Sears, M. R., Rea, H. H., Fenwick, J., Beaglehold, R., Gillies, A. J., Holst, P. E., O'Donnell, T. V., Rothwell, R. P., Sutherland, D. C. (1986) Deaths from asthma in New Zealand. *Archives of Disease in Childhood* 61(1): 6–10.

Sibbald, B., Anderson, H. R., Mc-Guigan, S. (1992) Asthma and employment in young adults. *Thorax* 47(1):19–24.

Silvers, A., Florence, B.T., Rourke, D.L., Lorimor, R.J. (1994) How children spend their time: a sample survey for use in exposure and risk assessments. *Risk Analysis* 14(6): 931–44.

Soto-Quiros, M., Bustamante, M., Gutierrez, I., Hanson, L.A.,Strannegard, I.L., Karlberg, J. (1994) The prevalence of childhood asthma in Costa Rica. *Clinical Experimental Allergy* 24 (12): 1130–36.

Spengler, J. D., Sexton, K. (1983) Indoor air pollution: a public health perspective. *Science.* 221:9–17.

Sporik, R., Holgate, S. T., Platts-Mills, T. A., Cogswell, J. (1990) Exposure to house-dust mite allergen (Der p I) and the development of asthma in childhood: a prospective study. *The New England Journal of Medicine* 323(8):502–7.

Stearns, S. C. (1992) *The Evolution of Life Histories.* New York: Oxford University Press.

Stephenson, L. (1987) *The impact of helminth infections on human nutrition: Schistosomes and soil-transmitted helminths.* New York: Taylor and Francis.

Sullivan, S. D., Weiss, K. B. (1993) Assessing cost effectiveness in asthma care:building an economic model to study the impact of alternative intervention strategies. *Allergy* 48(suppl. 17):146–52.

Sutton, B. J., Gould, H. J. (1993) The human IgE network. *Nature* 366(2): 421–28.

Szczeklik, A., Dropinski, J., Gora, P. F. (1993) Serum immunoglobulin E and sudden cardiac arrest during myocardial infarction. *Coronary Artery Disease* 4(11):1029–32.

Tambyraja, R. L. (1991) The prematurity paradox of the small Indian baby. *Indian Journal of Pediatrics* 58(4):415–19.

Taylor, W. R., and Newacheck (1992) Impact of childhood asthma on health. *Pediatrics* 90(5):657–62.

Turkeltaub, P. C., Gergen, P. J. (1991) Prevalence of upper and lower respiratory conditions in the US population by social and environmental factors: data from the second National Health and Nutrition Examination Survey, 1976 to 1980 (NHANES II). *Annals of Allergy* 67 (2): 147–54.

Turner, K. J., Fisher, E. H., Holt, P. G. (1982) Host age determines the effects of helminthic infestation upon allergic reactivity in rats. *Australian Journal of Experimental Biology and Medicine* 60:147–57.

U.S. Department of Health and Human Services (1984) *Air Power: Self-Management of Asthma through Group Education.* Washington, DC: National Institutes of Health.

U.S. Department of Health and Human Services (1991) Guidelines for the diagnosis and management of asthma. *National Asthma Education Program: Expert Panel Report.* Washington, DC: National Institutes of Health.

Venge, P., Shaw, M. (1989) Parasites

and allergy. *Allergologie* September: S96–99.

Warner, J. O. (1995) Review of prescribed treatment for children with asthma in 1990. *British Medical Journal* 311(7006):663–66.

Warr, G. W., Magor, K. E., Higgings, D. A. (1995) IgY:clues to the orgins of modern antibodies. *Immunology Today* 16(8):392–98.

Wegner, C. D., Gundel, R. H., Letts, L. G. (1993) Efficacy of monoclonal antibodies against adhesion molecules in animal models of asthma. *Agents Actions* (Suppl. 43):151–62.

Weiss, E. B., Stein, M., eds. (1993) *Bronchial Asthma*, 3rd ed. Boston, MA: Little, Brown and Company.

Weiss, K. B., Gergen, P. J., Crain, E. F. (1992) Inner-city asthma. The epidemiology of an emerging US public health concern. *Chest* 101(suppl. 6): 362S–367S.

Welliver, R. C., Sun, M., Rinaldo, D., Ogra, P. L. (1986) Predictive value of respiratory syncytial virus-specific IgE responses for recurrent wheezing following bronchiolitis. *Journal of Pediatrics* 109:776–80.

Wells, N. (1994) The economic and social impact of childhood asthma. In Christie, M., French, D. *Assessment of Quality of Life in Childhood Asthma*. Geneva: Harwood Academic; 27–34.

Willies-Jacobo, L. J., Denson-Lino, J. M., Rosas, A., O'Connor, R. D., Wilson, N. W. (1993) Socioeconomic status and allergy in children with asthma. *Journal of Allergy and Clinical Immunology* 92(4):630–32.

Wolf, A. M., Gortmaker, S. L., Cheung, L., Gray, H. M., Herzog, D. B., Colditz, G. A. (1993) Activity, inactivity, and obesity: racial, ethnic, and age differences among schoolgirls. *American Journal of Public Health* 83(11):1625–27.

Wood, P. R., Casey, R., Kolski, G. B., McCormick, M. C. (1985) Compliance with oral theophylline therapy in asthmatic children. *Annals of Allergy* 54(5):400–4.

Yadav, R., Yadav, S., Yadav, J. (1994) IgE levels of normal Indian children. *Indian Journal of Medical Sciences* 48(9):195–98.

6

EVOLUTIONARY PERSPECTIVES ON THE ONSET OF PUBERTY

CAROL M. WORTHMAN

[I]t would be a mistake to leap to the conclusion that because human immaturity makes possible high flexibility in later adjustment, anything is possible for the species. ... The technical-social way of life is a deep feature of the species adaptation. But we would err if we assumed a priori that man's inheritance places no constraint on his power to adapt. Some of the preadaptations can be shown to be presently maladpative.
                                   —J. Bruner (1972: 687)

Recent human history has witnessed worldwide, ongoing trans-formations of child growth and development. These changes are best documented by increasing height by age in children and decreasing ages at menarche. Increased size and accelerated maturation have always characterized privileged social groups or classes relative to those less privileged, but commencing about 130 years ago in a few western European countries, a pattern of accelerating development has swept across entire populations (Tanner, 1978, p. 152). Similar, though often more precipitous, shifts continue today in populations around the world, in parallel with socioeconomic and public health changes that strongly alter living conditions. These secular trends to accelerated growth and maturation are amply documented and widely discussed as markers of well-being (Tanner, 1990; van Wierigen, 1986), but their implications for human health have scarcely been explored. This analysis will apply an evolutionary and comparative approach to considera-tion of the causes and consequences of secular trends in the onset of puberty by presenting a series of studies on variation in pubertal timing, particularly with respect to sex differences, that demonstrate the impact of socioecological factors. Further, this chapter considers the clinical and public health implications of early maturation and suggests a number of challenges in the conceptualization of normal development, of developmental norms, and of sex differences in health needs and risks. The marked context dependency of human development points to a need for "social prescriptions" for healthy development.

Secular Trends in the Onset of Puberty

Estimates of pubertal timing used to track historical trends and population differences in maturation rates have been based primarily on age at menarche (Eveleth & Tanner, 1990, p. 161). Developmental studies document an enormous range of population variation in normative age at menarche, from population medians of around age 12.0–12.5 years in segments of urban postindustrial societies to 18.0–18.6 years in rural highland Papua New Guinea or high altitude Nepali groups (Eveleth & Tanner, 1990; Wood, 1994). The causes of this variation are not definitively established, although genetic variation apparently plays quite a small role (Ellison, 1981a).[1] Where social class or residence (urban/rural) predict significant differences in living environments, they also consistently predict earlier age at menarche in the better-off group (Bielicki, 1986). These and other socioecological factors appear to act as organizers or mediators for other variables that directly influence the pace of development and timing of puberty. The principal candidate factors are morbidity (timing, type, severity, number, and duration of infectious disease episodes), nutrition (including quality [sufficiency of macro- and micronutrients] and stability of food supply), and maternal-gestational conditions (Liestøl, 1982), but all are supported by associational evidence, and none has received definitive critical tests. Recent studies of Indian adoptees in Sweden (Proos et al., 1991), and of effects of early (Khan et al., 1995) or late (Maclure et al., 1991; Moisan et al., 1990) nutritional status all reflect the importance of early, including gestational, experience for timing of puberty.

Given the established importance of contextual factors for determining cross-population differences in the timing of puberty, variation in age at menarche within populations through time is unsurprising. Such secular trends reflect either positive or negative change in maturation rate, depending on whether conditions improve or worsen. Secular trends to decreased age at menarche were first pointed out in data on northern European populations by Tanner (Eveleth & Tanner, 1990; Tanner, 1962), who inferred a sustained decline at the rate of 4 months per decade over the previous 130 years. The small size and questionable representativeness of historical samples (Eveleth, 1986) confound the interpretation of these findings (Bullough, 1981). Calculation of how quickly age at menarche and, inferentially, puberty, can and did fall in these and other populations is complicated by the effects of sampling bias: age at menarche at any point in time differs among subgroups for reasons described above, and the impact of social-ecological change often falls unequally among subgroups. Less advantaged groups or classes exhibit greater developmental acceleration than previously advantaged groups when general population conditions improve (Padro, 1984; Singh & Malhotra, 1988; Tan-Boom et al., 1983). Recent work has compiled a substantial worldwide literature that leaves no doubt that marked secular trends do occur, but yet fails to identify the precise proximal causes of declining menarcheal age. A recent report concurs with the bulk of contemporary evidence, which amplifies but scarcely alters Tanner's early conclusions, by noting "that broad improvements in nutrition and in living standards are determinants of age at menarche" (Khan et al., 1995: 1091S).

In the past 50 years, worldwide improvements in public health and nutrition accompanied by declining mortality rates and increased life expectancy (World Bank, 1993) have been paralleled by accelerations in physical development, though change has occurred over different time periods and at different rates across populations. Some reports document secular trends roughly equivalent to Tanner's initial estimate, on the order of 0.39 years per decade in a cohort of rural Guatemalans spanning birth years 1962–1977 (Khan et al., 1995); 0.34 years per decade in middle-class Madrid suburb over 1935–1965 (Padro, 1984); or 0.63 years per decade in lower class, versus 0.30 in higher class, Indian schoolgirls surveyed 1974–1986 (Singh & Malhotra, 1988). Again, relatively privileged classes show earlier menarcheal ages and less decline with time (Wood, 1994, p. 438). Some studies indicate more precipitous reductions in later-maturing groups that undergo marked environmental improvements, though here difficulties in obtaining accurate age estimates increase the potential for measurement error in estimating median age at menarche. For instance, highland Papua New Guinea peoples with some of the latest documented menarcheal ages have also been reported to undergo very rapid developmental acceleration (Heywood & Norgan, 1992; Wood, 1992). Social change, including introduction of cash economy, schooling, and health care, as well as migration to urban centers, is occurring rapidly in Papua New Guinea (PNG) and provides an opportunity to observe how these changes affect pubertal timing.

In 1966–67, Malcolm (1970) studied physical development among the Bundi, Gende-speaking horticulturalists in northwest highland PNG, and reported a median age at menarche of 18.0 years. Seventeen years later, in 1983–84, Zemel and Jenkins (1989) returned to remeasure the same Bundi population represented by two groups: those who remained in the rural locales studied by Malcolm, and those who had migrated to towns. A dramatic decline in age of menarche had occurred in both groups, but was most marked in urban settings: age at menarche in rural girls was 17.2 years, and even 1.4 years earlier in urban girls, at age 15.8 years. In less than two decades, Bundi populations had undergone a 0.8 to 2.2-year decrease in the timing of menstrual onset, at average rates of 0.47 and 1.29 years per decade for rural and urban girls, respectively (Zemel et al., 1993). During this interval, schooling, wage labor, shops with imported foods (rice, fish, sugar, flour), and health care had been introduced to the rural area and were even more readily available in the town settings. Consistently, migrants have been found to be larger and developmentally advanced compared with nonmigrants from the same source population (Bogin, 1988, pp. 146-8); thus, the rural–urban difference may be confounded by selective migration of those who are in some way better off (Kaplowitz et al., 1993; Mascie-Taylor & Lasker, 1988).

Studies that measure growth, sex character development, and/or endocrine change at puberty indicate that onset of puberty is not as delayed in later-maturing populations as their late ages at menarche would suggest. For instance, a developmental study in a later-maturing rural East African population of Kikuyu showed that the median age at menarche of 15.9 years was due in part to a prolongation of puberty, and that boys and girls entered puberty just 0.8 and 1.8 years later than British children among whom age at menarche was 2.4 years earlier, at 13.5 years (Worthman, 1987). That is, time elapsed between onset of breast development and first menses (2.9 years) was 0.6 years longer than the 2.3-year lapse observed in a well-studied London population. Thus, it appears that early-maturing populations such as contemporary western European or American youth not only undergo onset of puberty and reproductive maturity earlier, but they also progress through puberty more quickly than their later-maturing ancestors and contemporaries. Concomitantly, they attain adult height much earlier: a pattern of prolonged growth characterizes later-maturing populations, in which females reach adult height only in the late teens, males in the early twenties (Martorell & Kettel Kahn, 1994). Growth prolongation, supported by slow gonadal maturation, allows achievement of greater adult height despite chronic moderate malnutrition or other morbidity.

Because of their association with improving conditions and health, accelerated maturation and reduced age at puberty are viewed as positive indicators of well-being (Bielicki, 1986; van Wierigen, 1986). Indeed, so strong is the association of growth and pubertal timing to environmental quality, that the former are taken as indicators of health or disadvantage (Martorell, 1996). Class or cross-population parity on these indicators are viewed as markers of equity (Brundtland et al., 1980; Lindgren, 1976). Nevertheless, early onset of puberty per se is not known to confer a health advantage; rather, the early health and environmental conditions that promote accelerated development and the degree of physical well-being reflected in pubertal timing also predict future physical health and longevity (Martorell, 1995). In addition, microsimulation analysis has demonstrated that timing of menarche, particularly delayed menarche, can affect lifetime fertility of women by altering the reproductive life span, but this factor is markedly less influential than are age at marriage or duration of lactational amenorrhoea (Campbell & Wood, 1988). Some observers have pointed out that earlier attainment of reproductive capacity at the same time that attainment of adult social status is delayed through expectations of schooling and occupational attainment has opened a window for untimely reproduction, which has resulted in the flurry of concern over schoolage pregnancy (Konner & Shostak, 1986).

Moreover, secular trends are not mediated through natural selection or genetic shifts (Ellison, 1981b), but are acute facultative responses to environmental factors experienced by developing individuals. These acute facultative responses to changing conditions of development are biologically constrained but exhibit rather extensive developmental plasticity. Despite the careful documentation of secular trends and their worldwide distribution, the effects of early maturation and rapid pubertal transition on psychosocial development, on mental health, and possible physical health risks have scarcely been investigated.

Evolutionary Perspectives

The causes and consequences of variable timing of puberty and reproductive maturation can be predicted from life history theory, a powerful component of evolutionary thinking which recognizes that, in fitness terms, differential reproductive success is the result of a much bigger story: what any creature must do is to "get a life."[2] Creatures, especially long-lived ones such as humans, must be designed to perform adequately in a changing world throughout their lives. Life history strategies are the characteristic ways that species allocate time and resources to growth and development, maintenance, and reproduction (Charnov, 1993; Stearns, 1976); these strategies are reflected in age-specific survival and reproduction rates (Promislow & Harvey, 1990). Examination of human life history strategy can therefore reveal ways in which humans are specifically adapted to survive and reproduce. Maintenance comprises costs of staying alive, or what is required to meet basal and behaviorally driven metabolic needs, repair and renew the body, and fend off predators and pathogens (Finch & Rose, 1995). Resources available above and beyond maintenance constitute productivity, and may be allocated either to growth or to reproduction. Productivity among primates is 60% lower than for other mammals (Charnov, 1993, p. 93), which implies that primates have relatively high maintenance costs and limited energy available for growth. Lengthened growth periods and slow growth rates allow primates to reach adult size despite this low productivity, and the strategy is feasible because mortality is sufficiently low to allow reproductive delay.

Humans have a distinctive developmental course characterized by childhood, a long period of moderate growth and nutritional dependency during which extensive learning and socialization take place. This unusual feature is paralleled by a unique pattern of prolonged juvenile dependence on sustained provisioning by adults (Bogin, 1994). The life history strategy of humans reflects evolutionary trends among primates that have strongly influenced life history organization—namely, the trends to intense sociality, cooperative subsistence activity, and cranial expansion with reliance on postnatal learning (Harvey & Clutton-Brock, 1985; Pereira & Fairbanks, 1993). These trends have been paralleled by prolongation of the dependent juvenile period with a concomitant slowing of growth rates, prolongation of growth with (in the sole case of humans) elaboration of a distinctive "childhood" phase, delay of reproductive maturation, marked pubertal acceleration of growth rates (growth spurt), increased adult body size, and increased longevity (Smith, 1992). Such developmental shifts are supported by other trends to intensifying postnatal care by parents, including increased paternal presence and maternal carrying, which contribute to low juvenile mortality and culminate in the uniquely human practice of sustained parental provisioning of young throughout the preadult period (Blurton Jones, 1993).

Timing of puberty provides a useful marker for comparative studies of human life history strategy because puberty is a pivotal point in life history (Bernardo, 1993), and onset of puberty is under direct biologic control. Puberty marks the end of childhood and the threshold of the reproductive career and thus the period during which allocation of productivity switches from growth to reproduction. Early maturation would be predicted under conditions of low mortality (or the environmental concomitants of low mortality—namely, good nutrition, low morbidity, and low stress, Chisholm, 1993), especially when these indicators remain consistently good throughout gestation, infancy, and childhood. Such experiences inform the organism that environmental quality is reliably high and that accelerated development and early reproduction will be sustainable and thus fit a low-risk strategy that increases fitness by reducing intergeneration time and increasing the reproductive life span. The question arises whether this explanation of early maturation adds anything to the more proximal and thus parsimonious one of simple absence of impairment (Bielicki, 1986; Dennett & Connell, 1988). By this interpretation, early maturation results from sufficiency of available resources and optimal conditions for ontogeny, and, obversely, from the absence of biological insults that impair the ability of the organism to achieve its full developmental program or potential.

Several lines of evidence contradict the completeness of the impairment explanation. First, periodic food shortfalls and pathogen pressure prevaded the environment of evolutionary adaptedness for humans (Barnes et al., this volume; Brown & Konner, 1987; Ewald, this volume). Comparative analysis has repeatedly shown mortality rates to be the major determinant of life history parameters, including timing of reproduction maturation (Charnov, 1993, p. 90; Promislow & Harvey, 1990). These points underscore the selective pressures for (and thus the likelihood that) humans evolved the capacity to facultatively adjust age at maturity to resolve the adaptive dilemma of variable environmental quality and to determine the optimal course of reproductive development.

Second, although environmental insult is clearly a component of growth deficits and delays (Martorell & Kettel Kahn, 1994), it does not fully account for developmental variation. For instance, an impairment model fails to explain why Indian and Bangladeshi girls adopted at *later* ages by Swedish parents would show earliest menarche (Proos et al., 1991): girls <3 years of age at adoption averaged 11.9 years at menarche, compared to 11.1 years in those adopted at ≥3 years of age. Pediatricians studying this population note that "the reasons for the earlier pubertal maturation are not clear" (Proos et al., 1991: 852), but the life history model would predict that girls experiencing persistent deprivation would react to a dramatic improvement in environmental quality by hastening reproduction in order to exploit a narrow window of resource opportunity. Finally, an impairment model does not fit the possible observation of slightly earlier menarche in girls growing up under conditions of good resource availability but domestic social stress, but a life history model predicts earlier reproduction consonant with reduced parental investment under these circumstance of poor gender relations and disrupted domestic function (Belsky et al., 1991; Chisholm, 1993; Moffitt et al., 1993).

The general point here is that the impairment and the adaptationist schemas are both likely operative in human developmental variation. The reaction norm, or distribution of phenotypes produced by a genotype across a spectrum of environmental conditions (Stearns, 1992, pp. 41, 143–4), slopes sharply for menarcheal age in humans, ranging from about 12.5 to about 18.5 years for group medians (Eveleth & Tanner, 1990). Humans apparently have evolved capacities that are sensitive to availability and reliability of crucial material and social resources in order to adjust pubertal timing to environmental quality. The biological bases of such capacities and of life history evolution in general are the subject of current intense interest, with applications not only to development but also to aging (Finch & Rose, 1995). Life history theory requires elaboration with respect to explanations of intraspecific rather than cross-species variation (Leroi et al., 1994). Although the comparative method can identify patterns of life history evolution (Promislow & Harvey, 1990; Ridley, 1993), identification of its causes requires within-species investigation (e.g., Falconer, 1990; Gebhardt & Stearns, 1993; Lenski et al., 1991) of proximal mechanisms that drive life history and underlie change (Finch & Rose, 1995). For instance, in humans, intraspecific variation runs counter to comparative life history analyses finding *negative* correlation of juvenile mortality with age at maturity (Charnov, 1993; Promislow & Harvey, 1990) because humans facultatively respond to environmental improvements reflected in mortality declines by decreasing the age at maturity (Ellison, 1981a; Liestøl, 1982). In productivity terms, improved circumstances mediate more favorable productivity returns, which in turn increase resources available for development and thus favor rapid growth and early maturation. Persistently good conditions and high productivity also favor early maturation by indicating that accelerated investment in reproduction is energetically viable.

A key feature of human life history is that development of children is pervasively affected by "the company they keep" (Chisholm, 1993, 1996; Worthman, 1993a). Socially, the context of development is created through conditions of everyday life and the care provided by adults, especially parents. Physiologically, childhood is created through specific patterns of endocrine regulation in which central mechanisms restrain gonadal and some adrenal function. Puberty results from a true developmental neuroendocrine "switch" in the central nervous system that de-represses and turns on gonadal development. The switch is initiated by changes in hypothalamic function that precipitate gonadal maturation and a final burst of growth (Grumbach et al., 1990). Humans also exhibit sexual bimaturism; that is, the sexes differ in timing and markers of reproductive maturation (Wiley, 1974). Biological sex differences in puberty are illustrated in figure 6.1, which shows the timing of pubertal events for American youth, as compiled from several sources. Most of the features of puberty are outlined here and illustrate the phenomenon of sexual bimaturism.

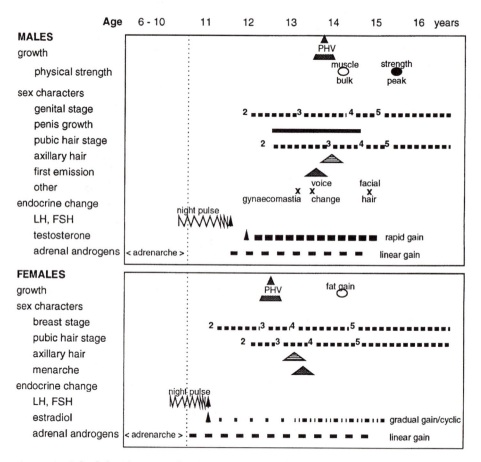

Figure 6.1. Schedule of events of puberty for both sexes, as manifested in a population with early maturation, such as in the United States, with a median age at menarche of 13.2 years. Note that these median ages of occurrence do not represent the large individual variation in timing and duration of any one state, process, or event, and that a given child is unlikely to experience precisely this chronology of pubertal development (based in part on Lee, 1980, and references cited in Worthman, 1993a).

   Onset of puberty is defined by the initial biological events of hormonal (gonadotropin, luteinizing hormone [LH], and follicle-stimulation hormone, [FSH]) and morphologic change (breast and genital growth). Based on these changes, girls in western populations enter puberty on average 6–12 months before boys. But most people, if asked, would say that boys undergo puberty much later than girls, by about 2 years. The reason for this perception lies in sex differences in the outward manifestation of puberty (Simmons & Blyth, 1987, pp. 38–45). Early pubertal changes, gonadal and hormonal, are not detectable by casual inspection. Attainments such as height gain, muscle bulking, facial hair, and breast development are more obvious. Note that males manifest several highly visible cues to maturity, but that these come relatively late in the course of puberty. Girls, on the other hand, show few but striking signs of maturation—breast development and peak height velocity—early in puberty. Consequently, although boys enter physiologic puberty only 6 months later than do girls in western populations, they are generally believed to mature much later than girls because of sex differences in timing of the conspicuous markers of maturation. Actually, boys manifest physiologic reproductive maturity (i.e., first emission) shortly after girls reach menarche. Figure 6.1 suggests the limitations imposed by using menarcheal age as a comparative measure of maturation: little is known about the physiology of normatively late-maturing populations and how it might differ, endocrinologically or metabolically, from that of the intensively studied earlier maturing ones in the United States or Europe, nor is it clear what variation in age at menarche reveals about variation in maturational timing in boys.

   Life history theory predicts that development in boys should be more sensitive to environmental quality than in girls because boys are under less time pressure to initiate a reproductive career (Stearns, 1992, pp. 123–8). That is, the reproductive performance of females is limited by time (from menarche to menopause); developmental delay may reduce that time and thereby reduce their relative reproductive output, or fitness. Males, on the other hand, are thought to experience less fitness cost from delaying development, both because their reproductive career is not truncated by menopause and because they have at least the theoretical prospect of increasing their reproductive output through multiple mates or wives. This logic has also supported the prediction that girls' development should be more canalized, or directed toward attaining maturity despite poor environmental conditions, so that boys are expected to delay development more than girls under unfavorable situations (Stinson, 1985). Life history theory thence predicts greater acceleration of maturation in boys than in girls when environmental circumstances improve.

   A major point as yet omitted here is that, although environmental quality and "the company they keep" determine the course of development and timing of puberty, adaptationist theory predicts that parental caretaking decisions will not always maximize advantage to the individual child (Hewlett, 1991; Trivers, 1974). Indeed, the parent's point of view is crucial in determining caretaking decisions that affect the parent's own resource allocation (Daly &

Wilson, 1995); such decisions may at times run directly contrary to the short-term interests of an individual child (e.g., Boone, 1988). Although parent and child interests are not entirely synonymous, parent–child conflict is muted by the extent to which, especially in humans, parental care makes a difference to child survival and success and, conversely, child demands impair parental ability to continue support into the future (for decades, in humans) (Bateson, 1994). Adaptationist theory pedicts that, given that resources are limited, parental care will be selective and that sex of child is often an important determinant of the kind of environment and care a child encounters (reviewed in Worthman, 1996); thus, analysis of sex differences in variation of pubertal timing can illuminate the importance of environmental quality in general, and parental care in particular, for shaping developmental schedules.

These three key points from evolutionary perspectives on timing of puberty—its acute responsiveness to variation in environmental quality, its evolved sex differences, and the influence of differential care—will be applied below to studies of the impact of different living conditions and rapid social transformations on timing and sex differences in puberty.

Adaptationist Studies of Pubertal Timing

We can test the differential effect of environmental change on development of boys and girls by examining the effects of "modernization" in late-maturing populations. In an earlier section, the case of rapidly decreasing menarcheal ages among Bundi of highland Papua New Guinea was mentioned as an illustration of this effect. Recall that life history theory predicts greater acceleration of maturation in boys than in girls when environmental circumstances improve (Stinson, 1985). Contrary to this prediction, data from a recent growth survey ($n = 471$; Zemel & Jenkins, 1989; Zemel et al., 1993) showed that environmental changes that accelerated development were accompanied among rural Bundi by greater maturational acceleration in girls than in boys: in the 17-year interval since Malcolm's survey (Malcolm, 1970), age of greatest pubertal height gain (peak height velocity [PHV]) declined more rapidly in girls than in boys (figure 6.2). Conversely, although at the second survey, in 1983–84, both sexes showed a secular trend to increased height at age 17 years (i.e., the cohort born at the time of the 1966–67 survey, and thus the oldest cohort closely reflective of interim conditions), girls showed yet greater gains in height at this age than did boys. Why might this be? The answer seems to lie in social changes that differentially affected the environmental quality encountered by either sex. The microenvironments experienced by individual children within a population and even within households can vary considerably, for stochastic as well as systematic reasons, such as differential parental care. Among traditional Bundi, as in many populations, preference is given to sons, and daughters are subject to relative neglect. Girls, then, benefited differentially from social modifications which included workload change with

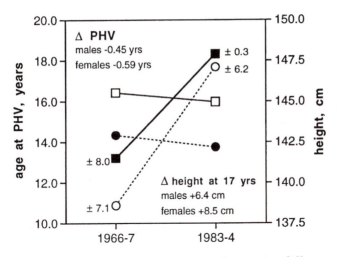

Figure 6.2. Secular trends in a late-maturing population. Sex differences in declines of age at peak height velocity observed in Bundi youth surveyed in 1966–1967 and 1983–1984. Age at peak height velocity (PHV) for males (□) and females (●), calculated by Zemel using the Preece-Baines model 1 equation (Zemel & Jenkins, 1989), is given along with mean height (± SD) by sex among 17 year old males ■ and females ○) measured at each survey. Survey differences in PHV and mean height of 17 year olds over the 17-year interval are noted.

schooling and domestic amenities, protein and fat enrichment of the diet, and acculturative changes in parental attitudes through changes in the perceived value of daughters relative to sons. Maturation rates of girls accelerated more than in boys, as reflected in their greater reduction in age at PHV and greater increase in height at 17 years, as growth of boys was also less complete than girls' at that age.

Similarly, expectations concerning sex bimaturism based on both clinical western data (Joffe, 1990; Wheeler, 1991) and evolutionary theory (Charnov, 1991, pp. 109–112) were violated by findings from the first intensive mixed longitudinal study ($n = 84$) that included endocrine measures on puberty in a late-maturing population. The study group was from a well-characterized rural population of Kikuyu in central highland Kenya which, while having reasonably good health and nutritional status, had late maturation. Median age at menarche in this population was 15.9 years ($n = 415$) (figure 6.3), and the relationship of endocrine to morphologic change in puberty was as seen elsewhere, except that adrenarche was temporally dissociated from puberty (Worthman, 1986, 1987). Menarche was late in part because puberty was prolonged; that is, the time between onset of breast development and first menses was, conservatively, 0.6 years longer than the 2.3 years observed in western girls (Marshall & Tanner, 1969). But the surprising and novel observation concerned the relative timing of puberty in boys and girls (Marshall & Tanner,

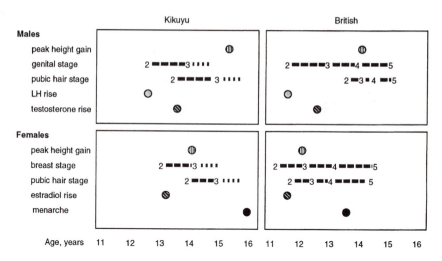

Figure 6.3. Schedule of pubertal events for both sexes in a later-maturing East African (Kikuyu) horticulturalist population (Worthman, 1987), compared with an earlier-maturing British (London) postindustrial one (Marshall & Tanner, 1969, 1970). Median age at onset and duration of stages of morphologic puberty (genital, breast, and pubic hair), from early (2) through adult (5), are plotted along with endocrine markers of onset of puberty (first average detectable daytime leutinizing hormone [LH] and gonadal steroids, testosterone in boys and estradiol in girls) and timing of growth spurt (average age of maximum height gain).

1969, 1970). Depending on the pubertal marker used (endocrine or morphological), boys appeared to enter puberty not after, but slightly before, or at the same time as, girls. Therefore, the timing of puberty can be differentially adjusted by sex within populations and must involve systematic differences in the microenvironments in which males and females grow up.

Evolutionary thinking can be of assistance here in considering why and under what conditions parents may provide variable quality of care to their children. Because virtually all parents have limited resources (time, energy, food, etc.), evolutionary theory predicts that parents will allocate those resources to optimize fitness returns *to the parents*; that is, parents' perception of both the costs to themselves and the survival and reproductive chances of children determines how they distribute care (Blurton Jones, 1993; Daly & Wilson, 1995; Hewlett, 1991). Consequently, parents are expected *not* to act always in the best interests of individual children, but to differentially provide care and resources based on their perception of future viability, costs, and benefits from each specific child. Sex-differentiated treatment of children is a specific category of selective parental care and is predicted to track parental expectations of the relative costs and fitness advantages by sex of child (Worthman, 1996). For instance, if boys appear less onerous to raise because they provide more resource benefits and incur fewer rearing costs to parents and/

or offer greater prospects for future reproductive success than do their sisters who marry, leave home early, and have lower potential reproductive output, then parents are expected to devote more resources to sons than daughters. Note that the resources available to parents, as well as costs and benefits of parenting, are largely determined by social organization and cultural practices.

Infanticide and aggressive neglect are extreme examples of selective parental care. Classic examples occur in India (Dickemann, 1979). Though such practices outraged the sensibilities of colonial personnel and currently unsettle western program officers, researchers, and caregivers whose ideal of parent–child relations emphasizes egalitarian and intensive parental care, they make sense in local and adaptationist terms. Certain castes and ethnic groups traditionally practiced female infanticide because of low marriage prospects for daughters (Dickemann, 1979), and sons were favored because of their high reproductive (Dickemann, 1979) or labor (Mamdani, 1972; Muhuri & Preston, 1991) value. Infanticide has long been illegal in India, but aggressive neglect of daughters has if anything been exacerbated by the recent inflation of dowry that makes the cost of arranging a suitable marriage for daughters nearly unmanageable for most parents. The more affluent the family, the more expensive and unlikely the prospect of an advantageous marriage for daughters. A recent study of child mortality in Bangladesh shows that parents "pursue a child survival strategy aimed at securing some balance between male and female offspring, albeit with a strong bias toward sons" (Muhuri & Preston, 1991: 415). Biased care is linked to resource availability: among rural Bangladeshi where labor value of sons is high, famine has been shown to result in greater erosion of nutritional status of girls than boys (Bairagi, 1986). Moreover, high socioeconomic status families showed greater male–female nutritional discrepancies than low socioeconomic status families, even in times of plenty, and famine had the stongest impact on high socioeconomic status girls. Researchers found these results "surprising and disturbing" (Bairagi, 1986: 314), but they are predictable from the adaptationist view that resource restriction will induce or aggravate selective parental allocation of (increasingly scarce) resources.

Little is known about developmental effects of differential care. I suggest that customary sex-differentiated child care contributed to the unusually late age at menarche in traditional Bundi and to the sex-reversed difference in pubertal timing among Kikuyu. In a collaborative study with Carol Jenkins of Hagahai, a small, group of forager-farmers in Papua New Guinea (population size 250), that was first contacted around 1983, we have begun to examine this issue more closely (Worthman et al., 1993). Hagahai live under strong environmental stress from disease and nutrition (Jenkins et al., 1989) and show pronounced sex differences in well-being (Jenkins, 1987). Demographic data collected since 1985 indicate that the sex ratio of males to females is already high at birth (126:100, compared to the human average of 105:100 [Hewlett, 1991; Sieff, 1990]) and reaches 175:100 by age 15 years. Infanticide and selective neglect contribute to differential female mortality. For instance, infant mortality is high, at 16% for boys and 21% for girls in the first year, but this is opposite from the usual sex difference. Preliminary analyses from this on-going study show little impact of preferential treatment on the timing of pu-berty. Serum LH patterns (figure 6.4) date average onset of puberty at age 12–13 years in girls and about 1 year later in boys. Hagahai are late-maturing, but the sex difference in timing is comparable to populations with minimal prefer-ential treatment, such as in the United States or Europe. Why? Life history theory predicts stronger canalization of development in girls than boys, but here girls are subject to poor treatment. First, pre- and early-contact Hagahai appear to have confined maltreatment of girls to infanticide and early neglect; once they survived, they were treated equivalently to boys. Then, Hagahai children experience powerful selection processes that allow survival of only those females who can successfully endure strong caloric, disease, work, and social stresses, whereas parents attempt to keep sons and give them less work and more freedom to forage for themselves. Girls who reach puberty are therefore a select group: they are survivors of much more stringent mortality than are boys. Here, social practices act to reduce the variation in females and diminish the developmental impact of environment, rather than exacerbate it, because girls die who cannot thrive under the adverse conditions with which they are faced.

Essentially, Hagahai parents allow or undertake a process of postnatal se-lection of children for survivorship ability and socially desirable traits. Such behavior is not uncommon (Daly & Wilson, 1995). Scheper-Hughes (1992) has

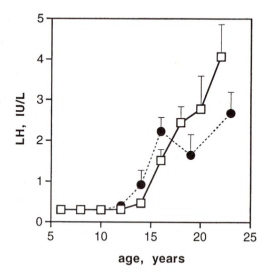

Figure 6.4. Timing and progression of puberty, as manifested by leutinizing hormone (LH) changes, among Hagahai, a later-maturing forager-horticulturalist population of fringe highland Papua New Guinea. Daytime LH (mean ± SE) is not significantly elevated over baseline in girls (●) or boys (□) until around ages 13 and 14.5 years, respectively. Data are cross-sectional; data points are connected to aid readability.

extensively documented selective maternal neglect among the very poor in a slum of northeast Brazil. There, infants who languish under the grim nutrition, sanitation, and housing conditions are allowed to perish because the very symptoms of infant malnutrition, diarrhea, and illness are taken as an unwillingness to live and an inability to cope. Mortality in Scheper-Hughes's sample was 36% by age 1 and 44% by age 5 (Scheper-Hughes, 1992: 307); death is seen as a relief for the weak who cannot deal with the harsh life they will have to face, and maternal care is concentrated on those who apparently can.

Sex-differentiated pubertal timing can also be less obviously the result of selective care than of more subtle general subsistence and child-rearing practices. We have undertaken a collaborative study of development among eastern Hadza (population size 750), a demographically successful group of foragers in northwest Tanzania under long-term study by Nicholas Blurton-Jones and co-workers (Blurton Jones, 1993; Blurton Jones et al., 1992). Hadza are interesting because foragers are thought to be more gender-egalitarian societies (Knauft, 1991; Smuts, 1992); indeed, stated sex preferences and overt practices of sex-differentiated care are absent among Hadza. Parental investment in child support is relatively low, for children of both sexes are expected to forage for themselves and obtain around half their diet by age 6 (Blurton Jones et al., 1989). Sex ratio is female biased, around 94:100 for both children under 5 and everyone under the age of 45 years (Blurton Jones et al., 1992), compared to juvenile sex ratios averaging 112:100 among foragers and 107:100 among horticulturalists and pastoralists (Hewlett, 1991). Early mortality, at about 28% to age 5 years, is comparable to that observed in other foragers under similar conditions (Hewlett, 1991).

Figure 6.5 shows the LH profile for the Hadza population ($n = 155$). Absolute ages are not available, so age must be reconstructed (see method in Blurton Jones et al., 1992). In the left panel, values are plotted by rated appearance of maturity. Blood LH concentrations are about what we would expect from endocrine profiles in American adolescents (McDade et al., 1995) and are statistically equivalent by sex at all maturity stages, except for late childhood. But plotting by reconstructed age (right panel) reveals that Hadza are late maturing, and pubertal onset in males is dramatically delayed relative to females, by more than 2 years. We have considered various factors that may explain these findings, including relative foraging ineffectiveness or inactivity in boys, covert differential provisioning of girls, sex-differentiated morbidity, and sampling bias. The last possibility is weakened by this being a fairly complete sample (76%) of the age-appropriate population, including those in school and those not in school. A definitive answer will require close examination of the behavioral ecological data collected by the site team, for other studies have concluded that one cannot rely on parental disclaimers of preferential treatment (Cronk, 1991). We can, however, say that this unusually large sex difference in timing of puberty likely represents yet another refutation of universality in biological function and developmental schedule in humans.

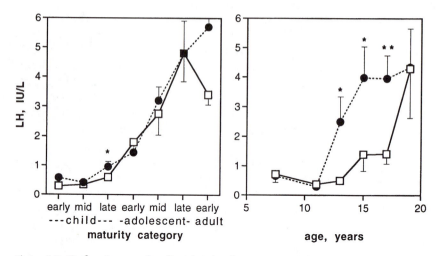

Figure 6.5. Endocrine marker (leutinizing hormones; LH) of pubertal onset and progression among 155 boys (□) and girls (●) of eastern Hadza, a Khoisan-speaking forager population of northwestern Tanzania. The left panel shows LH (mean ± SE) by maturation status coded from visual inspection; the right panel displays LH by extrapolated age. Significance of sex differences in LH denoted by $*p < .05$, $**p < .01$. Data are cross-sectional; data points are connected to aid readability.

A final set of findings concerning pubertal timing relates to an emerging insight from reproductive ecology—namely, that human populations may differ in the endocrine set-points for reproductive function (Ellison, 1994; Ellison et al., 1993). These differences likely have developmental bases, and, indeed, findings from our comparative studies of puberty suggest such bases. Late-maturing populations appear to manifest flatter trajectories of gonadotropin and gonadal hormone output over the course of puberty that result in lower levels of these hormones in adults than are reported in clinical studies of western populations (Worthman, 1998). Figure 6.4 shows such a pattern among Hagahai, among whom conditions are rather harsh and development is late. Characterization of reproductive hormone output is most readily achieved among males because in women, once menstrual cyclicity and then the reproductive cycle of pregnancy and lactation have commenced, endocrine status varies widely by reproductive status. Young adult Hagahai men ages 21–25 years show average serum concentrations of LH at 4.07 ± 3.50 IU/L and of testosterone at 307 ± 159 ng/dL (n= 19, mean ± SD), whereas serum measures (also performed in our laboratory) of American men the same ages show similar values of LH (average around 3.3 IU/L, range 1–8.3 IU/L) but double the testosterone (average around 750, range 550–2100 ng/dL). A final contrast is provided by our endocrine data from Amele, a fairly healthy, well-nourished population of subcoastal Papua New Guinea (Jenkins et al., 1984), among whom serum LH is 7.99 ± 2.98 U/L and testosterone is 667 ± 169 ng/ dL in men of the same ages ($n = 36$).

That both the regulatory hormone, LH, and the target tissue hormone, testosterone, vary in parallel across the New Guinea populations suggests that the differences are centrally mediated and that hypothalamo-gonadal regulatory set-points have been shifted, rather than that a peripheral effect such as gonadal insult or refractoriness is primarily responsible for this variation.[3] Such population variation fits within existing normal ranges for individual variation in western clinical populations and remains compatible with our existing models of reproductive physiology. That populations differ in hormone concentrations and endocrine regulation due to ontogenetic differences and that this also occurs on axes other than the reproductive (as suggested by birthweight effects on metabolic regulation [Law et al., 1995; Phillips et al., 1993; Yajnik et al., 1994], cardiovascular function [Barker et al., 1993a, b], and altered immune function [Godfrey et al., 1994]) are observations that require further investigation and replication. Should the observations prove robust, the notion that consequential functional differences emerge from ecologically sensitive divergences in developmental trajectories carries wide-reaching implications for our understanding of the significance of the secular trends and population variation in timing of puberty documented at the beginning of this chapter.

Clinical Implications

This chapter has examined population variation and sex differences in timing of puberty as a model for developmental ecology, or human development from an evolutionary–ecological perspective. The empirical points raised were the following (1) normative ages for developmental events vary widely among populations; (2) sex differences in onset of puberty also vary across populations; (3) inter- and intrapopulation variation is produced largely by specific social and ecological conditions; (4) parents and cultural practices may not act for the equal benefit of all children, for specific, predictable reasons based on resource availability and social–political structures; (5) social change, by definition, alters practices and resources and has differential impact between and within populations; and (6) differences in conditions of ontogeny in general, and variation in pubertal timing in particular, are paralleled by shifts in endocrine output and regulation on the reproductive and perhaps other axes.

What are the implications for health and development relevant to clinical practice? First, the fact that early maturation among entire populations is unusual in human history may lead us to consider the possible negative physical health consequences of early maturation. Changes in age at maturation represent shifts in life history allocation of resources among growth and development, reproduction, and maintenance, and we do not yet know whether early reproductive maturation is associated with other trade-offs in maintenance or development. As reviewed above, early maturation appears to be a facultative response to persistently high environmental quality that also predicts positive health and other functional outcomes. Insofar as it reflects well-being, early maturation can scarcely be viewed as a negative outcome. Yet early-maturing populations may also have upregulated output of gonadal steroids (testosterone, estradiol, progesterone) that, along with earlier maturation itself, increase the lifetime exposure to these potent hormones and raise the risk for reproductive cancers (Apter et al., 1989; Eaton et al., this volume). Psychobehavioral effects of greater gonadal steroid hormone output and, in the case of women, increased magnitude of cyclic variation and of menopausal decline in these hormones have not been investigated.

Second, although worldwide variation and population secular trends in physical development are well documented, the concomitant changes in psychosocial development and mental health remain unknown and unstudied. Variation in timing and course of puberty is as great within as between populations (Marshall, 1969; Marshall & Tanner, 1969, 1970): studies of American or western European adolescents have persistently though weakly linked timing of puberty, relative to peers, to psychobehavioral problems and long-term adjustment (Angold & Worthman, 1993; Buchanan et al., 1992). Such findings concern correlates of maturational timing within populations that are early maturing. Relevance to effects on psychosocial development of population-wide modal shifts in maturation timing is not established; minimally, comparative data on psychosocial development are needed from populations with quite different maturational schedules. Preferably, longitudinal data should be gathered on populations undergoing the secular trend to accelerated maturation and hence experiencing modal shifts toward earlier and more rapid pubertal transitions. As pediatricians and parents in the United States and many other parts of the world already know, the average 12 year old today is biologically not in the same developmental state as one of 50 or 100 years ago. The modal, or "normal," chronology as well as the intensity, or rate, of pubertal transitions has changed. Yet clinicians and the public alike are continually amazed at the psychological and behavioral precocities of contemporary youth and remain conditioned by schedules of childhood and youth that are dissonant with current patterns of maturation. This is an area of enormous concern for physicians, educators, and parents and really deserves attention.

Third, and related to these concerns, are issues of sex difference that receive little attention in the pediatric literature. In prior sections of this chapter, sex differences in course and content of puberty plus extensive sex differences in treatment of children, and thus in microenvironments of development, were related to variable degrees of sexual bimaturism across populations. These observations raise the question of sex differences in developmental vulnerabilities, requirements for optimal development, and health outcomes of specific conditions or changes in conditions of development. The recent initiatives on women's health reflect a belated recognition of sex differences in these areas, but work has concentrated on the last half of the life cycle and thus eschews the opportunity to consider developmental bases of sex-differentiated health risks in either sex. The ways in which gender organizes child care and socialization in relation to pediatric risk, health evaluation, and treatment has not been probed, yet an evolutionary perspective suggests that sex of the individual will powerfully influence social organization of context and the individual response to it.

Fourth, the phenomenon of developmental plasticity challenges the conceptualization of normative development in pediatrics (Zelazo & Barr, 1989). These issues are not particularly new for the international public health community concerned with establishing norms for growth (Freeman et al., 1995; Seth et al., 1990; Sullivan et al., 1991; Ulijaszek, 1994), but they remain under discussion and have yet to enter clinical pediatric discourse within countries such as the United States (Denniston, 1994; Hall, 1995; but see Cole, 1995). Briefly, that developmental variation reflects systematic interactions of genetic and localized human ecological variables, as attested in the pubertal variation documented here, confounds both the conceptual and quantitative definition of "ideal" function and formulation of developmental goals toward which pediatric practice and public health intervention should strive (e.g., Pelto & Pelto, 1989). What is "normal" or modal for one population may be optimal for that situation but yet fall on an extreme of the range of normal variation for another population. To what extent must local genetics and ecology (e.g., high altitude or latitude) be taken into account in the formulation of clinical norms? The challenge is to identify the range of population developmental variation outside of which a problem is indicated and remediation is necessary. The issue of whether international or local growth norms are most appropriate remains the subject of lively debate, but the question has not been systematically extended to pediatric issues outside of growth.

A substantial developmental literature on the relationship between ontogeny and context and on adaptation and plasticity from the molecular to the psychobehavioral level (Cairns et al., 1990; Changeux, 1985; Gottlieb, 1991; Oyama, 1985, 1988; Plomin & Daniels, 1987; Scarr & McCartney, 1983), has shown that, while the locus of ontogeny is the individual, ontogeny is a process coextensive with context. More simply, organisms are *designed* to rely on interactions with the environment to guide physical development and ongoing function (Worthman, 1993a). Implications for medicine are direct: maturational norms and pathogenesis are inseparable from context, and by extension, functional states and illness are not simply located in the person but have etiologies based in context. Patterns of human development and health are outcomes of the interaction between our evolved biological selves and the company we keep, or the specific social-ecological contexts in which we grow up and live. Demonstrations that infant and child anthropometric status strongly contribute to early mortality (Schroeder & Brown, 1994), that adult growth attainments are exquisitely sensitive to wealth inequity (Fogel, 1986), and that adult psychosocial competence is markedly diminished by developmental insults reflected by delayed maturation (Martorell, 1996; Martorell et al., 1996) all document the tight linkage of physical development with resource access and resource allocation. Evolutionary medicine considers that this linkage is mediated not only through direct impairments, but also through evolved responses of developmental reorganization and life-history redirection to cope with high-risk, high-mortalty circumstances. Accordingly, physicians may need to concern themselves more with writing "social prescriptions" for the structural sociopolitical and economic conditions fostering health, thus extending the logic of public health to prophylaxis and treatment of developmental insults. Family medicine and family therapy address contextual bases for etiology of dysfunction and consequently of treatment, but structural "social prescriptions" for healthy development need to go beyond the household level.

Fifth, the evolutionary adaptationist view emphasizes access to valuable resources as key to understanding development, function, and behavior; it predicts that distribution of resources, not just pathogens, has important effects on development and health (Chisholm, 1993). Further, evolutionary biologists increasingly recognize that, in a complex and changing world, organisms cannot develop and operate in a way that maximizes everything (Finch & Rose, 1995). Rather, they optimize and adapt through trade-offs and compromises in use of restricted time and resources: borrowing a phrase from developmental psychology (Scarr, 1992, but see Baumrind, 1993), we could call this "good enough" functioning. Developmental delay and pathogenesis can each be viewed either as a failure to adapt or cope with existing circumstances or as the outcome of trade-offs (or side-effects) in the process of coping with environmental challenges (Wolff, 1989). In a world of limited resources and enormous health needs, medicine, too, may have to develop an articulated view of "good enough" health and development. Such considerations sit uneasily with the deeply established and clearly desirable view that best is better, and that anything less indicates a problem to be remediated. Adaptationist thinking can complement the more usual epidemiologic (Schroeder & Brown, 1994), ethical, and economic (Fogel, 1994) perspectives to help reengage these thorny issues. The close associations among resource allocation, pubertal timing, adult competence, and human health make equity in access to resouces for child health and development a priority for preventive medicine.

Finally, because of our evolved pattern of slow development, children are dependent on parents for extended care. However, evolutionary theory predicts that parents, operating under resource constraints and with a view to the perceived costs to themselves and to the futures open to individual children, *do not and cannot* be expected to operate in a way that is best for each individual child from our western individual-centered notion of egalitarianism. We cannot assume a universality of parental goals, strategies, and motivations any more than we can assume a universality of biological function. This insight would allow us to take a more informed approach to famine relief, development, and public health efforts, which have often foundered on a failure to recognize local practices, perceptions, and priorities. Values of, perceptions by, and constraints on parents must form an important part of clinical and public health interventions designed to improve child well-being.

In sum, an evolutionary approach to timing of puberty has shown that contemporary early-maturing populations represent an outlier in the evolutionary history of human development as outcomes of unusually favorable populationwide and consistently sustained epidemiologic and nutritional circumstances. Biologically, these fortunate circumstances signal that a reallocation of developmental priorities is in order, which results in early maturation. Because the bulk of biomedical research occurs in these unusually favored, early-maturing populations, they inform our norms of development and adult function. Thus, for instance, we have been slow to recognize that, compared to later-maturing groups, these populations have prolonged exposure and increased output of gonadal steroids mediated by early maturation and that this contributes to their increasing rates of reproductive cancers. Clinicians may do well to recall that these privileged populations are developmental outliers that push the envelope of maturation rate for our species (Liestøl & Rosenberg, 1995) and be alert to identifying possible short- and long-term sequellae and costs of widespread early maturation. For instance, our notions of the epidemiology and processes of aging may be conditioned by their foundation in studies of early-maturing people (Finch & Rose, 1995).

Conclusion

This analysis overturns two cherished myths about evolution and the human condition. For one, evolutionary thinking is emphatically not about social Darwinism—that things are as they must be. It points unswervingly to the criticality of resource access for all dimensions of biological and social-behavioral functioning and highlights the role of social practices (including power, values, customs) that determine it (Chisholm, 1993; Worthman, 1993b). By emphasizing the centrality of resource distribution to understanding human development, health, and behavior, evolutionary analysis focuses our attention on the necessity for social change to effect change in development and health. Instead of confirming the social status quo, an evolutionary perspective provides a deeper reason for critically analyzing resource distribution or maldistribution and its consequences. Likewise, evolutionary medicine can provide critical evaluation of the conditions that generate health risk and, rather than to point to some return to a mythical adaptive golden age in our species behavioral-ecological history, can suggest contemporary and novel biosocial solutions to the challenges that our present circumstances raise for human adaptability and health.

Second, an evolutionary perspective is popularly associated with species-universal, shared features, and some fear that it promotes a deterministic, invariant view of human nature. This analysis has shown that, to the contrary, an adaptationist, evolutionary view also leads us to an understanding of the causes of our differences and reveals that those causes lie in the dynamic between individual and context that cross-cuts biological and social-ecological domains. Similarly, as medicine necessarily engages the enormous range of human diversity, it also must move beyond narrowly normative diagnostics and conceptualizations of the biologically ideal to consider the biosocial causes and consequences of biological variation in order to enhance the prevention and treatment of human suffering. The logic of evolutionary medicine is deftly caught in an early, prescient essay by Jerome Bruner (1972: 706):

> remember that the human race has a biological past from which we can read lessons for the culture of the present. We cannot adapt to everything, and in designing a way to the future we would do well to examine again what we are and what our limits are. Such a course does not mean opposition to change but, rather, using man's natural modes of adapting to render change both as intelligent and as stable as possible.

Notes

1. Genetic factors do play a significant role in within-group variation in menarcheal age. Several studies of Americans report mother–daughter correlations in age at menarche which average around 0.30 (Graber et al., 1995).

2. As the developmental biologist John Bonner put it, "what reproduces is the life cycle" (Bonner 1974: 14).

3. That the LH of young males in both New Guinea populations averages higher than the American, while testosterone is lower, suggests that some end-organ unresponsiveness could also be involved. Environmental (e.g., high altitude hypoxia) or morbid (e.g., filariasis) conditions may acutely impair testicular output, in addition to the developmental reprogramming of regulatory set-points of the hypothalamo-pituitary-gonadal axis, as suggested here.

References

Angold, A., and Worthman, C.M. (1993) Puberty onset of gender differences in rates of depression: a developmental, epidemiologic, and neuroendicrone perspective. *Journal of Affective Disorders* 29: 145–158.

Apter, D., Reinilla, M., and Vihko, R. (1989) Some endocrine characteristics of early menarche, a risk factor for breast cancer, are preserved into adulthood. *Journal of Clinical Endocrinology and Metabolism* 57: 82–86.

Bairagi, R. (1986) Food crisis, nutrition, and female children in rural Bangladesh. *Population and Development Review* 12: 307–315.

Barker, D. J. P., Martyn, C.N., Osmond, C., Hales, C.N., and Fall, C.H.D. (1993a) Growth in utero and serum cholesterol concentrations in adult life. *British Medical Journal* 307: 1524–1527.

Barker, D. J. P., Osmond, C., Simmonds, S.J., and Wield, G.A. (1993b) The relation of small head circumference and thinness at birth to death from cardiovascular disease in adult life. *British Medical Journal* 306: 422–426.

Bateson, P. (1994) The dynamics of parent-offspring relationships in mammals. *Trends in Ecology and Evolution* 9: 399–403.

Baumrind, D. (1993) The average expectable environment is not good enough: a response to Scarr. *Child Development* 64: 1299–1317.

Belsky, J., Steinberg, L., and Draper, P. (1991) Child experience, interpersonal development, and reproductive strategy: an evolutionary theory of socialization. *Child Development* 62: 647–70.

Bernardo, J. (1993) Determinants of maturation in animals. *Trends in Ecology and Evolution* 8: 166–173.

Bielicki, T. (1986) Physical growth as a measure of the economic well-being of populations. In F. Falkner & J. M. Tanner, eds., *Human Growth*, 2nd ed., pp. 283–305. New York: Plenum Press.

Blurton Jones, N. (1993) The lives of hunter-gatherer children: effects of parental behavior and parental reproductive strategy. In M. E. Pereira and L. A. Fairbanks, eds., *Juvenile Primates*, pp. 309–326. New York: Oxford University Press.

Blurton Jones, N., Hawkes, K., and O'Connell, J. (1989) Studying the costs of children in two foraging societies: implications for schedules of reproduction. In V. Standen and R. Foley, eds., *Comparative Socioecology*, pp. 367–390. Oxford: Blackwell.

Blurton Jones, N., Smith, L.C., O'Connell, J., Hawkes, K., and Kamuzora, C.I. (1992) The demography of the Hadza, an increasing and high density population of savannah foragers. *American Journal of Physical Anthropology* 89: 159–181.

Bogin, B. (1988) *Patterns of Human Growth*. Cambridge: Cambridge University Press.

Bogin, B. (1994) Adolescence in evolutionary perspective. *Acta Paediatrica Supplement* 406: 29–35.

Bonner, J. T. (1974) *On Development*. Cambridge: Harvard University Press.

Boone, J. L. (1988) Parental investment, social subordination and population processes among th 15th and 16th century Portugese nobility. In L. Betzig, M. Borgerhoff Mulder, and P. Turke, eds., *Human Reproductive Behavior: A Darwinian Perspective*, pp. 201–219. Cambridge: Cambridge University Press.

Brown, P. J., and Konner, M. (1987) An anthropological perspective on obesity. *Annals of the New York Academy of Sciences* 499: 29–46.

Brundtland, G. H., Liestøl, K., and Walløe, L. (1980) Height, weight, and menarcheal age of Oslo schoolchildren during the last 60 years. *Annals of Human Biology* 7: 307–322.

Bruner, J. (1972) Nature and uses of immaturity. *American Psychologist* 27: 687–708.

Buchanan, C. M., Eccles, J. S., and Becker, J. B. (1992) Are adolescents the victims of raging hormones: evidence for activational effects of hormones on moods and behavior at adolescence. *Psychological Bulletin* 111(1): 62–107.

Bullough, V. L. (1981) Age at menarche: a misunderstanding. *Science* 213: 365–366.

Cairns, R. B., Gariépy, J. -L., and Hood, K. E. (1990) Development, microevolution, and social behavior. *Psychological Review* 97: 49–65.

Campbell, K., and Wood, J.W. (1988) Fertility in traditional societies. In P. Diggory, M. Potts, and S. Teper, eds., *Natural Human Fertility: Social and Biological Determinants*, pp. 39–69. London: Macmillan.

Changeux, J. -P. (1985) *Neuronal Man*. L. Garey, trans. New York: Pantheon.

Charnov, E. L. (1991) Evolution of life history variation among female mammals. *Proceedings of the National Academy of Sciences, USA* 88: 1134–1137.

Charnov, E. L. (1993) *Life History Invariants*. Oxford: Oxford University Press.

Chisholm, J. S. (1993) Death, hope, and sex: life history theory and the development of alternative reproductive strategies. *Current Anthropology* 34: 1–24.

Chisholm, J. S. (1996) The evolutionary ecology of attachment organization. *Human Nature* 7: 1–38.

Cole, T. J. (1995) Conditional reference charts to assess weight gain in British infants. *Archives of Disease in Childhood* 73: 8–16.

Cronk, L. (1991) Preferential parental investment in daughters over sons. *Human Nature* 2: 387–417.

Daly, M., and Wilson, M. (1995) Discriminative parental solicitude and the relevance of evolutionary models to the analysis of motivational systems. In M. S. Gazzaniga, eds., *The Cognitive Neurosciences,* pp. 1269–1286. Cambridge, MA: MIT Press.

Dennett, G., and Connell, J. (1988) Acculturation and health in the highlands of Papua New Guinea. *Current Anthropology* 29: 273–299.

Denniston, C. R. (1994) Assessing normal and abnormal patterns of growth. *Primary Care* 21: 637–654.

Dickemann, M. (1979) Female infanticide, reproductive strategies, and social stratification: a preliminary model. In N. Chagnon and W. Irons, eds., *Evolutionary Biology and Human Social Behavior,* pp. 321–367. North Scituate, MA: Duxsbury Press.

Ellison, P. T. (1981a) Morbidity, mortality, and menarche. *Human Biology* 53: 635–643.

Ellison, P. T. (1981b) Threshold hypotheses, development age, and menstrual function. *American Journal of Physical Anthropology* 54: 337–40.

Ellison, P. T. (1994) Advances in human reproductive ecology. *Annual Review of Anthropology* 23: 255–275.

Ellison, P. T., Panter-Brick, C., Lipson, S. F., and O'Rourke, M. T. (1993) The ecological context of human ovarian function. *Human Reproduction* 8: 2248–2258.

Eveleth, P. B. (1986) Timing of menarche: secular trend and population differences. In J. B. Lancaster and B. A. Hamburg, eds., *School-Age Pregnancy and Parenthood: Biosocial Dimensions,* pp. 39–52. New York: Aldine de Gruyter.

Eveleth, P. B., and Tanner, J. M. (1990) *Worldwide Variation in Human Growth,* 2nd ed. Cambridge: Cambridge University Press.

Falconer, D. S. (1990) Selection in different environments: effects on environmental sensitivity (reaction norm) and on mean performance. *Genetic Research* 56: 57–70.

Finch, C. E., and Rose, M. R. (1995) Hormones and physiological architecture of life history evolution. *Quarterly Review of Biology* 70: 1–52.

Fogel, R. W. (1986) Physical growth as a measure of the economic well-being of populations: the eighteenth and nineteenth centuries. In F. Falkner and J. M. Tanner, eds., *Human Growth,* pp. 263–281. New York: Plenum Press.

Fogel, R. W. (1994) Economic growth, population theory and physiology: the bearing of long-term processes on the making of economic policy. Working Paper no. 4638. Cambridge, MA: National Bureau of Economic Research.

Freeman, J. V., Cole, T. J., Chinn, S., Jones, P. R., White, E. M., and Preece, M. A. (1995) Cross sectional stature and weight reference curves for the UK, 1990. *Archives of Disease in Childhood* 73: 17–24.

Gebhardt, M. D., and Stearns, S. C. (1993) Phenotypic plasticity for life history traits in *Drosophila melanogaster.* I. Effect on phenotypic and environmental correlations. *Journal of Evolutionary Biology* 6: 1–16.

Godfrey, K. M., Barker, D.J.P., and Osmond, C. (1994) Disproportionate fetal growth and raised IgE concentration in adult life. *Clinical and Experimental Allergy* 24: 641–648.

Gottlieb, G. (1991) Experiential canalization of behavioral development: theory. *Developmental Psychology* 27: 4–13.

Graber, J. A., Brooks-Gunn, J., and Warren, M.P. (1995) The antecedents of menarcheal age: heredity, family environment, and stressful life events. *Child Development* 66: 346–359.

Grumbach, M. M., Sizonenko, P. C., and Aubert, M. L., eds. (1990). *Control of the Onset of Puberty*. Baltimore, MD: Williams and Wilkins.

Hall, D. (1995) Monitoring children's growth. *British Medical Journal* 311: 583–584.

Harvey, P. H., and Clutton-Brock, T. H. (1985) Life history variation in primates. *Evolution* 39: 559–581.

Hewlett, B. S. (1991) Demography and childcare in preindustrial societies. *Journal of Anthropological Research* 47: 1–37.

Heywood, P. F., and Norgan, N. G. (1992) Human growth in Papua New Guinea. In R. D. Attenborough and M. P. Alpers, eds., *Human Biology in Papua New Guinea: The Small Cosmos*, pp. 234–248. Oxford: Clarendon Press.

Jenkins, C. (1987) Medical anthropology in the Western Schrader Range, Papua New Guinea. *National Geographic Research* 3: 412–430.

Jenkins, C., Dimitrakakis, M., Cook, E., Sanders, R., and Stallman, N. (1989) Culture change and epidemiological patterns among the Hagahai, Papua New Guinea. *Human Ecology* 17: 27–57.

Jenkins, C., Orr-Ewing, A., and Heywood, P. (1984) Cultural aspects of early childhood growth and nutrition among the Amele of lowland Papua New Guinea. *Ecology of Food and Nutrition* 14: 261–275.

Joffe, A. (1990) Adolescent medicine. In F. A. Oshi, C. D. DeAngelis, R. D. Feigan, and J. B. Warshaw, eds., *Principles and Practice of Pediatrics*, pp. 708–719. New York: Lippincott.

Kaplowitz, H., Martorell, R., and Engle, P. L. (1993) Rural-urban migrants in Guatemala. In L. M. Schell, M. T. Smith, and A. Bilsborough, eds., *Urban Ecology and Health in the Third World*, pp. 144–162. Cambridge: Cambridge University Press.

Khan, A. D., Schroeder, D. G., Martorell, R., and Rivera, J. A. (1995) Age at menarche and nutritional supplementation. *Journal of Nutrition* 125: S190–S196.

Knauft, B. M. (1991) Violence and sociality in human evolution. *Current Anthropology* 32: 391–428.

Konner, M. J., and Shostak, M. (1986) Adolescent pregnancy and childbearing: an anthropological perspective. In J. B. Lancaster and B. Hamburg, eds., *School-Age Pregnancy and Parenthood*, pp. 325–345. New York: Aldine.

Law, C. M., Gordon, G. S., Shiell, A. W., Barker, D. J. P., and Hales, C. N. (1995) Thinness at birth and glucose tolerance in seven-year-old children. *Diabetic Medicine* 12: 24–29.

Lee, P. A. (1980) Normal ages of pubertal events among American males and females. *Journal of Adolescent Health Care* 1: 26–29.

Lenski, R. E., Rose, M. R., Simpson, S. E., and Tadler, S. D. (1991) Long-term experimental evoluation in *Escherichia coli* I. Adaptation and divergence during 2000 generations. *American Naturalist* 138: 1315–1341.

Leroi, A. M., Rose, M. R., and Lauder, G. E. (1994) What does the comparative method reveal about adaptation? *American Naturalist* 143: 381–402.

Liestøl, K. (1982) Social conditions and menarcheal age: the importance of early years of life. *Annals of Human Biology* 9: 521–537.

Liestøl, K., and Rosenberg, M. (1995) Height, weight and menarcheal age of schoolgirls in Oslo—an update. *Annals of Human Biology* 22: 199–205.

Lindgren, G. (1976) Height, weight, and menarche in Swedish urban school children in relation to socioeconomic and regional factors. *Annals of Human Biology* 3: 501–528.

Maclure, M., Travis, L. B., Willett, W., and MacMahon, B. (1991) A prospective cohort study of nutrient intake and age at menarche. *American Journal of Clinical Nutrition* 54: 649–656.

Malcolm, L. A. (1970) Growth and development of the Bundi child of the New Guinea highlands. *Human Biology* 42: 293–328.

Mamdani, M. (1972) *The Myth of Population Control.* New York: Monthly Review.

Marshall, W. A. (1969) Individual variations in rate of skeletal maturation. *Nature* 221: 175–191.

Marshall, W. A., and Tanner. J. M. (1969) Variation in the pattern of pubertal changes in girls. *Archives of Disease in Childhood* 44: 291–303.

Marshall, W. A., and Tanner, J. M. (1970) Variations in the pattern of pubertal changes in boys. *Archives of Disease in Childhood* 45: 13–23.

Martorell, R. (1995) Promoting healthy growth: rationale and benefits. In P. Pinstrup-Anderson, D. Pelletier, and H. Alderman, eds., *Child Growth and Nutrition in Developing Countries. Priorities for Action*, pp. 15–31. Ithaca: Cornell University Press.

Martorell, R. (1996) The role of nutrition in economic development. *Nutrition Reviews* 54(4): S66–S71.

Martorell, R., and Kettel Kahn, D. G. (1994) Reversibility of stunting: epidemiological findings in children from developing countries. *European Journal of Clinical Nutrition* 48: S45–S57.

Martorell, R., Ramakrishnan, U., Schroeder, D. G., and Ruel, M. (1996) Reproductive performance and nutrition during childhood. *Nutrition Reviews* 54: S15–S21.

Mascie-Taylor, C. G., and Lasker, G. W., eds. (1988). *Biological Aspects of Human Migration.* Cambridge: Cambridge University Press.

McDade, T., Angold, A., Costello, E. J., Stallings, J. F., and Worthman, C. M. (1995) Physiologic bases of individual variation in pubertal timing and progression: report from the Great Smoky Mountains Study. *American Journal of Physical Anthropology* (suppl.) 20: 148.

Moffitt, T. E., Caspi, A., Belsky, J., and Silva, P.A. (1993) Childhood experience and the onset of menarche: a test of a sociobiological model. *Child Development* 63: 47–58.

Moisan, J., Meyer, F., and Gingras, S. (1990) A nested case-control study of the correlates of early menarche. *American Journal of Epidemiology* 132: 953–961.

Muhuri, P. K., and Preston, S. H. (1991) Effects of family composition on mortality differentials by sex among children in Matlab, Bangladesh. *Population and Development Review* 17: 415–434.

Oyama, S. (1985) *The Ontogeny of Information: Developmental Systems and Evolution.* Cambridge: Cambridge University Press.

Oyama, S. (1988) Stasis, development and heredity. In *Evolutionary Processes and Metaphors*, pp. 255–274. New York: John Wiley.

Padro, C. (1984) Secular change in menarche in women in Madrid. *Annals of Human Biology* 11: 165–166.

Pelto, G. H., and Pelto, P. J. (1989) Small but healthy? An anthropological perspective. *Human Organization* 48: 11–15.

Pereira, M. E., and Fairbanks, L. A., eds. (1993). *Juvenile Primates: Life History, Development, and Behavior.* New York: Oxford University Press.

Phillips, D., Barker, D., and Osmond, C. (1993) Infant feeding, fetal growth and adult thyroid function. *Acta Endocrinologica* 129: 134–138.

Plomin, R., and Daniels, D. (1987) Why are children in the same family so different from one another? *Behavioral and Brain Sciences* 10: 1–60.

Promislow, D. E., and Harvey, P. H. (1990) Living fast and dying young: a comparative analysis of life history variation among mammals. *Journal of Zoology* 220: 417–437.

Proos, L. A., Hofvander, Y., and Tuvemo, T. (1991) Menarcheal age and growth pattern of Indian girls adopted in Sweden. *Acta Paediatrica Scandinavia* 80: 852–858.

Ridley, M. (1993) *Evolution.* Boston: Blackwell Scientific.

Scarr, S. (1992) Developmental theories for the 1990s: development and individual differences. *Child Development* 63: 1–19.

Scarr, S., and McCartney, K. (1983) How people make their own environments: a theory of genotype/environment effects. *Child Development* 54: 424–435.

Scheper-Hughes, N. (1992) *Death without Weeping.* Berkeley: University of California Press.

Schroeder, D. G., and Brown, K. H. (1994) Nutritional status as a predictor of child survival: summarizing the association and quantifying its global impact. *Bulletin of the World Health Organization* 72: 569–579.

Seth, V., Rai, A., Gupta, M., Semwal, O. P., Patnaik, K. K., and Sundaram, K. R. (1990) Construction of growth reference standards for urban slum children in developing countries. *Indian Pediatrics* 27: 1081–1087.

Sieff, D. (1990) Explaining biased sex ratios in human populations: a critique of recent studies. *Current Anthropology* 31: 25–47.

Simmons, R. G., and Blyth, D. A. (1987) *Moving into Adolescence: The Impact of Pubertal Change and School Context.* New York: Aldine de Gruyter.

Singh, S. P., and Malhotra, P. (1988) Secular shift in menarcheal age of Patiala (India) schoolgirls between 1974 and 1986. *Annals of Human Biology* 15: 77–80.

Smith, B. H. (1992) Life history and the evolution of human maturation. *Evolutionary Anthropology* 1: 134–142.

Smuts, B. (1992) Male aggression against women: an evolutionary perspective. *Human Nature* 3: 1–44.

Stearns, S. (1976) Life history tactics: a review of the ideas. *Quarterly Review of Biology* 51: 3–47.

Stearns, S. C. (1992) *The Evolution of Life-Histories.* New York: Oxford University Press.

Stinson, S. (1985) Sex differences in environmental sensitivity during growth and development. *Yearbook of Physical Anthropology* 28: 123–147.

Sullivan, K., Trowbridge, F., Gorstein, J., and Pradilla, A. (1991) Growth references. *Lancet* 337: 1420–1421.

Tan-Boom, A., Othman, R., Butz, W. P., and DaVanzo, J. (1983) Age at menarche in Penninsular Malaysia: time trend, ethnic differentials, and association with ages at marriages and at first birth. *Malaysian Journal of Reproductive Health* 1: 91–108.

Tanner, J. M. (1962) *Growth at Adolescence,* 2nd ed. Oxford: Blackwell Scientific.

Tanner, J. M. (1978) *Fetus into Man: Physical Growth from Conception to*

Maturity. Cambridge, MA: Harvard University Press.

Tanner, J. M. (1990) Growth as a mirror of conditions in society. In G. W. Lindgren, eds., *Growth as a Mirror of Conditions in Society*, pp. 9–48. Stockholm: Stockholm Institute of Education Press.

Trivers, R. L. (1974) Parent-offspring conflict. *American Zoologist* 14: 240–264.

Ulijaszek, S. (1994) Between-population variation in pre-adolescent growth. *European Journal of Clinical Nutrition* 48 (suppl. 1): S5–S13.

van Wierigen, J. C. (1986) Secular growth changes. In F. Falkner and J. M. Tanner, eds., *Human Growth*, pp. 307–331. New York: Plenum Press.

Wheeler, M. D. (1991) Physical changes of puberty. *Endocrinology and Metabolism Clinics of North America* 20: 1–15.

Wiley, R. H. (1974) Effects of delayed maturation on survival, fecundity, and the rate of poulation increase. *American Naturalist* 108: 705–709.

Wolff, P. (1989) The concept of development; how does it constrain assessment and therapy? In P. R. Zelazo and R. Barr, eds., *Challenges to Developmental Paradigms*, pp. 13–28. Hillsdale, NJ: Lawrence Erlbaum.

Wood, J. W. (1992) Fertility and reproductive biology in Papua New Guinea. In R. D. Attenborough and M. P. Alpers, eds., *Human Biology in Papua New Guinea: The Small Cosmos*, pp. 93–118. Oxford: Clarendon Press.

Wood, J. W. (1994) *Dynamics of Human Reproduction.* New York: Aldine de Gruyter.

World Bank (1993) *World Development Report.* New York: Oxford University Press.

Worthman, C. M. (1986) Later-maturing populations and control of the onset of puberty. *American Jour-*

nal of Physical Anthropology 69: 282.

Worthman, C. M. (1987) Interactions of physical maturation and cultural practice in ontogeny: Kikuyu adolescents. *Cultural Anthropology* 2: 29–38.

Worthman, C. M. (1993a) Bio-cultural interactions in human development. In M. Pereira and L. Fairbanks, eds., *Juvenile Primates: Life History, Development and Behavior*, pp. 339–358. Oxford: Oxford University Press.

Worthman, C. M. (1993b) Comment on J. Chisholm, "Death, hope, and sex: life history theory and the development of reproductive strategies." *Current Anthropology* 34: 17–18.

Worthman, C. M. (1998) The epidemiology of human development. In C. Panter Brick and C. M. Worthman, eds. *Hormones, Health, and Behavior.* Cambridge: Cambridge University Press.

Worthman, C. M. (1996) Survivorship, selection, and socialization: biosocial determinants of sex ratios. In S. J. Ulijaszek and C. J. K. Henry, eds., *Long Term Consequences of Early Environments*, pp. 45–68. Cambridge: Cambridge University Press.

Worthman, C. M., Stallings, J. F., and Jenkins, C. L. (1993) Developmental effects of sex-differentiated parental care among Hagahai foragers. *American Journal of Physical Anthropology (suppl.)* 16: 212.

Yajnik, C. S., Fall, C. H. D., Vaidya, U., Pandit, A. N., Bavdelar, A., Bjat, D., Osmond, C., Hales, C. N., and Barker, D. J. P. (1994) Fetal growth and glucose and insulin metabolism in four-year-old Indian children. *Diabetic Medicine* 12: 330–336.

Zelazo, P. R., and Barr, R., eds. (1989) *Challenges to Developmental Paradigms.* Hillsdale, NJ: Lawrence Erlbaum.

Zemel, B. S., and Jenkins, C. (1989) Dietary change and adolescent growth among the Bundi (Gende-speaking) people of Papua New Guinea. *American Journal of Human Biology* 1: 709–718.

Zemel, B., Worthman, C. M., and Jenkins, C. (1993) Differences in endocrine status associated with urban-rural patterns of growth and maturation in Bundi (Gende-speaking) adolescents of Papua New Guinea. In L. M. Schell, M. T. Smith, and A. Bilsborough, eds., *Urban Health and Ecology in the Third World*, pp. 38–60. Cambridge: Cambridge University Press.

7

INCEST AVOIDANCE

Clinical Implications of the Evolutionary Perspective

MARK T. ERICKSON

Until recently, incest was believed to be extraordinarily rare, literally a one in a million occurrence. Research now shows, however, that it is far from uncommon; by one estimate, 1 in 20 women may be victims of father–daughter sexual abuse (Russell 1986). Research also shows that incest often has devastating psychological effects. Depression, anxiety disorders, substance abuse, borderline personality disorder, somatoform disorder, and dissociative disorders are all more frequently diagnosed in adults who experienced incest as children (Braun 1990; Chu and Dill 1990; Courtois 1979; Gelinas 1983; Herman 1981; Kluft 1985; Meiselman 1978; Morrison 1989; Putnam 1989; Schetky 1990; Shearer et al. 1990; Stone 1990; Summit and Kryso 1978).

Ironically, just as we have gained awareness of the alarming frequency of incest in contemporary society, we have also discovered that incest is naturally *avoided* by most animal species. Furthermore, anthropological studies now convincingly show that humans also have an innate, and hence evolved, capacity to avoid incest (McCabe 1983, Shepher 1971; Wolf 1970, 1993, 1995). In both humans and other species, findings show that innate incest avoidance depends, in large part, on the developmental conditions of early life. Species in which incest rarely occurs will become incestuous if the developmental conditions essential for avoidance are significantly disrupted (Agren 1984; Gavish et al. 1984; Yamazaki et. al. 1988). Many clinicians believe we are in the midst of an epidemic of incest. Characteristics of modern culture which interfere with familial bonding, such as scant involvement of the father in early parenting (Williams and Finklehor 1995), may be creating conditions for such an epidemic by disrupting our evolved propensity to avoid incest.

There have been few attempts to integrate the growing literature on the evolutionary biology of incest avoidance with the clinical findings on incest

(Erickson 1989, 1993; Parker and Parker 1986; Wolf 1993, 1995). The purpose of this chapter is to show how aspects of incest and innate incest avoidance can be better understood when viewed from the perspective of contemporary evolutionary biology, animal studies, and anthropology. I first present earlier hypotheses of incest avoidance and then review pertinent recent findings. I then discuss an evolutionary hypothesis which argues that familial bonding in early childhood functions to establish behaviors that are adaptive in the context of a high degree of genetic relatedness. These behaviors include attachment behavior and, in later life, preferential altruism or kin selection and incest avoidance (Erickson 1989, 1993; Wolf 1993, 1995). Finally, I discuss the importance of this perspective for the clinical sciences.

Freud and Westermarck

For most of this century it was widely believed that a cultural rule, an incest taboo, was essential, not only for inhibiting incest, but also for the survival of culture in general. Without a taboo it was thought that incest would be rampant, leading to the destruction of the family and society (Freud 1913; Levi-Strauss 1969; White 1948). This view rested on the crucial assumption that animals, and precultural humans, mated incestuously. Freud (1913) used this assumption to argue that humans inherit ancient incestuous tendencies which, because of a cultural taboo, are repressed. In *Totem and Taboo* he proposed that repression of incestuous impulses creates a universal neurosis, unique to our species. Freud called this neurosis the Oedipus complex.

A different view was presented by the Finnish anthropologist Edward Westermarck (1922). Aware of the harmful effects of close inbreeding, Westermarck believed that, through natural selection, humans had acquired an innate aversion to incest. Specifically, he proposed that an innate aversion to sexual intercourse develops between individuals who have lived in close proximity from early childhood. Because children are raised in close proximity to their family in virtually all traditional cultures, this hypothesis could explain incest avoidance. Widely rejected during his lifetime, Westermarck's hypothesis is now recognized as an important contribution to our understanding of incest avoidance (Demarest 1977; Erickson 1989, 1993; Fox 1980; Murray and Smith 1983; Parker and Parker 1986; Roscoe 1994; Shepher 1971; van den Berghe 1980; Wolf 1970, 1995).

Incest Avoidance in Nature

Biologists have long realized that close inbreeding increases expression of deleterious recessive genes and reduces heterozygotic vigor (Bateson 1983; Packer 1979), effecting a marked increase in mortality and morbidity in humans (Adams and Neel 1967; Carter 1967; Seemanova 1971) and other species (Packer 1979; Ralls and Ballou 1982a, b). Given this, incest avoidance could evolve by natural selection.

The first study to examine the prevalence of incest in nonhuman species was not carried out until the 1960s. Because of the long-held belief that incest was common in nature, researchers were surprised to find that mother–son incest was rare in rhesus monkeys, *Macaca mulatta* (Sade 1968). Subsequent studies have shown incest to be uncommon between mother–son, father–daughter, and sibling pairs in a variety of primate species (Baxter and Fedigan 1979; Enmoto 1978; Evans 1986; Goodall 1986; Murray and Smith 1983; Packer 1979; Paul and Kuester 1985; Pusey 1980; Scott 1984; Smuts 1985; Tokuda 1961). Completely reversing earlier belief, it now appears that, with few exceptions, incest is avoided throughout the animal kingdom (Bischof 1975; Murray and Smith 1983).

One of the more striking examples of innate incest avoidance comes from studies of the pilot whale, *Globicephala melas* (Amos et al. 1993). Adult male pilot whales swim within their mother's group or "pod." Here, they live in nearly continuous social contact with their mother, mature sisters, and other female kin. Despite this extraordinary proximity, genetic analysis of entire pods shows that incest is virtually nonexistent. Males are sexually active with females of other pods but apparently do not mate within the family.

Research suggests that close proximity to immediate kin, from early life, is critical for establishing incest avoidance in nonhuman species, just as Westermarck had proposed for humans decades earlier (Bramblett 1983; Gavish et al. 1984; Hoogland 1982). The prairie vole, *Microtus ochrogaster*, for example, rarely mates incestuously when reared naturally with biological siblings. If, however, siblings are experimentally separated at birth into foster litters, as adults they sexually avoid unrelated foster sibs but readily mate incestuously with their unfamiliar biological kin (Gavish et al. 1984). These data are strikingly similar to human anthropological and clinical findings suggesting that the biological processes underlying incest avoidance have been conserved during evolutionary history.

Given the importance of infantile sexuality in psychoanalytic theory, it should be noted that early sexual play is observed in other primate species. An infant male Japanese macaque, for example, may mount his mother or siblings (Hanby and Brown 1974). Such play is accepted impassively and disappears as maturity approaches.

Evolution and Human Incest Avoidance

The major obstacle to studying incest avoidance in humans has been the incest taboo. Virtually all societies share this culturally constituted rule which has made it difficult to isolate biological influences. What was needed, to test Westermarck's hypothesis in particular, were circumstances in which there existed close proximity between individuals during childhood without a later taboo on sexual interaction. Two important test cases will be reviewed, one in rural Taiwan and the other on communal farms (kibbutzim) in Israel (see also Bevc and Silverman 1993; McCabe 1983).

Simpua Marriage and Sexual Avoidance

In Taiwan, Stanford anthropologist Arthur Wolf studies a type of marriage called *simpua* marriage. In *simpua* marriage a couple is typically betrothed in infancy by parental arrangement. The "child bride" moves into the home of her future husband, usually before her first birthday, to be raised by his family. Couples are usually close in age and live in continuous association throughout childhood. They are married in their mid-teens in a ceremony often marked by the extreme reluctance of the couple to consummate their relationship (Wolf 1970).

Westermarck's hypothesis predicts that *simpua* couples should develop a mutual sexual aversion and, indeed, this seems to be the case. Marital records show that the divorce rate for *simpua* couples is much higher than that of other arranged marriages. Marital infidelity is far more common among *simpua* couples, and birth records reveal that the birth rate, per year of marriage, is about 30% lower for these couples than other arranged marriages. Wolf believes that the fecundity of *simpua* couples would have been even less were it not for the intense parental pressure to produce offspring. Remarkably, a number of *simpua* couples confessed to having never consummated their relationship despite years of marriage. They explained that their children were the products of extramarital affairs and were simply recorded as their own (Wolf 1970, 1995).

Wolf (1995) has studied more than 14,000 *simpua* couples. After submitting his findings to virtually every conceivable alternative explanation, he concludes that Westermarck's evolutionary hypothesis provides the most plausible explanation of his data. His findings indicate that proximity during the first 3 years of childhood are a sensitive phase for the development of incest avoidance in humans. Also of note is Wolf's finding that early close proximity often led to later sexual aversion, and not just sexual disinterest as might occur through habituation alone (Parker and Parker 1986). This aversion developed despite cultural influences which unambiguously and insistently encouraged sexual interaction.

Sexual Avoidance Between Kibbutz Peers

Until recently, children living on communal farms (kibbutzim) in Israel were raised in a unique manner. Infants were moved from their parents' home shortly after birth to be cared for by nurses in a "children's house." In this setting, children of similar age, from different families, were raised in stable heterosexual peer groups through high school graduation. Kibbutz peers lived in continuous close association, playing together, eating together, bathing together, and so forth. Contact with their family consisted only of evening visits (Shepher 1971; Talmon 1964).

Under these circumstances, Westermarck's hypothesis predicts that kibbutz peers should sexually avoid each other upon reaching adolescence. The results of several studies are supportive (Bettelheim 1969; Shepher 1971; Spiro 1965;

Talmon 1964). In the most extensive study, the late Israeli anthropologist, Joseph Shepher, followed the premarital sexual preferences of adolescents raised on the same kibbutz. He found no evidence of heterosexual activity between peer group members ($n = 65$), aside from one instance in which the boy had not moved into the kibbutz until late childhood, at age 10.

Shepher (1971) went on to analyze marriage records of kibbutz-raised individuals. In a large study ($n = 2769$), he found not a single case of marriage in which a couple had been associated in the same peer group throughout early childhood (0–6 years). In contrast, there were eight marriages between individuals who had spent considerable time together in the same peer group in later childhood (6–12 years) and nine marriages between individuals who had been together during most of their adolescence (12–18 years). Shepher and many other authors, but not all (e.g., Spiro 1965), have interpreted these data as supportive of Westermarck's hypothesis.

A Contemporary Evolutionary View of Incest Avoidance

The discovery that incest is rare in nature has undermined the long-held belief that incest avoidance is uniquely human and dependent on a cultural taboo. The sexual avoidance of *simpua* couples and kibbutz peers provides evidence that humans, like other species, innately avoid incest. Westermarck's original hypothesis is a good first approximation of how this process works, but evolutionary theory, anthropological research, and clinical findings now support a significant refinement of the hypothesis.

With regard to evolutionary theory, Westermarck's hypothesis rests on the reasonable assumption that an environmental marker, early childhood proximity, has been stably associated with kinship. With this stable association, it was possible for a reliable incest avoidance process to evolve. If there is more to establishing incest avoidance than just early close proximity, it is almost certainly some factor(s) that is also stably associated with kinship. A likely candidate is altruistic behavior.

Altruism, Attachment, and the Familial Bond Hypothesis

Altruistic behavior is common in nature and includes such responses as caring for young, mutual grooming, and defensive warning calls. Since Darwin, however, and until recently, altruism remained a theoretical dilemma. How can a behavior be adaptive if it decreases, even slightly, the ability of the altruist to survive? In 1964, biologist W. D. Hamilton provided an answer. A useful summation of his argument is that natural selection maximizes the abilities of individual organisms, *not species*, to gain genetic representation in future generations (Williams and Nesse 1991). Because individuals share a higher proportion of their genes with kin, as opposed to nonkin, an individual can, under many circumstances, maximize his or her own genetic representation, or *in-*

clusive fitness, by altruistically helping kin. Hamilton thus predicted that altruistic behavior should occur predominantly between closely related individuals. This prediction is now supported by data indicating that altruism in nature typically occurs between closely related individuals, usually between *immediate kin* (Arnold et al. 1996; Axelrod and Hamilton 1981; Vogel 1985). Altruism, like early proximity, is then stably associated with kinship and could therefore function as a marker of kinship in the evolution of incest avoidance.

At about the same time that biologists were beginning to understand the foundations of altruism, British psychiatrist John Bowlby (1969) began studies of attachment or bonding of children to parents. Bowlby believed that attachment is an innate process because it occurs in virtually all cultures at about the same time, it is widespread in primate species, and, once formed, the bond persists, even if attachment figures are abusive. Bowlby further observed that attachment occurs primarily between immediate kin in virtually all human cultures. Importantly, research shows that secure bonding develops in a responsive milieu (Ainsworth 1967). Responsivity to the needs of an infant or young child is, in essence, altruistic behavior.

With this in mind, a reasonable extension of Westermarck's hypothesis is to propose that *secure attachment* is crucial to fully establish incest avoidance (Erickson 1989, 1993; Wolf 1993, 1995). Secure attachment requires more than early close proximity, it also requires a responsive or altruistic milieu, in most cases created by a child's parents or older siblings. Succinctly, secure attachment may be a *kin recognition process* that canalizes several adaptive kin-directed behaviors including early attachment behavior and, later, preferential altruism and incest avoidance. As the quality of an attachment bond diminishes, the stability of a kinship relationship and the associated adaptive behaviors should also diminish.

It is noteworthy that Bowlby (1969) and Westermarck (1889) specified that close proximity in early childhood is essential for childhood attachment and incest avoidance, respectively. It now seems quite possible that Bowlby and Westermarck each described one aspect of a single, more inclusive phenomenon (Erickson 1989, 1993; Wolf 1993, 1995), which could be called familial bonding. The *familial bond* hypothesis makes the following predictions:

Prediction 1: Incest will be least likely when a secure bond or attachment exists between individuals. The most useful data for examining this prediction are from the kibbutz studies because they include descriptions of early peer relationships and because peers faced no social taboo on later sexual interaction. These data are supportive, as observers have commented on the apparently secure attachment between peers in early childhood and, as mentioned, their sexual avoidance in later life (Bettelheim 1969; Rabin 1965; Shepher 1971; Talmon 1964). Rabin (1965:3) writes, "Attachments . . . between infants are often formed very early in life. . . . The infant may watch his friend . . . babble at him and become extremely upset and morose when his friend happens to be moved to another room." As yet, there have been no formal studies of attachment between kibbutz peers, so the support for this prediction is tentative.

Of clinical importance are data that suggest that the incidence of sexual interaction between kibbutz peers may be significantly *lower* than the incidence of sibling incest in the United States (Meiselman 1978). As remarkable as this observation seems, it is consistent with the familial bond hypothesis because kibbutz peers seem to consistently form secure attachments in childhood. Secure attachment may not routinely occur between siblings in contemporary society because of various cultural practices that reduce early association. Of note, research on nonhuman primates also suggests that the strength of early attachment covaries with later incest avoidance (Packer 1979; Pusey 1980; Smuts 1985).

Prediction 2: Incest will be somewhat likely to occur between individuals who have an insecure or weak attachment. The familial bond hypothesis argues that individuals who are weakly or insecurely attached during childhood will subsequently experience immediate family members as being less "kinlike." The effect of this should be an increased probability of incest and a decrease in kin-directed altruism.

A profile of families in which incest has occurred is characterized by emotional deprivation and rejection during the childhood of all involved. In the case of father–daughter incest, the father typically comes from a background of emotional deprivation, having experienced maternal rejection and abandonment by his father. The daughter is typically starved for affection, and the mother usually has also had an emotionally deprived childhood (Herman 1981; Justice and Justice 1979). A similar pattern is evident in mother–son incest in which the mother was usually deprived of a nurturant relationship as a child (Justice and Justice 1979; Weinberg 1955). A son who attempts an incestuous relationship with his mother may do so because he has been severely rejected by her and perceives incest as a way to force her to be close to him (Weinberg 1955). Although the data concerning sibling incest is rather limited, a similar pattern seems to exist (Mrazek 1981).

Although alcoholism, drug abuse, and marital tension are often associated with incest, these seem to be precipitators of incest, not etiological factors (Weinberg 1955). A general image which emerges is one in which a rather extreme lack of nurturance characterizes the childhood of both perpetrators and victims of incest. A reasonable tentative conclusion is that individuals who developed markedly impaired attachments in childhood are more likely to engage in incest with parents, siblings, and/or offspring. A recurrent generational cycle of failed kin-directed altruism, including neglectful parenting and perturbed affiliative behavior (e.g., sexualized interaction with offspring), may then be established.

Prediction 3: Incest will be most likely to occur between kin who have not formed a mutual attachment bond. To my knowledge, the only occasion in which incest is clearly motivated by sexual passion on the part of both individuals occurs when kin are separated, during infancy or early childhood, and are then later reunited (Greenberg and Littlewood 1995; Weinberg 1955). Under conditions of early separation, an attachment would not develop, and kin, once united, would not experience sexual avoidance. The clinical literature, although limited, is supportive. Weinberg (1955:78), for example, writes, "Six pairs of siblings were separated in infancy. . . . Though intellectually aware of an incest taboo, each did not feel an aversion to the other as a sex partner" (see also Greenberg and Littlewood 1995). Anecdotally, the poet Lord Byron had what was described as a torrid and unabated affair with his half-sister Augusta. Byron had been separated from Augusta until he was fourteen (Marshand 1957).

It is of psychological interest that a recurrent story line in poems, novels, and plays in which incest is a theme is one of separation in early childhood, with a later incestuous reunion (Cory and Masters 1963; Rank 1992). Examples include *The Book Bag* by Somerset Maugham, *Pierre, or the Ambiguities* by Herman Melville, and *The Caryatids* by Isak Dinesen. The myth of Oedipus is by far the best known example. Oedipus is separated from his mother, Jocasta, at birth and much later reunites with her incestuously. From the perspective of the familial bond hypothesis, the Oedipus myth portrays a literal truth that early separation undermines natural incest avoidance. Parenthetically, literature such as Mary Shelly's *Frankenstein* also provides evidence that the reverse situation, proximity in early childhood, precludes later sexual attraction (see Price 1995).

Animal studies (such as that of the prairie vole earlier discussed) are also supportive. Species that avoid incest under normal rearing conditions of early close proximity become incestuous if experimentally separated during early life (Agren 1984; Gavish et al. 1984; Yamazaki et al. 1988).

Prediction 4: Incest avoidance and altruistic behavior will directly covary with the strength of early attachment. Because familial bonding is hypothesized as the foundation of both altruistic behavior and incest avoidance, these behaviors should covary in a direct manner. The kibbutz data support this prediction, indicating that secure early attachment, later sexual avoidance, and remarkably supportive relationships exist between peers (Bettelheim 1969; Shepher 1971). Shepher (1971:296) writes, "[Peers] are tied together by a very warm, friendly relationship. . . . they are always prepared to help each other, to share joys and sorrows." Research on other species also indicates that preferential altruism covaries with sexual avoidance. For example, in a study of Japanese macaques, *Macaca fuscata*, investigators observed no mating between immediate kin, whereas grooming, a form of altruistic behavior, was far more common between immediate kin than between unrelated animals (Baxter and Fedigan 1979).

The Psychology of Incest Avoidance

When kibbutz peers or *simpua* couples are asked why they are not sexually attracted, they frequently reply that they regard each other as brother or sister.

Psychologically, this is a remarkable comment. What is implied is the existence of a subjective experience of kinship which is distinct from, and precludes, a sexual affiliation.

Most earlier attempts to explain incest avoidance have assumed that childhood social affiliation is fundamentally, or at least significantly, sexual (Demarest 1977; Fox 1962; Freud 1953). Given this assumption, the dilemma of incest avoidance was to explain how sexual impulses of children might be diverted from kin before adulthood. Not surprisingly, it was hypothesized that some form of repression or aversive conditioning (e.g., castration fears: Freud 1953 or punishment in childhood: Demarest 1977) was critical. In my view, the significance of childhood sexuality has been seriously overemphasized (see also Chodoff 1966; Schrut 1994). Incest avoidance is more easily explained as an evolutionarily ancient tendency to bond, during early childhood, which then establishes a subjective experience of affiliation which is predominantly nonsexual. This "familial" form of affiliation is what kibbutz peers experience. Their subjective experience of kinship is ultimately traced to developmental and evolutionary pathways that are distinct from sexual affiliation.

Although familial and sexual bonds have many similarities (e.g., they are often enduring and involve pleasurable physical associations such as kissing and hugging), they are distinct in obvious ways. Familial attachments are formed, in part, because of proximity during a sensitive phase in early life. Sexual affiliations may develop if individuals come together beyond this early phase. In sexual affiliation there is usually a demand for sexual exclusivity. In familial bonds, the opposite is true, there exists a sexual aversion within the dyad (and I suspect a concern that each other find a suitable conjugal partner). Further, as evidenced by the *simpua* marriage and kibbutz findings, there appears to exist a distinct subjective experience or psychology of kinship (Bailey 1988), which given adequate early attachment is easily distinguished from sexual affiliation.

Consistent with this view, ethologists have stressed the necessity of distinguishing between sexual bonding and what has alternatively been termed attachment bonding, bonding motivation, or filial imprinting (Bischof 1972; Bowlby 1969; Lorenz 1935). Bishof (1972:16) writes, "Bonding motivation is one of the concrete specifications necessary for scientific clarity of the hazy term 'love.' It is in no way synonymous with sexual eroticism and is probably not even derived therefrom." A number of psychoanalytically oriented authors have also, contrary to traditional theory, argued or inferred that a nonsexual form of affiliation emerges between kin under normal developmental conditions (Bacal and Newman 1990; DeVos 1975; Doi 1989; Ferenczi 1949; Johnson 1993; Kohut 1984; Roscoe 1994; Suttie 1935). Both ethologists and psychoanalytic authors have, however, lacked an explanation of how a nonsexual form of affiliation could have evolved (cf. Roscoe 1994). The familial bond hypothesis provides an explanation. Attachment behavior, altruistic behavior, and incest avoidance may have coevolved during natural history because each social behavior is adaptively directed toward the same individuals—namely, immediate kin. A single developmental process, attachment bonding, could have acquired the function of canalizing all of these adaptive kin-directed behaviors (Erickson 1989, 1993; Wolf 1993, 1995). The existence of a biologically distinct form of affiliation, characterized by attachment behaviors, altruistic behaviors, and sexual avoidance, is then entirely plausible.

The developmental circumstance and psychological disposition which I believe typify a fully intact familial bond and distinguish it from sexual affiliation, is well depicted in the following comment on brother–sister relationships recorded in the Philippines by Yu and Liu (1980:215–216):

> Brothers and sisters are naturally deeper and su-od (closer) to one another. . . . Of course we are related by blood to our cousins, but somehow there exists a gap. Brothers and sisters really feel for one another, perhaps by virtue of the fact that we have lived with one another since we first saw light. . . . They are about the only ones that can help one another fully without thought of any remuneration or any strings attached.

Relationships as described above would, according to the familial bond hypothesis, display an extremely low incidence of incest.

What is of particular clinical significance is the fact that incest avoidance is not "hard wired." It is susceptible to being *misdirected*, as in *simpua* marriage or between kibbutz peers, or *stunted* in its emergence, as in families in which attachment is grossly impaired. Clinicians who work with families in which incest has occurred are often impressed by their grossly distorted, and often indiscriminately sexualized, affiliative behavior. Here, the adaptive intrapsychic boundaries, or the distinct domains (Tooby and Cosmides 1992) between familial and sexual experience appear inadequately developed. From a psychodynamic perspective, it could be argued that an unconscious confusion between familial and sexual forms of affiliation would manifest as an Oedipal-like conflict. Repression would then function as a necessary defense. Viewed in this way, the Oedipus complex is not universal, but rather a pathological phenomenon derived from impaired early attachment (cf. Badcock 1990).

Completing the Model

The data discussed thus far account for the sexual avoidance of an individual toward his or her parents or siblings, but it does not explain how a familial bond and incest avoidance is established within parents for their children. Probably the most important developmental process for parent-to-child bonding is familiarization. Familiarization simply refers to the recognition of phenotypic features through direct exposure. With mothers familiarization starts shortly after birth. Mothers typically adopt an *en face* orientation for close inspection of their infant. Within a very short time, mothers can reliably distinguish their offspring through either visual, olfactory, auditory, or tactile cues (see Porter 1991 for review). In one study, for example, 9 of 10 mothers correctly identified their newborns' odors after only 10–60 minutes of post-natal contact with the infant (Kaitz et al. 1988). Although the data are more limited, fathers also seem to have the ability to discriminate their offspring soon after birth. For example, fathers with an average of less than 7 hours of postnatal contact were able to recognize their offspring, blindfolded, by touch of the infant's hands alone (Kaitz et al. 1994).

In nonhuman species familiarization may be important to establish parent–offspring incest avoidance (Burda 1995). This appears to be the case for hu-mans as well. Nonincestuous fathers, for example, have been found to be sig-nificantly more involved in the home or in child care activities than incestuous fathers (Williams and Finkelhor 1995). Stepfathers, who generally spend much less time in early caretaking than the natural parents, are also known to be far more likely to engage in sexual abuse (Parker and Parker 1986; Russell 1984). However, if stepfathers *are* extensively involved in the early care of their step-daughters, they do not appear to be more sexually abusive than natural fathers (Parker and Parker 1986). Finally, parents who relinquish their offspring for adoption at birth and much later reunite appear to have no aversion to incest (Greenberg and Littlewood 1995).

Williams and Finkelhor (1995) suggest that caretaking enhances paternal feelings and capacities. Stated somewhat differently, the emergence of such feelings may reflect the development of a biologically based bond of kinship with offspring that precludes significant sexual feelings.

Innate incest avoidance thus appears to involve two temporally separate but linked developmental processes. First, in early childhood secure bonding establishes the capacity for a stable familial relationship. This manifests as an adaptive, nonsexualized, altruistically inclined affiliation with parents and siblings that tends to continue through the life span. Second, as an adult, even if one experienced secure bonding in childhood, some degree of early famil-iarization, or reciprocal bonding, with one's own children appears to be im-portant for establishing a stable parent–child bond and hence incest avoidance and an altruistically inclined relationship. Without this foundation, the nor-mal boundaries between familial and sexual forms of affiliation may become blurred, creating the potential for incest, abuse, and neglect.

Clinical Implications

Over the last 30 years, our understanding of the biology of incest avoidance and the harmful psychological effects of incest have advanced dramatically. It is now quite clear that the foundation of incest avoidance rests in a biological process and that understanding this process has clinical relevance. Clinical interventions can be broadly divided into preventive measures and treatment of individuals who have been victims of incest.

With regard to prevention, public health policy should be consistent with what we currently understand about the biology of incest avoidance. Unfor-tunately, we do not yet know how *much* early proximity is necessary to establish incest avoidance. This is a particularly important question given the common use of surrogate parental care in modern society, whether in the form of a nanny or day care. It appears that secure bonding can develop between child and parent even with the use of day care during early childhood, provided parents are responsive or attuned to the needs of their children (NICDH 1996; cf. Belsky 1990). In cases in which parents are not adequately responsive to offspring, where attachment gets off to a poor start, interventions are being developed to aid parents in acquiring the necessary parenting skills (Lieberman et al. 1991).

In recent years there has been an increased popular interest in what has been come to be known as "attachment parenting." In attachment parenting, parents maintain as much proximity to their infants and young children as is possible (Sears and Sears 1993). The mother breast-feeds and the father participates in the birth and in ongoing child care. Both parents are encouraged to "wear their baby" by carrying it in slinglike carriers. Parents may also decide to share a bed with their baby or young children. Although this latter recommendation may seem odd to those raised in westernized industrial societies, in virtually all other cultures babies sleep either with, or in immediate proximity to, their parents. This is almost certainly the sleeping arrangement in which our species evolved and may have adaptive advantages (see the chapter by McKenna in this volume). Attachment parenting is supported by clinical research that indicates that parental involvement in the ongoing care of their child reduces the likelihood of abuse (Williams and Finkelhor 1995).

With regard to the treatment of individuals who suffer from the effects of sexual abuse, significant progress has been made in recent years. A review of the literature is beyond the scope of this chapter. Several outstanding books are now available to assist clinicians and patients in the therapeutic work needed for recovery (see especially Allen 1995; Herman 1992).

For most of this century it was believed that incest was natural and that its avoidance was established by cultural factors. It now appears that the opposite is true. Incest is naturally rare, and its occurrence in humans may most often be due to the disruption of an evolved process of reciprocal familial bonding between kin. More research is needed to refine our understanding, but what is known at present has significance for clinicians as well as those who shape public policy.

References

Adams, M. S., and Neel, J. V. (1967) Children of incest. *Pediatrics* 40:55–62.

Agren, A. (1984) Incest avoidance and bonding between siblings in gerbils. *Behavioral Ecology and Sociobiology* 14:161–169.

Ainsworth, M. D. S. (1967) *Infancy in Uganda: Infant Care and the Growth of Love.* Baltimore, MD: Johns Hopkins University Press.

Allen, J. (1995) *Coping with Trauma: A Guide to Self Understanding.* Washington, DC: American Psychiatric Association Press.

Amos, B., Schlotterer, C., and Tautz,

D. (1993) Social structure of pilot whales revealed by analytical DNA profiling. *Science* 260:670–672.

Arnold, G., Quinett, B., Cornuet, J., Masson, C., Schepper, B., Estoup, A., and Gasqui, P. (1996) Kin recognition in honeybees. *Nature* 379:498.

Axelrod, R., and Hamilton, W. D. (1981) The evolution of cooperation. *Science* 211:1390–1396.

Bacal, H. A., and Newman, K. M. (1990) *Theories of Object Relations: Bridges to Self Psychology.* New York: Columbia University Press.

Badcock, C. R. (1990) Is the Oedipus complex a Darwinian adaptation? *Journal of the American Academy of Psychoanalysis* 18:368–377.

Bailey, K. G. (1988) Psychological kinship: Implications for the helping professions. *Psychotherapy* 25(1): 132–141.

Bateson, P. (1983) Optimal outbreeding. In P. Bateson, ed., *Mate Choice.* London: Cambridge University Press, p. 257–277.

Baxter, M. J., and Fedigan, L. M. (1979) Grooming and consort partner selection in a troop of Japanese monkeys (*Macaca fuscata*). *Archives of Sexual Behavior* 8:445–458.

Belsky, J. (1990) Parental and nonparental child care and children's socioemotional development: A decade in review. *Journal of Marriage and the Family* 52:885–903.

Bettleheim, B. (1969) *The Children of the Dream.* New York: Macmillan.

Bevc, I., and Silverman, I. (1993) Early proximity and intimacy between siblings and incestuous behavior: A test of the Westermarck theory. *Ethology and Sociobiology* 14:171–181.

Bischof, N. (1972) The biological foundations of the incest taboo. *Social Science Information* 2:7–36.

Bischof, N. (1975) Comparative ethology of incest avoidance. In R. Fox, ed., *Biosocial Anthropology.* New York: John Wiley & Sons.

Bowlby, J. (1969) *Attachment and Loss,* vol. 1. *Attachment.* New York: Basic Books.

Bramblett, C. A. (1983) Incest avoidance in socially living vervet monkeys. *American Journal of Physical Anthropology* 60:176–177.

Braun, B. G. (1990) Dissociative disorders as sequelae to incest. In R. P. Kluft, ed., *Incest-Related Syndromes of Psychopathology.* Washington, DC: American Psychiatric Association Press, p. 227–246.

Burda, H. (1995) Individual recognition and incest avoidance in eusocial common mole-rats rather than reproductive suppression by parents. *Experientia* 51:411–413.

Carter, C. O. (1967) Risk to offspring of incest. *Lancet* 1:436.

Chodoff, P. (1966) A critique of Freud's theory of infantile sexuality. *American Journal of Psychiatry* 123: 507–518.

Chu, J. A., and Dill, D. L. (1990) Dissociative symptoms in relation to childhood physical and sexual abuse. *American Journal of Psychiatry* 147:887–892.

Cory, D. W., and Masters, R. E. L. (1963) *Violation of Taboo: Incest of the Great Literature of Past and Present.* New York: Julian Press.

Courtois, C. A. (1979) The incest experience and its aftermath. *Victimology* 4:337–347.

Demarest, W. J. (1977) Incest avoidance among human and nonhuman primates. In S. Chevalier-Skolnikoff and E. F. Poirer, eds., *Primate Social Behavior.* New York: Garland Press, p. 323–342.

DeVos, G. A. (1975) Affective dissonance and primary socialization: Implications for a theory of incest avoidance. *Ethos* 3:165–182.

Doi, T. (1989) The concept of *amae* and its psychoanalytic implications. *International Review of Psychoanalysis* 16:349–354.

Enmoto, T. (1978) On social preference in sexual behavior of Japanese monkeys (*Macaca fuscata*). *Journal of Human Evolution* 7:283–293.

Erickson, M. T. (1989) Incest avoidance and familial bonding. *Journal of Anthropological Research* 45:267–291.

Erickson, M. T. (1993) Rethinking Oedipus: An evolutionary perspective of incest avoidance. *American Journal of Psychiatry* 150:411–416.

Evans, S. (1986) The pair bond of the common marmoset. In D. M. Taub and F. A. King, eds., *Current Perspectives in Primate Social Dynamics*. New York: Van Nostrand Reinhold.

Ferenczi, S. (1949) Confusion of tongues between the adult and the child. *International Journal of Psychoanalysis* 30:225–230.

Fox, J.R. (1962) Sibling incest. *British Journal of Sociology* 13:128–50.

Fox, J. R. (1980) *The Red Lamp of Incest*. New York: Dutton.

Freud, S. ([1913] 1953) Totem and taboo. In *Complete Psychological Works*, Standard ed., vol. 13. London: Hogarth Press, pp. 1–161.

Gavish, L., Hofmann, J. E., and Getz, L. L. (1984) Sibling recognition in the prairie vole, *Microtus ochrogaster. Animal Behavior* 23:362–366.

Gelinas, D. J. (1983) The persistent negative effects of incest. *Psychiatry* 46:312–332.

Goodall, J. (1986) *The Chimpanzees of Gombe*. Cambridge, MA: Belknap Press.

Greenberg, M., and Littlewood, R. (1995) Post-adoption incest and phenotypic matching: Experience, personal meanings and biosocial implications. *British Journal of Medical Psychology* 68:29–44.

Hamilton, W. D. (1964) The genetical evolution of social behavior, I and II.

Journal of Theoretical Biology 7:1–52.

Hanby, J. P., and Brown, C. E. (1974) The development of sociosexual behavior in Japanese macaques. *Behavior* 49:152–196.

Herman, J. L. (1981) *Father-Daughter Incest*. Cambridge, MA: Harvard University Press.

Herman, J. L. (1992) *Trauma and Recovery*. New York: Basic Books.

Hoogland, J. L. (1982) Prairie dogs avoid extreme inbreeding. *Science* 215:1639–1641.

Johnson, F. A. (1993) *Dependency and Japanese Socialization: Psychoanalytic and Anthropological Investigations into* Amae. New York: New York University Press.

Justice, B., and Justice, R. (1979)*The Broken Taboo*. New York: Human Sciences Press.

Kaitz, M., Good, A., Rokem, A. M., and Eidelman, A. I. (1988) Mothers' recognition of their newborn by olfactory cues. *Developmental Psychology* 20:582–591.

Kaitz, M., Shuri, S., Danziger, S., Hershko, Z., and Eidelman, A. I. (1994) Fathers can also recognize their newborns by touch (brief report). *Infant Behavior and Development* 17:205–207.

Kluft, R. P. (1985) *Childhood Antecedents of Multiple Personality*. Washington, DC: American Psychiatric Press.

Kohut, H. (1984) *How Does Analysis Cure?* A. Goldberg and P. Stepansky, eds. Chicago: University of Chicago Press, pp. 13–33.

Levi-Strauss, C. (1969) *The Elementary Structures of Kinship*. Boston, MA: Beacon Press.

Lieberman, A.F., Weston, D. R., and Pawl, J. H. (1991) Preventive intervention and outcome with anxiously attached dyads. *Child Development* 62:199–209.

Lorenz, K. ([1935] 1970) *Studies in Animal and Human Behavior.* Cambridge, MA: Harvard University Press.

Marshand, L. A. (1957) *Byron: A Biography.* New York: Knopf.

McCabe, J. (1983) FBD marriage: Further support for the Westermarck hypothesis of the incest taboo? *American Anthropologist* 85:50–69.

Meiselman, K. C. (1978) *Incest: A Psychological Study of Causes and Effects.* San Francisco, CA: Jossey-Bass.

Morrison, J. (1989) Childhood sexual histories of women with somatization disorder. *American Journal of Psychiatry* 146:239–241.

Mrazek, P. K. (1981) The nature of incest: A review of contributing factors. In P. B. Mrazek and C. H. Kempe, eds., *Sexually Abused Children and Their Families.* Elmsford, NY: Pergamon Press, pp. 87–107.

Murray, R. D., and Smith, E. O. (1983) The role of dominance and intrafamilial bonding in the avoidance of close inbreeding. *Journal of Human Evolution* 12:481–486.

NICDH (1996) Infant child care and attachment security: Results of the National Institute of Child Health and Human Development study of early child care. Presented at the International Conference on Infant Studies, Providence, RI, April 20, 1996.

Packer, C. (1979) Inter-troop transfer and inbreeding avoidance in *Papio anubis. Animal Behavior* 27:1–36.

Parker, H., and Parker, S. (1986) Father-daughter sexual abuse: An emerging perspective. *American Journal of Orthopsychiatry* 56:531–549.

Paul, A., and Kuester, J. (1985) Intergroup transfer and incest avoidance in semi-free ranging Barbary macaques (*Macaca sylvanus*) at Salem (FRG). *American Journal of Primatology* 8:317–322.

Porter, R. H. (1991) Mutual mother-infant recognition in humans. In P. G. Hepper, ed., *Kin Recognition.* Cambridge: Cambridge University Press, pp. 413–432.

Price, J. S. (1995) The Westermarck trap: A possible factor in the creation of Frankenstein. *Ethology and Sociobiology* 16:349–353.

Pusey, A. E. (1980) Inbreeding avoidance in chimpanzees. *Animal Behavior* 28:543–552.

Putman, F. W. (1989) *The Diagnosis and Treatment of Multiple Personality Disorder.* New York: Guilford Press.

Rabin, I. A. (1965) *Growing Up on a Kibbutz.* New York: Springer.

Ralls, K., and Ballou, J. (1982a) Effects of inbreeding on juvenile mortality in some small mammal species. *Laboratory Animals* 16:159–166.

Ralls, K., and Ballou, J. (1982b) Effects of inbreeding on infant mortality in captive primates. *International Journal of Primatology* 4:491–505.

Rank, O. ([1912] 1992) *The Incest Theme in Literature and Legend.* G. C. Richter, trans. Baltimore, MD: Johns Hopkins University Press.

Roscoe, P. B. (1994) Amity and aggression: A symbolic theory of incest. *Man* 29:49–76.

Russell, D. E. H. (1984) The prevalence and seriousness of incestuous abuse: Stepfathers vs. biological fathers. *Child Abuse and Neglect* 8:15–22.

Russell, D. E. H. (1986) *The Secret Trauma: Incest in the Lives of Girls and Women.* New York: Basic Books.

Sade, D. S. (1968) Inhibition of son-mother mating among free ranging rhesus monkeys. *Science and Psychoanalysis* 12:18–38.

Schetky, D. H. (1990) A review of the literature on the long-term effects of childhood sexual abuse. In R. P. Kluft, ed., *Incest-Related Syndromes of Adult Psychopathology.* Washington, DC: American Psychiatric Press, pp. 35–54.

Schrut, A. H. (1994) The Oedipus complex: Some observations and questions regarding its validity and universal existence. *Journal of the American Academy of Psychoanalysis* 22:727–751.

Scott, L. M. (1984) Reproductive behavior of adolescent female baboons in Kenya. In M. F. Small, ed., *Female Primates: Studies by Women Primatologists*. New York: Alan R. Liss, pp. 95–99.

Sears, W., and Sears, M. (1993) *The Baby Book*. Boston, MA: Little Brown.

Seemanova, E. (1971) A study of children of incestuous matings. *Human Heredity* 21:108–128.

Shearer, S. L., Peters, C. P., Quaytman, M. S., and Ogden, R. L. (1990) Frequency and correlates of childhood sexual and physical abuse histories in adult female borderline patients. *American Journal of Psychiatry* 147:214–216.

Shepher, J. (1971) Mate selection among second generation kibbutz adolescents and adults: Incest avoidance and negative imprinting. *Archives of Sexual Behavior* 1:293–307.

Smuts, B. B. (1985) *Sex and Friendship in Baboons*. New York: Aldine-de-Gruyter.

Spiro, M. E. (1965) *Children of the Kibbutz*. New York: Schocken Books.

Stone, M. H. (1990) Incest in the borderline patient. In R. P. Kluft, ed., *Incest-Related Syndromes of Adult Psychopathology*. Washington, DC: American Psychiatric Press, pp. 183–204.

Summit, R., and Kryso, J. (1978) Sexual abuse of children: A clinical spectrum. *American Journal of Orthopsychiatry* 48:237–251.

Suttie, I. D. (1935) *The Origins of Love and Hate*. London: Kegan Paul.

Talmon, S. (1964) Mate selection in collective settlements. *American Sociological Review* 29:491–508.

Tokuda, K. (1961) A study of sexual behavior in the Japanese monkey troop. *Primates* 3:1–40.

Tooby, J., and Cosmides, L. (1992) The psychological foundations of culture. In J. H. Barkow, L. Cosmides, and J. Tooby, eds., *The Adapted Mind: Evolutionary Psychology and the Generation of Culture*. Oxford: Oxford University Press, pp. 19–136.

van den Berghe, P. L. (1980) Incest and exogamy: A sociobiological reconsideration. *Ethology and Sociobiology* 1:151–162.

Vogel, C. (1985) Helping, cooperation and altruism in primate societies. In B. Holldobler, and M. Lindauer, eds., *Experimental Behavioral Ecology and Sociobiology*. New York: Sinauer, pp. 375–389.

Weinberg, S. K. (1955) *Incest Behavior*. Secaucus, NJ: Citadel.

Westermarck, E. A. ([1889] 1922) *The History of Human Marriage, vol. 2*. New York: Allerton Press.

White, L. A. (1948) The definition and prohibition of incest. *American Anthropologist* 50:416–435.

Williams, G. C., and Nesse, R. M. (1991) The dawn of Darwinian medicine. *The Quarterly Review of Biology* 66(1):1–22.

Williams, L. M., and Finklehor, D. (1995) Paternal caregiving and incest: Test of a biosocial model. *American Journal of Orthopsychiatry* 65(1):101–113.

Wolf, A. P. (1970) Childhood association and sexual attraction: A further test of the Westermarck hypothesis. *American Anthropologist* 72:503–515.

Wolf, A. P. (1993) Westermarck redivivus. *Annual Review of Anthropology* 22:157–175.

Wolf, A. P. (1995) *Sexual Attraction and Childhood Association: A Chi-*

nese Brief for Edward Westermarck. Stanford, CA: Stanford University Press.

Yamazaki, G. K., Beauchamp, G. K., Kupriewski, D., Bard, J., Thomas, L., and Boyce, E. A. (1988) Familial imprinting determines H-2 selective mating preferences. *Science* 240: 1331–1332.

Yu, E., and Liu, W. T. (1980) *Fertility and Kinship in the Phillipines.* South Bend, IN: University of Notre Dame Press.

8

EVOLUTIONARY OBSTETRICS

WENDA R. TREVATHAN

Many current challenges to human health and well-being can be accounted for by the inconsistencies between the environments in which we live today and those in which human physical, emotional, and psychological needs evolved. An extreme example is the contrast between childbirth as experienced by our ancestors and by women in industrialized nations delivering their infants in modern hospital labor and delivery suites. In particular, the technology associated with childbirth and the social environment in which birth takes place are two areas of difference between past and present that will be examined in this chapter. Contemporary criticisms of "high-tech birthing" are presented,[1] and ways in which birth technology could be made to align with a woman's emotional needs are suggested. My argument is that the roots of social support during labor and delivery are as ancient as the human species itself, and that one of the reasons for dissatisfaction with the way in which childbirth is practiced in many industrialized nations is the failure of the medical system to acknowledge and work with the evolved emotional needs of women at this time. Providing social and emotional support during labor and delivery may improve both physical and emotional outcomes and lead to more positive perceptions of the birth experience.

Birth Technology: Past and Present

As with any area of health care, one of the major differences between past and present childbirth practices involves changes in the use of technology. Table 8.1 summarizes practices of contemporary birth technology in the United States, comparing each with probable historical practices in maternity care. Current hospital deliveries ("twenty-first century birth technology") may in-

Table 8.1 Technology of childbirth

21st Century birthing technology	Intended to	Implicated in	Low-tech alternative	Reference
Intravenous oxytocin	Augment or induce labor	High dystocia incidence; high rates of cesarean section	Manual breast stimulation; waiting	Curtis (1993)
Electronic fetal heart monitor	Monitor fetal heart rate; reduce perinatal mortality and morbidity	High rates of cesarean section and maternal infection	Visual and aural monitoring (intermittent)	Shy et al. (1990); Freeman (1990)
Medications to reduce sensations of pain	Reduce stress on the laboring woman	Fetal stress; numerous side effects for mother and infant	Reassuring presence; breathing techniques	Brackbill et al. (1984)
Enema, shaving of pubic hair	Reduce risk of infection	Increased risk of infection; discomfort	No enema (women often have natural diarrhea in labor), no shaving	Davis-Floyd (1992); Brackbill et al. (1984)
Lying down in labor	Ease of monitoring labor progress	Interruption of blood flow and oxygen to fetus; discomfort; increase length of labor, decreased strength of contractions	Walking; movement at will	Davis-Floyd (1992)

Lithotomy position for delivery	Ease of assisting in delivery, especially if instruments are used	Lengthens second stage of labor; less effective pushing	More upright posture, such as squatting or sitting	Sleep et al. (1989)
Episiotomy	Enlarge vaginal canal; reduce chances of perineal tears, uterine prolapse, etc.	Pain and discomfort; infection	Perineal massage; slow, controlled delivery of head	Banta and Thacker (1982); Mynaugh (1991)
Cesarean section	Alleviate problems from dystocia and fetal distress	Excessive expense of maternity care; related morbidity and mortality	Reduce incidence of cesarean section deliveries	Marieskind (1989)

clude one or more of the following practices: induction or augmentation of labor with intravenous oxytocin, use of electronic fetal heart-rate monitors, medication to reduce pain, maintenance of a "sterile field," supine posture during labor and delivery, episiotomy, use of forceps, and cesarean section. Each of these procedures is meant to reduce mortality and morbidity and/or to facilitate the ease with which health care providers assist women in labor and delivery. Unfortunately, several of the routine and expected practices in hospital deliveries are occasionally linked to increases in morbidity and mortality (see Brackbill et al., 1984, for a review). Coupled with general feelings of dissatisfaction on the part of recipients of maternity care (see, e.g., Kyman, 1991), concerns about overreliance on technology have led to calls for changes in the ways normal deliveries are managed in the United States, and increasing numbers of requests for "natural childbirth," which are referred to in table 8.1 as "low-tech alternatives."

Until the early twentieth century in the United States, and in many cultures today, pregnancy care, labor, and delivery were the province of women, with midwives providing the needed services. Most scholars of the history of medicine cite the invention of the forceps in 1588 by Peter Chamberlen as the beginning of the transformation of childbirth from a home-based "healthful" event supervised by midwives to a hospital- or clinic-based *medical* event under the control of medical specialists (Mitford, 1992). That transformation from home to hospital was not complete, however, until the middle of the twentieth century (Devitt, 1977). Only 5% of all American births occurred in hospitals in 1900, but by 1970, virtually all births in the United States were in hospitals (Wertz and Wertz, 1989). In 1994, 99% of U.S. births occurred in hospitals (Guyer et al., 1995). Rationale for the claim that hospitals were the normal and expected place for childbirth included safety, sterility, ease of monitoring, efficiency, and proximity to emergency equipment. These rationale were behind the emergence of what Mitford (1992) calls "the American way of birth."

One issue that concerns modern obstetrics is time. Any review of an obstetric text will highlight the expected length of gestation and of normal labor and delivery and will note the points at which the bounds of the norm are exceeded, calling for intervention. This intervention is usually designed to induce labor or to speed up labor or delivery.[2] Common methods used include amniotomy (rupture of the amniotic sac and membranes) and administration of intravenous oxytocin to induce uterine contractions. Neither of these practices is without risks. Amniotic fluid helps to cushion the fetal head during early contractions and serves to protect the umbilical cord. Once the membranes are ruptured, risks of infection increase. Exogenous oxytocin administration has been implicated in fetal distress, uterine rupture, extremely painful and powerful contractions, increased maternal stress and anxiety, and increased reliance on pain medication (Davis-Floyd, 1992). The "low-tech" option is to wait until labor begins and to avoid dependence on arbitrary time limits for its progression, unless there are clearer indicators of the need for intervention.

One of the primary functions of birth attendants, past and present, is to monitor the health and state of the mother and fetus. In the past this required spending time with the laboring woman, listening to fetal heart tones, and talking with the woman to assess her status. Internal and external electronic fetal monitors (EFM) were designed to improve the ability to detect abnormalities in fetal heart rate that indicate fetal distress and that require intervention, thereby reducing perinatal morbidity and mortality. Although their use has become almost routine in hospital deliveries in the United States,[3] there is no evidence that EFMs have contributed to a reduction in morbidity and mortality (Brackbill et al., 1984; Freeman, 1990; Prentice and Lind, 1987). In fact, they have consistently led to an increase in rates of cesarean section, which is associated with greater morbidity and mortality for mothers and infants (Brackbill et al., 1984). One reason for the increase in cesarean section rates associated with EFM use is that the monitors detect subtle changes in fetal heart rate during uterine contractions, changes that are often normal responses to the contractions (Lagerkrantz and Slotkin, 1986). Lagerkratnz and Slotkin (1986:103) observe that normal labor can "cause alterations in heart rates that might well be misinterpreted as signals of fetal distress." These misinterpretations often lead to decisions to deliver the infant by cesarean section. Furthermore, most studies conclude that EFMs are no more effective at detecting fetal distress than manual monitoring with a fetoscope (Haverkamp and Orleans, 1983). Thus, the low-tech alternative of watchful waiting is at least as effective as the high-tech option, at far lower expense and without restricting a woman's freedom of movement in labor.

Another way in which modern deliveries in the United States differ from those in the past is the position the woman assumes for delivery. The lithotomy position, commonly used for delivery in the United States, is one in which a woman lies flat on her back with her legs up in stirrups. This position is apparently good for the birth attendant, but may be less than optimal for the parturient. Studies by Caldeyro-Barcia (1979) suggest that an upright position (standing, squatting, sitting) is better than the supine position because the cervix dilates more effectively and contractions are stronger and more efficient. One reason for these differences is that the expulsive efforts of the contractions work with gravity in dilating the cervix and in delivering the infant. Furthermore, when a woman is upright, the presenting part (usually the back of the infant's head, the occiput) bears most of the force of the contractions. The occipital bone is the most developed of the cranial plates at birth and can best absorb this stress (D'Esopo, 1941). When a woman is lying down for delivery, the force of the infant's body must be absorbed by the more fragile frontal bones that lie against her sacrum. If the fetus is lying with the back of the head against the mother's sacrum (i.e., in a posterior position), the stress will be applied to the base of the skull, close to the spinal cord (D'Esopo, 1941). Babies delivered by their mothers in the upright position show fewer abnormal heart rate patterns and have higher Apgar scores at birth (Sleep et al., 1989).

Women in many cultures of the world deliver in an upright position. Naroll and coworkers (1961) reviewed birth position in 76 non-European cultures and found that the upright position was stated or implied for 62. Upright delivery postures includes sitting, squatting, kneeling, and standing. Often advocated by midwives is the squatting position, which increases the diameter of the pelvic outlet by as much as 0.5–2.0 cm. (Golay et al., 1993; Russell, 1969). In a study of 200 squatting births compared with 100 semirecumbent births, squatting women required significantly less oxytocin stimulation ($p<.01$) and fewer episiotomies ($p<.0001$) (Golay et al., 1993).

Unlike women in many cultures, however, most American women are not accustomed to performing daily activities while squatting and find it difficult to deliver in this position without assistance. Thus, although it may be the position with the longest history (most nonhuman primates deliver in the squatting position), it is often not compatible with twenty-first century lifestyles in the United States. Sitting, however, is something that Americans do especially well. Many hospitals have incorporated the use of a birthing stool into their delivery options, or women are simply allowed to deliver in a sitting or reclining position on a bed. These semi-upright positions probably represent a reasonable compromise between the ideal position for delivery and one which is practical.

The potential for contamination is of great concern in modern obstetrics. One of the major causes of neonatal and maternal mortality throughout human history has been sepsis. The source of "childbed fever" was first recognized by Semmelweis in 1846, who observed that mortality was much higher for women delivered by physicians than for those attended by midwives (Mitford, 1992). Physicians often came to deliveries following their ministrations to sick people or following autopsies, without washing their hands, thus transferring infectious agents to otherwise healthy women. Most physicians were angered by the suggestion of Semmelweis that they were the sources of childbed fever (see Mitford, 1992, for a review). Notably, in the last century, deaths from childbed fever were much higher in hospital maternity wards than among home births. At one point in the late nineteenth century, it was even recommended that in order to avoid the dangers of sepsis associated with hospital births, all births should take place at home (Oakley, 1984).[4] In the 1930s, the drug prontosil was found to be useful in the battle against childbed fever, although the decline in virulence of the infection appears to have preceded the use of the drug (Oakley, 1984). Perhaps childbed fever is another example of coevolutionary processes leading to reduced virulence of a disease (see Ewald, this volume). Today the problem has been almost entirely eliminated in some parts of the world through the use of antibiotics and improved hygiene measures.

Despite the fact that childbed fever has not been a major concern since the middle of this century, and despite the evidence that home environments in the United States are no more dangerous as sources of infections than hospitals (indeed, the opposite may be the case; see Brackbill et al., 1984, for a review), concern for contamination (sterile technique) during the birth process has assumed major importance in contemporary hospital births. This concern about contamination is the rationale behind restricting access to the laboring woman, using hospital gowns, administering an enema before delivery and shaving the pubic area ("prepping"). Each of these (nonmedical personnel, street clothes, diarrhea, pubic hair) is seen as a potential source of contamination. Davis-Floyd (1992), Brackbill et al. (1984), and others argue that there is no scientific evidence that any of these procedures reduce infection. Itching and other discomforts while pubic hair is growing back contribute to problems in the postpartum period. Enemas are uncomfortable and do not appear to shorten labor, as is often claimed (Romney and Gordon, 1981). Exposure to fecal matter is likely to occur, whether an enema is used or not.

Many women interviewed by Davis-Floyd (1992) reported feelings of discomfort and embarrassment resulting from the practices to which they were subjected upon entering the hospital, most especially being separated from their families, being dressed in hospital gowns, being given enemas, and being shaved. Considering the near absence of evidence that any of these effectively reduce infection, it is reasonable to argue against their routine use in normal labor and delivery. In this case, the alternatives of no separation from family, comfortable clothing, no shaving, and no enemas seem acceptable.

Another routine practice associated with deliveries in the United States is episiotomy, the cutting of the perineum (the tissue and skin between the vaginal opening and the rectum), which has been cited as the most commonly performed surgery on women in the United States (Rothman, 1993). According to the National Center for Health Statistics, the episiotomy rate in the United States in 1990 was 55.8 per 100 (Labrecque et al., 1994). In a study of practices associated with low-risk deliveries in five hospitals in the United States, episiotomy rates ranged from 53.7% to 74.5% (Hueston et al., 1995). The purpose of this minor surgical procedure is to enlarge the vaginal opening and ease the passage of the infant from the birth canal. Other rationale for this practice include reducing the incidence of uncontrolled tearing of the perineum and preventing later "pelvic relaxation." Although it has been demonstrated that episiotomy prevents anterior lacerations, none of the other claimed benefits have been supported (Woolley, 1995). In fact, episiotomies were found to increase maternal blood loss, anal sphincter damage, and the depth of posterior perineal lacerations, all of far more serious consequence than the claimed benefits (Woolley, 1995). Adding this to the pain associated with the cutting, suturing, and subsequent healing leads many to suggest that episiotomies should not be done routinely in hospital deliveries (Reynolds, 1993). This is not to suggest that uncontrolled tearing of the perineum should be tolerated. Women in the past and in many other cultures today probably spent significant portions of their lives in the squatting position and thus had stronger perineal tissue than most women do today. Thus, the risk of tearing is probably higher today. Nevertheless, many midwives believe that tearing can be reduced with perineal massage, lubrication, and exercise (Labrecque et al., 1994). Use of a delivery position other than lithotomy significantly reduces the need for episiotomy (Lydon-Rochelle et al. 1995).

A final aspect of modern obstetrics that was certainly lacking in the past is the ability to deliver an infant by cesarean section. Much has been written about the increase in the rate of surgical deliveries in the United States and in other industrialized countries (Francome and Savage, 1993; Nielsen, 1986; Nielsen et al., 1994; Taffel et al., 1992). In the United States, the proportion of babies delivered by cesarean section in 1993 was 21.8% (Guyer et al., 1995).[5] I will not add to this literature except to say that the increased morbidity related to surgical delivery is often unappreciated because it persists for months or even years after a woman has left the hospital, or it may be viewed as minor in the eyes of medical accounting. Nielsen (1986) lists several postoperative complications, primarily infections such as endometritis, urinary tract infection, wound infection, peritonitis, and pelvic abscess (risks ranged from 13% to 65%). More subtle problems have also been reported. For example, a study of 588 women who delivered by cesarean section in a Glasgow teaching hospital in 1986 revealed that three months after delivery, 35% believed that they had not yet fully recovered and 28% felt less healthy than they had before pregnancy (Hillan, 1992). Many reported backaches (55%), constipation (49%), and depression (38%). Although these may not seem like life-threatening situations, they still have an effect on the lives and experiences of individual women, their infants, and their families.

Obviously, there is no reasonable alternative to justified and necessary surgical delivery. Death for both mother and infant were the most likely results of a number of risky situations in our ancestral past, including cephalopelvic disproportion (head too large to pass through the maternal pelvis), placenta previa (the placenta covering the cervical opening), and prolapsed umbilical cord (when the cord precedes the infant in delivery). In these cases, the twenty-first century high tech alternative is welcomed. Considering the risks associated with this major surgery, however, there are a number of situations in which the ancient practices of waiting, walking, and offering emotional support may prove more beneficial.

Emotional Support in Labor and Delivery

Among the by-products of advances in the technology of childbirth has been an increase in the distance between caregivers and birthing women, an increase in the number of specialists attending women, and a decrease in the amount of time each attendant is able to devote to each woman. In other words, a major result has been a change in the social environment of birth. Several of these changes have met with resistance on the part of women giving birth in hospitals. Michaelson (1988) cites a 1957 article in the *Ladies Home Journal* decrying the dehumanizing nature of hospital birth, which was followed by hundreds of letters from women expressing their negative reactions to the lack of concern on the part of those who attended them in childbirth. The women who wrote the letters were generally positive about the decrease in mortality and morbidity,[6] but were dissatisfied that "the safe efficiencies had become a kind of industrial production" (Wertz and Wertz, 1977:173). Many expressed dissatisfaction with how impersonal this most significant event in their lives was regarded by the people who attended them.

In the first half of this century, it is possible that there was more social interaction and personalized care than we see today, because, for example, fetal heart tones were monitored with a manual fetoscope, requiring nurses to make tactile contact with a laboring woman several times an hour. Moreover, nurses or physicians visually assessed labor progress and cervical dilation stage, often by simply talking with the laboring woman and observing her behavior (see e.g., recommendations of Myles, 1975).[7] Their frequent presence likely provided more information to the laboring woman, potentially alleviating some of her concerns and anxieties. In contrast, a recent study reported a near absence of physical comfort measures provided by nurses during labor and delivery (Hodnett and Osborn, 1989). The authors of this study note that their results confirm anecdotal accounts of "changes in the intrapartum nursing role from one emphasizing hands-on comfort to one relying on technology, including pharmacologic methods of pain relief" (Hodnett and Osborn, 1989: 182).

Devices such as electronic fetal heart monitors often substitute for the jobs previously performed by nurses. But use of these devices may fail to provide the personal human touch, including answering questions of laboring women. McNiven and her colleagues (1992) note that the evolution of obstetric care in the United States is almost synonymous with the evolution of technology and physician-directed interventions (McNiven et al. 1992). They report that the primary relationships during labor and delivery seem to be those between individuals and electronic devices and the technicians that work with the machines.

In an Australian study of 790 women, Brown and Lumley (1994) found that women who reported dissatisfaction with the information they were given by physicians, midwives, and nurses during labor and delivery were four to six times more likely to report overall dissatisfaction with the quality of care they received. Women who rated caregivers negatively on kindness and understanding were four to five times more likely to rate overall care negatively. There is also evidence that caregivers are not aware that they are not being supportive of women during labor and delivery. McKay and Smith (1993) found, for example, that caregivers' perceptions of quality of support and communication was higher than the perceptions of the recipients of their support. Even when the caregivers viewed videotapes of their activities during labor and delivery, they felt that they had given adequate care and support. Clearly what is defined as helpful by one group (caregivers) is not perceived as such by the other (laboring women). Considering all of the demands on nurses and physicians during labor and delivery (e.g., working with and interpreting the electronic devices noted above), it is understandable that they feel that they have given maximum support to the women in their care. Furthermore, physicians and nurses are usually responsible for several patients at one time and cannot provide the continuous support often desired by laboring women.

Dissatisfaction with medical care occasionally leads to legal action, which has the effect of increasing the use of technology during birth because insurers request or require it (Shearer and Eakins, 1988). Shearer and Eakins (1988:32) point out that one reason for an increase in malpractice suits is that physicians do not spend enough time with patients: "Why are midwives and family practitioners not sued at the same rate as obstetricians? The answer is that people do not sue their friends, their trusted resources. . . . Midwives respect the sanctity of birth, and they spend the time to become trusted friends and resources for parents."

Dissatisfaction of women about their childbirth experiences has spawned numerous books[8] and has resulted in a number of changes in the ways in which hospitals and physicians treat this event. Changes include allowing husbands and other family members to be with women during labor and delivery, offering alternatives to delivering in a supine position on a delivery table, honoring numerous individual requests from laboring women and their partners, encouraging women to use fewer drugs during labor and delivery, and providing more homelike atmospheres for normal deliveries. But still, for many women, there remains the unmet need of emotional support during labor and delivery.

What Can the Evolutionary Perspective Tell Us?

Why might women need emotional support during labor and delivery? After all, for almost all mammals, birth is a solitary affair (see Trevathan, 1987, for a review of birth in several mammalian species). The most typical mammalian response to the increasing intensity of contractions is for the laboring female to seek an isolated spot in which to deliver her infant, brood, or litter. Even among social species, such as monkeys and apes, the female typically moves to the periphery of the group to deliver (see, e.g., Altmann, 1980; Goodall and Athumani, 1980). These reported responses suggest that the contractions of labor are a signal to seek solitude. In contrast, the typical human response to increasing intensity of contractions is to seek companionship. Why would there be a difference in the meaning of labor contractions? Another way of putting it is to ask: At what point in our evolutionary history did the signal content of active labor change from "seek isolation" to "seek companionship"?

A reasonable proposition, given the evolutionary history of the human species, is that natural selection favoring bipedal locomotion among human beings ultimately led to changes in the way females interpreted and responded to intense labor contractions (Trevathan, 1987). This does not mean to suggest that difficult and painful birth are unique to bipedal humans, however; birth is a physical challenge for most primates (see Schultz, 1949). An important characteristic of the primate order is a large ratio of brain or head size to body size. This means that the passage of the fetal head through the maternal pelvis

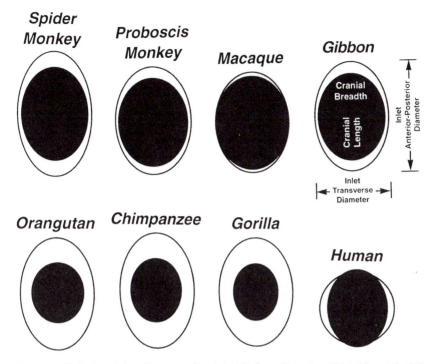

Figure 8.1. Relationship of maternal pelvis (dark outlines) and fetal head (solid dark ovals) (after Schultz, 1949).

is generally a tight squeeze (figure 8.1). Because of this, mortality from ce-phalopelvic disproportion is not insignificant in primate species such as mar-mosets, squirrel monkeys, baboons, and macaques (Leutenegger 1981). The modern great apes appear to be exceptions to this phenomenon, because the pelvic canal is not so narrow in comparison with the neonatal head (Leute-negger 1972), but it is likely that the last common ancestor of humans and great apes had a pelvis/neonatal head ratio similar to that of modern monkeys and gibbons (Ward, 1994). (Selection for increased body size in great apes occurred after the divergence of the ape and human lines.) But modern pri-mates still accomplish birth without assistance, so it is probable that our pre-hominid ancestors did so as well.

Approximately 5 million years ago, selection began to favor the anatomical and behavioral changes that led to bipedal walking in hominids (Conroy, 1990; Lovejoy, 1988). Whatever the "cause" or benefits of this new mode of loco-motion, it resulted in fundamental changes in the way birth occurred. In quad-rupedal species like monkeys (and in the probable quadrupedal ancestor of humans), the entrance and exit of the birth canals have their greatest breadth in the front-to-back (sagittal) dimension (figure 8.2). The infant, whose head is also largest in the sagittal dimension, passes straight through the birth canal

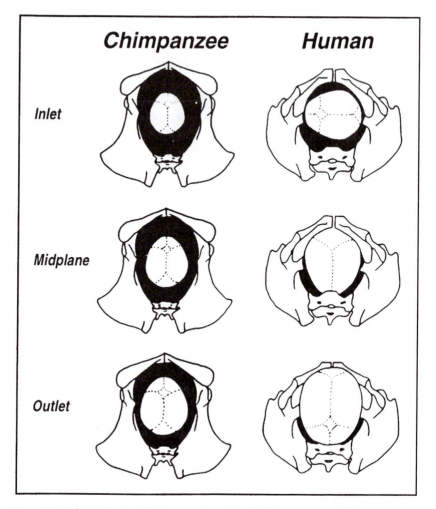

Figure 8.2. Pelvic inlet, midplane, and outlet of chimpanzee and human.

with the head extended (figure 8.3) (Stoller, 1995a,b). Stoller (1995a,b) has presented evidence that when the monkey fetal head is extended (with the back of the head against the top of the vertebral column), the mandible first contacts the bottom of the pelvis and rotates under the pubic arch, thus emerging in a position referred to in obstetric texts as "occiput posterior." This term refers to the fact that the base of the fetal skull (the occiput) is against the back (posterior) of the mother's pelvis. (Stoller points out that the emergence of the infant monkey is more appropriately referred to as a "face presentation" because the occiput posterior position of human obstetric texts assumes a flexed fetal head, one in which the chin is pressed against the throat.) This means

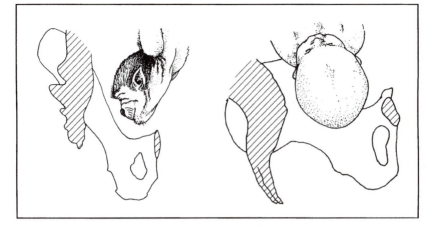

Figure 8.3. Lateral view of fetal descent in baboon (with head extended) and human (with head flexed).

that the baby monkey emerges facing toward the front of the mother's body and that she can reach down with her hands and guide it from the birth canal, or it can crawl up toward her nipples, unassisted (for a review, see Rosenberg and Trevathan, 1996; Trevathan, 1987).

One result of the evolution of bipedalism is that the birth canal is twisted in the middle so that the inlet is broadest in the transverse dimension, the outlet in the sagittal dimension. Thus, the maximal breadths of the entrance and exit are perpendicular to each other. The relevant fetal dimensions are also perpendicular: the head is largest in the sagittal dimension, the broad, rigid shoulders in the transverse dimension. According to Stoller (1995b), the passage of these broad, rigid shoulders through the deep bony pelvis requires that the neck be flexed. (Monkey shoulders are not as broad as those of apes and humans [Schultz, 1949], so they do not have this same effect on the birth process, and the head remains extended.) This flexion, coupled with the re-structured bony birth canal, means that the human infant must undergo a series of rotations to pass through the birth canal without hindrance (figure 8.4). When the head is flexed during delivery, the occiput contacts the bottom of the pelvis first and rotates under the pubic arch (Stoller, 1995b). Therefore, in most cases, the human infant emerges with the occiput against the mother's pubis ("occiput anterior" in obstetric texts) and thus faces away from the mother when it is born. (Once the head has delivered, the infant's body rotates to enable the shoulders to pass through the pelvic outlet, so the baby turns to face to the side at this point.)

This tendency for the human infant to be born facing away from the mother is the change that I argue had the greatest impact in transforming birth from a solitary to a social event (figure 8.4E). This position hinders a mother's ability

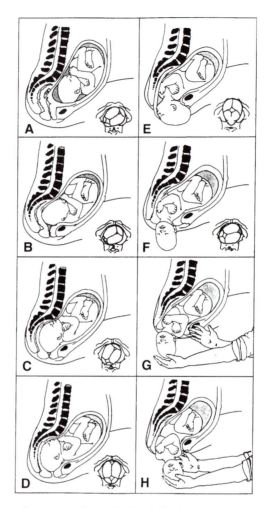

Figure 8.4. Series of rotations through which the human infant must pass during the birth process.

to reach down and clear a breathing passageway for the infant and to remove the cord from around the neck if it interferes with breathing or continued emergence. In most deliveries, if she attempts to guide the infant from the birth canal, she risks pulling it against the body's angle of flexion, perhaps damaging nerves and muscles in the process (Trevathan, 1988). It is reasonable to argue that simply having another person in the vicinity who could assist in this final stage of delivery would have reduced mortality. (Today, unattended births are usually associated with higher mortality than attended births [Edwards, 1973; Ohio State Medical Society, 1975]). Thus, it can be argued that with the origin of bipedalism, risks of mortality from unattended birth became greater than the risks associated with having others in the vicinity of the birthing female. I suggest that early hominid females who sought assistance or companionship at the time of delivery had more surviving and healthier offspring than those who continued the ancient mammalian pattern of delivering alone. Thus, the evolutionary process itself first transformed birth from an individual to a social enterprise, with the accompanying underlying emotions that motivated the behavior in the first place.

With increased brain size in the genus *Homo* (beginning about 2 million years ago), the already tight fit between infant head and maternal pelvis became even tighter, although one of the compromises to the conflict between selection for large brains and narrow birth canals was to delay most brain growth to the postnatal period. The result of this adaptation is that human infants today are less developed and more altricial at birth than most primate infants (see table 8.2). Giving birth to more helpless altricial infants, unable to assist themselves during delivery, posed more challenges to ancestral hominid females, adding to the advantages of having another person present at delivery.

It is not likely that early hominids consciously sought assistance at birth to reduce mortality, but merely that they felt a range of emotions that caused this positive result. The more immediate need (proximate cause) was probably to reduce fear, pain, and other emotional stress. Another consequence of increased brain size may have been an increased awareness or consciousness about the dangers of birth. Associated with this consciousness are the emotional signals, whatever their meaning, which led to seeking companionship during the later stages of labor and delivery (see also Hays, 1996).

Certainly there is much variation in the ways women interpret labor contractions, depending on cultural context and individual experiences. Intense labor contractions may lead to fear, doubt, confusion, joy, excitement, or un-

Table 8.2 Neonatal brain weight as a percentage of adult brain weight for selected primate species (data from Harvey and Clutton-Brock, 1985)

Species	% of adult weight
Galago *(Galago senegalensis)*	48
Howler monkey *(Alouatta palliata)*	56
Spider monkey *(Ateles geoffroyi)*	58
Rhesus macaque *(Macaca mulatta)*	57
Yellow baboon *(Papio cyocephalus)*	43
White-handed gibbon *(Hylobates lar)*	47
Orangutan *(Pongo pygmaeus)*	41
Common chimpanzee *(Pan troglodytes)*	31
Gorilla *(Gorilla gorilla)*	45
Human *(Homo sapiens)*	31

certainty. But, except in rare and unusual circumstances, the behavior resulting from the signal is usually seeking another person. Underlying these emotions, and perhaps contributing to their expression, are painful contractions,
extreme exertion, hormones such as prolactin and oxytocin (see Insel, 1992,
for a review of the effects of oxytocin on maternal behavior), and the presence
of the newborn infant.

Contemporary Studies of the Effects of Social and Emotional Support in Childbirth

Recent studies consistently demonstrate the positive effects of social and emotional support during labor and delivery (Hofmeyr et al., 1991; Kennell et al.,
1991; Klaus et al., 1992; Sosa et al., 1980; Wolman et al., 1993). In one study
(Hodnett and Osborn, 1989), the categories of support recalled by mothers as
being most meaningful to them during labor were (1) physical comfort (strok-
ing, massaging, assisting with movement, offering fluids), (2) emotional support (reassuring, quietly talking, giving encouragement, being continuously
present), (3) information (coaching with breathing, giving advice, interpreting
assessments of medical staff, explaining procedures), and (4) advocacy (interpreting the laboring woman's needs to medical staff, supporting her decisions).
All of these contributed to the women reporting more positive feelings about
their experiences and the people who cared for them.

Not only does social support contribute to more positive feelings about the
birth experiences, but there is evidence that biomedical outcome measures are
improved as well. For example, Klaus and his colleagues (1992) summarized
randomized trials conducted in hospitals in four different countries. They
evaluated the effectiveness of emotional support provided by women whose
job was solely to provide support consisting of praising and reassuring the
laboring women, providing physical contact such as rubbing and holding, explaining procedures, and being continuously present. Women delivering in
selected hospitals were randomly assigned either to receive social support or
to receive the routine procedures offered by the hospital. A summary and
analysis of five independent studies revealed that the effect of social support
included a reduction in the caesarean section rate by as much as two-thirds,
reduction in the length of labor by approximately 25%, and reduction in the
use of forceps by as much as 82%. Six weeks after the birth, the effects were
still apparent, represented by increased breast-feeding,[9] more time spent with
the infant, less anxiety, lower scores on a depression scale, higher self-esteem,
and more positive feelings about partners and infants.

Thornton and Litford (1994) reviewed the British practice known as "active
management of labor," which includes early amniotomy, use of oxytocin to
accelerate labor, and continuous support by a labor companion; the use of this
protocol has been demonstrated to reduce the incidence of cesarean section.
The reviewers conclude, however, that "the effective ingredient seems to be

the presence of a companion in labour rather than the performance of amniotomy or administration of oxytocin" (Thornton and Litford, 1994:368).

A South African study of the effects of companionship during birth on postpartum depression found that those who received extra companionship had higher self-esteem scores ($p<.001$) and lower postpartum depression and anxiety ratings ($p<.001$) 6 weeks after delivery compared with those who experienced the routine hospital care without extra social support (Wolman et al., 1993). This research is consistent with previous evidence that social support during pregnancy, delivery, and postpartum is the single most important factor in preventing postpartum depression and such negative sequelae as physical and emotional health disorders and problems in the early mother–infant relationship that can be long-lasting. In the study cited above, women in the support group were more likely to be breast-feeding their infants exclusively (Hofmeyr et al., 1991) If we accept that successful breast-feeding is desirable (see, e.g., Stuart-Macadam and Dettwyler, 1995), then this suggests that social support in labor and delivery may not only have positive effects on the woman herself, but also on the long-term health of her infant.

In their study of Guatemalan women, Sosa et al. (1980) found that women who had received social support during labor and delivery were awake longer after delivery ($p<.02$) and stroked ($p<.09$), smiled at ($p<.01$), and talked to ($p<.01$) their babies more than those who received the routine hospital care regimen. These authors suggest that providing social support has implications for the ease with which mothers and infants attach and thus for long-term family health and relationships (see Klaus and Kennell, 1982, for a review). In a Houston study of deliveries, the cesarean-section rate for those who received extra emotional support during labor and delivery ($n = 212$) was reduced by 10% and the epidural rate by 75% (Kennell et al., 1991). Extrapolating from these figures, it is estimated that maternity costs in the United States could be reduced by $2 billion if every woman had a support person with her in labor (Klaus et al., 1992). It seems that these welcomed changes would come at very little additional costs per delivery and at great reductions in obstetric costs per capita.

It is quite common for studies like those reviewed above to cite the extensive cross-cultural evidence that emotional support from another woman is an almost-universal component of maternity care today and one that has a long history. Kennell and his colleagues, for example, note that the practice of providing support during labor is "centuries old" (Kennell et al., 1991). An evolutionary perspective, however, informs us that this type of support is not, in any sense, a recent historical phenomenon. As argued above, an evolutionary perspective argues that dependence on assistance at birth is a characteristic of human behavior that results from bipedal upright walking, which evolved approximately 5 million years ago. Its very antiquity leads to the suggestion that social and emotional support in labor may be as crucial for optimal health of mother and infant as physical and medical assistance.

Who can provide this crucial support? Cross culturally and historically, normal pregnancy, birth, and the postpartum period take place surrounded by supportive friends and family in familiar environments (see, e.g., Jordan, 1993; Kay, 1982; MacCormack, 1982; Trevathan, 1987). But for many women giving birth today, their closest support persons may be their husbands. Are husbands likely to provide the needed support for women in labor and delivery? In most of the studies reviewed above that reported positive effects of companionship during labor and delivery, the support person was *not* the husband. Although his presence may be important to the woman, it is not likely that he can provide the informational and emotional support described above because of his own emotional investment in his wife and the birth of their child. In fact, his needs at this time may often equal those of his wife, and he may not be able to give the support she needs. Several studies of the effects of husband's presence on a woman's experience with labor and delivery report that, under certain circumstances, the husband's presence actually exacerbates the stresses of labor and delivery (see, e.g., Katz, 1993; Kennell et al., 1991; Nolan, 1995). Keinan et al. (1992), for example, found that for multiparas who scored low on anxiety rating, the presence of the husband resulted in higher tension than did his absence. They also note, however, that 85% of the women in their study reported that they wanted their husbands to be present for reasons other than stress reduction. This suggests that husbands play a different role from that of the birth assistants (see also Bertsch et al., 1990).

Solutions

If, indeed, there is a mismatch between the evolved needs of women during labor and delivery and the contemporary medical system designed to meet those needs, this potential mismatch can be easily remedied. The addition of a labor companion to the obstetric teams in hospital deliveries could likely lead to a decrease in complications of labor and delivery, decrease in postpartum depression, increase in patient satisfaction, increase in breast-feeding, decrease in obstetric costs, and decrease in malpractice suits.

Most labor companions who provide social and informational support are not ordinarily professionally trained as midwives, nurses, or psychologists. In the studies cited above, most of the supporting women had previously given birth and, upon interview, showed evidence of being able to be empathetic toward laboring women (see, e.g., Hofmeyr et al., 1991). The labor companions were also not usually part of the hospital staff, and most were drawn from the same community as the laboring women, although they were usually unknown to the women.

How important is it that the labor support people be trained? Should they be paid for their services? In some of the studies reviewed by Chalmers and Wolman (1993), the women acting as labor companions volunteered their services; in other cases, they were paid as much as $200 per delivery (Kennell et al., 1991). In the latter study (Kennell et al., 1991), the labor companions also received 3 weeks of specialized training in the process of normal labor, obstetric procedures, hospital policies, and supportive techniques, although there is evidence that labor support can be provided by untrained women. Indeed, in a survey of 19 studies of the effect of social support in labor, Chalmers and Wolman (1993:6) conclude that "Of all the different types of labor support which have been studied, the most impressive, consistent, and methodologically sound results have been obtained for support given by lay, untrained female supporters."

A professional service class of birth counselors and companions would be a contemporary response to the deep-rooted social needs of twenty-first century women. This profession could absorb the functions of childbirth educators, monitrices, lactation counselors, and *doulas* (see Raphael, 1973), all of which have recently appeared in many Western societies to replace family-centered support groups of times past. Compatible with the Western emphasis on training and education, professional birth counselors could receive training in reproductive biology, reproductive psychology, midwifery, lactation, child development, and grief counseling. Most of these topics could be covered in a semester-long course in community colleges similar to "nanny" classes currently offered.

Childbirth education should be part of the routine services provided by the birth counselor and, following delivery, lactation and other postpartum support could continue to be provided. In the event of a stillbirth, neonatal death, or malformation, the birth counselor could also be the person to provide initial grief counseling and other similar support that the mother and father needed. Although the baby would be taken for pediatric exam early in life, most of the routine child-care advice could be provided by the birth counselor. This opportunity for continuity in care would probably help to fill the void and uncertainty that many contemporary women feel when they are pregnant and experiencing first-time motherhood.

Conclusion

Oakley (1983) has used the concepts of "hard" and "soft" outcomes to describe the results of labor and delivery. The usual measures of morbidity and mortality assessments are the hard outcomes; "psychosocial morbidity" refers to the soft outcomes. Both may be equally important in evaluating the success of parturition, and they are complementary. If we accept the studies demonstrating the positive effects of having someone present at delivery to provide emotional support, it seems that we can have the best of both worlds: the welcomed reduction in mortality provided by many modern obstetric procedures and the increased positive feelings provided by the ancient practice of having emotional support from another woman.

We have behind us a heritage of 5 million years of hominid evolution, during which time having someone present to help during the final stages of delivery, especially with challenges to neonatal respiration, probably made the difference between life and death for many hominid mothers and infants. With encephalization and the origins of consciousness, awareness of vulnerability heightened the emotional impact of birth, which probably led the normally gregarious human female to seek companionship at this time. In other words, women experience heightened emotions at birth which lead them to seek companionship, which, in turn, leads to the ultimate outcome of lowered mortality and greater reproductive success for the women who behave in this manner. These emotions of childbirth are among the human adaptations to the obstetrical complications of bipedalism, and their roots are ancient.

At its best, evolutionary medicine links knowledge of human evolutionary history and cross-cultural studies with recent developments in biomedical and clinical research. Those of us who adopt this perspective argue that only by understanding the human condition in both its evolutionary and historical/cultural context can we help to bring about changes in contemporary medical practice that positively affect not only morbidity and mortality, but also human social and emotional fulfillment.

Notes

1. For a more thorough review of contemporary birth technology, see Brackbill et al. (1984) and Davis-Floyd (1992).

2. For example, Oxorn and Foote (1975) cite the following as the bounds of normal labor: First stage, 6–18 hours in primiparas (first-time mothers) and 2–10 hours in multiparas; second stage, 30 minutes to 3 hours in primiparas, 5–30 minutes in multiparas.

3. The Monthly Vital Statistic Report #44 (1996: 15–16) reports that EFM was used in 80% of all live births in 1994.

4. Oakley (1984:33) quotes Duncan (1870, p. 110): "The mortality of maternity hospitals is said to be so great that it is expedient, indeed absolutely necessary, to close them entirely."

5. Cesarean section rates were first available from national birth certificates in 1989. That year, 22.8% of births were delivered by cesarean sec-

tion, so there has actually been a slight decline in rates since 1989.

6. Indeed, between 1950 (when standards for safe medical practices in obstetrics were established by the American College of Obstetrics and Gynecologists) and 1975, there was an 8- to 10-fold reduction in maternal mortality and an approximate 50% reduction in infant mortality in the United States (Aubry, 1977). One important element in the standards was that birth should take place in hospitals.

7. Here is an example of advice given to midwives in Myles's (1975: 228) classic text, under the heading "Emotional Support During Labour":

Loneliness breeds fear. The comforting companionship of the midwife who will listen, explain, encourage and assure, or keep silent as required, is of inestimable value to the woman at this time. When labour is established the midwife should remain in constant attendance.

It is essential for the peace of mind of most women that they be kept informed regarding the progress they are making. Women respond magnificently to a word of praise, and being given reasons or explanation.

8. Examples of books calling for changes in the way in which childbirth is treated in the United States include Gilgoff (1978), Romalis (1981), Oakley (1984), Inch (1984), Brackbill et al. (1984), Sagov et al. (1984), Davis-Floyd (1992), and Mitford (1992).

9. For example, in one study 51% of those with social support at delivery were breast-feeding at 6 weeks compared with only 29% of those who did not receive the extra support (Hofmeyr et al., 1991).

References

Altmann, J. (1980) *Baboon Mothers and Infants.* Cambridge, MA: Harvard University Press.

Aubry, R. (1977) The American College of Obstetricians and Gynecologists (ACOG) standards for safe childbearing. In L. Stewart and D. Stewart, eds., *Twenty-First Century Obstetrics Now! vol. 1*, pp. 15–26. Marble Hill, MO: NAPSAC.

Banta, D., and Thacker, S. (1982) The risks and benefits of episiotomy: a review. *Birth* 9:25–30.

Bertsch, T. D., Nagashima-Whalen, L., Dykeman, S., Kennell, J. H., and McGrath, S. (1990) Labor support by first-time fathers: direct observations with a comparison to experienced doulas. *Journal of Psychosomatic Obstetrics and Gynecology* 11:251–260.

Brackbill, Y., Rice, J., andYoung, D. (1984) *Birth Trap: The Legal Low-Down on High-Tech Obstetrics.* St. Louis, MO: C. V. Mosby.

Brown, S., and Lumley, J. (1994) Satisfaction with care in labor and birth: a survey of 790 Australian women. *Birth*: 21:4–13.

Caldeyro-Barcia, R. (1979) The influence of maternal position on time of spontaneous rupture of the membranes, progress of labor, and fetal head compression. *Birth and the Family Journal* 6:7–15.

Chalmers, B. and Wolman, W. (1993) Social support in labor—a selective review. *Journal of Psychosomatic Obstetrics and Gynecology* 14:1–15.

Conroy, G. C. (1990) *Primate Evolution.* New York: W. W.Norton.

Curtis, P. (1993) Oxytocin and the augmentation of labor. *Human Nature* 4: 351–366.

Davis-Floyd, R. E. (1992) *Birth as an American Rite of Passage.* Berkeley: University of California Press.

Devitt, N. (1977) The transition from home to hospital birth in the United States, 1930–1960. *Birth and the Family Journal* 4:47–58.

D'Esopo, D. A (1941) The occipito-posterior position: its mechanism and treatment. *American Journal of Obstetrics and Gynecology* 38:937–957.

Edwards, M. E. (1973) Unattended home birth. *American Journal of Nursing* 73:1332–1335.

Francome, Colin, and Savage, Wendy (1993) Caesarean section in Britain and the United States 12% or 24%: is either the right rate? *Social Science and Medicine* 37:1199–1218.

Freeman, R. (1990) Intrapartum fetal monitoring—a disappointing story.

New England Journal of Medicine 322:624–626.

Gilgoff, A. (1979) *Home Birth: An Invitation and a Guide*. Granby, MA: Bergin and Garvey.

Golay, J., Vedam, S., and Sorger, L. (1993) The squatting position for the second state of labor: effects on labor and on maternal and fetal well-being. *Birth* 20:73–78.

Goodall, J., and Athumani, J. (1980) An observed birth in a free-living chimpanzee (*Pan troglodytes schweinfurthii*) in Gombe National Park, Tanzania. *Primates* 21:545–549.

Guyer, B., Strobino, D. M., Ventura, S. J., and Singh, G. K. (1995) Annual summary of vital statistics- 1994. *Pediatrics* 96:1029–1039.

Harvey, P. H., and Clutton-Brock, T. H. (1985) Life history variation in primates. *Evolution* 39:559–581.

Haverkamp, A. D., and Orleans, M. (1983) An assessment of electronic fetal monitoring. D. Young, ed., *Obstetrical Intervention and Technology in the 1980s*. New York: The Haworth Press.

Hays, B. M. (1996) Authority and authoritative knowledge in American birth. *Medical Anthropology Quarterly* 10:291–294.

Hillan, E. M. (1992) Short-term morbidity associated with cesarean delivery. *Birth* 19:190–194.

Hodnett, E. D., and Osborn, R. W. (1989) A randomized trial of the effects of monitrice support during labor: mothers' views two to four weeks postpartum. *Birth* 16:177–183.

Hofmeyr, G. J., Nikodem, V. C., Wolman, W. L., Chalmers, B. E. and Kramer, T. (1991) Companionship to modify the clinical birth environment: effects on progress and perceptions of labour and breastfeeding. *British Journal of Obstetrics and Gynecology* 98:756–764.

Hueston, W. J., Applegate, J. A., Mansfield, C. J., King, D. E. and McClaflin, R. R. (1995) Practice variations between family physicians and obstetricians in the management of low-risk pregnancies. *Journal of Family Practice* 40:345–351.

Inch, S. (1984) *Birthrights: What Every Parent Should Know about Childbirth in Hospitals*. New York: Pantheon Books.

Insel, T. R. (1992) Oxytocin—a neuropeptide for affiliation: evidence from behavioral, receptor autoradiographic, and comparative studies. *Psychoneuroendocrinology* 17:3–35.

Jordan, B. (1993) *Birth in Four Cultures*. Prospect Heights, Il.: Waveland Press.

Katz, V. L. (1993) Two trends in middle class birth in the United States. *Human Nature* 4:367–382.

Kay, M. A. (1982) *Anthropology of Human Birth*. Philadelphia, PA: F. A. Davis.

Kennell, J., Klaus, M., McGrath, S., Robertson, S., and Hinkley, C. (1991) Continuous emotional support during labor in a US hospital. *Journal of the American Medical Association* 265:2197–2201.

Keinan, G., Ezer, E., and Feigin, M. (1992) The influence of situational and personal variables on the effectiveness of social support during childbirth. *Anxiety Research* 4:325–337.

Klaus, M. H., and Kennell, J. H. (1982) *Parent-Infant Bonding*. St. Louis, MO: C.V. Mosby.

Klaus, M., Kennell, J., Berkowitz, G., and Klaus, P. (1992) Maternal assistance and support in labor: father, nurse, midwife, or doula? *Clinical Consultations in Obstetrics and Gynecology* 4:211–217.

Kyman, W. (1991) Maternal satisfaction with the birth experience. *Journal of Social Behavior and Personality* 6:57–70.

Labrecque, M., Marcoux, S., Pinault, J-J, Laroche, C., and Martin, S. (1994) Prevention of perineal trauma by perineal massage during pregnancy: a pilot study. *Birth* 21:20–25.

Lagerkrantz, H., and Slotkin, T.A. (1986) The "stress" of being born. *Scientific American* 254:100–107.

Leutenegger, W. (1972) Functional aspects of pelvis morphology of simian primates. *Journal of Human Evolution* 3:201–222.

Leutenegger, W. (1981) Encephalization and obstetrics in primates with particular reference to human evolution. In E. Armstrong and D. Falk, eds., *Primate Brain Evolution: Methods and Concepts*, pp. 85–92. New York: Plenum.

Lovejoy, O. (1988) Evolution of human walking. *Scientific American* 259: 118–125.

Lydon-Rochelle, M. T., Albers, L. and Teaf, D. (1995) Perineal outcomes and nurse-midwifery management. *Journal of Nurse-Midwifery* 40:13–18.

MacCormack, C. P. (1982) *Ethnography of Fertility and Birth*. New York: Academic Press.

Marieskind, H. I. (1989) Cesarean section in the United States: has it changed since 1979? *Birth* 16:196–202.

McKay, S., and Smith, S.Y. (1993) What are they talking about? Is something wrong? Information sharing during the second stage of labor. *Birth* 20:142–147.

McNiven, P., Hodnett, E., and O'Brien-Pallas, L. (1992) Supporting women in labor: a work sampling study of the activities of labor and delivery nurses. *Birth* 19:3–8.

Michaelson, K. L. (1988) Childbirth in America: a brief history and contemporary issues. In K. L. Michaelson, ed., *Childbirth in America*, pp. 1–32. South Hadley, MA: Bergin and Garvey.

Mitford, J. (1992) *The American Way of Birth*. New York: Dutton.

Myles, M. F. (1975) *Textbook for Midwives*. Edinburgh: Churchill Livingstone.

Mynaugh, P. A. (1991) A randomized study of two methods of teaching perineal massage: effects on practice rates, episiotomy rates, and lacerations. *Birth* 18:153–159.

Naroll, Frada, Naroll, Raoul, and Howard, Forrest H. (1961) Position of women in childbirth. *American Journal of Obstetrics and Gynecology* 82:943–954.

Nielsen, T. F. (1986) Cesarean section: a controversial feature of modern obstetric practice. *Gynecological and Obstetrical Investigations* 21:57–63.

Nielsen, T. F., Olausson, P. O., and Ingemarsson, I. (1994) The cesarean section rate in Sweden: the end of the rise. *Birth* 21:34–38.

Nolan, M. (1995) Supporting women in labour: the doula's role. *Modern Midwifery* 5:12–15.

Oakley, A. (1983) Social consequences of obstetric technology: the importance of measuring "soft" outcomes. *Birth* 10:99–108.

Oakley, A. (1984) *The Captured Womb: A History of the Medical Care of Pregnant Women*. Oxford: Basil Blackwell.

Ohio State Medical Society Committee on Maternal Health. (1975) Maternal deaths following unattended delivery. *The Ohio State Medical Journal* 71:627–628.

Oxorn, H., and Foote, W. R. (1975) *Human Labor and Birth*, 3rd ed. New York: Appleton-Century-Crofts.

Prentice, A. and Lind, T. (1987) Fetal heart rate monitoring during labor—too frequent intervention, too little benefit. *Lancet* 2:1375–1377.

Raphael, D. (1973) *The Tender Gift: Breastfeeding*. New York: Schockin Books.

Reynolds, J. L. (1993) The final fatal

blow to routine episiotomy. *Birth* 20: 162–163.

Romalis, S. (1981) *Childbirth: Alternatives to Medical Control*. Austin: University of Texas Press.

Romney, M. and Gordon, H. (1981) Is your enema really necessary? *British Medical Journal* 282:1269.

Rosenberg, K. and Trevathan, W. (1996) Bipedalism and human birth: the obstetrical dilemma revisited. *Evolutionary Anthropology* 4:161–168.

Rothman, B. K. (1993) Episiotomy. In B. K. Rothman, ed., *Encyclopedia of Childbearing: Critical Perspectives*, pp. 126–127. Phoenix, AZ: The Oryx Press.

Russell, J. G. B. (1969) Moulding of the pelvic outlet. *Journal of Obstetrics and Gynaecology of the British Commonwealth* 76:817–820.

Sagov, S. E., Feinbloom, R. I., Spindel, P., and Brodsky, A. (1984) *Home Birth: A Practitioner's Guide to Birth Outside the Hospital*. Rockville, MD: Aspen Systems.

Schultz, A. H. (1949) Sex differences in the pelves of primates. *American Journal of Physical Anthropology* 7: 401–424.

Shearer, M. H. and Eakins, P. S. (1988) Commentary and response: the malpractice insurance crisis and freestanding birth centers. *Birth* 15:30–32.

Shy, K. K., Luthy, D. A., Bennett, F. C., Whitfield, M., Larson, E. B.,Van Belle, G., Hughes, J. P., Wilson. J. A., and Stenchever. M. A. (1990) Effects of electronic fetal-heart-rate monitoring, as compared with periodic auscultation, on the neurologic development of premature infants. *New England Journal of Medicine* 322: 588–593.

Sleep, J., Roberts, J., and Chalmers, I. (1989) Care during the second stage of labour. In I. Chalmers, M. Enkin, and M. J. N. C. Keirse, eds., *Effective Care in Pregnancy and Childbirth*, vol. 2, pp. 1129–1144. New York: Oxford University Press.

Sosa, R., Kennell, J., Klaus, M., Robertson, S., and Urrutia, J. (1980) The effect of a supportive companion on perinatal problems, length of labor, and mother-infant interaction. *The New England Journal of Medicine* 303:597–600.

Stoller, M. (1995a) The obstetric pelvis and mechanism of labor in nonhuman primates. *American Journal of Physical Anthropology* (suppl.) 20: 204.

Stoller, M. (1995b) The obstetric pelvis and mechanism of labor in nonhuman primates. Ph.D. dissertation, University of Chicago.

Stuart-Macadam, P., and Dettwyler, K. A. (1995) *Breastfeeding: Biocultural Perspectives*. Hawthorne, NY: Aldine de Gruyter.

Taffel, S. M., Placek, P. J., and Kosary, C. L. (1992) U. S. cesarean section rates 1990: an update. *Birth* 19:21–22.

Thornton, J. G., and Litford, R. J. (1994) Active management of labour: current knowledge and research issues. *British Medical Journal* 309: 366–369.

Trevathan, W. R. (1987) *Human Birth: An Evolutionary Perspective*. Hawthorne, NY: Aldine de Gruyter.

Trevathan, W. R. (1988) Fetal emergence patterns in evolutionary perspective. *American Anthropologist* 90:19–26.

Ward, C. V. (1994) Locomotor and obstetric constraints in hominoid pelvic evolution. *American Journal of Physical Anthropology* (suppl.) 18: 204.

Wertz, R. W., and Wertz, D.C. (1977) *Lying-In: A History of Childbirth in America*. New York: Free Press.

Wolman, L., Chalmers, B., Hofmeyr, J., and Nikodem, V.C. (1993) Postpartum depression and companionship in the clinical birth environment: a randomized, controlled study. *American Journal of Obstetrics and Gynecology* 168:1388–1393.

Woolley, R. J. (1995) Benefits and risks of episiotomy: a review of the English-language literature since 1980. Part I. *Obstetrics and Gynecology Survey* 50:806–820.

9

DARWINIAN MEDICINE AND THE EMERGENCE OF ALLERGY

KATHLEEN C. BARNES

GEORGE J. ARMELAGOS

STEVEN C. MORREALE

A relatively recent view within the paradigm of Darwinian medicine is that allergy is an adaptive response because it serves a defensive function in which cellular attack by toxins is minimized and potentially harmful substances are diluted and expelled from the body. Advocates of this particular hypothesis question the theory that IgE-mediated, immediate hypersensitivity is part of an earlier adaptive response to helminthic infestations, which in the absence of exposure to parasites (the antigens for which the system arguably may have evolved to combat) overreacts to common proteins (e.g., pollens, dust mite allergen) and subsequently manifests as allergic disease. We argue that the rise of allergy can in fact be best understood in part from this broader evolutionary perspective, and that IgE, one of the key players involved in the immediate-type allergic response, originated in mammals chronically exposed to helminthic parasites as an adaptive response. Within our theoretical framework, we contend that the immunological response probably did not produce widespread adverse health effects because of the co-evolution of a check-and-balance mechanism by which the adverse effects of the hypersensitive state were controlled; specifically, IgG-blocking antibodies and possibly other mediators in the cytokine network minimized the systemic effects of the allergy cascade but at the same time allowed for a control of the worm load in chronically infested individuals. Thus, this mechanism was advantageous to both the host and pathogen, creating an acceptable level of symbiosis between the two.

In *The Dawn of Darwinian Medicine*, Williams and Nesse (1991:2) decry the lack of attention that medicine pays to evolutionary biology, contending that the "application of evolutionary principles to medical problems" has already demonstrated that advances would be expedited "if medical professionals were as attuned to Darwin as they are to Pasteur." Darwinian medicine, in their view, applies evolutionary theory to particular problems by constructing hypotheses that can be tested by comparative and experimental methods. The key to Darwinian medicine is the adaptationist program, in which unsuspected adaptive processes "can be searched for and, if found, described" (Williams and Nesse 1991:3); for example, they argue that fever, morning sickness, sleep disorders, menstruation, and allergy are among the "medical problems" that can be understood from this perspective. Puzzling over the function of IgE antibody production, they state:

> IgE antibody seems to do almost nothing, at least in modern industrial countries, except cause allergy. It would appear that we evolved this special IgE machinery for no better reasons than to punish random individuals for eating cranberries or wearing wool or inhaling during August (Nesse and Williams 1994:160).

In the context of Darwinian medicine, they surmise:

> We can be sure that the capacity for an allergic reaction is a defense against some kind of danger, or else the underlying mechanism, the immunoglobulin-E (IgE) part of the immune system, would not exist. It is perhaps conceivable that our IgE system is a remnant of a system that was useful for other species, but this is unlikely because systems of this complexity degenerate quickly if they are not maintained by natural selection and even more quickly if they cause any harm. It is much more likely that the IgE system is somehow useful (Nesse and Williams 1994: 159).

We argue that the theory of allergy as an adaptive process as formulated by Williams and Nesse (1991; Nesse and Williams 1994), Ewald (1980, 1993, 1994), and Profet (1988, 1991, 1992, 1995) is conceived from a perspective that narrowly defines adaptation in a biological system as one that is perfectly attuned to the insults that threaten survival. Our evolutionary perspective considers adaptation as a *temporal* evolutionary process that often involves tradeoffs that may not represent a perfect adjustment to the insult. If the selective pressure for a specific adaptive feature is reduced, natural selection will not necessarily lead to a quick degeneration of the system. Adaptations are built upon existing systems, and they do not always respond to challenges in a perfect manner. Furthermore, once there is a reduction in the selection for a particular trait, it may be maintained because of the inertia that exists in a cascade of complex features, including redundancies, that are part of the adaptive system.

Evolution of hominids, environmental changes, and evolution of cultural systems have altered the disease ecology, and these changes have had an impact on infectious, chronic, and allergic disease in human populations. We use an evolutionary framework to describe the rise of allergy in human populations as the result of a series of epidemiological transitions,[1] with special consideration of the role of extracellular parasites as a driving force for this process.[2] This chapter focuses particularly on the mechanics of the IgE-mediated process, which is a common denominator for both atopy and the human host's response to helminthiasis.

The Epidemiological Transition and Allergy

As early hominids expanded their niche and increased their dietary breadth, they likely consumed new foods and became an intermediate host to more species of helminths. As they evolved, hominids would therefore have carried the pathogens and parasites that plagued their ancestors. Consequently, these pathogens and parasites, or "heirloom species" (Sprent 1969a), would have evolved with the hominids. Kliks (1983) notes that helminths easily made this trip because they are less environmentally and host restricted. Humans are, in fact, definitive, intermediate, or accidental hosts[3] for more than 25 species of helminths (Warren et al. 1990).

The shift from gathering-hunting to primary food production represents such a dramatic change in disease ecology that we consider it the first epidemiological transition. The shift to agriculture increased sedentism and with it an increase in parasitic disease spread by contact with animals and other humans and their waste. Products of domestic animals that often cause allergic responses today, such as milk, dander, and components of dust kicked up by animals, also contributed to the emergence of new zoonotic diseases (e.g., anthrax, Q fever, brucellosis, tuberculosis) with the domestication of animals during this first epidemiological transition (Lappe 1994). Agricultural practices also increased contact with parasites not dependent on a vector. Irrigation (contact with schistosomal cercariae) and the use of feces as fertilizer (infection with intestinal flukes) would have been a frequent source of infection (Cockburn 1971).

The pronounced relationship between humans and parasites continued until the second epidemiological transition, in which acute infectious diseases are controlled and there is a rise in chronic, noninfectious, degenerative diseases. Allergy is just one example of a chronic disease associated with the second epidemiological transition (Corruccini and Kaul 1983). Even though many human populations throughout the world have benefited from the control of infectious disease, vast segments of the earth's inhabitants are still exposed to infectious disease and parasites. Consider, for example, that approximately 300 million people are infected with schistosomes (UNDP 1990), and more than 1 billion are infected with nematodes (Bundy and Medley 1992). It can be argued, therefore, that there is continued selective pressure for a protective mechanism(s) against helminthiasis among a substantial proportion of the contemporary human population.

Defining Allergic Disease

"Allergy" is a common medical term (Greek *allos* = other) that originally referred to an altered reaction, either beneficial or harmful, within the immune system. A more contemporary and specific description of allergy is a hyper-, or abnormal, sensitivity to a substance that, for most individuals, is innocuous (Note that "allergy" and the term "atopy" are used interchangeably.) The process of "becoming allergic" is not entirely clear, but by definition begins after sensitization to a certain *allergen* (a substance that elicits an allergic response). During sensitization, a period of time which can vary from weeks to decades, *antibodies* (a certain class of molecules produced by B lymphocytes, which are a type of white blood cell that are part of the adaptive immune system) are generated that are specific to the allergen(s) to which an individual has been exposed. There are five major classes of antibodies referred to as *immunoglobulins*, and they are, in order of concentration, IgG, IgA, IgM, IgD, and IgE. Each type of immunoglobulin has a specific function in the immune system, and IgE is most relevant in allergic disease.

Technically there are four types of hypersensitivity. True allergy falls under the category of type I or immediate-type hypersensitivity, which is also referred to as *IgE-mediated* hypersensitivity. A common and well-known example of type I hypersensitivity is an intense reaction to an allergen such as pollen. Type II hypersensitivity refers to antibody-dependent cytotoxic hypersensitivity, which occurs when antibody which is directed against certain target cells (e.g., incompatible blood cells) interacts with molecules and effector cells of the immune system (e.g., platelets, neutrophils, eosinophils, macrophages, monocytes) in order to destroy the target cells. A good example of type II hypersensitivity is the response to blood cells following an incompatible blood transfusion. Type III hypersensitivity, or immune complex-mediated hypersensitivity, is most typically manifested as *serum sickness*, which results from circulating immune complexes that are deposited in tissues which can lead to inflammatory diseases (e.g., arthritis). Type IV, or delayed-type hypersensitivity (DTH), is usually characterized as *allergic contact dermatitis,* such as poison ivy or poison oak, or as a *graft rejection.* Unlike types I–III, which are antibody-mediated forms of hypersensitivity, type IV hypersensitivity is mediated by *T cells* and macrophages and does not involve the IgE antibody. T cells are a second type of lymphocyte which recognize different regions of antigens than the regions recognized by B cells.

Allergy can be manifested as asthma, allergic rhinitis, urticaria, eczema (dermatitis), or anaphylaxis. The overall effect of IgE antibodies, released by plasma cells and then bound to receptors on mast cell surfaces (located in connective tissue on epithelial surfaces including the respiratory, urinary, and gastrointestinal tracts and the skin), is not unlike that seen in helminthiasis (see Capron and Capron 1994; Sutton and Gould 1993; Zanders et al. 1992 for review). After IgE antibodies encounter the antigen to which they are specific, the antibodies crosslink with each other, and there is rapid activation of mast cells to degranulate and release an array of anaphylaxis mediators, including histamine, platelet-activating factor, leukotrienes, prostaglandin D, and proinflammatory cytokines. However, unlike helminthiasis, in which IgE antibodies are directed at the worm and its by-products, the allergic response is directed at seemingly innocuous substances (e.g., pollens, components of house dust).

The general manifestation of this overreaction of mediators includes bronchial constriction, vascular dilatation, increased vascular permeability, and an increase in mucus secretions, which lead to associated symptoms such as wheezing, coughing, itching, sneezing, vomiting, and diarrhea.

Anaphylaxis is the most severe form of allergic reaction. It is a life-threatening immunological event that results from the generation and release of these inflammatory mediators and their effects on various organs. Numerous IgE-mediated agents can induce an anaphylactic response, including antibiotics, (e.g., penicillin), foreign protein agents (e.g., venoms), therapeutic agents, and foods (notably milk, eggs, shellfish, and nuts). Fatal anaphylaxis from a bee sting was first recorded some 4500 years ago in Egypt (Cohen and Samter 1992), and Hymenoptera venom remains one of the most common causes of anaphylaxis today, yet causes only about 40 deaths per year in the United States (Marquardt and Wasserman 1993). Because a very small amount of substance, or venom, is injected from one or two stings, it is believed that most reactions to hymenoptera venom are genuinely allergic rather than toxic (Valentine 1992).

Common Allergens

It is impossible to engage in any discourse on the evolution of allergy without first addressing the question of what makes an allergen an allergen. Over the past several decades a tremendous amount of work has been conducted in the field of allergen research. In general, most antigens that evoke an allergic response and that have been purified so far are proteins or glycoproteins that range in molecular weight from 3 to 70 kDa, and they are usually acidic (Lind and Løwenstein 1988). Most helminth allergens are about the size and charge of inhalant allergens, such as pollens and animal danders, although one exception is the very large allergen (360 kDa) from the large intestinal nematode *Ascaris*, commonly referred to as roundworm (O'Donnell and Mitchell 1978). Unlike inhalant allergens, however, which must be small enough to pass through mucosal membranes (because they are inhaled) in order to activate mast cells, helminth allergens are dropped directly on the tissues where the worm is located or migrating, such as the gut, so size limitation is less of a factor (Kennedy 1992).

In recent years, the most common allergens affecting humans have been purified, characterized, and studied intensively (e.g., Arruda et al. 1990; Burge 1989; Chua et al. 1988; de Weck and Todt 1988; Platts-Mills et al. 1985, 1989; Pollart et al. 1991; Schou et al. 1991; van Metre et al. 1988). The two major groups of indoor allergens found in the domestic environment are arthropod-borne allergens (especially the house dust mites *Dermatophagoides pteronyssinus* and *D. farinae*) and allergens from domestic pets (e.g., cats) (Colloff et al. 1992; Platts-Mills et al. 1992). Mites of the genus *Dermatophagoides* are considered to be the primary indoor allergens worldwide (Platts-Mills and de Weck 1989; Platts-Mills et al. 1992; Sporik et al. 1990). Mold and pollen allergens are also important, but represent a smaller proportion of the total do-mestic allergen level, and exposure to them occurs primarily outdoors and in the occupational setting (Colloff et al. 1992).

Sensitivity to many potential allergens can be measured either by the skin-prick test or by the measurement of specific IgE antibodies in the blood (e.g., radioallergosorbent test, enzyme-linked immunosorben assay). People who are hypersensitive to a particular allergen can reduce their morbidity by al-lergen avoidance—for example, by reduction of house dust mites in the home (Platts-Mills and Chapman 1987; Platts-Mills et al. 1982). In fact, so much information has been collected in this area of clinical immunology that at-risk levels of some of the most common allergens (e.g., house dust mites, cat) have been established (Hamilton et al. 1992; Platts-Mills and de Weck 1989; Sporik et al. 1990).

Therefore, it seems that one of the most significant consequences of mod-ernization is the creation of a microenvironment that increases our exposure to peridomestic and domestic arthropods and other pests. As we remove our-selves from climatic elements by constructing buildings that protect us from the elements, we have created an ideal habitat for a vast array of pests. Urban entomology has in fact shown us that household infestations of most pests increase with the degree of urbanization (mosquitoes, Gratz 1973; cock-roaches, Koehler et al. 1987; and rodents, McNeill 1976).

Epidemiologic Patterns

Allergy is a relatively common disease, afflicting some 50–60 million Ameri-cans, or approximately 1 in 5 adults and children (AAFA 1996). The general consensus is that prevalence, morbidity, and mortality are increasing (Platts-Mills et al. 1997; Vogel 1997; Anonymous 1998), and better diagnostics and awareness cannot completely account for the overall increase. After allergic rhinitis, asthma is the most common chronic condition, affecting about 5–8% of the general population in the United States and in other developed coun-tries; however, this figure varies considerably according to ethnicity and na-tionality.

Curiously, there is a lower prevalence of atopy in less-developed countries than in developed countries (Cookson 1987; Waite et al. 1980). A close look at changes in the environment of contemporary populations and the preva-lence of atopic disease among them supports our argument about the relation-ship between allergy and the second epidemiological transition. That is, pop-ulations in underdeveloped regions with a high prevalence of helminthiasis tend to demonstrate a lower prevalence of asthma than populations whose better-developed infrastructure has resulted in the elimination of helminths and an increase in chronic disease (Godfrey 1975; Lynch 1987, 1992; Turner 1980). One hypothesis that has arisen from this finding is that antibodies nor-mally targeted at fighting helminth allergens do so at such a rate that there is literally a saturation of these antibodies to the degree that there can be no response to innocuous allergens such as pollens. However, in the absence of this saturation, the immune system is free to overreact to these otherwise "unimportant" allergens. To this end, helminthiasis and allergy are mutually exclusive (Bazaral et al. 1973).

However, evidence has been presented that suggests that parasites actually cause asthma (Joubert et al. 1980; Tullis 1970). Tullis (1970) proposed this counter-hypothesis and observed that Canadian asthmatics were more likely to be infested with gastrointestinal nematodes than nonasthmatics. The study was subsequently criticized on the ground of mistaken identification of fecal parasite ova (Grove 1982) and on the lack of association between asthma and intestinal parasites in other populations (Van Dellen and Thompson 1971). Lynch (1992:59–60) conducted a more thorough investigation taking into consideration ethnicity, socioeconomic status, and disease severity and found that "allergic reactivity of tropical populations could depend on the intensity, and possibly type, of helminthic infection." The conclusion from this finding is that one is more likely to develop asthma in the absence of helminthiasis, or during phases of helminthiasis in which the person is only suffering from mild disease, in which case there is much less antibody directed at the parasite than in full-blown parasitic disease. In short, helminthiasis and allergy are probably not mutually exclusive, but allergy is much less likely to occur in severe helminthic disease than in mild helminthic disease.

Populations that undergo rapid modernization, in terms of the introduction of objects that promote common allergens (e.g., foam and spring mattresses, upholstered furniture, carpeting, air-tight homes, domestic pets) also demonstrate rapid increases in asthma prevalence regardless of parasite status. The first and most convincing study was conducted by Dowse et al. (1985), who studied the incidence of asthma in eight South Fore villages in the Eastern Highlands of Papua New Guinea. There had been an alarming increase in prevalence of asthma over a 10-year period; the overall rate was 39 and 46 times higher than rates noted in 1972 and 1975, respectively, in the same villages and was much higher than rates reported from other highland regions in New Guinea. The increase was attributed in part to the introduction of wool blankets to the villagers; the sudden and profound exposure to house dust mites embedded in the blankets resulted in the expression of allergy.

The effects of housing on sensitivity to common inhalant allergens has been demonstrated elsewhere. Barnes and Brenner (1996) found a relationship between allergy to cockroach antigens and better built homes in the Dominican Republic and concluded that the masonry house was more conducive to the proliferation of large cockroach populations than the cheaper, wood-frame houses because the masonry homes provided a better habitat for cockroaches, allowing for a greater build-up of cockroach allergen. The same association was found between better-built homes and higher house dust mite allergen concentrations in Barbados (Barnes et al. 1997) probably because the indoor plumbing associated with masonry homes contributes to higher humidity levels that are conducive to house dust mite proliferation than the somewhat drier environment present in wood homes.

As a population undergoes modernization, extreme changes in the environment and technology produce an array of organic and inorganic substances. These substances are ubiquitous and are inhaled and consumed at home and at the workplace. In situations where class differences emerge, there is often a continuous effort to "move up" socially and "improve" the home environment by adding to it nonconsumable goods. This creates an environment in which allergens become problematic. If, in the process of modernization, parasitic disease is reduced or eliminated, the risk of allergic disease increases even more dramatically.

The Role of Genetics

Much of our understanding of the genetics of IgE and its protective role against helminths derives from studies of the genetics of allergic disease. Numerous studies of twins and multigenerational families have demonstrated that genetic factors are involved in total serum IgE concentration (Bazaral et al. 1974; Edfors-Lubs 1971; Gerrard et al. 1978; Marsh et al. 1974; Martinez et al. 1994; Meyers et al. 1987). To date, genes have been implicated for IgE responsiveness in a number of chromosomal regions (Barnes et al. 1996; CSGA 1997; Daniels et al. 1996; Huang et al. 1991; Marsh et al. 1982, 1990, 1994; Meyers et al. 1994; Moffatt et al. 1994; Ohe et al. 1995; Postma et al. 1995; Rosenwasser et al. 1995; Sandford et al. 1993; Shirakawa et al. 1994; Song et al. 1996; Turki et al. 1995; Zwollo et al. 1991).

It comes as no surprise, given a long symbiotic history with parasites, that the mammalian inflammatory network has evolved mechanisms, both unique and redundant, that function to maximize host well-being in the event of parasitic infestation. There are redundancies, for example, within the genes encoding the complement cytokines or cell-adhesion molecules and their receptors (Paul and Seder 1994; Springer 1990). We assume that mutations have occurred but that their effects are modulated by the interactive nature of the complex system. Thus, we accept a priori that there will be multiple mutations of multiple genes controlling the inflammatory response (including regulatory elements of these genes) and that these mutations will have variable and interdependent effects on the expression of the different allergic disease phenotypes. The expression of these "inflammation genes" will also vary according to the environmental context.

Some of the early molecular studies of genes controlling the allergic response focused on the HLA-D subregions (DR, DQ, DP) of the human major histocompatibility complex (MHC), located on chromosome 6p21.3 (Marsh et al. 1982, 1987, 1990. Barnes and March 1998). This region was an obvious group of candidate genes for an immune response that is specific—that is, for the recognition of specific allergens, such as the most important allergen of a certain pollen. It was hypothesized that the specific immune response involved the interaction of a single MHC class II molecule with a single major T-cell epitope of the antigen. Based on this assumption, it was expected that unrelated subjects would share a role for *HLA-D* genes in specific immune responses. Analyses were simplified by studying a very restricted phenotype: immune responsiveness to the major allergen of short ragweed pollen (Amb a 5). Marsh and his associates (1982) were able to establish that the class II restriction was limited to allergic people carrying a specific HLA genotype in 95% of the cases. But allergic people who were not sensitive to Amb a 5 had a frequency of this particular HLA genotype similar to the Caucasian population at large (~22%). Over the past several years, numerous studies have confirmed these findings, and the HLA-D association has been extended to other allergens, such as giant ragweed and perennial rye, and also to house dust mite allergens (Blumenthal et al. 1992; Freidhoff et al. 1988; O'Hehir et al. 1991; Roebber et al. 1985).

Prevalence of atopy has been shown to be higher in non-European descendants living in developed countries (Davis et al. 1961; Marsh et al. 1980; Orgel et al. 1974; Worth 1962), possibly because those individuals carry a greater propensity for the production of IgE and at the same time are exposed to a vast array of inhalant allergens. Under our paradigm, the selective advantage for a haplotype predisposed to producing high levels of IgE, a key component in the mechanisms for destroying invading helminths, would be highest in a population chronically exposed to helminths, or in subpopulations that were, until recently, chronically exposed.

Both genetic and epidemiological studies of human susceptibility/resistance to helminthiasis support the notion that natural selection is in fact operating to maintain the IgE immune network as a protective measure against parasites. Perhaps the most convincing data for the protective role of the human immune system in helminthiasis are from studies that suggest a genetic propensity for susceptibility/resistance to parasitic infection. The importance of predisposition to helminthiasis has been established as an epidemiological factor for whipworms, roundworms, hookworms, and schistosomes (Haswell-Elkins et al. 1987; Schad and Anderson 1985; Tingley et al. 1988). Certainly, predisposition may depend on factors such as nutrition, behavior, or the environment, but the role of genetics cannot be underestimated. For example, one study concluded that protective IgG4 antibodies, a subset of IgG antibodies, to *Onchocerca volvulus* were significantly higher in Ecuadorian Indians compared to blacks of African origin living in the same area and that ophthalmological complications were less pronounced in the Indian population (Kron et al. 1993). Furthermore, certain individuals are more likely to suffer from rapid and severe reinfection following chemotherapy (e.g., oxaminiquine) than others (Dessein et al. 1988; Tingley et al. 1988).

Perhaps the most-relied upon model for understanding both the genetics and epidemiology of helminthic infection and its relationship to IgE is the disease associated with three species of the trematode *Schistosoma* (*S. mansoni, S. haematobium,* and *S. japonicum*), schistosomiasis. The schistosoma parasite is transmitted to humans via free-swimming cercariae, which emerge from snails, the intermediate host. Once the parasite penetrates the human skin, it may grow and reproduce inside the human host for years, in either the genitourinary tract or gastrointestinal tract (depending upon the species), producing chronic disease associated with either of those systems (see Hagan et al. 1992; Jordan and Webbe 1982 for review).

Early studies of at-risk subjects—people living in regions endemic for *S. mansoni*—found that infection intensities were largely dependent on the intrinsic susceptibility/resistance to infection (Butterworth et al. 1985; Dessein et al. 1988; Katz et al. 1978). Abel and co-workers (1991) detected a codominant major gene responsible for human susceptibility/resistance of infection to *S. mansoni* among 20 Brazilian multigenerational families. More recently, the same group of scientists reported evidence for linkage to susceptibility of *S. mansoni* infection and a major gene on chromosome 5q31-q33 (Marquet et al. 1996), which is the same region that linkage to high IgE antibodies among allergic subjects was previously reported (Marsh et al. 1994; Meyers et al. 1994; Ohe et al. 1995; Postma et al. 1995; Roesnwasser et al. 1995; Song et al. 1996; Turki et al. 1995). Similarly, the same group who found the first evidence of linkage to atopy on Chromosome 11g13 has found that a variation of FCERIB, the candidate gene in that region, regulates IgE levels in a group of Australian Aborigines living in a region endemic for helminth infections (Palmer et al. 1997). It would not be surprising to find tremendous overlap in the many genes that are no doubt contributing to both allergic disease and helminthic disease.

Thus, there is clearly a genetic component in the development of high IgE levels. This is perhaps best illustrated in epidemiologic studies, where it has been shown that levels of mite allergens defined as clinically 'high' (i.e., high enough that they are a risk factor for allergies in people with a family history of allergies) do not cause allergies in most people (Lau-Schadendorf and Wahn 1991; Platts-Mills et al. 1989; Sporik et al. 1990). This finding implies that only genetically susceptible individuals become allergic or asthmatic and only when adequately and persistently exposed to certain allergens, most notably dust mites, but also to pollens, molds, and animal danders.

The Role of Helminthic Infection in the Evolution of IgE

IgE and Immunoecology

Nesse and Williams (1994:159) purport it is unlikely that "our IgE system is a remnant of a system that was useful for other species" because a useless or harmful mechanism (vis-a-vis allergy) would be quickly eliminated under natural selection. A review of the evolution of the human immune system illustrates that, whatever the selective forces, this complex system evolved long before the emergence of humans and independently of their cultural ways. Of particular interest here is the humoral arm immunity, or antibodies—specifically *immunoglobulins*. The five classes of immunoglobulins are present in all eutherian mammals. IgE is believed to have emerged most recently, evolving within the past 300 million years (Hadge and Ambrosius 1984). Given the relatively recent emergence of modern humans (~100,000 years ago)[4], the complete elimination of such a fundamental process in the immune system, even if it occasionally overreacts, seems unlikely in such a short period of time.

We agree with Nesse and Williams that the "IgE system" is not likely to be a "remnant" of a system that is no longer beneficial, but the basis of our agreement is the overwhelming amount of data collected over the years that has suggested that IgE confers at least some benefit to the mammalian host (humans included) in a relatively predictable manner in response to parasites. Some of the most convincing evidence of the IgE–helminthic relationship is derived from animal studies. High levels of serum IgE actually minimize the number of parasites in a nonhuman animal host chronically exposed to large numbers of parasites (Dessein et al. 1981; Hsu et al. 1974); similar findings have recently been made in humans (Dunne et al. 1992; Hagan et al. 1991). In other words, serum IgE antibodies and the degree of parasitic infestation are inversely correlated in both animals and humans. This observation is further supported by the finding that selective removal of IgE reduces the protection against schistosomes (and presumably other helminths) in animal models (see Capron et al. 1987).

Because the first appearance of mammals was toward the end of the Triassic period of the Mesozoic era, we can assume that, potentially, the symbiotic relationship of mammals and parasites spans nearly 200 million years. The explosive divergence and expansion of mammals with the onset of the Cenozoic era some 65 million years ago must have allowed for the creation of new niches for both host and parasite at a staggering rate, contributing to the complexity and, subsequently, the stability of parasites and a refinement of their *immunoecology*.[5] It is little wonder that these fast-growing "infrapopulations" of parasites have developed an immunological mechanism that is complex and not entirely understood if we consider the contrast in the pace of host–parasite *microecological* changes (e.g., microhabitat of the host, such as the immune system) compared to the much slower *macroecological* changes, such as the environment in which humans live, human food ways, and epidemiological transitions (Price 1990). It is noteworthy that "the large number of parasite generations that occur in vivo would be the equivalent to thousands of evolutionary years for humans and other large animals with long generation times" (Seed 1993:470). Indeed, given this long symbiotic history, and "because of the rapid changes that occur in the microenvironment during an infection, one has parasite infrapopulations under dramatic physical, physiological, and immunological selective pressures" (Seed 1993:470–471), a number of feedback mechanisms have developed that maintain parasite survival and, at the same time, minimize damage to the host.

Antibody-dependent Mechanisms of the Host

Having established that there is indeed an irrefutable relationship between IgE antibodies and extracellular parasitic disease, we now ask the questions, How is IgE antibody regulated, and how does it benefit the host? Basal serum IgE is relatively low when compared to quantities of the other four immunoglobulins; levels of IgE typically range from 10 to 100 ng/ml, or less than 0.001 of the body's total immunoglobulins (Johansson and Bennich 1985). Elevated IgE occurs primarily in two clinical conditions: helminthic infestations and allergy. Serum IgE levels are typically 10 times higher in people with allergy than in nonallergics, and 10 times higher in people with parasitic infections than in those with allergies (Sher and Ottesen 1988). Contemporary human populations that are chronically infested with parasites exhibit some of the highest titers of serum IgE in response to parasite stimuli (e.g., helminths) compared to other common allergens (Bazaral et al. 1974; Ito et al. 1972; Johansson et al. 1968; Sher and Ottesen 1988).

All extracellular parasites stimulate a humoral and cell-mediated immune response in humans against antigens produced from the parasite throughout its various life cycles (Capron et al. 1982; Dessaint et al. 1975; Jarrett and Bazin 1974; Orr and Blair 1969). The key players in the humoral immune response are IgE antibodies, polymorphonuclear granulocytes such as basophils and eosinophils, and tissue cells such as macrophages and mast cells. The initial IgE-mediated response to the helminth is one of *nonspecific*, polyclonal activation of B cells. In other words, the antibodies are not directed toward any particular antigen or part of the parasite, but to the parasite in general. Subsequently, the response becomes antigen-specific (Hussain et al. 1981; Jarrett et al. 1976), directed toward specific aspects of the helminth, and it is this component of the host's response which has been linked with protection of the host (Hagan et al. 1991). In studies in which parasite-specific IgE antibodies were transferred from one animal to another (*passive transfer*), and the animal receiving the antibodies was exposed to the helminth, protective immunity was evident (Capron and Dessaint 1985; Verwaerde et al. 1987), confirming the conclusion that IgE antibodies are capable of providing protection against parasitic infestations.

When a helminth enters the host, soluble antigens from the parasite diffuse across the host's intestinal mucosa. These antigens are transported to the lymph nodes, where an IgE-mediated response occurs. IgE antibody is then released systemically, where it can attach to cells bearing specific receptors, such as mast cells, located in virtually every tissue in the body, including the intestinal mucosa. Upon contact with the parasitic antigen, the mast cells degranulate and release their pharmacological mediators such as histamine, leukotrienes, prostaglandins, and cytokines, which attract eosinophils and other cells to the site. This cascade of reactions functions to damage and expel the parasite (Brothell 1972; Capron et al. 1975; see also Capron and Capron 1994; Maizels et al. 1993).

In addition to humoral immunity, or IgE-mediated, type I hypersensitivity, in the fight against helminths, a second line of defense has evolved in the human immune system which is also antibody-mediated and is the primary mechanism for the killing of parasite larvae: *antibody-dependent, cell-mediated cytotoxicity* (ADCC). As described here, ADCC, or type II hypersensitivity, occurs when antibody binds to antigen and leads to killer cell activity, phagocytosis (the engulfment of material), or the lysis (killing) of an organism by cell mediators, in this case *complement* (proteins in the serum that control inflammation, the activity of phagocytes, and the attack on foreign cell surfaces). Unlike humoral immunity, which depends on the antibody IgE, ADCC is an IgG-dependent process. Antibodies cause damage to the parasite either by interacting with complement or by interracting with inflammatory cells, or effector cells (e.g., platelets, eosinophils, macrophages).[6] In short, type I hypersensitivity and type II hypersensitivity act in concert to defend the host (human) against extracellular parasites.

Of particular interest in this chapter is the evidence that suggests that human helper T cells can be divided into T_H1 and T_H2 subsets, wherein T_H1 cells are cytolytic and produced in response to intracellular infections (e.g., bacteria, viruses) and T_H2 cells are produced in response to extracellular infections (e.g., helminths) and allergens (Mosmann et al. 1986; Romagnani 1991, 1992). T_H1 cells secrete a ctyokine called interferon-gamma (IFN-γ) and also interleukin-2 (IL-2) (and a host of other cytokines), and T_H2 cells secrete the cytokine IL-4 (and, similarly, other cytokines). *Cytokines* are molecules that mediate intereactions between cells. Interleukin-4 is a critical cytokine in the production of IgE, and it also promotes T cell, mast cell, and macrophage growth (see reviews in Maizels et al. 1993; Paul and Ohara 1987; Zanders et al. 1992). Interleukin-4 is also a major player in the recruitment of eosinophils (Bochner and Schleimer 1994 Mochizukietal, 1998). It is thought that IFN-γ and IL-4 work in opposition, or as a negative feedback mechanism for each other (Romagnani 1992; Romagnani et al. 1990). But this system is not absolute, and IFN-γ is not the only control for IL-4. Earlier in this chapter we indicated that the human immune system has developed a number of redundancies; one example of redundancy is that IL-4 is also produced by activated human basophils and perhaps mast cells (Bradding et al. 1992; MacGlashan et al. 1994). Typically in helminth infections, the ratio of T_H1/T_H2 is unbalanced in favor of the T_H2 profile (Del Prete et al. 1991; Mahanty et al. 1992; Sher and Coffman 1992; Zwingenberger et al. 1991). It is beyond the scope of this chapter to describe in full the complicated and ever-unfolding network of cytokines and other factors involved in this process, but suffice it to say that a yet-to-be fully described mechanism is involved that is seemingly common to both parasitic disease and atopic disease whereby IL-4-producing T_H2 T-cells are selected, which then leads to the production of parasite-specific IgE (Romagnani 1991).

The response to helminths is less than perfect, and the parasite has evolved several clever responses to the host's weapons of attack. Although it has been argued that a truly effective defense mechanism would eliminate the worm burden completely, it is important to note that there are several means by which immune-mediated resistance is conferred to certain hosts. There does not appear to be a single common antihelminth method by which the human immune system defends itself; rather, "the critical T-cell population may differ in each parasite combination and in each tissue site" (Maizels et al. 1993, 802). This once again highlights the complex nature of the parasite's immunoecology and lends further support for a long history of coexistence.

Parasite Feedback Mechanisms

We have just illustrated how the immune response to parasites is complex, being characterized by a number of feedback mechanisms and redundancy, as well as regulatory networks that are possibly independent of the T_H1 and T_H2 pathways. Indeed, as noted by Capron et al. (1992:6), the "heterogeneity" of different components of IgE as well as the different types of cellular responses "may be considered to account for the differences in the consequences of IgE antibody-dependent reactions, from immune defense against helminth parasites to allergy." A broad evolutionary approach in understanding the origin of IgE is supported if we analyze several of the elaborate feedback mechanisms that have seemingly evolved to benefit the parasite. In their review on evasive mechanisms of helminth parasites in the human host, Maizels and co-workers (1993:804) note that, given the elaborate immune system that has evolved in humans, we should expect that clever evasive mechanisms would have evolved in helminth parasites so that they might "counteract T_H2 responses." After all, the success of extracellular parasites in terms of their pervasiveness throughout the developing world "suggests a very dynamic evolution where continued selection enables the parasites to evade or modulate the immunological attack in at least some fraction of the host population (the'wormy' individuals)" (Maizels et al. 1993:804).

Immunological Modulation

Because contacts between the vector and the vertebrate host may be infrequent, it is important for the parasite to maintain chronic infections. Most human parasitic infections persist for years and often decades and must therefore not overwhelm the host if the parasite is to be maintained. Therefore, it is not surprising that parasitic infection seldom induces a potent sterilizing immunity in the vertebrate host; in fact, parasites that are poorly adapted to the host and produce self-limiting infections induce the most efficient protective immune responses (e.g., *Leishmania major*; Sher and Ottesen 1988).

There is good evidence to suggest that many parasites have evolved a balanced, homeostatic relationship with the vertebrate immune system which permits the host to defend itself against potentially lethal levels of infection while maintaining a viable parasite population at the same time. *Concomitant immunity* is one such method, and it is frequently observed in schistosomiasis (Hagan et al. 1992; Sher and Ottesen 1988). Concomitant immunity is a response by which adult schistosomes induce an immune response that limits (but does not eliminate) subsequent infection of the host by infective larvae of the parasite but that does not cause rejection of the adult worms. To this end, concomitant immunity is a sort of "voluntary reduction" of reproduction. Similarly, many parasitic protozoa induce partially effective immune responses that limit the extent of their own replication in the host (Sher and Ottesen 1988).

Although responses may become immunosuppressed during chronic infection, usually the initial encounter between parasites and host results in immune responsiveness. This almost universal finding suggests that it may be in the parasite's interest to be immunogenic, either to initiate an immunoregulatory chain of events that will eventually result in suppression or, alternatively, to protect the host against the potentially lethal effects of overinfection or reinfection (Sher and Ottesen 1988).

IL-10, a Downregulator of IFN-γ

Interleukin-10 is another cytokine which, like IL-4, inhibits the production of IFN-γ (Chomarat et al. 1993; Mosmann and Moore 1991; Romagnani 1991; Sher et al. 1991). The potential importance of IL-10 in parasitic infections has been illustrated by demonstrating a primarily T_H2 type response in mice infected with helminths (*N. brasiliensis, S. mansoni*, Mosmann and Coffman 1989; for review, see Zanders et al. 1992). Despite the observation of the clonal expansion in animals infected with *N. brasiliensis* (Street et al. 1990), an increased production of both IL-4 and IL-10 and a reduction of IFN-γ has been observed in infected mice (Mosmann and Coffman 1989). However, IFN-γ is important for microbiocidal activity because it induces macrophage activation for cytotoxicity against human parasites such as *S. mansoni* (Gazzinelli et al. 1992; Oswald et al. 1992). Interleukin-10 functions as a downregulator by blocking the production of tumor necrosis factor-alpha (TNF-α), which is an important costimulatory factor for the activation of IFN-γ (Oswald et al. 1992b). In this respect IL-10 is different from IL-4, a cytokine that can inhibit macrophage cytotoxicity but does not cause a reduction in TNF-α (Oswald et al. 1992b).

The role of IL-10 is unique and suggests a dual function for this cytokine in terms of parasite killing and parasite maintenance. Interestingly, sequence homology has been observed for IL-10 and IL-4 in both mice and humans, suggesting that "cytokine genes may have been subverted by viruses during evolution as an immunosuppressive strategy" (Zanders et al. 1992:87–88). In addition to the subset of CD4+ T cells, IL-10 is produced by mast cells, Ly-1+ B cells, and macrophages (Oswald et al. 1992a).

IgG-blocking Antibody and Nonspecific IgE Mechanisms

Perhaps the most notorious mechanism that mediates the impact of the IgE reaction is the "blocking" effect of parasite-specific antibodies. Individuals chronically exposed to helminths produce antigen-specific IgG-blocking antibodies of the subclass IgG4, and, like IgE, IgG4 is upregulated by IL-4 (King and Nutman 1993; King et al. 1993), although production can be inhibited by IFN-γ (Sutherland et al. 1993). IgG4 possibly functions by interfering with complement activation and by competing with worm-specific IgE, thereby blocking mast cell degranulation (Iskander et al. 1981; Hussain et al. 1992; Hajan et al. 1991). Interestingly, the amount of IFN-γ needed to block the synthesis of

IgE is lower than the amount needed to stop the synthesis of IgG4 (Hagan et al. 1991); in other words, the worm can stimulate IFN-γ to minimize the negative effect of IgE without jeopardizing the positive effect of IgG4. Egg laying by the parasite is believed to be a significant influence in the production of IgG4 (Grogan et al. 1996).

A study was conducted with volunteers infected with hydatidosis (*Echinococcus granulosus*), which is a small tapeworm of dogs that can cause serious and life-threatening cysts, or "hydatids," in the human liver, lung, and brain. High IgE producers showed an increase in antigen-driven IL-4 and IL-10, but a decrease in IFN-γ, whereas high IgG4 producers showed an increase in IL-4, IL-10, *and* IFN-γ, suggesting that the T_H1 cell subset is, in addition to the T_H2 subset, also regulating the immune response (Riganò et al. 1995). The ratio of IgE to IgG4 appears to be crucial in determining susceptibility to reinfection; for example, adolescents in Brazil with low IgE specific for *S. mansoni* but who had high IgG4 levels were 100 times more likely to be reinfected than their counterparts with high specific IgE levels (Rihet et al. 1991). Similar findings were reported for *S. haematobium* in the Gambia (Hagan et al. 1991).

In summary, the immune system by which the human host battles parasitic disease is highly developed and specialized and depends on the humoral response, which calls for the generation of specific IgE. The worm-specific IgE signals the ADCC response, which generates killer cell activity and phagocytosis via effector cells such as eosinophils, macrophages, and platelets. IgE, the key player, is synthesized and controlled by an elaborate network, the T_H1/T_H2 system, which is downregulated and upregulated by cytokines IFN-γ and IL-4, respectively, among many others. Alternatively, the invading parasite is protected, to some degree, by several clever, highly evolved mechanisms, most notably blocking IgG4 antibodies. The induction of IgG4 is one means by which the parasite has evolved a defense against specific IgE. However, there are a variety of responses to helminths and a continuum of helminthic disease expression. For example, the more successful human host will produce higher levels of specific IgE with which to kill the worm, but the immune system of a host who succumbs to greater worm loads may be less proficient at producing sufficiently high levels of worm-specific IgE. We maintain that allergy is indeed the price we must pay for the elaborate protection against extracellular parasites. We make this assumption based on epidemiological and genetic data that suggest that protection against helminths pre-dates, and in some cases, may even preclude, atopic disease.

How Does the Worm Theory Compare to Other Paradigms?

As discussed earlier, alternative hypotheses regarding the evolution of IgE have been put forth. One approach is derived from the work of Profet (1991), who suggests that the allergic reaction functions as an immunological defense against "toxins." Profet extends the entity of "toxin" beyond extracellular par-

asites to include what Schultes and Hofman (1987:10) define as any "nonnu-
tritional substance that exerts a biodynamic effect on the body." Illustrating
the ubiquitous nature of toxins, Profet (1991:28) explains that "all plants produce toxins" as a means of self-defense, including those that are commonplace
in our diet (e.g., bananas, oranges, spinach), and she also cites the "toxic defenses" of such nonplant organisms as insects, marine animals, and reptiles,
as well as the better-known toxins in the inorganic environment such as lead
and nickel. Many of these substances potentially bind covalently to the DNA
of target cells and are possibly mutagenic and carcinogenic.

To explain how the body can mount an IgE-mediated response to an array
of antigens smaller than 2 kDa, Profet provides a detailed discussion on the
covalent binding of low-molecular-weight toxins to a protein, notably serum
proteins (e.g., albumin). She correctly asserts that, because of their small size,
toxins can only elicit an IgE response when covalently bound to carrier proteins (and are therefore referred to as *haptens*); however, this argument is
weakened as she contends that "the connection between covalent binding,
toxicity, and allergenicity is especially strong in contact allergy" (Profet 1991:
31). The problem with this supposition is that contact allergy is *not* IgE-
mediated. Rather, it is one of the four types of hypersensitivity (type IV), and
is most typically manifested locally as a T cell-mediated dermatitis, peaking
approximately 48 hours after coming in contact with a particular antigen, such
as poison ivy. The antigens that induce contact allergy are typically haptens,
as described by Profet, but they evoke a predominantly "epidermal phenomenon," or reaction of the skin, rather than a *systemic* reaction, and an epidermal reaction in no way compares to a more life-threatening acute response
such as anaphylaxis.

Profet primarily focuses on plant compounds and venoms as the most important allergens, but as we described in detail in the section "Common Allergens," the most important allergens, according to worldwide epidemiologic
studies, are the indoor allergens, such as house dust mites, and outdoor pollens, such as birch or ragweed pollens. In fact, it is not the ingested (e.g., plant
compounds) or injected (e.g., bee stings) allergens that are the most important
etiological agents in allergy. Only 2–4% of children suffer from true food allergies (and the prevalence among adults is much lower; Sampson and Metcalfe 1992), and the prevalence of atopy related to food and especially Hymenoptera venom cannot be compared to that of house dust mites, for
example. Besides, even though toxic plant compounds no doubt were frequently encountered among early foraging humans, we can be certain that
proscriptions of the more poisonous plants, as well as other potential food
sources that were toxic, significantly minimized exposure and/or ingestion
(Harris and Ross 1987; Rozin 1981).

In short, the "toxin" theory presented by Profet argues that allergy is a last
line of defense against harmful environmental toxins, especially venoms and
secondary compounds found in food plants. She also concludes that, whereas
nonimmunological defenses (e.g., vomiting and diarrhea) are by themselves
insufficient in defending against a specific toxin, when co-opted with the im-

munological mechanisms of allergy, the two types of defenses synergistically prevent toxins from entering the bloodstream and other organs by rapidly expelling it from the body. Because toxic substances are commonly allergenic, Profet reasons, the chemicals released by the body's mast cells during an IgE antibody-mediated allergic reaction initiate vomiting, diarrhea, coughing, tearing, sneezing, or scratching. Those responses dilute and expel the toxic substance that triggered the reaction. Profet extends her analysis to suggest that those allergic reactions often lead to the development of aversions to the substances that have triggered those unpleasant responses. For example, "morning sickness" prevents pregnant women from consuming foods that contain toxins that could harm the fetus (Profet 1988, 1992, 1995). The toxin hypothesis implies, therefore, that atopy should not be viewed as a disease or disorder, but rather as a protective element of our physiology. But if toxins are truly ubiquitous and the human body is constantly challenged with their expulsion by means of the usual IgE antibody-mediated response, wouldn't we all be in a constant state of "vomiting, diarrhea, coughing, tearing, sneezing, or scratching"?

If the allergic response is necessary to "hinder the allergen from entering the bloodstream and circulating to the various organs" (Profet 1991:26), then why does only a subset of the population (atopics) produce IgE levels significantly higher than the majority (nonatopics; see Burrows et al. 1989)? Furthermore, if allergy only affects 20–30% of the population, then how do the other 70–80% battle these ubiquitous toxins? Interestingly, one of the hallmark symptoms of asthma, an important manifestation of allergic disease, is wheezing, although wheezing is rarely mentioned by Profet as a typical or adaptive response. We have difficulty with the notion that narrowing of the airways to the point of respiratory arrest is functional. A major problem with the toxin theory is that it is based on an oversimplification of the role of IgE in the human immune system, and it ignores the function of IgE in the context of a complex humoral and cell-mediated immune response.

Profet (1991) and others have argued that there is abundant evidence that IgE production is not the result of an adaptation to counter the effects of helminthic infestation. Some have rightly argued that too few studies have taken into consideration nutritional status, genetic predisposition, psychosocial variables, and the duration and degree of exposure to allergens (Masters and Barrett-Connor 1985). Another problem with the helminth hypothesis is that all parasites probably do not behave in the same way. For example, a certain parasite, *Heligmosomoides polygyrus*, suppresses *overall* immunity in the host, but other nematodes do not do this (Pritchard et al. 1984). Some parasites are probably more allergenic than others; that is, they can induce higher levels of IgE, or cause greater mast cell degranulation (Pritchard 1992b). Profet (1991) believes that many allergic responses are consistent with both the toxin and helminthic hypothesis. The allergic reaction (sneezing, coughing, tearing, vomiting) can expel toxins and dislodge helminths. Both toxins and helminths induce IgE production and have receptors on eosinophils that results in degranulation of mast cells.

Lappe (1994) has proposed a "gatekeeper" model in which allergies in general and asthma in particular are selected for to prevent toxins from reaching the human tissue. Lappe argues that the rise in asthma as an adaptation is tied to the domestication of grains 14,000 years ago. The risks of inhaling moist grain contaminated by products such as molds (*Aspergillusspp*) and fungal spores was the selective force in the development of asthma (Lappe 1994). Lappe believes that asthma originated as a means of shutting down the respiratory system enough to inhibit inhaling molds or fungi and that the present exploding prevalence of atopy is an overreaction caused by an environment packed with industrial pollutants and people living in contemporary houses that are built so tightly that they lack proper ventilation and are loaded with allergens (Lappe 1994). While the argument that allergies emerged in response to exposure to allergens found in grains on the basis of total IgE levels is interesting, we contend that human helminthic infections, which clearly predate domestication of grains, and which result in total IgE levels 100-fold higher than normal levels compared to 10-fold higher levels found in atopic disease such as asthma, are a more potent "toxin" than grains.

Certainly one unexplainable feature of the worm theory in terms of the evolution of IgE is the interesting relationship that has been observed between allergy and cancer. There is abundant evidence that cancer and allergy are inversely related (Cockroft et al. 1979; Eisenbud 1987; Golumbek et al. 1991; Hallgren et al. 1981; McWhorter 1988; Morales and Bruce 1975; Rosenbaum and Dwyer 1977; Vena et al. 1985). There appears to be some mechanism in the T_H2 cytokine profile that limits the growth and spread of malignant lesions (Golumbek et al. 1991). But it has also been suggested that the ultimate effect on tumors by the T_H2 cytokine profile is possibly a suboptimal response and may not be adequate to eliminate the tumor (Yamamura et al. 1993). In an experimental study using basal cell carcinoma as a model, Yamamura and co-workers (1993) concluded that the response to malignancy was typified by the T_H2 cytokine profile, whereas the response to a benign tumor (seborrheic keratosis) reflected the T_H1 profile (1993). They speculated that "the Type 2 cytokine response induces immunosuppression at the tumor site and in this way contributes to the pathogenesis of [basal cell carcinoma]" (Yamamura 1993: 1009). Others have proposed that antitumor immunity is possibly suppressed by the release of cytokines of the T_H2 type (Bloom et al. 1992).

The antitumor properties of the T_H1-type cytokine, IL-12, which induces IFN-γ, is of particular interest in the treatment of cancer (Tahara and Lotze 1995; Zeh et al. 1993; Zitvogel et al. 1995). Interferon-gamma, the key player in the T_H1 profile, along with IL-2, are believed to be critical in the activation of cytotoxic T lymphocytes and natural killer cells, which in turn lyse tumor targets.

In short, it appears that the role of the opposing cytokine profiles (T_H1 versus T_H2) depends largely on the stage of the malignancy. That is, participation of different cytokines in fighting cancer will vary along the continuum of disease; perhaps the T_H2 profile is beneficial in preventing malignancies, whereas the T_H1 profile is useful in fighting malignancies in the event they develop.

Moreover, it is likely that the expression of these profiles will differ according to the *type* of tumor. Although much of the data regarding cytokine profiles and cancer are based on experimental models, there is an undeniable amount of evidence for an inverse relationship between allergy and cancer.

Profet (1991) addresses the inverse relationship between cancer and allergy and has hypothesized that, by protecting tissue from acute toxicity, allergy might also defend against mutagens and carcinogens. This theory logically implies that prevention of cancer and mutation is more beneficial than the potentially harmful, temporarily debilitating, and often deadly risk of anaphylactic shock. An intuitive argument is that populations with greater numbers of individuals with allergy-prone (hypersensitive) immune systems should be protected from carcinogens and mutagens, vis-a-vis lower rates of cancer because their "overactive" immune system should rid them of toxins, but there is no firm epidemiological data to support this notion. Although one might make a case that the trade-off of allergic disease is prevention of cancer, we prefer to view this trade-off as more of a circumstantial benefit of atopy because the development of our complex, humoral immune network that conferred protection from parasites among early hominids certainly pre-dated selective pressure for a system that confers some protection from malignancies, given that early hominids rarely survived beyond an age at which cancer would be of epidemiological significance. Only recently in our human history has life expectancy been extended to a point at which cancer affects a large enough proportion of the population to the degree that a mechanism (e.g., the T_H2 cytokine profile) would be selected that would protect us from cancer. In summary, while there does appear to be strong evidence that the T_H2 cytokine profile, which is the driving force behind atopic disease, does confer a biological advantage such as protection against cancer among the contemporary population, it must be borne in mind that the proportion of that population for whom there is genetic susceptibility to overexpress the humoral response is only some 20–30%.

Conclusion

Although Darwinian medicine is not precluded from formulating a systematic analysis of a particular adaptive force, many of the explanations appear to support a single cause for the adaptive response. It may be the reliance on an adaptionist program that seeks the single-most positive explanation for a phenomenon. They are asynchronic functional adaptations that belie its name: Darwinian or evolutionary medicine. We believe a perspective that considers adaptive features as evolving is essential for Darwinian medicine to become a robust theoretical tool. Furthermore, the assumption that all aspects of an adaptive system are going to be 100% positive needs to be questioned. While neatness counts in other realms, evolution may not be one of them.

Adaptation often results in trade-offs that may compromise an individual's adjustment to his or her environment. Our approach conceptualizes adaptation in an ever-changing human disease ecology. It is unfortunate that some contemporary theories regarding the evolution of allergy present an advocacy approach that sees the accumulation of evidence as the proof of a scientific position, rather than testing alternative hypotheses in the fashion of "strong inference" (Platt 1964).

The goal of Darwinian medicine as proposed by Nesse, Williams, and Profet is not just a theoretical perspective but has practical application. Perhaps we may be faced with the decision to suppress or not to suppress allergy. If the allergic reaction is "non-life-threatening and non-debilitating responses may be better left untreated" (Profet 1991:52). But allergy is an important chronic disease and appears to be an emerging disease of the second epidemiological transition. An understanding of its evolution and etiology could enable us to begin control or prevent a continued increase in prevalence in both developed countries and those undergoing development.

Nesse and Williams (1994:159) have suggested that allergy is analogous to a "smoke-detector" that exists for many years, occasionally sounding a false alarm. They justify the maintenance of the smoke detector in spite of the false alarms because it can save one's life in the event of a major fire, and they contend that, although allergy may elicit certain unpleasantries along the way, its upkeep (in terms of a selective trait) is worth the effort.

Puzzled by the two constructs of allergy as either an "immunological mistake" and/or a defense against parasitic worms, Profet seeks a more satisfactory, "functional" explanation for the apparent aberration—allergy—and reasons that "the evolutionary persistence of the allergic capability, despite its physiological costs, implies the existence of an adaptive benefit . . . that outweighs the costs" (1991:24–25). We cannot discount helminths as that "functional" explanation in light of the fact that 70% of the contemporary population resides in the developing world, and given that helminths remain one of the most common types of infection in that environment, infecting one-third of the population (Warren and Mahmoud 1984). Certainly the some 200 million people infected with schistosomes (Capron and Capron 1994) benefit from the immune system's highly evolved strategy against the worm, as did our many ancestors before us.

Acknowledgments

We wish to thank Bruce Bochner and Shau-Ku Huang for their review of and comments on the manuscript.

Notes

1. The concept of "epidemiological transition" was originally used to describe the rise of chronic disease as infectious diseases came under control (Corruccini and Kaul 1983). We have broadened the concept of epidemiological transition to include changes in human disease ecology during the last

4 million years. Changes in disease ecology resulted in three epidemiological transitions. The first occurred with the shift to agriculture. The second occurred with the control of infectious disease in many developed societies. We are entering the third epidemiological transition with the reemergence and emergence of antibiotic-resistant disease on a global scale.

2. Parasites can be divided into "intracellular" and "extracellular," the former including bacteria, viruses, and protozoa, and the latter including helminths. Helminths of medical importance to humans include trematodes ("flukes") and nematodes ("roundworms"). Intracellular parasites reproduce within the host, and infections are typically acute and short-lived. Extracellular parasites possess a more complicated life cycle, reproducing sexually in the host but multiplying outside of the host. Extracellular parasitic infections are chronic and long-lasting. The term "parasites" in the context of this paper refers to extracellular parasites, unless otherwise indicated.

3. "Host" is defined as the local environmental media ("inanimate" reservoir) or, in the case of humans, the living organism ("living" reservoir) that supports the survival and reproduction of a pathogen. The "definitive" host is one in which the pathogen lives and reproduces. The "intermediate" host is one in which the immature stages of the pathogen's development occur. An "accidental" host is one on which the pathogen is not necessarily dependent, yet is capable of supporting the pathogen's various life cycles.

4. Figures are based on the single African origin model (Delson 1988; Stringer and Andrews 1988; Waddle 1994), which proposes that anatomically modern *Homo sapiens*, after evolving in Africa between 100,000 and 200,000 years ago, migrated throughout Europe and Asia, replacing the archaic groups in these regions.

5. The term "immunoecology" derives from the study of immunoparasitology, defined by Seed (1993:470) as "[the study of] the mechanisms by which multiple host cell populations interact to influence the distribution and abundance of infrapopulations or infracommunities of parasites in a defined host environment."

6. Eosinophils phagocytize and kill ingested microorganisms, although it is their degranulation activity that is most effective against helminths. Following stimulation from T cells, eosinophils kill schistosome larvae in vitro by depositing toxic granule proteins (Butterworth 1984) and reactive oxygen intermediates such as hydrogen peroxide (Yazdanbakesh et al. 1987). Further evidence for the protective role of eosinophils has been found in hypoeosinophilic animals in which immunity to *Trichinella spiralis* (Grove et al. 1977) and *Strongyloides venezuelensis* (Korenaga et al. 1991) is compromised. Macrophages are another type of phagocyte that function by engulfing worm particles and destroying the parasite. Their role in killing schistosomes is documented in both murine (Esparza et al. 1987; James and Glaven 1989) and human (Cottrell et al. 1989) studies. Inhibition of mast cell proliferation has resulted in a failure to expulse *Heligmosomoides polygyrus* from the gut (Dehlawi et al. 1987).

References

Abel, L., F. Demenais, A. Prata, A.E. Souza, and A. Dessein. (1991) Evidence for the segregation of a major gene in human susceptibility/resistance to infection by *Schistosoma mansoni. American Journal of Human Genetics* 48:959–970.

Anonymous. (1998). Surveillance for Asthma—United States, 1960–1995. *MMWR—Morbidity and Mortality Weekly Report* 47(SS–1):1–28.

Anonymous. (1997) The Collaborative Study on the Genetics of Asthma: a genome-wide search for asthma susceptibility loci in ethnically diverse populations. *Nature Genetics* (in press). 15(4):389–392.

Arruda, L.K., T.A.E. Platts-Mills, J.W. Fox, and M.D. Chapman. (1990) *Aspergillus fumigatus* allergen I, a major IgE-binding protein, is a member of the mitogillin family of cytotoxins. *Journal of Experimental Medicine* 1972:1529–1532.

AAFA. (1996) Allergy Facts. Asthma & Allergy Foundation of America Maryland Chapter. Towson, MD: Asthma & Allergy Foundation of America.

Barnes, K.C., and R.J. Brenner. (1996). Quality of housing and allergy to cockroaches in the Dominican Republic. *International Archives of Allergy and Immunology* 109:68–72.

Barnes, K.C., E. Fernández-Caldas, W.L. Trudeau, D.E. Milne, and R.J. Brenner. (1997) Spatial and temporal distribution of house dust mite (Astigmata: Pyroglyphidae) allergens Der p 1 and Der f 1 in Barbadian homes. *Journal of Medical Entomology* 34:212–218.

Barnes, K.C., R.J. Brenner, E. Fernandez-Caldas, W.L. Trudeau, R.P. Naidu, T. Roach, M.E. Howitt. (1993). Spatial and temporal distribution of *Der p* I and *Der f* I in Bar-

badian homes: Implications for the incidence of asthma. *Journal of Allergy and Clinical Immunology* 91(1): 356.

Barnes K.C., Marsh D.G. (1998). The genetics and complexity of allergy and asthma. *Immunology Today* 19(7):325–332.

Barnes, K.C., J.D. Neely, D.L. Duffy, L.R. Freidhoff, D.R. Breazeale, C. Schou, R.P. Naidu, P.N. Levett, B. Renault, R. Kucherlapati, S. Iozzino, E. Ehrlich, T.H. Beaty, and D.G. Marsh. (1996) Linkage of asthma and total serum IgE concentration to markers on chromosome 12q: evidence from Afro-Caribbean and caucasian populations. *Genomics* 37:41–50.

Bazaral, M., H.A. Orgel, and R.N. Hamburger. (1973) The influence of serum IgE levels of selected recipients, including patients with allergy, helminthiasis and tuberculosis, on the apparent P-K titre or reaginic serum. *Clinical Experimental Immunology* 14:117–125.

Bazaral, M., H.A. Orgel, and R.N. Hamburger. (1974) Genetics of IgE and allergy: serum IgE levels in twins. *Journal of Allergy and Clinical Immunology* 54:288–304.

Bloom, B.R., P. Sagame, and B. Diamond. (1992) Revisiting and revising T suppressor cells. *Immunology Today* 13:131–136.

Blumenthal, M., D. Marcus-Bagley, Z. Awdeh, B. Johnson, E. Ynuis, and C.A. Alper. (1992) HLA-DR2 [HLA-B7, SC31, DR2], and HLA-B8, SC01, DR3] haplotypes distinguish subjects with asthma from those with rhinitis only in ragweed pollen allergy. *Journal of Immunology* 148 (2): 411–416.

Bochner, B.S. and R.P. Schleimer. (1994). The role of adhesion molecules in human eosinophil and basophil recruitment. *Journal of Allergy and Clinical Immunology* 94:427–438.

Bousquet, J., F. Hatton, P. Godard, and F.B. Michel. (1987) Asthma mortality in France. *Journal of Allergy and Clinical Immunology* 80:389–394.

Bradding, P., I.H. Feather, P.H. Howarth, R. Mueller, J.A. Roberts, K. Britten, J.P. Bews, et al. (1992). Interleukin 4 is localized to and released by human mast cells. *Journal of Experimental Medicine* 176(5):1381–1386

Brothwell, D. (1972) The question of pollution in earlier and less developed societies. In P.R. Cox and J. Peel, eds., *Population and Pollution.* London: Academic Press, pp. 15–27.

Bundy, D.A.P., and G.F. Medley. (1992) Epidemiology and population dynamics of helminth infection. In R. Moqbel, ed., *Allergy and Immunity to Helminths.* London: Taylor & Francis. 17–37

Burge, H.A. (1989) Airborne allergenic fungi. *Immunology and Allergy Clinics of North America* 9:307–319.

Burney, P.G.J. (1986) Asthma mortality in England and Wales: evidence for a further increase, 1974–1984. *Lancet* 2:323–326.

Burrows, B., F.D. Martinez, M. Haonen, R.A. Barbee, and M.G. Cline. (1989) Association of asthma with serum IgE levels and skin-test reactivity to allergens. *New England Journal of Medicine* 320:271–277.

Butterworth, A.E. (1984) Cell-mediated damage to helminths. *Advances in Parasitology* 23:143–235.

Butterworth, A.E., M. Capron, J.S. Cordingley, P.R. Dalton, D.W. Dunne, H.C. Kariuki, D. Koech, M. Mugambi, J.H. Ouma, M.A. Prentice, B.A. Richardson, T.K. ArapSiongok, R.F. Sturrock, and D.W. Taylor. (1985) Immunity after treatment of human schistosomiasis mansoni. II.

Identification of resistant individuals, and analysis of their immune responses. *Transactions of the Royal Society of Tropical Medicine and Hygiene* 79: 393–408.

Capron, A., and J.P. Dessaint. (1985) Effector and regulatory mechanisms in immunity to schistosomes: a heuristic view. *Annual Review of Immunology* 3:455–476.

Capron, A., J.P. Dessaint, and M. Capron. (1992) Allergy and immune defence: common IgE-dependent mechanisms or divergent pathways. In R. Moqbel, ed., *Allergy and Immunity to Helminths.* London: Taylor & Francis. pg 1–16.

Capron, A., J.P. Dessaint, M. Capron, and H. Bazin. (1975) Specific IgE antibodies in immune adherence of normal macrophages to *Schistosoma mansoni* schistosomules. *Nature* 253: 474–475.

Capron, A., J.P. Dessaint, M. Capron, A.M. Ouma, and A.E. Butterworth. (1987) Immunity to schistosomes: progress toward vaccine. *Science* 238:1065–1072.

Capron, A., J.P. Dessaint, M. Capron, M. Joseph, and G. Torpier. (1982) Effector mechanisms of immunity to schistosomes and their regulation. *Immunological Reviews* 61:41–66.

Capron, M., and A. Capron. (1994) Immunoglobulin E and effector cells in schistosomiasis. *Science* 264: 1876–1877.

Chomarat, P., M. Rissoan, J. Banchereau, and P. Miossec. (1993) Interferon inhibits interleukin-10 production by monocytes. *Journal of Experimental Medicine* 177:523–527.

Chua, K.Y., G.A. Stewart, W.R. Thomas, R.J. Simpson, R.J. Wilworth, T.M. Plozza, and K.J. Turner. (1988) Sequence analysis of cDNA coding for a major house dust mite allergen, Der p 1. Homology with cysteine proteases. *Journal of Experimental Medicine* 167:175–182.

Cockburn, T.A. (1971) Infectious disease in ancient populations. *Current Anthropology* 12:45–62.

Cockroft, D.W., G.J. Klein, R.E. Donevan, and G.M. Copland. (1979) Is there a negative correlation between malignancy and respiratory atopy? *Annals of Allergy* 43:345–347.

Cohen, S.G., and M. Samter, eds. (1992) *Excerpts from Classics in Allergy*, 2nd ed. Carlsbad, CA: Symposia Foundation.

Colloff, M.J., J. Ayre, F. Carswell, P.H. Howarth, T.G. Merrett, E.B. Mitchell, M.J. Walshaw, J.O. Warner, J.A. Warner, and A.A. Woodcock. (1992) The control of allergens of dust mites and domestic pests: a position paper. *Clinical and Experimental Allergy* 22:1–28.

Cookson, J.B. (1987). Prevalence rates of asthma in developing countries and their comparison with those in Europe and North America. *Chest* 91: 97s–103s.

Corruccini, R.S., and S.S. Kaul. (1983) The epidemiological transition and the anthropology of minor chronic non-infectious diseases. *Medical Anthropology* 7:36–50.

Cottrell, B.J., C. Pye, A.M. Glauert, and A.E. Butterworth. (1989) Human macrophage-mediated cytotoxicity of *Schistosoma mansoni*. Functional and structural features of the effector cells. *Journal of Cell Science* 94:733–741.

Daniels, S.E., S. Bhattacharrya, A. James, N.I. Leaves, A. Young, M.R. Hill, J.A. Faux, G.F. Ryan, P.N. le Söuef, G.M. Lathrop, A.W. Musk, and W.O.C.M. Cookson. (1996) A genome-wide search for quantitative trait loci underlying asthma. Nature 383:247–250.

Davis, L.R., R.H. Marten and I. Sarkany. (1961) Atopic eczema in European and Negro West Indian infants in London. *British Journal of Dermatology* 73:410–414.

Dehlawi, M.S., D. Wakelin, and J.M. Behnke. (1987) Suppression of mucosal mastocytosis by infection with the intestinal nematode *Nematospiroides dubius. Parasite Immunology* 9:187–194.

Del Prete, G., M. De Carli, C. Mastromauro, R. Biagiotti, D. Macchia, P. Falagiani, M. Ricci, and S. Romagnani. (1991) Purified protein derivative of *Mycobacterium tuberculosis* and excretory-secretory antigen(s) of *Toxocar canis* expand in vitro human T cells with stable and opposite (type 1 T helper or type 2 T helper) profile of cytokine production. *Journal of Clinical Investigation* 88:346–350.

Delson, E. (1988) Paleoanthropology. One source not too many. *Nature* 332:206.

Dessaint, J.-P., M. Capron, D. Bout, and A. Capron. (1975) Quantitative determination of specific IgE antibodies to schistosome antigens and serum IgE levels in patients with schistosomiasis (*S. mansoni* and *S. haematobium*). *Clinical and Experimental Immunology* 20:427–436.

Dessein, A.J., M. Begley, C. Demeure, D. Caillol, J. Fueri, M. Galvao dos Reis, Z.A. Andradae, A. Prata, and J.C. Bina. (1988) Human resistance to *Schistosoma mansoni* is associated with IgG reactivity to a 37-kDa larval surface antigen. *Journal of Immunology* 140:2727–2736.

Dessein, A.J., W.L. Parker, S.L. James and J.R. David. (1981) IgE antibody and resistance to infection. *Journal of Experimental Medicine*. 153:423–436.

de Weck, A., and A. Todt, eds. (1988) *Mite Allergy: A World-Wide Problem*. Brussels: UCB Institute of Allergy.

Dowse, G.K., K.J. Turner, G.A. Stewart, M.P. Alpers, and A.J. Woolcock. (1985) The association between *Dermatophagoides* mites and the increasing prevalence of asthma in village communities within the Papua New Guinea highlands. *Journal of Allergy and Clinical Immunology* 75: 75–83.

Dunne, D.W., A.E. Butterworth, A.J.C. Fulford, M.C. Kariuki, J.G. Langley, J.H. Ouma, A. Capron, R.J. Pierce, and R.F. Sturrock. (1992) Immunity after treatment of human schistosomiasis mansoni: association between IgE antibodies to adult worm antigens and resistance to reinfection. *European Journal of Immunology* 22: 1483–1494.

Dunne, D.W., A.M. Grabowska, A.J.C. Fulford, A.E. Butterworth, R.F. Sturrock, D. Koech, and J.H. Ouma. (1988) Human antibody responses to *Schistosoma mansoni*: the influence of epitopes shared between different life cycle statges on the response to the schistosomules. *European Journal of Immunology* 18:123–131.

Edfors-Lubs, M.-L. (1971) Allergy in 7000 twin pairs. *Acta Allergology* 26: 249–285.

Eisenbud, M. (1987) Carcinogenicity and allergenicity. *Science* 23:1613.

Esparza, I., A. Männel Ruppel, W. Falk, and P.H. Krammer. (1987) Interferon gamma and lymphotoxin or tumor necrosis factor act synergistically to induce macrophage killing of tumor cells and schistosomula of *Schistosoma mansoni. Journal of Experimental Medicine.* 166:589–594.

Evans, R. (1987) Recent observations reflecting increases in mortality from asthma. *Journal of Allergy and Clinical Immunology* 80:377–379.

Ewald, P. (1980) Evolutionary biology and the treatment of signs and symptoms of infectious disease. *Journal of Theorethical Biology* 86: 169–176.

Ewald, P. (1993) The evolution of virulence. *Scientific American* April:86–93.

Ewald, P.W. (1994) *Evolution of Infectious Disease*. New York: Oxford University Press.

Freidhoff, L.R., E. Ehrlich-Kautzky, D.A. Meyers, A.A. Ansari, W.B. Bias, and D.G. Marsh. (1988) Association of HLA-DR3/Dw3 and total serum immunoglobulin E level with human immune response to *Lol p* I and *Lol p* II allergens in allergic subjects. *Tissue Antigens* 31:211–219.

Gazzinelli, R.T., I.P. Oswald, S.L. James, and A. Sher. (1992) IL-10 inhibits parasite killing and nitrogen oxide production by IFN-g-activated macrophages. *Journal of Immunology* 148:1792–1796.

Gergen P.J., and K.B.Weiss. (1990) Changing patterns of asthma hospitalization among children: 1979 to 1987. *Journal of the American Medical Association* 264:1688–1692.

Gerrard, J.W., D.C. Rao, and N.E. Morton. (1978) A genetic study of immunoglobulin E. *American Journal of Human Genetics* 30:46–58.

Godfrey, R.C. (1975) Asthma and IgE levels in rural and urban communities of the Gambia. *Clinical Allergy* 5: 201–207.

Golumbek, P.T., A.J. Lazenby, H.I. Levitsky, L.M. Jaffee, H. Karasuyama, M. Baker, and D.M. Pardoll. (1991) Treatment of established renal cancer by tumor cells engineered to secrete interleukin-4. *Science* 254:713–716.

Gratz, N.G. (1973) Mosquito-borne disease problems in the urbanization of tropical countries. *CRC Critical Reviews in Environmental Control* 3: 455–495.

Grogan, J.L., P.G. Kremsner, G.J. van Dam, W. Metzger, B. Mordmüller, A.M. Deelder, and M. Yazdanbakhsh. (1996) Antischistosome IgG4 and IgE responses are affected differentially by chemotherapy in children versus adults. *Journal*

of Infectious Disease 173:1242–
1247.

Grove, D.I. (1982) What is the relationship between asthma and worms?
Allergy 37:139–148.

Grove, D.I., A.A.F. Mahmoud, and K.S.
Warren. (1977) Eosinophils and resistance to *Trichinella spiralis. Journal of Experimental Medicine* 145:
755–759.

Hadge, D., and H. Ambrosius. (1984)
Evolution of low molecular weight
immunoglobulins. IV. IgY-like immunoglobulins of birds, reptiles and
amphibians, precursors of mammalian IgA. *Molecular Immunology* 21:
699–707.

Hagan, P., V.J. Blumenthal, D. Dunn,
A.J.G. Simposn, and H.A. Wilkins.
(1991) Human IgE, IgG4 and resistance to reinfection with *Schistosoma haematobium. Nature* 349:243–
245.

Hagan, P., U.J. Blumenthal, D. Dunn,
and H.A. Wilkins. (1992) A protective role for IgE in human schhisto-
somiasis. In R. Moqbel, ed., *Allergy
and Immunity to Helminths.* London:
Taylor & Francis, pp. 94–106.

Hallgren, R., H. Arrendal, K. Hiesche,
G. Lundquist, E. Nou, and O. Zetter-
strom. (1981) Elevated serum immunoglobulin E in bronchial carcinoma:
Its relation to the histology and
prognosis of the cancer. *Journal of
Allergy and Clinical Immunology* 67:
398–406.

Hamilton, R.G., M.D. Chapman, T.A.E.
Platts-Mills, and N.F. Adkinson.
(1992) House dust aeroallergen measurements in clinical practice: a
guide to allergen-free home and
work environments. *Immunology
and Allergy Practice* 14:96–112.

Harris, M., and E. Ross. (1987) *Food
and Evolution: Toward a Theory of
Human Food Habits.* Philadelphia,
PA: Temple University Press.

Haswell-Elkins, M.R., D.B. Elkins, and
R.M. Anderson. (1987) Evidence for

a predisposition in humans to infection with Ascaris, hookworm, Enter-
obius and Trichuris in a South Indian fishing community.
Parasitology 95:323–328.

Hsu, S.Y.L., J.F. Hsu, G.D. Penick, G.L.
Lust, and J.W. Osborne. (1974) Der-
mal hypersensitivity to schistosome
cercariae in rhesus monkeys during
immunization and challenge. *Journal of Allergy and Clinical Immunology.* 54:339–349.

Huang, S.K., P. Zwollo, and D.G.
Marsh. (1991) Class II MHC restriction of human T-cell responses to
short ragweed allergen, *Amb a* V.
European Journal of Immunology 21:
1469–1473.

Hussain, R., R.G. Hamilton, V. Kumaraswami, N.F. Adkinson, and E.A.
Ottesen. (1981) IgE response in human filariasis. I. Quantitation of fi-
laria specific IgE. *Journal of Immunology* 127:1623–1629.

Hussain, R., R.W. Poindexter, E.A. Ottesen. (1992). Control of allergic
reactivity in human filariasis. Predominant localization of blocking
antibody to the IgG4 subclass. *Journal of Immunology* 148:2731–2737.

Ishizaka, A., Y. Sakiyama, M. Nakanishi, K. Tomizawa, E. Oshika, D. Kojima, Y. Taguchi, E. Kandil, and S.
Matsumoto. (1990) The inductive effect of interleukin-4 on IgG4 and IgE
synthesis in human peripheral blood
lymphocytes. *Clinical and Experimental Immunology* 79:392–396.

Iskander, R., P.K. Das, and R.C. Aalberse. (1981) IgG4 antibodies in
Egypt patients with schistosomiasis. *International Archives of Allergy
and Applied Immunology* 66:200–
207.

Ito, K., T. Sawada, and S. Shigefus.
(1972) Increased serum IgE level in
individuals infected with *Schistosoma japonicum, Wucheria ban-
crofti* or hookworm, and the changes
by treatment in schistosomiasis. *Jap-*

anese Journal of Experimental Medicine 42:115–123.

James, S.L., and Glaven, J.J. (1989) Macrophage cytotoxicity against schistosomula of *Schistosoma mansoni* involves arginine-dependent production of reactive nitrogen intermediates. *Journal of Immunology* 143: 4208–4212.

Jarrett, E.E.E., and H. Bazin. (1974) Elevation of total serum IgE in rats following helminth parasite infection. *Nature* 25:613–614.

Jarrett, E.E., Haig, D.M., and H. Bazin. (1976) Time course studies of igE production in N. brasiliensis infection. *Clinical and Experimental Immunology* 24:346–351.

Johansson, S.G.O., and H.H. Bennich. (1985) Advances in the biology of IgE. In E.B. Weiss, M.S. Segal, and M. Stein, eds., *Bronchial Asthma. Mechanisms and Therapeutics*, 2nd ed. Boston, MA: Little, Brown and Company.

Johansson, S.G.O., T. Melbin, and B. Vahlquist. (1968) Immunoglobulin levels in Ethiopian preschool children with special reference to high concentrations of immunoglobulin E (IgND). *Lancet* May 25:1(7552)1118–1121.

Jordan, P., and G. Webbe. (1982) *Schistosomiasis. Epidemiology, Treatment and Control*. London: William Heinemann Medical Books.

Joubert, J.R., D.J. van Schalkwyk, and K.J. Turner. (1980) *Ascaris lumbricoides* and the human immunogenic response. Enhanced IgE-mediated reactivity to common antigens. *South African Medical Journal* 57: 409–412.

Katz, N., F. Zicker, R.S. Rocha, and V.B. Oiveira. (1978) Reinfection of patients in schistosomiasis mansoni endemic areas after specific treatment. *Revista do Instituto Medical Tropical de* Sao Paulo 20: 273–278.

Kennedy, M.W. (1992) Genetic control of the antibody response to parasite allergens. In R. Moqbel, ed., *Allergy and Immunity to Helminths*. London: Taylor & Francis. pp. 63–80.

King, C.L., S. Mahanty, V. Kumaraswami, J.S. Abrams, J. Regunathan, K. Jayaraman, E.A. Ottesen, and T.B. Nutman. (1993) Cytokine control of parasite-specific anergy in human lymphatic filariasis. *Journal of Clinical Investigation* 92:1667–1673.

King, C.L., and T.B. Nutman. (1993) IgE and IgG subclass regulation by IL-4 and IFN-γ in human helminth infections. *Journal of Immunology* 151: 458–465.

Kliks, M.M. (1983) Parasitology: on the origins and impact of human-helminth relationships. In N.A. Croll and J.H. Cross, eds., *Human Ecology and Infectious Disease*. New York: Academic Press.

Koehler, P.G., R.S. Patterson, and R.J. Brenner. (1987) German cockroach (Orthoptera: Blattellidae) infestations in low-income apartments. *Journal of Economic Entomology* 80: 446–450.

Korenaga, M., Y. Hitoshi, N. Yamaguchi, Y. Sato, K. Takatsu, and I. Tada. (1991) The role of interleukin-5 in protective immunity to *Strongyloides venezuelensis* infection in mice. *Immunology* 72:502–507.

Kron, M.A., B. Sisley, R.H. Guderian, Ch.D. Mackenzie, M. Chico, H. Jurado, and J. Rumbea Guzman. (1993) Antibody responses to *Onchocerca volvulus* in Ecuadorian Indians and Blacks. *Tropical Medical Parsitology* 44:152–154.

Lappe, M. (1994). *Evolutionary Medicine: Rethinking the Origins of Disease.*San Francisco, CA: Sierra Club Books.

Lau-Schadendorf, S., and U. Wahn. (1991) Atopic diseases in infancy. The German multicenter atopy study (MAS-90). *Pediatric Allergy and Immunology* 2:63–69.

Lind, P., and H. Løwenstein. (1988) Characterization of asthma-associated allergens. In *Bailliere's Clinical Immunology and Allergy, vol. 2*. London: Balliere-Tindall. pp. 67–89.

Lynch, N.R. (1987) Immediate hypersensitivity (allergic) reactions to intestinal helminthic infections. In *Bailleres Clinical Tropical Medicine and Communicable Diseases, vol. 2*. London: Bailliere-Tindall; pp. 573–593.

Lynch, N.R. (1992) Influence of socioeconomic level on helminthic infection and allergic reactivity in tropical countries. In R. Moqbel, ed., *Allergy and Immunity to Helminths*. London: Taylor & Francis, pp. 52–62.

MacGlashan, D., J.M. White, S. Huang, S.J. Ono, and L.M. Lichtenstein. (1994) Secretion of IL-4 from human basophils. The relationship between IL-4 mRNA and protein in resting and stimulated basophils. *Journal of Immunology* 152:3006-3016.

Mahanty, S., J.S. Abrams, C.L. King, A.P. Kimaye, and T.B.J. Nutman. (1992) Parallel regulation of IL-4 and IL-5 in human helminth infections. *Immunology* 148:3567–3571.

Maizels, R.M., D.A.P. Bundy, E. Selkirk, D.F. Smith, and R.M. Anderson. (1993) Immunological modulation and evasion by helminth parasites in human populations. *Nature* 365:797–805.

Malefyt, R.W., J. Abrams, B. Bennett, C. Figdor, and J.E. de Vries. (1991) Interleukin 10 (IL-10) inhibits cytokine synthesis by human monocytes: an autoregulatory role of IL-10 produced by monocytes. *Journal of Experimental Medicine* 174:1209–1220.

Marquardt, D.L., and S.I. Wasserman. (1993) Anaphylaxis. In E. Middleton Jr., C.E. Reed, E.F. Ellis, N.F. Adkinson, Jr., J.W. Yunginger, and W.W. Busse, eds., *Allergy. Principles and Practice, vol. 2*, 4th ed. Baltimore, MD: Mosby, pp. 1525–1536.

Marquet, S., L. Abel, D. Hillaire, H. Dessein, G. Kalil, J. Feingold, J. Weissenbach, and A.J. Dessein. (1996) Genetic localization on chromosome 5q31–q33 of a locus predisposing to high infections by *Schistosoma mansoni*. *Nature Genetics* 14: 181–184.

Marsh, D.G., W.B. Bias, and K. Ishizaka. (1974) Genetic control of basal serum immunoglobulin E level and its effect on specific reaginic sensitivity. *Proceedings of the National Academy of Sciences, USA* 71: 3588–3592.

Marsh, D.G., S.H. Hsu, M. Roebber, E. Ehrlich-Kautzky, L.R. Freidhoff, D.A. Meyers, M. K. Pollard, and W.B. Bias. (1982) HLA-Dw2: a genetic marker for human immune responses to short ragweed pollen allergen Ra5. I. Response resulting primarily from natural antigenic exposure. *Journal of Experimental Medicine* 155:1439–1451.

Marsh, D.G., L.R. Freidhoff, E.E. Kautzky, W.B. Bias, and M. Roebber. (1987) Immune responsiveness to *Ambrosia artemisiifolia* (short ragweed) pollen allergen Amb a VI (Ra6) is associated with HLA-DR5 in allergic humans. *Immunogenetics* 26: 230–236.

Marsh, D.G., D.A. Meyers, and W.B. Bias. (1980) The epidemiology and genetics of atopic allergy. *New England Journal of Medicine* 305(26): 1551–1559.

Marsh D.G., J.D. Neely, D.R. Breazeale, B. Ghosh, L.R. Freidhoff, E. Ehrlich-Kautzky, C. Schou, G. Krishnaswany, and T.H. Beaty. (1994) Linkage analysis of *IL4* and other chromosome 5q31.1 markers and total serum immunoglobulin E concentrations. *Science* 264:1152–1156.

Marsh, D.G., P. Zwollo, S.K. Huang, B.

Ghosh, and A.A. Ansari. (1990) Molecular studies of human response to allergens. *Cold Spring Harbor Symposim on Quantitative Biology* 54: 459–470.

Martinez, F.D., C.J. Holberg, M. Halonen, W.J. Morgan, A.L. Wright, and L.M. Taussig. (1994) Evidence for Mendelian inheritance of serum IgE levels in Hispanic and non-Hispanic white families. *American Journal of Human Genetics* 55:555–565.

Masters, M., and E. Barrett-Connor. (1985) Parasites and asthma—predictive or protective? *Epidemiology Reviews* 7:49–58.

McNeill, W.H. (1976) *Plauges and Peoples.* Garden City, NY: Anchor Books.

McWhorter, W.P. (1988) Allergy and risk of cancer: a prospective study using NHANES I followup data. *Cancer* 62:451–455.

Melewicz, F.M., J.M. Plummer, and H.L. Spiegelberg. (1982) Comparison of the Fc receptors for IgE on human lymphocytes and monocytes. *Journal of Immunology* 129:563–569.

Meyers, D.A., T.H. Beaty, L.R. Freidhoff, and D.G. Marsh. (1987) Inheritance of serum total IgE (basal levels) in man. *American Journal of Human Genetics* 41:51–62.

Meyers DA, D.S. Postma C.I.M. Panhuysen, J. Xu, P.J. Amelung, R.C. Levitt, and E.R. Bleeker. (1994) Evidence for a locus regulating total serum IgE levels mapping to chromosome 5. *Genomics* 23:464–470.

Mochizuki, M., J. Bartels, A.I. Mallet, E. Christophers, and J.M. Schroder. (1998). IL-4 induces eotaxin: a possible mechanism of selective eosinophil recruitment in helminth infection and atopy. *Journal of Immunology* 160(1):60–68.

Moffatt, M.F., M.R. Hill, F. Cornélis, C. Schou, J.A. Faux, R.P. Young, A.L. James, G. Ryan, P. Le Souef, A.W. Musk, et al. (1994) Genetic linkage of T-cell receptor a/d complex to specific IgE responses. *Lancet* 343: 1597–1600.

Morales, A., and A.W. Bruce. (1975) Allergy and cancer. *Urology* 6:78–80.

Moro-Furlani, A.M., and H. Krieger. (1992) Familial analysis of eosinophilia caused by helminthic parasites. *Genetic Epidemiology* 9:185–190.

Mossman, T.R., H. Cherwinski, M.W. Bond, M.A. Giedlin, and R.L. Coffman. (1986) Two types of munine helper T cel clone I: definition according to the profiles of lymphokine activities and secreted proteins. *Journal of Immunology* 136: 2348–2357.

Mosmann, T.R., and R.L. Coffman. (1989) TH1 and TH2 cells: different patterns of limphokine secretion lead to different functional properties. *Annual Review of Immunology* 7:145–173.

Mosmann, T.R., and K.W. Moore. (1991) The role of IL-10 in crossregulation of Th1 and Th2 responses. *Immunology Today* 12:A49– A53.

Nesse, R.M., and G.C. Williams. (1994) *Why We Get Sick: The New Science of Darwinian Medicine.* New York: Times Books.

O'Donnell, I.J., and G.F. Mitchell. (1978) An investigation of the allergens of *Ascaris lumbricoides* using a radioallergosorbent test (RAST) and sera from naturally infected humans: comparison with an allergen for mice identified by a passive cutaneous anaphylaxis test. *Australian Journal of Biological Science* 31:459–487.

Ohe, M., M. Munakata, N. Hizawa, A. Itoh, I. Doi, E. Yamaguchi, Y. Homma, and Y. Kawakami. (1995) Beta$_2$-adrenergic receptor gene polymorphism and bronchial asthma. *Thorax* 50:353–359.

O'Hehir, R.E., R.D. Garman, J.L. Green-stein, and J.R. Lamb. (1991) The specificity and regulation of T-cell responsiveness to allergens. *Annual Review of Immunology* 9:67–95.

Orgel, J.A., M.A. Lenoir, and M. Bazaral. (1974) Serum IgG, IgA, IgM, and IgE levels and allergy in Filipino children in the United States. *Journal of Allergy and Clinical Immunology* 53(4): 213–222.

Orr, T.S., and A.M. Blair. (1969) Potentiated reagin response to egg albumin and conalbumin in *Nippostrongylus brasiliensis*-infected rats. *Life Sciences* 8:1073.

Oswald, I.P., T.A. Wynn, A. Sher, and S.L. James. (1992) Interleukin 10 inhibits macrophage microbicidal activity by blocking the endogenous production of tumor necrosis factor a required as a costimulatory factor for interferon γ-induced activation. *Proceedings of the National Academy of Sciences, USA* 89:8678–8680.

Paul, W.E., and J.H. Ohara. (1987) B cell stimulatory factor-1/interleukin 4. *Annual Review of Immunology* 5: 427–459.

Paul, W.E., and R.A. Seder. (1994) Lymphocyte responses and cytokines. *Cell* 76: 241–251.

Platt, J.R. (1964) Strong inference. *Science* 146:347–353.

Platts-Mills, T.A.E., and M.D. Chapman. (1987) Dust mites: immunology, allergic disease, and environmental control. *Journal of Allergy and Clinical Immunology* 80:755–775.

Platts-Mills, T.A.E., and A.L. de Weck. (1989) Dust mite allergens and asthma—a worldwide problem. *Journal of Allergy and Clinical Immunology* 83:416–427.

Platts-Mills, T.A.E., P.W. Heymann, M.D. Chapman, T.F. Smith, and S. Wilkins. (1985) Mites of the genus *Dermatophagoides* in dust from the houses of asthmatic and other allergic patients in North America: development of a radioimmunoassay for allergen produced by *D. farinae* and *D. pteronyssinus*. *International Archives of Applied Immunology* 77: 163–165.

Platts-Mills, T.A.E., E.B. Mitchell, P. Nock, E.R. Tovey, H. Moszoro, and S.R. Wilkins. (1982) Reduction of bronchial hyperreactivity during prolonged allergen avoidance. *Lancet* 2:675–678.

Platts-Mills, T.A.E., W.R. Thomas, R.C. Aalberse, D. Vervloet, and M.D. Chapman. (1992) Dust mite allergens and asthma: report of a second international workshop. *Journal of Allergy and Clinical Immunology* 89: 1046–1060.

Platts-Mills T.A., D. Vervloet, W.R. Thomas, R.C. Aalberse, M.D. Chapman. (1997). Indoor allergens and asthma: report of the Third International Workshop. *Journal of Allergy and Clinical Immunology* 100(6 Pt 1): S2–24.

Platts-Mills, T.A.E., and A.L. de Weck. (1989). Dust mite allergens and asthma—a worldwide problem. *Journal of Allergy and Clinical Immunology* 83:416–427.

Pollart, S.M., T.F. Smith, E.C. Morris, L.E. Gelber, T.A.E. Platts-Mills, and M.D. Chapman. (1991) Environmental exposure to cockroach allergens: analysis with monoclonal antibody-based enzyme immunoassays. *Journal of Allergy and Clinical Immunology* 87:505–510.

Postma D.S., E.R. Bleecker, P.J. Amelung, K.J. Holroyd, J. Xu, C.I.M. Panhuysen, D.A. Meyers, R.C. Levitt. (1995) Genetic susceptibility to asthma—bronchial hyperresponsiveness coinherited with a major gene for atopy. *New England Journal of Medicine* 333:894–900.

Price, P.W. (1990) Host populations as resources defining parasite community organization. In G.W. Esch,

A.O. Bush, and J.M. Aho, eds., *Parasite Communities: Patterns and Processes*. London: Chapman & Hall.

Pritchard, D.I. (1993) Immunity to helminths: is too much IgE parasite-rather than host-protective? *Parasite Immunology* 15: 5–9.

Pritchard, D.I., N.H.M. Ali, and J.M. Behnke. (1984) Analysis of the mechanisms of immunodepression following heterologous antigenic stiumlation during concurrent infection with *Nematospiroides dubius*. *Immunology* 51:633–642.

Pritchard, D.I., S. Kumar, and P. Edmonds. (1992) Souble(s) CD23 levels in the plasma of a parsitized population from Papua New Guinea. *Parasite Immunology* 15(4):205–208.

Profet, M. (1988) The evolution of pregnancy sickness as protection to the embryo against Pleistocene teratogens. *Evolutionary Theory* 8:177–190.

Profet, M. (1991) The function of allergy: immunological defense against toxins. *Quarterly Review of Biology* 66:23–62.

Profet, M. (1992) Pregnancy sickness as adaptation: a deterrent to maternal ingestion of teratogens. In J.H. Barkow, L. Cosmides, and J. Tooby, eds., *The Adapted Mind*. New York: Oxford University Press, pp. 327–365.

Profet, M. (1995) *Protecting Your Baby to Be*. Addison-Wesley.

Riganò, R., E. Profumo,G. Di Felice, E. Ortona, A. Teggi, and A. Siracusano. (1995) In vitro production of cytokines by peripheral blood mononuclear cells from hydatid patients. *Clinical and Experimental Immunology* 99:433–439.

Rihet, P., C.E. Demeure, A. Bourgeois, A. Prata, and A.J. Dessein. (1991) Evidence for an association between human resistance to *Schistosoma mansoni* and high anti-larval IgE levels. *European Journal of Immunology* 21:2679–2686.

Roebber, M., D.G. Kapper, L. Goodfriend, W.B. Bias, S.H. Hsu, D.G. Marsh. (1985) Immunochemical and genetic studies of *Amb. t.* V (Ra5G), an Ra5 homologue from giant ragweed pollen. *Journal of Immunology* 134:3062–3069.

Romagnani, S. (1991) Type 1 T helper and type 2 T helper cells: functions, regulation and roles in protection and disease. *International Journal of Clinical Laboratory Research* 21:152–158.

Romagnani, S. (1992) Human TH1 and TH2 subsets: regulation of differentiation and role in protection and immunopathology. *International Archives of Allergy and Immunology* 98:279–285.

Rosenbaum, J.T., and J.M. Dwyer. (1977) The role of IgE in the immune response to neoplasia: a review. *Cancer* 39:11–20.

Rosenwasser, L.J., D.J. Klemm, J.K. Dresback, H. Inamura, J.J. Mascali, M. Klinnert, and L.L. Borish. (1995). Promoter polymorphisms in the chromosome 5 gene cluster in asthma and atopy. *Clinical and Experimental Allergy* 25 (suppl. 2): 74–78.

Rozin, P. (1981) Human food selection: the interaction of biology, culture and individual experience. In L.M. Barker, ed., *Psychobiology of Human Food Selection*. Westport, CT: AVI Press, pp. 181–205.

Sampson, H.A., and Metcalfe, D.D. (1992) Food allergies. *Journal of the American Medical Association* 268: 2840–2844.

Sandford AJ, T. Shirakawa, M.F. Moffatt, S.E. Daniels, C. Ra, J.A. Faux, R.P. Young, Y. Nakamura, G.M. Lathrop, and W.O.M. Cookson. (1993) Localization of atopy and β subunit of high-affinity IgE receptor (FceRI) on chromosome 11q. *Lancet* 341:332–334.

Schad, G.A., and R.M. Anderson.

(1985) Predisposition to hookworm infection in humans. *Science* 228: 1537–1540.

Schou, C., E. Fernandez-Caldas, R.F. Lockey, and H. Lowenstein. (1991) Environmental assay for cockroach allergens. *Journal of Allergy and Clinical Immunology* 87:828–834.

Schultes, R.E., and A. Hofman. (1987) *Plants of the Gods: Origins of Hallucinogenic Use*. New York: Alfred van der Marck.

Seed, J.R. (1993) Immunoecology: origins of an idea. *Journal of Parasitology*. 79:470–471.

Sher, A.F., and R.L. Coffman. (1992) Regulation of immunity to parasites by T cells and T cell-derived cytokines. *Annual Review of Immunology* 10:385–409.

Sher, A., D. Fiorentino, P. Caspar, E. Pearce, and T. Mosmann. (1991) Production of IL-10 by CD4+ T lymphocytes correlated with down-regulation of Th1 cytokine synthesis in helminth infection. *Journal of Immunology* 147:2713–2716.

Sher, A., and E. Ottesen. (1988) Immunoparasitology. In M. Samter, D.W. Talmage, M.F. Michael, K.F. Austen, and H.N. Claman, eds., *Immunological Diseases*, 4th ed., vol. I. Boston, MA: Little, Brown and Company, pp. 923–943.

Shirakawa T, A. Li, M. Dubowitz, J.W. Dekker, A.E. Shaw, J.A. Faux, C. Ra, W.O.M. Cookson, and J.M. Hopkin. (1994) Association between atopy and variants of the β subunit of the high-affinity immunoglobulin E receptor. *Nature Genetics* 7:125–129.

Song, Z., V. Casolaro, R. Chen, S.N. Georas, D. Monos, and S.J. Ono. (1996) Polymorphic nucleotides within the human IL-4 promoter that mediate overexpression of the gene. *Journal of Immunology* 156:424–429.

Sporik, R., S.T. Hogate, T.A.E. Platts-Mills, and J.J. Cogswell. (1990) Expsoure to house-dust mite allergen

(*Der p* I) and the development of asthma in childhood. *New England Journal of Medicine* 323:502–507.

Sprent, J.F.A. (1969a) Evolutionary aspects of immunity of zooparasitic infections. In G.J. Jackson, ed., *Immunity to Parasitic Animals*. New York: Appleton.

Springer, T. (1990) Adhesion receptors of the immune system. *Nature* 346: 425.

Street, N.E., J.H. Schumacher, T.A.T. Fong, H. Boss, D.F. Fiorento, J.A. Leverah, and T.R. Mossmann. (1990) Heterogeneity of mouse T helper cells. Evidence from bulk cultures and limiting dilution cloning for precursors of TH1 and TH2 cells. *Journal of Immunology* 144:1629–1639.

Stringer, C.B., and P. Andrews. (1988) Genetic and fossil evidence for the origin of modern humans. *Science* 239:1263–1264.

Sutherland, M., K. Blaser, and J. Pene. (1993). Effects of interleukin-4 and interferon-gamma on the secretion of IgG4 from human peripheral blood mononuclear cells. *Allergy* 48(7):504–510.

Sutton, B.J., and H.J. Gould. (1993) The human IgE network. *Nature* 366: 421–428.

Tahara, H., and M.T. Lotze. (1995) Antitumor effects of interleukin–12 (IL-12): applications for the immunotherapy and gene therapy of cancer. *Gene Therapy* 2:96–106.

Tingley, G.A., A.E. Butterworth, R.M. Anderson, H.C. Kariuki, D. Koech, M. Mugambi, J.M. Ouma, et al. (1988) Predisposition of humans to infection with *Schistosoma mansoni:* evidence from the reinfection of individuals following chemotherapy. *Transactions of the Royal Society of Tropical Medicine and Hygiene* 82: 448–452.

Tullis, D.C.H. (1970) Bronchial asthma associated with intestinal parasties.

New England Journal of Medicine 282:370–373.

Turki, J., J. Pak, S.A. Green, R.J. Martin, and S.B. Liggett. (1995) Genetic polymorphisms of the β_2-adrenergic receptor in nocturnal and nonnocturnal asthma: evidence that Gly16 correlates with the nocturnal phenotype. *Journal of Clinical Investigation* 95: 1635–1641.

Turner, K.J. (1980) Is the prevalence of allergy related to parasitic disese? In A. Oehling, ed., *Advances in Allergology and Immunology*. Oxford: Pergamon Press, pp. 279–287.

UNDP/World Bank/WHO. (1990) Schistosomiasis. In *Tropical Diseases: Progress in Research*. Geneva: World Health Organization.

Valentine, M.D. (1992) Anaphylaxis and stinging insect hyerpsensitivity. *Journal of the American Medical Association* 268:2830–2833.

Van Dellen, R.G., and J.H. Thompson. (1971) Absence of intestinal parasties in asthma. *New England Journal of Medicine* 285:146–148.

Van Metre, T.E., Jr., D.G. Marsh, N.F. Adkinson, Jr., A. Kagey-Sobotka, P.S. Norman, A. Khattignavong, and G.L. Rosenberg. (1988) Immunotherapy for cat asthma. *Journal of Allergy and Clinical Immunology* 82: 1055–1068.

Vena, J.E., J.R. Bona, T.E. Byers, E.J.R. Middleton, M.K. Swanson, and S. Graham. (1985) Allergy-related diseases and cancer: an inverse association. *American Journal of Epidemiology* 122:66–74.

Verwaerde, C. M. Joseph, M. Capron, R.J. Pierce, M. Damonneville, F. Velge, C. Auriault, and A. Capron. (1987) Functional properties of a rat monoclonal IgE antibody specific for *Schistosoma mansoni. Journal of Immunology* 138:4441–4446.

Vogel, G. (1997). Why the rise in asthma cases? *Science* 276:1645.

Waddle, D.M. (1994) Matrix correlation tests support a single origin for modern humans. *Nature* 368:452–454.

Waite, D.A., E.F. Eyles, S.L. Tonkin, and T.V. O'Donnell. (1980) Asthma prevalence in Tokelauan children in two environments. *Clinical Allergy* 10:71–75.

Warren, K.S., D.A.P. Bundy, R.M. Anderson, A.R. Davies, D.A. Henderson, D.T. Jamison, N. Prescott, and A. Senft. (1990) Helminth infections. In D.T. Jamison, and W.H. Mosley, eds., *Disease and Disease Cotnrol in Developing Countries*. Washington, DC: World Bank.

Warren, K.S., and A.A.F. Mahmoud. (1984) *Tropical and Geographic Medicine*. New York: McGraw-Hill, pp. 1–1159.

Williams, G.C., and R.M. Nesse. (1991) The dawn of Darwinian medicine. *Quarterly Review of Biology* 66:1–22.

Worth, R.M. (1962) Atopic dermatitis among Chinese infants in Honolulu and San Francisco. *Hawaii Medical Journal* 22: 31.

Yamamura, M., R.L. Modlin, J.D. Ohmen, and R.L. Moy. (1993) Local Expression of antiinflammatory cytokines in cancer. *Journal of Clinical Investigation* 91:1005–1010.

Yazdanbakhsh, M., P.-C. Tal, C.J.F. Spry, G.J. Geich, D.J. Roos. (1987) Synergism between eosinophil cationic protein and oxygen metabolites in killing of schistosomula of *Schistosoma mansoni. Immunology* 138: 3443–3447.

Zanders, E.D., C. Harris, S.J. Buckham, D.J. Quint. (1992) Regulation of IgE synthesis in allergy and helminth infections. In R. Moqbel, ed., *Allergy and Immunity to Helminths*. London: Taylor & Francis, pp. 81–93.

Zeh, H.J., S. Hurd, W.J. Storkus, and M.T. Lotze. (1993) Interleukin-12 promotes the proliferation and cytolytic maturation of immune effectors: implications for the immunotherapy of cancer. *Journal of Immunotherapy* 14:155–161.

Zitvogel, L., H. Tahara, P.D. Robbins, W.J. Storkus, M.R. Clarke, M.A. Nalesnik, and M.T. Lotze. (1995) Cancer immunotherapy of established tumors with IL-12. Effective delivery by genetically engineered fibroblasts. *Journal of Immunology* 155:1393–1403.

Zwingenberger, K., A. Hohmann, M. Cardoso de Brito, and M. Rittere.

(1991) *Scandinavian Journal of Immunology* 34:243–251.

Zwollo, P., E. Ehrlich-Kautzky, A.A. Ansari, S.J. Scharf, H.A. Erlich, and D.G. Marsh. (1991) Molecular studies of human immune response genes for the short ragweed allergen, Amb a V. Sequencing of HLA-D second exons in responders and nonresponders. *Immunogenetics* 33:141–151.

10

USING EVOLUTION AS A TOOL FOR CONTROLLING INFECTIOUS DISEASES

PAUL W. EWALD

The Long Lag

At first glance it may seem surprising that the integration of evolution with
the health sciences is occurring at the end of the twentieth century, a century
and a half after Darwin provided the framework for understanding the "why"
questions of biology. The health sciences, after all, are essentially applied areas of biology with a focus on the well-being of individuals of one species. A
peek below the surface, however, resolves this paradox. The health sciences
have been quick to incorporate areas of biology as soon as the first clear-cut
example of an effective intervention is demonstrated. Within a decade of Jenner's demonstration of an effective vaccine against smallpox in 1796, vaccination was accepted as a viable part of the medical arsenal, and has been so
now for nearly for two centuries. The demonstration of penicillin's effectiveness quickly transformed chemical therapy from an intervention of questionable value or even ridicule into a mainstay of modern medicine. The improved
recovery of patients in response to disinfection of surgical tools, the surgical
environment, and the patient prior to surgery led to a similarly rapid adoption
of disinfection procedures. Interventions that prevent or cure disease through
the treatment of individuals are especially prone to be rapidly accepted because effectiveness can be readily observed. A person who was vaccinated
against smallpox did not develop smallpox. A person who was treated with
penicillin before the emergence of antibiotic resistance rapidly recovered from
an otherwise serious disease. Because the benefits of such treatments were
readily observable, the health sciences almost immediately recognized the importance of the biological disciplines that fostered these interventions in their
primitive and modern phases—namely, cell biology, microbiology, biochemistry, and molecular biology.

The acceptance of interventions that alter interactions above the level of the individual patient has had a more plodding history. Strong evidence of the importance of looking at infectious disease at the level of human populations became available in the mid-nineteenth century, a few decades after universal acceptance of smallpox vaccination. By comparing cholera incidence rates among London residents receiving water from different sources, John Snow broadened the scale of inquiry to this population level, showing that cholera outbreaks in London were largely waterborne; Ignaz Semmelweis similarly traced the transmission of infections in the University of Vienna hospital, implicating transmission between patients via attendants. Although we now recognize these studies as the beginning of epidemiology, the study of disease at the population level, the value of epidemiological interventions took much longer to be incorporated into standard health science practice than those focusing on individual patients. Fifty years after Snow's discoveries, many developed countries, such as the United States, still were providing contaminated water supplies in most of their cities. Health benefits of water purification were more difficult to assess because they were not as clear cut as smallpox vaccination or penicillin treatment. Even after water supplies were purified, people would continue to become infected because the organisms that cause cholera and other water-borne diseases could be transmitted at least to some degree by other routes (e.g., personal contact). Eventually, however, the tremendous importance of controlling infectious disease by altering population-level processes was recognized by predictable statistical associations between interventions such as water purification and disease reduction.

The place of epidemiology in modern health sciences is an indication that the importance of disease processes at the population level has been recognized and integrated. But today's health sciences have not yet fully incorporated still larger scale contributions of biology, those of evolutionary changes that occur during epidemiological processes.

The general lack of attention to evolutionary effects of human activities on pathogen characteristics is intriguing because the influence of human activities on one pathogen characteristic has been recognized for a half century. That characteristic is antibiotic resistance. (For brevity, I use the term *antibiotics* broadly to subsume chemical against the entire spectrum of parasites from viruses to multicellular eukaryotes.) The recognition of the evolution of antibiotic resistance in response to the use of antibiotics, however, is consistent with the tendency for the health sciences to recognize effects most readily when they are apparent at the level of the treated individual. When antibiotic therapy began failing, the importance of an evolutionary process became glaringly apparent to clinicians.

An evolutionary epidemiologist would argue that the demonstrated effects of human activity on antibiotic resistance is a specific example of a broader potential for human activities to influence pathogen characteristics. Some of these other characteristics may be even more important to the health sciences than antibiotic resistance. This chapter focuses on effects of human activities on one such characteristic: the inherent virulence of a pathogen. By "inherent virulence," I mean the tendency for a pathogen to be harmful relative to other pathogens in the same host. Thus, although the virulence of an infection is host dependent, the concept of inherent virulence concerns differences between pathogens in their contribution to the virulence of the infection.

Although evolution of the inherent virulence of pathogens has until recently received much less attention than the evolution of antibiotic resistance, it is arguably more important. The health sciences are concerned with antibiotic resistance only if that resistance can be coupled with virulence. Virulence is thus a prerequisite for damage caused by resistance. Interventions that control the evolution of virulence may thus obviate the problem of antibiotic resistance in two ways. First, if pathogens are avirulent, the presence or absence of resistance is of no direct clinical relevance (although transmission of resistance genes to virulent pathogens could be important). Second, because pathogens with reduced virulence will tend to cause disease less often, such mild pathogens should tend to have reduced exposure to antibiotics, and hence should be less prone to evolve antibiotic resistance.

If we can eradicate a pathogen from the human population using conventional strategies, as was done with smallpox, then evolutionary epidemiology would be of relatively little importance. But over the two centuries that have passed since Jenner's discovery of a smallpox vaccine, this goal has been achieved for only one pathogen: the smallpox virus. In spite of the wealth of knowledge that has been generated from the disciplines of cell biology, biochemistry, and molecular biology and their integration with the germ theory of disease, the only pathogenic adversary that has been eradicated was eradicated by a vaccine generated and effectively used decades before these disciplines existed. The reasons for this limited success involve a variety of economic, political, and sociological processes in addition to biological characteristics, such as high mutation rates, that may make some pathogens particularly difficult to eradicate. Whatever the combination of reasons, the record of the past century and the persistence of the obstacles to eradication lead to a sober realization: most of our pathogens will probably be living with us for centuries to come.

This prospect necessitates a shift in goals away from eradication of pathogenic adversaries toward use of interventions as evolutionary tools to mold pathogenic adversaries into organisms that coexist benignly with us. The failure to fully integrate evolutionary principles into intervention strategies has led us into a state of continual and expensive arms races with our pathogens. These evolutionary arms races are exemplified by the urgency to develop new antibiotics before the disease organisms develop measures to counter them, and by the annual race to produce the appropriate influenza vaccines during the months before each flu season. The evolutionary approach offers the hope of avoiding arms races by making damaging pathogens benign or by preventing the organisms from being so damaging in the first place.

In this chapter I first summarize the current understanding of the environmental factors that cause pathogens to evolve toward different levels of virulence. I then suggest ways in which this understanding could be used as a tool for making pathogens evolve toward lower inherent virulences. Finally, I describe studies that will allow us to determine both the accuracy of our current understanding of virulence and the feasibility of using epidemiological interventions as evolutionary tools. Application of this approach to AIDS and other sexually transmitted diseases is discussed in chapter 11.

Evolution of Virulence: The Traditional View

The last century and a half is replete with misunderstandings and misapplications of Darwin's theory of evolution. The misapplication to virulence began to take the shape of dogma during the 1930s, at the same time that the evolutionary synthesis of Darwin's theory with genetics was occurring (Fisher 1930; Haldane 1932). The misapplication stemmed from a misunderstanding of the level at which natural selection occurred. Disease organisms were presumed to evolve toward a state of peaceful coexistence with their hosts because peaceful coexistence was in the best long-term interest of both parasite and host (Smith 1934; Zinsser 1935; Swellengrebel 1940). From that time until the late 1970s, this view was promulgated as fact by leading health scientists (Dubos 1965; Thomas 1972; Burnet and White 1972). A few health scientists expressed reservations (Andrewes 1960; Coatney et al. 1971) because they believed that the logic was unsound or thought that some mild symptoms such as sneezes and coughs could be maintained indefinitely because they facilitate the spread of disease. A few others (Ball 1943; Cockburn 1963) disagreed adamantly, but were ignored, perhaps because they did not have an alternative theoretical framework to explain the variation in virulence.

Such a theoretical framework was generated by ecologists and evolutionary biologists during the last quarter of the twentieth century through two independent branches of inquiry: (1) mathematical modeling focused on the general trade-offs between reproductive benefits that parasites obtain from exploitation of hosts and the transmission costs that parasites accrue as a result of the virulance associated with that exploitation (Fine 1975; Levin & Pimentel 1981; Anderson & May 1982; Levin et al. 1982), and (2) comparative studies focused on the evolutionary functions of symptoms and specific categories of transmission that in theory should be associated with high or low virulence (Ewald 1980, 1983). Both branches of this theoretical framework independently recognized the error inherent in the traditional view, emphasizing that natural selection does not necessarily favor long-term stability of host–parasite associations. Rather, natural selection favors characteristics that allow their bearers to survive and reproduce better than competitors. If exploitation of the host generates competitive superiority, that exploitation will be favored by natural selection, even if it is associated with severe disease and increased chances of death. According to this argument, virulence results from an evolutionary trade-off between the benefits and costs that the parasite accrues from different levels of host exploitation. Increased host exploitation benefits the parasite through greater conversion of host resources into parasite reproduction. But increased exploitation may impose evolutionary costs on the parasite by making the host a less effective vessel for obtaining resources from the environment and for transmitting the parasite to susceptible hosts.

Before the development of this framework, most medical experts interpreted severe disease as a state of maladaptation between host and parasite. They therefore had nothing to say about the ways in which evolution might be used as a tool for controlling infectious diseases. The evolutionary framework, in contrast, offers a wide range of possibilities for understanding why some disease organisms are more virulent than others and for using this understanding to identify ways that pathogens might be guided evolutionarily toward a more benign state. One set of possibilities derives from a focus on host mobility.

Host Mobility, Vectors, and Sanitation

Associations Between Virulence and Transmission Modes

Like the traditional view of host–parasite coevolution, the modern view identifies host illness as a potential liability for the pathogen. When pathogens rely on the mobility of their current host to reach uninfected individuals, the illness caused by intense exploitation of the host typically reduces the potential for transmission. The modern view differs from the traditional view, however, in weighing these costs of intense exploitation against the competitive benefits that the pathogen accrues from the exploitation. The modern view proposes that high virulence can be evolutionarily stable if the costs incurred by parasites from the host damage are particularly low and/or the benefits obtained by the parasites from exploitation are particularly high (figure 10.1). Thus, if host immobilization has little negative effect on transmission, then pathogen variants that exploit the host so intensely that the host is immobilized will get the benefits of exploitation while paying a relatively small decrement in transmission. Parasites that do not rely on host mobility for transmission should therefore evolve to be more virulent than those that do rely on host mobility. In this way the diversity of virulence found among host–parasite associations is explained according to modes of transmission rather than being dismissed as variation in the state of parasite adaptation to the host.

The trade-off between the benefits of increased host exploitation and the costs of reduced host mobility has led to several predictions: (1) Because biting arthropods can transmit parasites effectively from immobilized hosts, arthropod-borne parasites should evolve to a higher level of virulence than directly transmitted parasites. (2) Similarly, diarrheal disease organisms transmitted by water should evolve to relatively high levels of virulence because effective transmission can occur even when infected hosts are immobilized: attendants remove and wash contaminated clothing and bedsheets, contaminating water supplies, which then can infect large numbers of uninfected in-

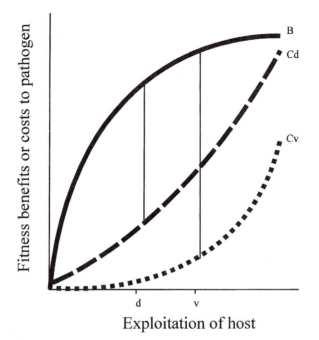

Figure 10.1. Cost–benefit trade-off associated with increased host exploitation. Increased host exploitation is assumed to be a cause of increased virulence. The curve B designates the fitness benefits to the pathogen, which eventually plateau as a function of exploitation of the host, reflecting the leveling off of the probability of infection per contact. The curve Cd designates the fitness costs to the pathogen associated with directly transmitted pathogens, which should rise as increased exploitation of host causes reductions in host mobility. The curve Cv represents the fitness costs to vectorborne pathogens. It rises less steeply at levels of exploitation that cause immobilizing, nonlethal illness because a vectorborne pathogen's reliance on a person's mobility is less than that of a directly transmitted pathogen. The maximum net fitness benefit to the pathogen (vertical bars) occurs at a higher level of exploitation for vectorborne than for directly transmitted pathogens (designated by v versus d on the x axis). See Ewald (1983) for a discussion of effects of vectorborne transmission on the benefit curve and the effect of host defenses on the outcome of host/parasite coevolution.

dividuals. (3) In hospitals, nurses and physicians can transmit pathogens from one immobilized patient to uninfected individuals, favoring relatively exploitative and hence virulent pathogens. (4) A variation on the theme occurs when parasites rely on the mobility of uninfected individuals to move to the location at which pathogens were released from infected hosts. Durability in the external environment allows for such transmission from immobile infected hosts (termed "sit-and-wait" transmission; Ewald 1987); durable pathogens are thus predicted to be more virulent than labile pathogens, which must rely on mobility of infected hosts for person-to-person transmission.

Each of these hypotheses has been tested by determining whether the variable that is hypothesized to influence virulence is in fact significantly associated with virulence. In each case the expected associations were confirmed (table 10.1, figures 10.2–10.4). Collectively this evolutionary framework explains the diversity of virulence found among human parasites according to modes of transmission rather than simply as variation in the state of parasite adaptation to the host. According to this framework, the agents of malaria, typhus, yellow fever, and sleeping sickness have evolved highly virulent relationships with their hosts because they are vectorborne (figure 10.2). The agents of classical cholera, typhoid fever, and the worst of the dysentery bacteria (*Shigella dysenteriae* type 1) are highly virulent because they are waterborne (figure 10.3). The agents of smallpox, tuberculosis and diphtheria are severe because they are durable in the external environment. The lethal hospital outbreaks that plagued Semmelweis's hospitals, the *Escherichia coli* that killed high percentages of infected babies in industrialized countries during the first half of the twentieth century (figure 10.4), and the recent outbreaks of particularly severe streptococci, staphylococci, enterococci and numerous gram-negative hospital-acquired bacteria may be similarly explained by high levels of attendant-borne transmission (table 10.1; Ewald 1994).

Some confusion has resulted from misrepresentations of these and related arguments. The trade-offs mentioned above are not based solely or even primarily on the idea that "increases in the rate of transmission favor increased virulence" (Levin 1996, p. 96). Also, they are not based on the idea that changes in the general level of hygiene in an area will cause an evolutionary change in virulence (van Baalen & Sabelis 1995). The argument does not "assume that a microparasite's virulence is constrained solely by the need to keep the host alive" (Levin 1996, p. 96). Nor does it represent an alternative to the idea that disease manifestations may be nonadaptive side effects of infection, a category that Levin (1996) calls "coincidental evolution" (p. 97). The idea that morbidity, mortality, and disease manifestations in general can be nonadaptive side effects has been stated repeatedly throughout the development of the comparative approach to disease manifestations (Ewald 1980, 1983, 1994).

Altering Virulence by Altering Transmission

Each of the preceding associations between transmission characteristics and virulence suggests ways in which epidemiological interventions that reduce the spread of infections can be used as evolutionary tools to mold the organism into a milder form. According to the theory presented above, vectorborne pathogens have evolved greater virulence than non-vectorborne pathogens because vectors can transmit pathogens from immobilized hosts. Interventions that prohibit transmission from immobilized hosts should alter the competitive balance in favor of the more benign variants by disproportionately restricting

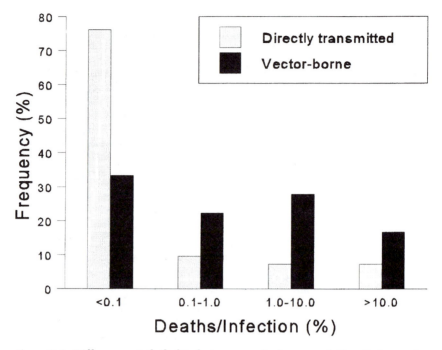

Figure 10.2. Differences in lethality between vectorborne and directly transmitted pathogens (calculations as in Ewald 1983).

transmission from the people who might otherwise transmit the most virulent variants. If houses and hospitals are made mosquito-proof, for example, a person immobilized by illness would be unlikely to act as a source of transmission. Vector-proof housing therefore raises the cost of virulence incurred by the parasite, making the cost–benefit trade-off for vectorborne pathogens more like that for directly transmitted pathogens (figure 10.5). If such improvements were made on a broad geographic scale, pathogen variants that affected people adversely enough to cause them to remain in their vector-proof structure would achieve virtually no propagation during the incapacitating illness.

Although mosquito-proof housing is not as glamorous as some of the high-tech candidates for controlling the spread of vectorborne pathogens (e.g., state-of-the-art vaccines and antimalarials), even partial improvements in restricting mosquitoes is strongly associated with a reduction in frequency of infection (Gamagemendis et al. 1991; Ko et al. 1992). Field experiments are now needed to determine whether investments in mosquito proofing cause evolutionary reductions in pathogen virulence.

The most lethal agent of malaria, *Plasmodium falciparum*, may be the most important test case of this idea. It currently causes more deaths than any other vectorborne pathogen, and it is very difficult to control by conventional means, partly because it can rapidly evolve resistance to antimalarials. The protective immunity generated by a mild *P. falciparum* strain against distantly related lethal strains (Fandeur et al. 1992) indicates that any favoring of transmission of mild variants would be enhanced by immunological resistance. Because this immunity appears to be more long-lived than previously thought (Deloron & Chougnet 1992), mild strains of *P. falciparum* circulating in a human population should act like a vaccine against more virulent strains.

The available information about variation in virulence within *Plasmodium* species supports the idea that a *Plasmodium* species could evolve to a milder state in response to a reduction in opportunities for transmission from immobile hosts. *Plasmodium vivax* tends to be milder in northern latitudes when vectors are more sporadically available (Ewald 1994). The unusual mildness of *P. falciparum* infections in an area with sporadic transmission of *P. falciparum* (Elhassan et al. 1995) also accords with the possibility that *P. falciparum*'s virulence varies in response to the local potential for vectorborne transmission, and it may therefore also evolve reduced virulence in response to a reduction in opportunities of transmission from immobile hosts.

Provisioning of uncontaminated drinking water should similarly drive the evolution of diarrheal pathogens toward benignness because those variants that immobilize hosts depend on waterborne transmission more than those

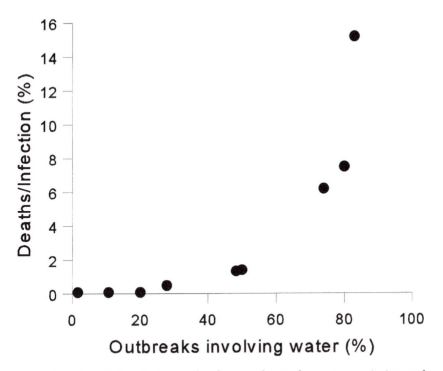

Figure 10.3. Association between the degree of waterborne transmission and lethality of bacterial diarrheal diseases of humans. Each data point represents a different bacterial species or subspecific type. (For calculations see Ewald 1991a.)

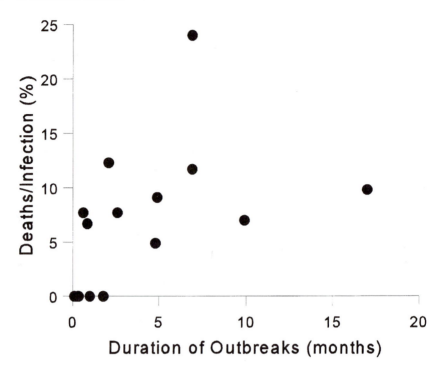

Figure 10.4. Association between lethality of hospital outbreaks of pathogenic *Escherichia coli* and the duration of the outbreak. (Data and calculation methods are given in Ewald 1988, 1991b.)

that are sufficiently mild to allow transmission by direct contact. The geographic distributions of diarrheal bacteria and protozoa with different levels of virulence accord with this argument, as do the changes in prevalence of these variants as access to uncontaminated water supplies increases (Ewald 1991a, 1994, 1995).

These temporal and geographic variations in virulence of diarrheal pathogens draw attention to the need for two categories of field studies: (1) descriptive studies of "natural experiments" in which changes in pathogen virulence are monitored after the pathogens have spread between source and recipient areas that differ in their degrees of water purity and (2) experimental studies that monitor changes in diarrheal pathogens of differing virulence and the prevalence of virulence genes before and after uncontaminated water supplies are introduced. The evolutionary theory predicts that the inherent virulence of diarrheal pathogens will decline in the areas with improved water supplies relative to control areas where such improvements have not been introduced.

Blocking attendant-borne transmission in hospitals should similarly shift the competitive balance in favor of the variants that are continually entering hospital environments from the outside community. The more conditions in the outside community favor mild variants (e.g., because water supplies are

Table 10.1 Lethality of human pathogens, their dependence on host mobility for transmission, and interventions that may favor evolution toward a lower level of pathogen virulence

Characteristic of pathogens allowing transmission from immobile hosts	Lethality per infection	Interventions favoring evolution toward benignness	Reference
Arthropod-borne	Higher among arthropod-born pathogens than among directly transmitted pathogens	Vector-proof housing	Ewald (1983)
Waterborne	Positively correlated with tendencies for waterborne transmission	Clean drinking water	Ewald (1991a)
Attendant-borne	Positively correlated with degree of attendant-borne cycling	Improved attendant hygiene	Ewald (1988)
Durability in the external environment	Positively correlated with durability	Destruction of pathogens at transmission sites	Walther & Ewald (1997)

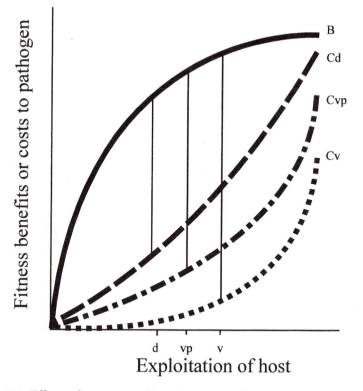

Figure 10.5. Effects of vector-proof housing on cost–benefit trade-off associated with increased host exploitation. The curve Cvp corresponds to the costs to a vectorborne pathogen associated with vector-proof housing, which reduces the potential for transmission from immobile hosts. By requiring a high level of mobility for transmission, vector-proof housing should favor those pathogens that have a lower level of exploitation (vp is less than v on the x axis). B, fitness benefits to pathogen; Cd, fitness costs to directly transmitted pathogens; Cv, fitness costs to vectorborne pathogens.

uncontaminated), the greater the potential for maintaining mild variants in institutions. Studies of compliance of doctors and nurses to guidelines for handwashing show much room for improvement, but they also show the effectiveness of interventions designed to obtain such improvement (e.g., Raju & Kobler 1991). Studies have documented the epidemiological effects of improved attendant hygiene on frequency of infection but have not investigated experimentally the evolutionary effects of such interventions on the inherent virulence of the pathogens that cause the infections (e.g., Easmon et al. 1981; Bentley 1990). Now that the relevance of evolutionary processes and virulence determinants are better understood, investigators and funding agencies are in a good position to integrate this evolutionary dimension into such studies.

By analogy, inhibition of sit-and-wait transmission by durable pathogens should also disfavor those pathogen variants that could benefit from high levels of host exploitation. This intervention would first identify "hot-beds" of sit-and-wait transmission and then eliminate the potential for transmission of durable variants in these areas. Now that the smallpox virus has been eradicated, the tuberculosis bacterium (*Mycobacterium tuberculosis*) is the most damaging sit-and-wait pathogen of humans. For *M. tuberculosis*, enclosed areas with poor rates of fresh air exchange, such as homeless shelters, hospitals, and homes for the elderly, can become hot-beds of transmission because they bring particularly vulnerable, susceptible people together with others who have particularly florid infections (Ehrenkranz and Kicklighter 1972; Nardell et al. 1986; Beck-Sagué et al. 1992; Dooley et al. 1992; Edlin et al. 1992; Pearson et al. 1992; Bellin et al. 1993; Paul et al. 1993). By requiring a high rate of air exchange in such areas with ventilation designs that draw air currents away from individuals and out to the external environment (where *M. tuberculosis* is quickly diluted to less dangerous concentrations and destroyed by ultraviolet light), the potential for transmission from immobilized individuals could be reduced, thus disproportionately reducing the transmission of virulent variants. Such interventions have proven effective in reducing the transmission of *M. tuberculosis* in hospital settings, where the potential for transmission of durable pathogens from immobile hosts is enhanced by the tendency for particularly sick individual to be brought in for treatment. As with the preceding examples of interventions, the expected evolutionary effects would simply make an intervention that is already considered beneficial on traditional epidemiological grounds even more beneficial. By raising estimates of the cost effectiveness of such interventions, evolutionary considerations may often elevate the interventions above cost-effectiveness thresholds necessary for initial or additional investment.

One of the benefits of using epidemiological interventions as evolutionary tools is that a given intervention can simultaneously lower the damage done by a broad range of pathogens. Screening against vectors, for example, is predicted to decrease the virulence not only of the agents of malaria but also those of other diseases like dengue. Uncontaminated water supplies should decrease the virulence of the diarrheal pathogens whether bacterial, viral, or protozoal, and whether their pathogenic mechanism is characterized by invasiveness (as with *Shigella* or *Entamoeba*) or fluid loss (as with *Vibrio cholerae*). Reduction in attendant-borne transmission should favor milder variants of the standard pathogens that are already causing severe problems in institutional settings, as well as new pathogens, which have repeatedly emerged and are often both severe and resistant to antibiotics. Modification of ventilation to reduce the spread of sit-and-wait respiratory tract pathogens such as *M. tuberculosis* in hot-bed areas should also reduce the potential for airborne transmission of other pathogens in those areas, particularly durable pathogens that have evolved or could evolve relatively high levels of virulence.

Virulence-Antigen Vaccines

Vaccines can also be used as evolutionary tools to favor evolution toward benignness. As successful as vaccines have been over the last two centuries, smallpox virus remains the only human pathogen that has been eradicated by a vaccine. This lone success story indicates that eradication through vaccination is not so easily achieved as was thought in the wake of the smallpox vaccination campaign.

Vaccines are now developed and chosen without consideration of their evolutionary effects. They are chosen on the basis of their safety, cost, ease of administration, and efficacy at immunologically protecting individuals from infection. Because vaccine protection can be narrow and pathogens can be variable, vaccination programs may favor reductions in their own effectiveness. Influenza viruses regularly generate variants that are poorly controlled by recently used vaccines, making selection of variants for vaccine development an annual guessing game. Similarly, the currently circulating measles viruses appear to have diverged from the measles viruses circulating when measles vaccination was first initiated: measles vaccines inhibit recently circulating measles viruses less strongly than the measles viruses of previous decades (Tamin et al. 1994). The changing composition of hepatitis B virus in a population subjected to heavy usage of vaccines and immunoprophylaxis suggests that this virus is also evolving around our immunological defenses against it (Oon et al. 1995).

These data suggest that our efforts at immunological interventions can alter the predominant variant in the pathogen population. The virulence-antigen strategy prescribes how to use this process to our advantage—specifically, how to make the variants that are favored by interventions benign.

The virulence-antigen strategy requires that policy makers resist the temptation to use vaccines that inhibit the broadest possible spectrum of existing variants because doing so suppresses mild as well as dangerous variants. Instead, antigens should be restricted to those that make viable mild organisms harmful. Such virulence-antigen vaccines should provide exceptionally strong long-term control because they disproportionately suppress the virulent variants, leaving behind the mild variants, which can augment disease control by competing with any remaining virulent variants. Because the mild forms will share antigens with the severe forms, the immunity generated against the mild variants should allow mild forms to be used like a free, live vaccine to protect those who are not vaccinated and those who develop a weak response to the administered vaccine. These effects may be particularly important when vaccine coverage temporarily wanes, a change that has occurred often in recent decades even in countries with high-quality health care (Hewlett 1990; Pinchichero et al. 1992). Evolutionary reductions in the virulence of circulating organisms will increase the chances that mild organisms will cause infections during such lapses. Also, if virulent organisms arise by mutation or enter from other areas, they will be less able to spread because of the increased resistance attributable to mild natural infections.

Diphtheria Vaccination

The evolutionary effects of virulence-antigen vaccines are not simply hypo-thetical. The standard diphtheria vaccine conforms to the virulence-antigen criteria. The active component of the vaccine is a toxoid derived from the diphtheria toxin. When *Corynebacterium diphtheriae* is faced with a shortage of iron, it begins producing this toxin, which causes death of nearby host cells. By this mechanism *C. diphtheriae* may generate nutrients for itself when nutrient availability is low. *C. diphtheriae* therefore appears to benefit from toxin production when it is infecting a person who has not been vaccinated against it.

In a person immunized with diphtheria toxoid, however, the toxin is im-potent. The bacterium is paying about 5% of its protein budget (Pappenheimer and Gill 1973) to make a product that does not work. In a vaccinated popu-lation this waste of resources should put the toxigenic variants at a competitive disadvantage. Accordingly, diphtheria has vanished from areas with long-standing, thorough vaccination programs. *C. diphtheriae* is still present but is dominated by variants that do not produce toxin. The most detailed evaluation of the association between application of diphtheria toxoid vaccine and *C. diphtheriae*'s evolution toward benignness came from a 15-year study in Ro-mania (Pappenheimer 1982). A steady increase in the number of vaccinated individuals was associated with a steady fall in toxigenicity; the majority of strains were toxigenic at the beginning of the vaccination effort, but few were toxigenic at the end (figure 10.6; Pappenheimer 1982).

In terms of reductions in morbidity and mortality per unit of investment, the diphtheria vaccination program is second only to the smallpox program. This disproportionate success in control without eradication is as expected from the virulence-antigen strategy because control of disease results not only from the manufactured vaccine but also from the remaining toxinless *C. diphtheriae*. Because these toxinless variants still share antigens with the tox-igenic variants, immunity against the former should inhibit the latter.

Although health scientists have suggested that the diphtheria vaccination caused an evolutionary shift from toxigenic to toxinless *C. diphtheriae* (Uchida et al. 1971; Pappenheimer & Gill 1973; Pappenheimer 1982; Chen et al. 1985), they did not take the next logical step: recognition that the success of the diphtheria vaccine may be attributable to the evolutionary shift toward mild-ness and that this process could serve as a model for achieving more powerful control of disease in other vaccination programs.

The records on diphtheria also illustrate the stability of control by a virulence-antigen vaccine. In recent decades the proportion of people with protective immunity against diphtheria toxin has not been particularly high. In the United States about three-quarters of children and one-quarter of adults have protective immunity (Chen et al. 1985). If the diphtheria vaccine pro-tected equally against all strains of *C. diphtheriae*, the large numbers of un-protected people might have generated a resurgence in diphtheria, much like the resurgence of other respiratory tract diseases, for which mild variants have

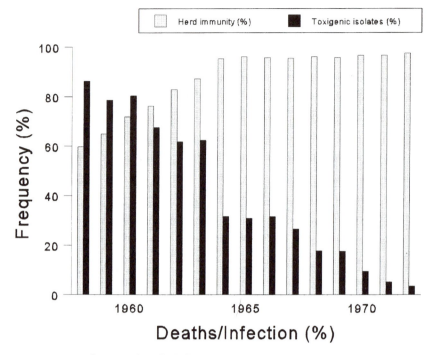

Figure 10.6. Evolution of *C. diphtheriae* toward benignness, as evidenced by a decrease in the percentage of isolates that were toxigenic, during an extensive vaccination effort in Romania (data from Pappenheimer 1982).

not been circulating to protect against resurgence. For example, even though diphtheria and pertussis vaccination have occurred at virtually the same rate, whooping cough has commonly recurred in areas where vaccination efforts have diminished; diphtheria has recurred substantially only in areas that have experienced more dramatically reduced rates of vaccination.

More generally, most major vaccination programs, such as those for measles, rubella, and mumps, parallel the whooping cough experience; they have been dealt setbacks from vaccine-caused disease, outbreaks that spread among unvaccinated people, or reinfections as vaccine-induced immunity fades (Hewlett 1990; Forsey et al. 1992; Hilleman 1992; Pinchichero et al. 1992; Weiss 1992). The incidence of measles in the United States, for example, rose dramatically during the 1980s, with infections often occurring in previously vaccinated adults (Poland & Jacobson 1994). This increase occurred in spite of a program initiated by the U.S. Public Health service to eliminate measles from the United States by 1982 (Centers for Disease Control 1978).

The *Haemophilus* Campaign

Only one other vaccination program has relied on a virulence-antigen vaccine: the program to control *Haemophilus influenzae*. Like diphtheria vaccination,

Haemophilus vaccination has been unusually successful (Shapiro 1993; Booy et al. 1994). Although *H. influenzae* has been a major cause of bacterial meningitis in recent decades, this association has been virtually eliminated where vaccination efforts have been rigorous (Eskola et al. 1992; Shapiro 1993; Booy et al. 1994).

Haemophilus vaccines have relied on the polysaccharide antigen of the type b capsule (polyribosyl ribotolphosphate; PRP) because type b *H. influenzae* (Hib) has been responsible for the majority of *Haemophilus* invasive disease. Simple PRP vaccines were introduced in the mid-1980s and replaced during the late 1980s and early 1990s with polysaccharide conjugate vaccines, which linked polysaccharide to carrier proteins such as the diphtheria toxoid. These conjugate vaccines generated better, longer lasting immunity, particularly among very young children. During this vaccination program, rates of invasive disease dropped precipitously, often even more than expected as a direct result of protection by the vaccine (Eskola et al. 1992; Peltola et al. 1992; Broadhurst et al. 1993; Murphy et al. 1993a, b; Shapiro 1993; Booy et al. 1994). But such a decline is just the effect one would expect from a virulence-antigen vaccine—a lower-than-expected frequency of disease in vaccinated as well as unvaccinated groups, even though the overall decline in mild infection appears negligible.

If the extraordinary success of the *Haemophilus* vaccination program is attributable to the evolutionary effects of using virulence antigens, *H. influenzae* vaccines should have favored evolution toward less virulent *H. influenzae* variants. The evidence is consistent with this assumption. A multistate study by the U.S. Centers for Disease Control showed that the absolute frequency of invasive disease caused by type b strains declined between 1986 and 1989 as PRP vaccines were being introduced, but the frequency of disease caused by other capsular serotypes and nontypeable strains did not (Wenger et al. 1992). Similarly, a study of schoolchildren showed that use of conjugate vaccine reduced the frequency of Hib carriage, confirming that fewer sources of Hib were present after vaccination (Murphy et al. 1993a). These findings suggest that the composition of *H. influenzae* has evolved toward a lower relative frequency of the particularly harmful type.

The stability of protection afforded by virulence-antigen vaccines depends on the organism's potential to evolve other virulence mechanisms. If a variety of virulence-generating exploitation mechanisms already exists or are readily created by mutations or genetic recombinations, then the disproportionate success of a virulence-antigen vaccine could be short-lived because the reduction in the dominant competitor may open the niche for competitors that exploit hosts through the use of a different virulence antigen.

Various virulence-enhancing components occur among Hib and non-b strains. A compound called haemocin, for example, is associated with increased virulence and increased competitiveness (LiPuma et al. 1992). The presence of these virulence components indicates that the current success of Hib vaccines will probably diminish over time as the inhibition of Hib opens the door for other *Haemophilus* variants to invade an open niche. To maintain the success of *Haemophilus* vaccination over the long term, we may need to work now to allow timely incorporation of virulence antigens like haemocin into future generations of *Haemophilus* vaccines.

These considerations indicate that as the long-term effects of *Haemophilus* vaccines become documented, the vaccination program may serve as a model for the stability of other virulence antigen vaccination programs. On the positive side, the findings may illustrate the degree of success generated from vaccines that guard against the majority of virulent variants. The findings may also, however, illustrate the potential for reduced long-term efficacy that may result from use of antigens that fail to protect against an originally small proportion of virulent variants and the need for continual monitoring of variants and incorporation of additional components into vaccines, even when vaccination programs seem to have achieved an extremely high level of success.

Potential Applications for Virulence-Antigen Vaccines

The virulence-antigen strategy may be applicable to any pathogen for which the determinants of virulence can generate an effective immune response and are not essential for reproduction and transmission. Because many of the components that contribute to virulence in bacteria are relatively easy to identify, the virulence-antigen strategy will probably first be applied intentionally to these components. Candidate antigens include the pneumolysins of *Streptococcus pneumoniae* (a common cause of pneumonia), and the pertussis toxin of *Bordetella pertussis* (the agent of whooping cough).

Application of the approach to viruses is inherently more difficult because the way in which viruses generate virulence is so intertwined with the functioning of the host cell. When the precise functions of viral products are understood, however, the virulence-antigen strategy should be applicable to viral vaccines as well.

The virulence-antigen strategy draws attention to the need for detailed knowledge about the molecular determinants of virulence, down to the variation that can result from a single mutation. This kind of variation is particularly applicable to pathogens such as HIV, for which there is a high mutation rate and considerable viral evolution within each host. For such pathogens, virulence-antigen vaccines may prevent the evolutionary emergence of virulent variants within hosts when administered to already infected hosts. HIV is the most likely candidate for the first of this kind of therapeutic vaccines both because the knowledge of HIV's virulence antigens is outpacing knowledge about virulence antigens of other viruses and because of the great need for effective treatments. The application of the virulence antigen strategy to sexually transmitted viruses is considered further in chapter 11.

Because virulence-antigen vaccines do not change any underlying environmental factors that may be favoring high levels of pathogen virulence, they may be best applied in conjunction with other interventions that favor evolution toward benignness by altering the relevant environmental factors. Virulence-antigen vaccines against cholera, for example, could be based on the cholera toxin, which is a primary determinant of virulence for *Vibrio cholerae*. The recent pandemic associated with the new O139 variant of *V. cholerae* (Albert et al. 1993), however, suggests that an immune response to cholera toxin alone is not sufficient to suppress highly toxigenic variants, the O139 variant invaded areas in which other variants of *V. cholerae*, which were producing the same toxin, were endemic. Other antigenic differences between O139 variants and the previously dominant variants of *V. cholerae* (Sengupta et al. 1994) presumably allowed the O139 variant to spread in the presence of a high level of acquired immunity to these other *V. cholerae*. In this case, a successful use of a virulence antigen vaccine would probably need to be coupled with water purification to favor the benign variants.

When the details of transmission favor a relatively benign association, a two-pronged attack is less necessary. Such is the case with *H. influenzae*, which is a directly transmitted pathogen that is relatively nondurable in the external environment. *H. influenzae* causes severe damage in only a tiny proportion of infected people. The task for a virulence-antigen vaccine is therefore much more feasible in this situation than it would be when transmission favors higher virulence (e.g., as in areas with waterborne transmission of *V. cholerae*).

Evolutionary Epidemiology, Uncertainty, and Ethics

At first glance one might think that the use of epidemiological interventions as evolutionary tools is not an innovation because we are already using such interventions. But this first impression is misleading. The decision to invest in any intervention is not made in isolation of other possible interventions. The amount of the investment will depend on the relative cost effectiveness of the various interventions and how the cost effectiveness varies as a function of the total investment.

These evolutionary considerations add a new dimension to disease diagnosis: not only does the causative organism need to be identified, but the variants of these organisms need to be monitored at both the individual and the population level. This integration of evolution represents a fundamental change in the health sciences. Twentieth-century epidemiology developed from the nineteenth century focus on species of pathogens to serological and biochemical variants, and finally to molecular variants characterized by nucleic acid similarities and differences. The first steps toward integrating principles of adaptation to disease control broadened the analysis to consider the evolution of antibiotic resistance. The central argument presented in this chapter is that this integration of evolution into the health sciences can be broadened beyond the realm of antibiotic resistance. Resistant variants can reduce our ability to control disease with antibiotics. Antigenic variants can reduce our ability to prevent disease with vaccines. Pathogens can differ in the degree to which they exploit hosts, and these variants can be transmitted to different degrees depending on how our activities alter our environments. The framework offered in this chapter and in chapter 11 addresses ways in which evolutionary processes can be harnessed in these situations to work for us instead of against us.

To resolve these issues, large-scale experiments need to be run on human populations. Such experiments are ethically acceptable, even if they disprove the hypotheses—even if reduction of transmission from immobilized people had no evolutionary effects, introduction of vector-proof housing, clean water or improved hygiene of attendants would have positive epidemiological effects.

Noting that theory about evolutionary molding of pathogens by epidemiological interventions has not generated a track record of success like that of molecular phylogenetics, Bull (1994) has suggested that the evolutionary molding of pathogens will be a less significant application of evolution to the health sciences than molecular phylogenetics. His arguments, however, overlook three issues.

First, available evidence indicates that such interventions have already molded pathogen characteristics over the past century. *C. diphtheriae* has evolved toward benignness in concert with diphtheria vaccination (figure 10.6). The most dangerous diarrheal pathogens have been replaced by more benign pathogens in areas where uncontaminated water supplies were provided (Ewald 1994). The evolution of antibiotic resistance offers ample evidence that human interventions can predictably and rapidly affect the evolution of our pathogens.

Second, definitive studies will require tracking pathogen virulence in experimental and control populations of humans. A preponderance of nonexperimental evidence may be necessary, however, before financial support for such large-scale experiments will be granted. At that point the high costs of the experiments can be justified as investments in disease control for those populations receiving it. The necessary investments in the synthesis of molecular phylogenetics with the health sciences were minuscule by comparison. That synthesis required only the sequencing of pathogen variants and quantitative comparisons of these sequences. Once the techniques of DNA and phylogenetics had been developed by researchers outside the health sciences, these studies could be done expeditiously, and their value could be readily appreciated.

Finally, evolutionary molding and molecular phylogenetics are likely to become so intertwined that a comparison of one with the other will be increasingly inane. It may be like debating whether Darwin's ideas about the origin of species is more important than his principles of natural selection. Both are critically important to the theory of evolution. The proposed test of the evolutionary effects on HIV virulence of a low potential for sexual transmission in Japan gives a sense of how these two applications of evolution to the health sciences will probably be used in concert (see Chapter 11, figure 11.3).

Until the critical studies are conducted, the usefulness of interventions as evolutionary tools will remain uncertain. The resolution of this matter will be protracted relative to the resolution of arguments about the value of nonevolutionary interventions such as vaccination, antibiotic treatment, and sanitary improvements. This expectation may seem surprising because the rate of technological change has accelerated over the last two centuries, but the rate of the resolution depends on the time scale of the processes being studied and the time scale over which research support for the necessary studies is made available. Unlike the evolution of antibiotic resistance, detection of evolution toward benignness will involve expansion of epidemiology, the health science discipline having the largest temporal scale, to a still larger scale.

The definitive experimental tests will involve tracking epidemiological trends over evolutionary time and will require comparisons of different track records with each other to assess alternative explanations. We can therefore expect that if these tests are begun promptly, decades will be required before the value of evolutionary tools for controlling epidemiological problems can be assessed. Given the prematurity of such a assessment, a more appropriate focus is on evaluating particular predictions by specific tests that are more controlled statistically and experimentally than the tests that have been conducted to date. The final judgment we can leave to future historians.

Acknowledgments
This study was supported by a grant from Leonard X. Bosack and Bette M. Kruger Charitable Foundation, a George E. Burch Fellowship in Theoretic Medicine and Affiliated Sciences awarded by the Smithsonian Institution, and an Amherst College Faculty Research Award.

References

Albert, M. J., Ansaruzzaman, M. Bardhan, P. K. Faruque, A. S. G. Faruque, S. M. Islam, M. S. Mahalanabis, D. Sack, R. B. Salam, M. A. Siddique, A. K. Yunus, M. D. and Zaman K. (1993) Large epidemic of cholera-like disease in Bangladesh caused by *Vibrio cholerae* 0139 synonym Bengal. *Lancet* 342:387–390.

Anderson, R. M., and May, R. M. (1982) Coevolution of hosts and parasites. *Parasitology* 85:411–26.

Andrewes, C. H. (1960) The effect on virulence of changes in parasite and host. In G. E. W. Wolstenholme and C. M. O'Connor, eds., *Virus Virulence and Pathogenicity*; pp. 34–39. Boston, MA: Little, Brown and Company.

Ball, G. H. (1943) Parasitism and evolution. *American Naturalist* 77:345–64.

Beck-Sagué, C., Dooley, S. W., Hutton, M. D., Otten, J., Breeden, A., Crawford, J. T., Pitchenik, A. E., Woodley, C., Cauthen, G., and Jarvis, W. R. (1992) Hospital outbreak of multidrug-resistant *Mycobacterium tuberculosis* infections. Factors in transmission to staff and HIV-infected patients. *Journal of the American Medical Association* 268: 1280–86.

Bellin, E. Y., Fletcher, D. D., and Safyer, S. M. (1993) Association of tuberculosis infection with increased time in or admission to the New York city jail system. *Journal of the*

American Medical Association 269: 2228–2231.

Bentley, D. W. (1990) Clostridium difficile-associated disease in long-term care facilities. *Infectious Control and Hospital Epidemiology* 11: 432–38.

Booy, R., Hodgson, S., Carpenter, L., Mayonwhite, R. T., Slack, M. P. E., Macfarlane, J. A., Haworth, E. A., Kiddle, M., Shribman, S., Roberts, J. S. C., and Moxon, E. R. (1994) Efficacy of *Haemophilus influenzae* type b conjugate vaccine PRP-T. *Lancet* 344:362–366.

Broadhurst, L. E., Erickson, R. L., and Kelley, P. W. (1993) Decreases in invasive *Haemophilus influenzae* diseases in US army children, 1984 through 1991. *Journal of the American Medical Association* 269:227–231.

Bull, J. J. (1994) Perspective—virulence. *Evolution* 48:1423–1427.

Burnet, F. M., and White, D. O. (1972) *Natural History of Infectious Disease*, 4th ed. Cambridge: Cambridge University Press.

Centers for Disease Control. (1978) Goal to eliminate measles from the United States. *Morbidity and Mortality Weekly Report*. 27:391.

Chen, R. T., Broome, C. V., Weinstein, R. A., Weaver, R., and Tsai, T. F. (1985) Diphtheria in the United States, 1971–81. *American Journal of Public Health* 75:1393–97.

Coatney, G. R., Collins, W. E., Mc-Wilson, W., and Contacos, P. G. (1971) *The Primate Malarias*. Washington, DC: U.S. Government Printing Office.

Cockburn, A. (1963) *The Evolution and Eradication of Infectious Diseases*. Baltimore, MD: Johns Hopkins University Press.

Deloron, P., and Chougnet, C. (1992) Is immunity to malaria really short-lived? *Parasitology Today* 8:375–78.

Dooley, S. W., Villarino, M. E., Lawrence, M., Salinas, L., Amil, S., Rullan, J. V., Jarvis, W. R., Bloch, A. B., and Cauthen, G. M. (1992) Nosocomial transmission of tuberculosis in a hospital unit for HIV-infected patients. *Journal of the American Medical Association* 267:2632–34.

Dubos, R. (1965) *Man Adapting*. New Haven, CT: Yale University Press.

Easmon, C. S. F., Hastings, M. J. G., Clare, A. J., Bloxham, B., Marwood, R., and Rivers, R. P. A. (1981) Nosocomial transmission of group B streptococci. *British Medical Journal* 283:459–61.

Edlin, B. R., Tokars, J. I., Grieco, M. H., Crawford, J. T., Williams, J., Sordillo, E. M., Ong, K. R., Kilburn, J. O., Dooley, S. W., Castro, K. G., Jarvis, W. R., and Holmberg, S. D. (1992) An outbreak of multidrug-resistant tuberculosis among hospitalized patients with the acquired immunodeficiency syndrome. *New England Journal of Medicine* 326: 1514–21.

Ehrenkranz, N. J., and Kicklighter, J. L. (1972) Tuberculosis outbreak in a general hospital: evidence of airborne spread of infection. *Annals of Internal Medicine* 77:377–82.

Elhassan, I. M., Hviid, L., Jakobsen, P.H., Giha, H., Satti, G.M.H., Arnot, D.E., Jensen, J B., and Theander, T G. (1995) High proportion of subclinical *Plasmodium falciparum* infections in an area of seasonal and unstable malaria in Sudan. *American Journal of Tropical Medicine and Hygiene* 53:78–83.

Eskola, J., Peltola, H., Kayhty, H., Takala, A. K., and Makela, P. H. (1992) Finnish efficacy trials with *Haemophilus influenzae* type b vaccines. *Journal of Infectious Disease* 165: S137–38.

Ewald, P. W. (1980) Evolutionary biology and the treatment of signs and symptoms of infectious disease.

Journal of Theoretical Biology 86: 169–76.

Ewald, P. W. (1983) Host-parasite relations, vectors, and the evolution of disease severity. *Annual Review of Ecology and Systematics* 14:465–85.

Ewald, P. W. (1987) Pathogen-induced cycling of outbreak insect populations. In P. Barbosa and J.C. Shultz, eds., *Insect Outbreaks: Ecological and Evolutionary Processes*; pp. 269–86. New York: Academic Press.

Ewald, P. W. (1988) Cultural vectors, virulence, and the emergence of evolutionary epidemiology. *Oxford Surveys of Evolutionary Biology* 5:215–45.

Ewald, P. W. (1991a) Waterborne transmission and the evolution of virulence among gastrointestinal bacteria. *Epidemiology and Infections* 106:83–119.

Ewald, P. W. (1991b) Transmission modes and the evolution of virulence, with special reference to cholera, influenza and AIDS. *Human Nature* 2:1–30.

Ewald, P. W. (1994) *Evolution of Infectious Disease*. New York: Oxford University Press.

Ewald, P. W. (1995) The evolution of virulence: a unifying link between parasitology and ecology. *Journal of Parasitology* 81:659–669.

Fandeur, T., Gysin, J., and Mercereau-Puijalon, O. (1992) Protection of squirrel monkeys against virulent *Plasmodium falciparum* infections by use of attenuated parasites. *Infections and Immunology* 60:1390–96.

Fine, P. E. F. (1975) Vectors and vertical transmission: an epidemiological perspective. *Annals of the New York Academy of Sciences* 266:173–94.

Fisher, R. A. (1930) *The Genetical Theory of Natural Selection*. Oxford: Clarendon.

Forsey, T., Bentley, M. L., Minor, P. D., and Begg, N. (1992) Mumps vaccines and meningitis. *Lancet* 340: 980.

Gamagemendis, A. C., Carter, R., Mendis, C., Dezoysa, A. P. K., Herath, P. R. J., and Mendis, K. N. (1991) Clustering of malaria infections within an endemic population: risk of malaria associated with the type of housing construction. *American Journal of Tropical Medicine and Hygiene* 45:77–85.

Haldane, J. B. S. (1932) *The Causes of Evolution*. New York: Longmans, Green.

Hewlett, E. L. (1990) *Bordetella* species. In G. L. Mandell, R. G. Douglas, and J. E. Bennett, eds., *Principles and Practice of Infectious Disease*; pp. 1757–62. New York: Wiley.

Hilleman, M. R. (1992) Past, present, and future of measles, mumps, and rubella virus vaccines. *Pediatrics* 90: 149–53.

Ko, Y. C., Chen, M. J., and Yeh, S. M. (1992) The predisposing and protective factors against dengue virus transmission by mosquito vector. *American Journal of Epidemiology* 136:214–20.

Levin, B. R. (1996) Evolution and maintenance of virulence in microparasites. *Emerging Infectious Diseases* 2:93–102.

Levin B.R., Allison A.C., Bremermann H.J., Cavali-Sforza L.L., Clarke B. C., Grentzel-Beyme R., Hamilton W.D., Levin S.A., May R.M., and Thieme H.R. (1982) Evolution of parasites and hosts. In R.M. Anderson and R.M. May, eds., *Population Biology of Infectious Diseases*; pp. 213–43. Berlin: Springer Verlag.

Levin, S., and Pimentel, D. (1981) Selection of intermediate rates of increase in parasite-host systems. *American Naturalist* 117:308–15.

LiPuma, J. J., Sharetzsky, C., Edlind, T. D., Stull, T. L. (1992) Haemocin production by encapsulated and nonencapsulated *Haemophilus influenzae*.

Journal of Infectious Disease 165: S118–19

Murphy, T. V., Pastor, P., Medley, F., Osterholm, M. T., and Granoff, D. M. (1993a). Decreased *Haemophilus* colonization in children vaccinated with *Haemophilus influenzae* type b conjugate vaccine. *Journal of Pediatrics* 122:517–23.

Murphy, T. V., White, K. E., Pastor, P., Gabriel, L., Medley, F., Granoff, D. M., and Osterholm, M. T. (1993b) Declining incidence of *Haemophilus influenzae* type b disease since introduction of vaccination. *Journal of the American Medical Association* 269:246–48.

Nardell, E., McInnis, B., Thomas, B., and Weidhaas, S. (1986) Exogenous reinfection with tuberculosis in a shelter for the homeless. *New England Journal of Medicine* 315:1570–1575.

Oon, C. J., Lim, G. K., Ye, Z., Goh, K. T., Tan, K. L., Yo, S. L., Hopes, E., Harrison, T. J., and Zuckerman, A. J. (1995) Molecular epidemiology of hepatitis b virus vaccine variants in Singapore. *Vaccine* 13:699–702.

Pappenheimer, A. M. (1982) Diphtheria: studies on the biology of an infectious disease. *Harvey Lectures* 76: 45–73.

Pappenheimer, A. M., and Gill, D. M. (1973) Diphtheria. *Science* 182: 353–58.

Paul, E. A., Lebowitz, S. M., Moore, R. E., Hoven, C. W., Bennett, B. A., and Chen, A. (1993) Nemesis revisited—tuberculosis infection in a New York city men's shelter. *American Journal of Public Health* 83:1743–45.

Pearson, M. L., Jereb, J. A., Frieden, T. R., Crawford, J. T., Davis, B. J., Dooley, S. W., and Jarvis, W. R. (1992) Nosocomial transmission of multidrug-resistant *Mycobacterium tuberculosis*: a risk to patients and health care workers. *Annals of Internal Medicine* 117:191–96.

Peltola, H., Kilpi, T., and Anttila, M. (1992) Rapid disappearance of *Haemophilus influenzae* type b meningitis after routine childhood immunization with conjugate vaccines. *Lancet* 340:592–94.

Pichichero, M. E., Francis, A. B., Blatter, M. M., Reisinger, K. S., Green, J. L., Marsocci, S. M., and Disney, F. A. (1992) Acellular pertussis vaccination of 2-month-old infants in the United States. *Pediatrics* 89:882–87.

Poland, G. A., and Jacobson, R. M. (1994) Failure to reach the goal of measles elimination. Apparent paradox of measles infections in immunized persons. *Archives of Internal Medicine* 154:1815–20.

Raju, T. N. K., and Kobler, C. (1991) Improving handwashing habits in the newborn nurseries. *American Journal of Medical Science* 302:355–58.

Sengupta, T. K., D. K. Sengupta, G. B. Nair, and A. C. Ghose. (1994) Epidemic isolates of *Vibrio cholerae* 0139 express antigenically distinct types of colonization pili. *FEMS Microbiology Letters* 118:265–71.

Shapiro, E. D. (1993) Infections caused by *Haemophilus influenzae* type b. The beginning of the end? *Journal of the American Medical Association* 269:264–66.

Smith, T. (1934) *Parasitism and Disease*. Princeton, NJ: Princeton University Press.

Swellengrebel, N. H. (1940) The efficient parasite. In *Proceedings of the Third International Congress of Microbiology*; pp. 119–27. Baltimore, MD: Waverly.

Tamin, A., Rota, P. A., Wang, Z. D., Heath, J. L., Anderson, L. J., and Bellini, W. J. (1994) Antigenic analysis of current wild type and vaccine strains of measles virus. *Journal of Infectious Disease* 170:795–801.

Thomas, L. (1972) Notes of a biology-watcher: germs. *New England of Journal of Medicine* 247:553–55.

Uchida, T., Gill, D. M., and Pappenheimer, A. M. (1971) Mutation in the structural gene for diphtheria toxin carried by temperate phage β. *Nature New Biology* 233:8–11.

van Baalen, M., and Sabelis, M. W. (1995) The scope for virulence management: a comment on Ewald's view on the evolution of virulence. *Trends in Microbiology* 3:414–16.

Walther, B. A., and Ewald, P. W. (1997) Pathogen survival in the external environment and the evolution of virulence. *The Quarterly Review of Biology* (submitted).

Weiss, R. (1992) Measles battle loses potent weapon. *Science* 258:546–47.

Wenger, J. D., Pierce, R., Deaver, K., Franklin, R., Bosley, G., Pigott, N., and Broome, C. V. (1992) Invasive *Haemophilus influenzae* disease: a population-based evaluation of the role of capsular polysaccharide serotype. *Journal of Infectious Disease* 165:S34–S35.

Zinsser, H. (1935) *Rats, Lice and History*. Boston, MA: Little, Brown and Company.

11

EVOLUTIONARY CONTROL OF HIV AND OTHER SEXUALLY TRANSMITTED VIRUSES

PAUL W. EWALD

An Evolutionary Perspective on Sexually Transmitted Viruses

At a press conference on April 23, 1984, the secretary of the U.S. Department of Health and Human Services, Margaret Heckler, forecasted the imminent development of a vaccine to prevent AIDS. Focusing on the newly developed process for mass production of HIV, she said "we also believe that the new process will enable us to develop a vaccine to prevent AIDS in the future. We hope to have such a vaccine ready for testing in approximately two years" (Heckler 1984). Those two years came and went over a decade ago. Between then and now a disturbing realization has emerged: HIV is one of the most formidable pathogenic adversaries that the health sciences have ever confronted. One of the things that makes it so formidable is its great potential for evolutionary change; it is a continually moving target. This potential for evolutionary change also makes HIV a prime example of the need for an integration of evolutionary biology with medicine.

The strategies used against HIV have involved both traditional and novel approaches, but all have been based on nonevolutionary mechanisms of control. The most intensely investigated strategies involve (1) antiviral compounds, such as AZT (azidodideoxythymidine), (2) vaccines to guard against infection or suppress the progression of infection to AIDS, and (3) behavioral modifications to reduce the frequency of infection. When such interventions leave in their wake a subset of the pathogens that are genetically different from those that were present before intervention, the intervention is acting as an agent of evolutionary change. An evolutionary perspective emphasizes that such effects need to be understood and, if possible, used to our advantage, if we are to invest wisely in measures to control AIDS and other sexually transmitted diseases.

This chapter considers HIV's evolutionary responses to the selection pressures that humans impose on it. I have three specific goals: (1) to assess the degree to which evolutionary theory can explain HIV's characteristics, (2) to identify critical information that needs to be acquired for better understanding the evolutionary consequences of medical interventions, and (3) to consider how insights generated for HIV might offer solutions for controlling other sexually transmitted viruses. My more general goal is to integrate the categories of interventions mentioned above into a three-pronged strategy for evolutionary control of the damage caused by HIV and other sexually transmitted viruses.

Antiviral Treatment of HIV

The Resistance Problem

The early work on anti-HIV drugs generated an optimistic view of their potential for effective therapy. Two years after the discovery of HIV, Mitsuya et al. (1985) reported anti-HIV effects of AZT, which inhibits HIV's reverse transcriptase, the molecule that transcribes a DNA copy of HIV's genes from the original RNA strand. Within another 2 years, AZT was shown to be effective in prolonging life during the late symptomatic phase of infection (Fischl et al. 1987). By that time a variety of anti-HIV drugs were being studied, increasing optimism that more effective drugs and combinations of drugs could push back the negative effects of HIV, leading to continued improvement in the length and quality of life for HIV-infected patients. Within another 2 years, however, HIV with resistance to AZT had been isolated, and causative mutations identified (Larder and Kemp 1989; Larder et al. 1989).

Since then, HIV has educated us about its versatility at generating resistance to antivirals. Resistance-conferring mutations not only vary in the degree of resistance conferred, but their effects are compounded when they co-occur (see table 11.1 for resistance to AZT); moreover, additional resistance-conferring mutations are still to be identified. Resistance in one isolate, for example, was more than an order of magnitude greater than that of the most resistant combination of the identified resistance-conferring mutations. It had 11 mutations in the reverse transcriptase molecule, but only two were among the identified resistance-conferring mutations (Muckenthaler et al. 1992).

Resistance to ddI (dideoxyinosine, another inhibitor of reverse transcriptase) also can be generated by a variety of mutations (St. Clair et al. 1991; Shafer et al. 1994, 1995; Shirasaka 1993, 1995). In fact, HIV has evolved resistance to every antiviral that has been adequately studied (Nunberg et al. 1991; Richman 1991, 1992; Fitzgibbon et al. 1992; Gu et al. 1994; Condra et al. 1995; Tisdale et al. 1995). This facility at evolving resistance has led to the virtual

Table 11.1 Quantifications of AZT resistance associated with specified amino acid substitutions introduced by site-directed mutagenesis into the gene coding for reverse transcriptase

Substitutions introduced[a]	Increase in IC_{50}
Met41→Leu	4-fold
Thr215→Tyr	16-fold
Met41→Leu & Thr215→Tyr	60-fold
Asp67→Asn, Lys70→Arg & Thr215→Tyr	31-fold
Met41→Leu, Asp67→Asn, Lys70→Arg & Thr215→Tyr	179-fold
Asp67→Asn, Lys70→Arg, Thr215→Tyr & Lys219→Gln	121-fold
Asp67→Asn, Lys70→Arg, Thr215→Phe & Lys219→Gln	147-fold

The number next to each amino acid refers to the amino acid site on the reverse transciptase molecule. IC_{50} refers to the AZT concentration associated with a 50% reduction in the number of viral plaques observed in the assay. Data from Kellam et al. (1992).
[a] Amino acid abbreviations: Arg, arginine; Asn, asparagine; Asp, aspartic acid; Gln, glutamine; Leu, leucine; Lys, lysine; Met, methionine; Phe, phenylalanine; Thr, threonine; Tyr, tyrosine.

abandonment of hopes for prolonged control of HIV infections using any single antiviral.

Combination therapy imposes a higher hurdle for HIV. Joint usage of AZT and ddI seemed particularly promising (Gao et al. 1992a; Cohen et al. 1995). Because the mutation that conferred resistance to ddI also resulted in sensitivity to AZT, researchers were hopeful that the two could be used in combination to postpone the emergence of resistance, or in sequence, with ddI treatment causing a return to sensitivity to AZT. HIV, however, has proven itself quite able to develop resistance to both drugs in vivo and in vitro (Japour et al. 1991; Gao et al. 1992b; Shafer et al. 1994). When ddI and AZT were administered jointly, the percentage of patients generating the common substitution (at amino acid site 74 of the reverse transcriptase molecule), was reduced from about 65% to 8%, (rows 3 and 2, respectively, of table 11.2), but novel substitutions that conferred resistance to both antivirals arose in another 8% of the patients, (row 1 of table 11.2), and the number of patients developing resistance only to AZT was not significantly altered by combination therapy (rows 4 and 5 of table 11.2). The higher hurdle imposed by joint administration of these two drugs seems insufficiently high to prohibit emergence of variants that are resistant to therapeutic dosages, although the joint administration does appear to lengthen the time before this emergence (table 11.3).

New combinations of antivirals have recently provided a resurgence of optimism, particularly because some can reduce the density of HIV RNA in the plasma of at least some patients by an order of magnitude or more (Richman 1996). Patients treated in the preliminary studies showed not only reversals

Table 11.2 Evolution of resistance to AZT and ddI in response to monotherapy and combination therapy (data from Shafer et al. 1994)

Reverse transcriptase sites with amino acid substitutions	Treatment	Resistance	Frequency of resistance
62, 75, 77, 116, and/or 151	AZT + ddI	AZT + ddI	2 of 24 patients
74	AZT + ddI	ddI	2 of 24 patients
74	ddI	ddI	17 of 26 patients
41, 67, 70, 215, and/or 219	AZT + ddI	AZT	10 or 24 patients
41, 67, 70, 215, and/or 219	AZT	AZT	8 of 26 patients

of immunological and virological indicators and a delay of disease progression, but clinical improvement during the late symptomatic period of infection (Richman 1996). Much of the current optimism focuses on the possibility of treating patients as soon as possible after infection because the lower numbers of HIV at the outset of infection should limit its potential for generating resistance mutations (Richman 1996).

The previous experiences with AZT, however, show that the best effects of a particular antiviral therapy will tend to occur upon first usage, before resistance begins to accumulate in the viral population. Considering that many patients will be receiving combination therapy after they are carrying a high viral load, we can expect the success of combination therapy to erode, even if the outlook for some early-treated patients may be dramatically improved.

Table 11.3 Emergence of antiviral resistance during the course of individual HIV infections

Antiviral	Stage of patient during initiation of treatment	Development of resistance after onset of treatment	Reference
AZT	AIDS	Slight by 6 months; Moderate by 1 year; High by 2 years	Larder et al. (1989)
AZT	Asymptomatic	Moderate by 2 years; High by 3 years	Richman (1993), Boucher (1992)
ddI	AIDS	Moderate by 1 year	St. Clair et al. (1991)
ddI and AZT	Asymptomatic	Slight by 2 years; Moderate by 3 years	Shirasaka et al. (1993, 1995), Shafter et al. (1995)
Nevirapine/ L-697,661	AIDS	1–2 months	Kappes et al. (1992), Richman (1993)

HIV's facility at evolving resistance to any particular antiviral would not be so damaging if the resistant forms were always competitively inferior to the sensitive forms when the antiviral is not present. Then one could expect a return to sensitivity after cessation of antiviral treatment. Some resistant variants do show a rapid return to sensitivity. HIV that are resistant as a result of an amino acid substitution at site 70 on the reverse transcriptase molecule, for example, have been displaced by sensitive types within a few months after cessation of AZT treatment (Boucher et al. 1992). Substitutions at site 215, however, can persist for more than a year after cessation of AZT therapy, even with simultaneous ddI therapy, which was thought to decrease the competitive abilities of AZT-resistant strains, as mentioned above (Boucher et al. 1992; Smith et al. 1994).

The slow return to sensitivity raises the prospects of a steadily worsening resistance problem in human populations. If the resistant variants are readily transmissible (i.e., if they can be readily transmitted to a partner during sexual contact), then their increased frequency in one patient as a result of treatment will generate increased frequencies in previously untreated patients. As a consequence, the overall clinical value of antivirals will decrease over time, even in previously untreated patients. This possibility has now been evaluated for AZT. Resistant viruses have been detected in the genital fluid of AZT-treated patients who had resistant HIV in their blood (Wainberg et al. 1993), in adult patients who had not previously received AZT therapy (Erice et al. 1993; Mohri et al. 1993), and in an infant born to an AZT-treated mother (Masquelier et al. 1993). Considering the relative stability of the resistance conferring substitution at site 215, it is not surprising that viruses with this substitution appear to be frequently transmitted (Fitzgibbon et al. 1993; Wahlberg et al. 1994). The limited information concerning transmission of HIVs that are resistant to other antivirals indicates that they, too, can be transmitted (e.g., Najera et al. 1994).

This realization of HIV's potential for rapidly evolving resistance, for slowly evolving a return to sensitivity, and for having transmissible resistant variants has sobered the early optimism about using antivirals as magic bullets for the long-term control of HIV infections. This realization clarifies the need for clever applications of antivirals, for strategies that exploit the short-term effectiveness of antivirals while reducing the negative impact of HIV's evolutionary responses. The use of antivirals needs to consider four evolutionary processes: (1) the time scales over which HIV evolves resistance to different antivirals, which will influence the duration of time over which particular antivirals can be used, (2) effects of various resistance mechanisms on the competitive abilities of the resistant HIV, which will influence the rate of return to sensitivity after termination of antiviral therapy, (3) variation in the different molecular mechanisms of resistance to a particular antiviral, which will influence the predictability of evolving a given resistance mechanism, and (4) variation in the potential for the transmission of resistant variants to new hosts, which will influence the prevalence of resistance in the general population and the time at which resistance occurs during the course of individual infections. The relevance of these processes can be understood by considering how the evolution of resistance can be incorporated into evolutionary strategies for handicapping HIV and by the consequences of the evolution of resistance for the timing of treatment. These two issues are considered below.

The Resistance Handicap Strategy

Although evolution of resistance typically reduces the degree to which a pathogen can be controlled, some forms of evolved resistance might handicap a pathogen by being especially costly for the pathogen to maintain. The handicaps that resistance imposes on viruses are evidenced by the rates of evolutionary return to sensitivity when the antiviral treatment is ended. The more rapid return to sensitivity of some resistant HIV variants than others suggests that some of the resistant configurations of HIV proteins impose a cost on the HIVs that encode for them. The information about resistance to AZT and the return to sensitivity after cessation of AZT treatment, however, indicates that some resistant variants, particularly those containing a substitution at site 215 of the reverse transcriptase molecule are constrained little, if at all, by resistance. The AZT experience therefore does not engender optimism about handicapping HIV reproduction through evolution of antiviral resistance.

Other antivirals may not be so easily countered, perhaps because they impose different kinds of handicaps. One of the most promising uses of antiviral therapy against HIV involves a handicap of HIV's mutation rate. Theoretical considerations suggest that high mutation rates benefit HIV by allowing it to stay ahead of immune responses. If HIV's ability to generate mutations were handicapped, the immune system might be better able to control HIV infections. HIV evolves resistance to 3TC (2'-deoxy-3'-thiacytidine) within a few months (Wainberg et al. 1995, 1996b), but 3TC seems to suppress HIV's damage even after the evolution of this resistance. The reverse transcriptase of 3TC-resistant strains transcribe with a 5- to 10-fold lower mutation rate, which may contribute to 3TC's positive therapeutic effect even after the evolution of 3TC resistance (Wainberg et al. 1996a, b). On the basis of these results, Wainberg (personal communication) has proposed that 3TC could be particularly effective if used at the beginning of infection. At that time, 3TC might strongly delay the onset of AIDS by reducing HIV's potential to generate the variants that cannot be controlled by the patient's immune system and hence delay the eventual erosion of the immune system. This application of the resistance handicap strategy should be especially important for neonatal infections because they can be identified at the onset of infection through identification of infections in their mothers. Used at this time the joint control by 3TC and the immune system should inhibit HIV particularly strongly because the mutation rate of the resistant HIV is restricted, as is the total population of HIV in the patient that can generate mutations. Over the long run, however, we can expect some mutations to generate resistant reverse transcriptases that are more prone to mutation because mutation proneness should involve reduced rather than increased precision, and there are more ways to generate reduced precision than increased precision. When mutation rate begins to rise, reducing control of HIV by 3TC, cessation of 3TC therapy should reduce the chances that 3TC-resistant viruses would evolve both increased mutation proneness and the increased replication rate that characterizes HIV infections before AIDS (replication rate refers to the production rate of new virions from a parent virus). In the absence of 3TC, the continued resistance handicap that the 3TC-resistant viruses experience due to low mutation rate should cause the viral population in the patient to be taken over by mutation-prone, 3TC-sensitive viruses, which have the best chance of staying ahead of the immune system in the absence of 3TC. When that takeover occurs, 3TC treatment can be restarted.

This possibility of handicapping a virus by forcing it to evolve toward a lower mutation rate raises an interesting aspect of the resistance handicap strategy: If pathogens will reevolve sensitivity, continued application of the antiviral might be necessary to keep the virus in a handicapped state. At first glance, this need would seem to create an ethical problem: increasing usage of antivirals to benefit the patient increases the overall incidence of new infections with resistant viruses. The apparent ethical problem vanishes, however, if resistant viruses need continued treatment to persist in competition with nonresistant viruses.

The resistance handicap strategy is similar to combination therapy, when simultaneous application of antivirals restricts HIV's ability to evolve resistance to each drug. The preceding application of the handicap strategy differs from combination therapy, however, in an important aspect: the characteristic being handicapped is HIV's ability to respond to attack from the immune system rather than its ability to reproduce during exposure to more than one antiviral. The ability of a mutation-restricted HIV to evolve resistance to the ever-changing defenses of the immune system should be less than the ability of a mutation-prone HIV to evolve resistance to one or a few unchanging antivirals. This contention is supported by the years over which HIV is held in check by the immune system before the onset of AIDS. By taking away HIV's main device for staying ahead of immune responses (its high mutation rate), the resistance handicap strategy should be more effective than simple combination therapy in achieving AIDS-free time for the patient.

As additional antivirals become available, the application of this information to timing of transmission will become increasingly important. Hints that a resistance handicap may be imposed by an antiviral can be obtained by observing whether signs of clinical or immunological deterioration are halted or slowed even after resistance evolved (as discussed for 3TC above). Resistance handicaps may be fairly common. Preliminary studies indicate, for example, that HIV evolves resistance to the protease inhibitor indinavir sulfate within 6 months after initiation of therapy, but a positive clinical effect persists afterward (Condra et al. 1996; J.H. Condra, personal communication). HIV variants possessing high level resistance to other protease inhibitors have proteases with reduced catalytic activity and are rapidly outcompeted by sensitive strains when cultured in the absence of the proteases (Croteau et al. 1997).

The preceding considerations draw attention to the need for continuing surveillance to identify new resistance mutations and to determine the degree to which and the means by which such mutations handicap the organisms. Variations in the degree to which different antivirals handicap HIV bear on these decisions, particularly with regard to the timing of treatment.

Timing of Antiviral Treatment

HIV's facility for evolving resistance to antivirals has led researchers to shift their focus from searching for an antiviral that will provide long-term control to assessments of the proper timing of antiviral treatment and the alternative sets of antivirals for combination therapy (Rutherford 1994). On the basis of high levels of viral replication early during infection and mathematical models of viral growth and immune responses, researchers have advocated intense antiviral treatment as early as possible during infection (Nowak et al. 1991; Ho 1995). The preceding considerations, however, emphasize that a treatment decision should take into account the handicap imposed by resistance mutations. Those antivirals that generate resistance-induced handicaps should be used early because the intact immune system will be better able to control a handicapped HIV and because cessation of treatment with such antivirals should lead to reevolution of sensitivity. This generalization is particularly true for antivirals that handicap HIV by decreasing its mutability, as discussed earlier for 3TC.

Those antivirals that readily generate resistance without handicaps, such as AZT, should be used later, during or just before AIDS. The late usage of such antivirals may allow the temporary period over which these drugs are effective to coincide with the period when the viruses are most damaging when control of the virus is most important. Because these viruses will not be handicapped after cessation of treatment, they will not readily reevolve sensitivity and will therefore be particularly dangerous to the outside community. But late usage will reduce this danger because it narrows the window of time during which transmission of the unhandicapped, resistant viruses can occur and restricts this window to the overtly symptomatic phase of infection when sexual contact between susceptibles and an infected patient is less likely. Overtly ill patients tend to be less motivated for sexual contact, avoided by susceptibles, and, by being more aware of their positive status, less likely to risk infecting others. In short, late use of antivirals that do not generate a resistance handicap allows control of HIV when the need is great while reducing the risk of transmission of resistant HIV that will not readily reevolve sensitivity.

The variability in HIV's evolutionary responses to different antivirals is not yet clear. Only one antiviral, AZT, has been studied sufficiently to get a sense of the spectrum of viral resistance. Currently, we can say that for antivirals that generate resistance as AZT does, certain guidelines for timing of treatment should apply. As already discussed, treatment with AZT immediately after infection is not advisable, especially when alternative antivirals, such as 3TC, appear to be more suited to early control. In this context combination therapy may take on a new meaning: the use of antivirals that impose resistance handicaps, such as 3TC and perhaps indinavir sulfate, during the early years of infection, followed by antivirals with little potential for imposing handicaps, such as AZT, later during each infection. If antivirals that do not impose handicaps are to be used late in AIDS, exactly when should they be used? The answer requires a detailed consideration of the evolution of HIV within each host.

This success of AZT among AIDS patients (Fischl et al. 1987, 1990; Swanson & Cooper 1990; Moore et al. 1991) and its suppression of viral replication during the asymptomatic period before AIDS (DeWolf et al. 1988), generated optimism that early AZT treatment could postpone the onset of AIDS. The first test of this idea was conducted by the AIDS Clinical Trial Group (ACTG); it confirmed a positive effect of AZT treatment before the onset of AIDS (Volberding et al. 1990). The protocol of this study was used as the basis for current guidelines for AZT use in the United States. A delay in the onset of symptoms occurred only in the AZT-treated group that had < 500 CD4 lymphocytes/mm^3 blood. This threshold density was therefore used as a primary criterion for treatment (Friedland 1990; Hecht et al. 1990; National Institute of Allergy and Infectious Diseases 1990; FDA 1991; Kess et al. 1991). Among healthy, uninfected people, densities of CD4 lymphocytes are variable but are generally near 1000/mm^3; they fall below 100/mm^3 in people dying from AIDS.

Evolutionary considerations indicated that many people with CD4 cell counts < 500 should not be receiving AZT (Ewald 1994a). One of these evolutionary considerations was apparent to health scientists at the time these guidelines were being enacted: several experts cautioned that evolution of AZT resistance and side effects of AZT could negate its beneficial effects over the entire course of an infection (National Institute of Allergy and Infectious Diseases 1990; Ruedy et al. 1990; Swart et al. 1990; Volberding et al. 1990; Graham et al. 1991). In response to these concerns, European researchers advocated more restricted use of AZT. The British Department of Health, for example, recommended treatment when CD4 counts are between 200 and 500/mm^3 and rapidly falling ("Clinical trials of zidovudine" 1989; Swart et al. 1990; Gazzard 1992).

The dialogue, however, overlooked a second evolutionary process that occurs during the course of infection. HIV evolves increased replication rates during infections, probably as a result of within-host competition among HIV variants; these increases are associated with greater pathology and advancement of the progression to AIDS (Cheng-Mayer et al. 1988; Levy 1989; Tersmette et al. 1989; Schellekens 1992; Connor & Ho 1994). Accordingly, if an antiviral is effective late during an infection, it is controlling HIVs that appear to be more dangerous than those controlled earlier during the infection. Because HIV evolves resistance during the early, asymptomatic stage of an infection (table 11.3), early treatment may provide protection against less dangerous early viruses at the expense of protection against more dangerous late viruses whenever resistant viruses do not reevolve sensitivity during the intervening years. As described above, some of the HIV variants that are resistant to AZT (e.g., those containing a substitution at site 215) reevolve sensitivity slowly if at all after cessation of AZT therapy. These arguments indicate that HIV evolves increased virulence coincident with the evolution of relatively stable AZT resistance during the course of an infection. For AZT, therefore, early treatment is contraindicated. Each antiviral needs to be analyzed similarly to evaluate the value of treatment at early stages of infection. Rejection of such early treatment still leaves open the question of when later during infection treatment should be initiated. Again the AZT experience is informative.

The delay of AIDS documented in the ACTG study (Volberding et al. 1990) provided a basis for an evolutionarily based prediction about the best time to initiate AZT treatment (Ewald 1994a). Because this prediction was formulated in March 1991, several years before the data to evaluate it were available, it provides an opportunity to assess the predictive value of an evolutionary approach to determine timing of treatment. The AZT-induced delay in the onset of AIDS owes its statistical significance entirely to patients who would have become symptomatic during the study period but did not because they were taking AZT. These patients composed about 4% of the total number of treated patients (table 11.4). Because the average duration of treatment was about 1 year, the results of the study indicate that this percentage of individuals would have become symptomatic during that year without AZT treatment. Because AZT resistance tends to be only partially developed at about 1 year of treatment (table11.3), these results justified AZT treatment for those individuals who would otherwise become symptomatic with within a year (the 8% indicated by the placebo group in table 11.4). For about half of these patients, severe symptoms are postponed, and AZT will still be at least partially effective at the time when AIDS would have developed without treatment.

For the other remaining asymptomatic individuals with CD4 cell counts < 500, evolutionary considerations predicted a different effect. Without AZT treatment, only a small percentage of these individuals would be expected to develop AIDS within 2 years (Fahey et al. 1986). The development of resistance in asymptomatic patients after a few years of AZT treatment (table 11.3) led me to conclude that most of these patients should not be treated (Ewald 1994a). The key prediction was that the measured effectiveness of AZT in postponing AIDS would diminish because a greater proportion of the treated people would have had their treatments initiated 2 or more years before the normal onset of AIDS; as the treatment of these patients continued, the patients increasingly would have resistant viruses during the later years of infection when they most need effective control of the more virulent viruses.

Studies published since this initial ACTG study have supported this prediction of diminishing effectiveness of AZT. The first of these studies involved patients that had entered the chronic, symptomatic phase of infection (Hamilton et al. 1992). Those patients whose treatment had begun at the beginning of this phase were compared with those whose treatment had begun about 1 year after the beginning of this phase. No difference in survival was found

Table 11.4 Effects of timing of AZT treatment during the asymptomatic and chronic, symptomatic phases of HIV infections

Study group	Treatment group	Disease progression	Approximate treatment duration (years)	Reference
ACTG	Asymptomatic with CD4 cell counts ≤ 500; treatment vs. placebo	Progression to AIDS: Placebo = 8% Treated = 4%	≈ 1	Volberding et al. (1990)
	Asymptomatic with CD4 cell counts ≥ 500; immediate vx. deferred until CD4 cell counts < 500	Progression to AIDS or death Immediate = 4% Deferred = 4%	Immediate ≈ 3 Deferred ≈ 2	Volberding et al. (1995)
Veterans Affairs	Treated at onset of symptoms vs. middle of AIDS	No difference in survival	Onset ≈ 2 Middle = 1	Hamilton et al. (1992)
Concorde	Asymptomatic with CD4 ≤ 500; treatment vs. placebo (includes some deferred treatment)	Progression to AIDS or death: Placebo = 7% Treated = 5% Placebo = 29% Treated = 32%	≈ 1 ≈ 3	Seligmann et al. (1994)
Euro-Australian Collaborative Group	Asymptomatic, 80% with T4 < 400/mm³	Progression to AIDS: Placebo = 11% Treated = 3% Placebos = 14% Treated = 14%	≈ 1 ≈ 2	J. Mulder et al. (1994)

(table 11.4). AIDS experts considered the accelerated deterioration of early-treatment patients during the last year of their infections to be paradoxical (Corey & Fleming 1992; Fackelmann 1992), but the results make sense in light of the preceding evolutionary analysis—early treatment of symptomatic infections should initially delay the progression to AIDS because HIV is held in check by AZT treatment. Symptomatic patients whose treatment is begun later should initially pay the price of more rapid progression before initiation of treatment. But when treatment is initiated, these late-treatment patients should reap the benefits of more effective control by AZT because their HIV will be less resistant. Because evolution toward increased replication rate should have continued during early AZT treatment, the HIV in early-treatment patients should have a greater intrinsic replication rate at the onset of late treatment than at the onset of early treatment. The viruses in the late-treatment patients should also have this potential for more rapid growth when the late treatment is begun (because they too have been replicating over the period between the beginning of early treatment and the beginning of late treatment). But because this more dangerous virus population is still sensitive to AZT, it can be held in check temporarily by AZT treatment. The initial postponement of AZT among early-treatment patients is thus offset by improved effectiveness of AZT among late-treatment patients at a time when HIV has evolved a greater potential for replication and is hence more dangerous. The early-treatment patients thus start to show a better clinical course than the late-treatment patients, but this initial advantage dissipates soon after the late treatment begins.

Recognition of this trade-off associated with early treatment does not warrant a shift away from early treatment of symptomatic infections because the early treatment of symptomatic patients delays progression during the first year of the symptomatic phase. The trade-off underlying these results did, however, offer some support for the evolutionary argument about the disadvantages of treatment years before the normal onset of the chronic, symptomatic phase.

Data yielding direct tests of this argument were published subsequently, after people who received such early treatment had been monitored for several years (table 11.4). These tests support the argument that early treatment can be justified only for individuals who would otherwise be within a year of the onset of AIDS. The French and British Concorde study showed a slightly lower rate of progression among AZT-treated patients after approximately 1 year of treatment, but no difference was apparent after 3 years of follow-up (table 11.4). The Euro-Australian Collaborative study showed a similar trend: the slower progression to AIDS among AZT-treated patients that was observable after 1 year of follow-up had vanished by the end of the second year of follow-up (table 11.4).

The 5-year follow-up of the patients in the original ACTG study who had CD4 cell counts > 500 (and who therefore would tend to be several years away from the chronic, symptomatic phase) showed a similar result. Patients treated immediately did not have a longer interval before AIDS (or death without documentation of AIDS) than patients whose treatment was deferred until CD4 cell counts dropped below 500 (table 11.4).

In retrospect, researchers have proposed that this kind of effect is "what one would expect" based on the toxicities of AZT (Egger et al. 1994). I disagree. Toxicities of AZT alone are not sufficient to predict that the positive effect of early treatment would be completely outweighed by a long-term negative effect associated with early treatment. Specifically, the negative effects of AZT toxicity (manifested particularly by reduced leukocyte counts; Koch et al. 1992; Schmitz et al. 1994) do not tend to accelerate as a function of duration of treatment. To the contrary, these negative effects tend to be most apparent during the first year of treatment (Koch et al. 1992; Swanson et al. 1994). Yet it is during the first year of treatment that the net effects of AZT are most positive.

This evolutionary argument predicts that the overall positive value of AZT treatment will eventually become negative because patients will harbor resistant viruses when they most need control, but will still have paid the price of AZT toxicity. Although this hypothesis will be difficult to test because of the ethical problems associated with the proper controlled experiments, the data set derived from the most prolonged treatment of asymptomatic patients is consistent with this prediction: after 3 years of follow-up, the progression of treated patients to AIDS in the Concorde study was, if anything, marginally faster than that of placebo patients (table 11.4).

The controversy over timing of AZT treatment has recently been called "moot" on the basis of the new combination therapy treatments (Richman 1996). But a dismissal of this controversy opens the door for a repetition of past errors. In particular, it overlooks the importance of determining which of the drugs should be used in combination at the outset of infection and which toward the later stages of infection. According to the evolutionary arguments presented here, combination treatments that cause resistance handicaps, particularly those that do so by reducing mutation rate, should be used early, and combinations that do not induce resistance handicaps should be used late. Although early combination therapy is often effective, it is unclear from the current evidence whether this effectiveness is attributable primarily to the reduction of mutation rate imposed by one or more of the antivirals. One of the most dramatic effects, for example, occurred when 3TC and indinavir were combined with AZT; the combination reduced HIV RNA in the plasma to below detectable levels (Richman 1996). As indicated above, 3TC and indinivir may be particularly effective because they impose resistance handicaps.

The therapeutic usefulness of these insights depends on the accuracy with which progression of asymptomatic infections to the chronic, symptomatic phase can be predicted. Combinations of various indicators should allow this progression to be accurately predicted at various points in time after the onset of infection and before the onset of AIDS (Farzadegan et al. 1996; earlier studies reviewed by Ewald 1994a). One of the most promising recent developments is the quantification of unspliced cellular RNA, which can identify individuals who will progress rapidly to AIDS (Michael et al. 1995a). The precise timing of initiation of therapy will depend on the rates at which resistance to the particular antivirals develop, the level of antiviral resistance existing at the onset of therapy, the degree of resistance handicap, and whether patients prefer a longer AIDS-free time or a longer total duration of survival (Ewald 1994a).

A decade of intensive research on anti-HIV drugs suggests that HIV will be with us regardless of what antivirals we use. We may not be able to create a magic bullet, but we can expect more than a squirt gun. If we use antivirals wisely, with an eye toward discovering the best time during the infection cycle for their administration, we should have something like tear gas and rubber bullets—antivirals that will not eradicate HIV but will reduce its damage by altering its activities. Use of handicapping antivirals early during infection and nonhandicapping antivirals later may extend the asymptomatic period and overall survival by several years. If that benefit can be coupled with evolutionarily astute uses of vaccines and behavioral interventions, the damage caused by HIV may be delayed further.

Vaccination Against HIV

Traditional approaches to vaccination probably will not work well against HIV. Experimental studies on related viruses suggest that benign strains of HIV can be developed for use as vaccines (Baba et al. 1995), but such benign viruses would probably prove too dangerous because HIV's high mutation rate and long duration of infection should allow benign HIVs to evolve increased virulence (i.e., increased capacity to harm the host). Some researchers have suggested that complete deletion of genes might circumvent this problem; benign vaccine viruses may be created, for example, by deleting accessory genes, such as *nef* (Almond et al. 1995), but there are many evolutionary routes to increased viral replication. Deletion of a gene such as *nef* may reduce the virulence of a vaccine virus, but so long as the virus is replicating in a mutation-prone way, as *nef*-deleted HIV can do (McNearney et al. 1995), the virus's evolutionary potential is left in tact. The generation of the variants with increased replication rates (through mutations in the undeleted genes) could drive the evolutionary return to increased replication and hence to increased virulence during the course of an infection. Even a slight increase in virulence may be sufficient to negate the usefulness of a prophylactic live vaccine where unvaccinated people have small probabilities of being infected.

Another approach is to extract immunogenic components of HIV for use in vaccines. Such component vaccines will be safer, but HIV's potential for generating variation will probably make them only marginally effective at preventing infection. Such marginal effectiveness may still be sufficient to warrant widespread use (Bloom 1996), but the expectation of marginal and diminishing effectiveness emphasizes the need to use strategies that capitalize on incomplete protection.

Although protection against a subset of the existing variants may be seen as a liability for standard vaccine strategies, such partial protection forms the basis of the virulence-antigen strategy (Ewald 1996, this volume). Present knowledge about determinants of HIV virulence is too incomplete to target particular variants for use in virulence-antigen vaccines, but even this incomplete knowledge offers a glimpse of future opportunities. Therapeutic vaccines (i.e., vaccines administered to infected patients as therapy) could target the particularly virulent variants regularly generated during the course of an infection. For example, evolution of increased virulence within hosts is associated with an increasing proportion of syncytium-inducing (SI) variants. These variants replicate rapidly though a process involving fusion of infected with uninfected cells, which generates syncytia (i.e., cells with a common cytoplasm and cell membrane). If patients are given a vaccine composed of antigens unique to the SI variants, the emergence of SI variants should be preferentially inhibited. Insofar as the SI variants tend to be more harmful than non-SI strains, such a therapeutic vaccine should tend to have a disproportionate suppressive effect on harmful variants. Using such a vaccine early when the immune system is still effective should help delay the onset of AIDS. Because there are undoubtedly many other determinants of virulence among HIV strains, the effectiveness of such virulence-antigen vaccines at postponing AIDS would depend on the degree to which different virulence antigens can be incorporated into a vaccine—the more complete the spectrum of virulence antigens incorporated, the better the suppression of virulent variants within the patient both through the direct effects of the vaccine and the indirect protective effects of the strains remaining after vaccination.

If most of the virulence antigens could be incorporated into a vaccine, the remaining strains would thus generate an immune response much as a live vaccine would, without the ethical unacceptability of a prophylactic live vaccine. For a virulence-antigen therapeutic vaccine to be ethically acceptable, it needs to meet the medical criterion (e.g., prolonging survival or increasing AIDS-free time) better than the next best alternative for the particular patient. The protection from a prophylactic live vaccine, in contrast, must be discounted by the negative effects of the vaccine in those who would not have been infected had they not received the vaccine. Because HIV infections tend to occur at relatively low frequency, these negative effects may often overwhelm the beneficial effects of the vaccine. Even when the net effect of a vaccine is positive at the population level, ethical considerations at the level of individual patients may limit the degree to which vaccination should be recommended for uninfected individuals.

Therapeutic vaccines have been discredited to some extent because they do not seem to be particularly good at eliminating infection. Incomplete protection, however, seems a realistic goal for therapeutic vaccines (Burke 1993). This issue is particularly relevant to HIV vaccines because HIV's mutability is a barrier to complete elimination of infection. Virulence-antigen therapeutic vaccines do not require complete elimination of infection. Rather they need to protect preferentially against the antigens that contribute to a high level of virulence. The possibility of virulence-antigen therapeutic vaccines against AIDS emphasizes the need for continued work to decipher existing virulence mechanisms. It also requires the monitoring for HIV variants that have generated new virulence mechanisms so that the spectrum of antigens in virulence-antigen vaccines can be continually revised.

An analogous application would target protein variants that allow HIV to resist other control strategies. If reverse transcriptases that contain the substitution at site 215 could be made immunologically active, generating, for example, a cytotoxic T cell response (i.e., stimulation of T-cells that kill cells infected with this variant), then incorporation of such antigens into a virulence-antigen vaccine could guard against the HIV variants that are able to cause damage because of their resistance to AZT. Such a strategy might allow AZT (and other antivirals that generate resistant variants without handicaps) to be effective longer.

Evolution of Virulence among Sexually Transmitted Pathogens

General Requirements for Sexual Transmission

Sexual transmission imposes special requirements on pathogens for persistence within hosts. The frequency of transmissible contact with different susceptible hosts is relatively low. For most sexually transmitted pathogens in most populations of humans, the standard acute infection strategy would therefore be a failure. Imagine, for example, that people changed sexual partners once per year. If a pathogen caused an acute infection that was terminated by an immune response or host death within a week or two, it would have little chance of being transmitted. To "break even," a sexually transmitted pathogen must be transmissible for a period of time that extends into the time of the next sexual partnership. To prosper, the pathogen needs to be transmissible for periods that span more than one change in sexual partners. This requirement of prolonged infectiousness means that sexually transmitted pathogens may often need tropisms for particular types of host cells that allow survival for long periods in their hosts and transmission from their hosts, all in the presence of activated immunological defenses.

Evolutionary effects of changes in sexual behavior on virulence may be strongly influenced by the cell tropisms that were present before the change in behavior. Increased potential for sexual transmission should favor those pathogen variants that reproduce more extensively soon after the onset of infection. If the preexisting tropisms were for infection of nonessential cell types, then this selection for earlier reproduction could be expected to have relatively small effects on virulence. If, for example, people began changing partners every few days, such a sexually transmitted pathogen should evolve to resemble a typical acute respiratory tract infection that relies on host mobility for

transmission. This relatively benign situation corresponds to sexually transmitted pathogens such as *Neisseria gonorrhea* and *Chlamydia trachomatis*,
which tend to infect mucosal tissues, and therefore tend to have relatively
minor negative effects on the survival of adult hosts even when reproducing
rapidly.

If, however, the tropisms involve cells that are more important for host
survival, the damage associated with increased levels of host exploitation may
have more critical effects. HIV, for example, has a tropism for CD4 lymphocytes, which are critical regulators of immune responses. Although a high level
of replication in these CD4 cells can be tolerated for short time periods, it
eventually leads (by mechanisms that are still being clarified) to the decima-
tion of CD4 cells and the collapse of the immune system.

HIV Virulence and the Potential for Sexual Transmission

The argument about evolutionary forces and tissue tropisms predicts that HIVs
should be more virulent in areas where the potential for sexual transmission
is greater. This prediction is supported by differences between the two major
phylogenetic categories of HIV, which are designated HIV-1 and HIV-2.
Throughout the AIDS pandemic, East and Central Africa were important sites
of epidemic spread of HIV-1 but not of HIV-2. West Africa, on the other hand,
has been home to HIV-2 for at least several decades, but has been invaded by
HIV-1 since the early 1980s. Sociological studies indicate that for a variety of
historical, sociological, and economic reasons, the potential for sexual transmission has been higher in the eastern than in the western regions of sub-
Saharan Africa. Behavioral surveys, for example, indicate that although
women became sexually active at about the same age in Senegal and Kenya,
the number of different partners was higher in Kenya (Pison et al. 1993; Daly
et al. 1994; see also Ewald 1991, 1994a).

Such survey data are prone to error, however, for several reasons: unwillingness among participants to admit sexual activity, regional biases, and short-
term behavioral changes that may make the potential for sexual transmission
at one time and place different from the overall potential to which the pathogens under study have been exposed. Aside from the sources of error, the
potential for sexual transmission is difficult to quantify because it may depend
on many other variables that are difficult to quantify, such as mean rates of
partner change, variability of partner change rates among people, behavioral
interactions between high-and low-risk people, and geographic movement of
groups.

One way of coping with these difficulties is to use surrogate variables that
should respond to these various influences in a manner analogous to the response of HIV. One such surrogate is the age prevalence of another retrovirus:
the human T-cell lymphotropic virus type I (HTLV-I). HTLV-I tends to be endemic in many regions, including many parts of Africa. It can be transmitted
vertically from mother to offspring as well as sexually. This vertical transmission can create a core of endemic infection from which sexual transmission

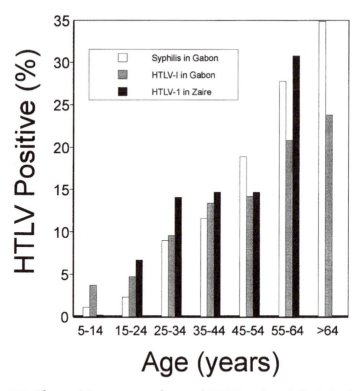

Figure 11.1. The positive age prevalence of HTLV and syphilis infections in Gabon and Zaire (data from LeHesran et al. 1994).

can occur. Because seropositives rarely if ever revert to seronegativity, the distribution of infection in a population provides a record of past infection for that population. HTLV-I causes death (usually by triggering lymphoma or leukemia), but does so in only about 1 of 20 infections, typically after many decades of infection. In Japan, for example, the median age of death from HTLV-induced cancer is about 60 years, and virtually all of these deaths occur in people who were infected via their mothers as neonates (Sugiyama et al. 1986; Tajima & Ito 1990).

The characteristics of HTLV should cause its prevalence to increase in a cohort as the cohort ages. If the potential for sexual transmission is relatively stable over time, a snapshot at any given time should show an age prevalence that rises in accordance with the amount of sexual transmission. In areas with low potential for sexual transmission, the prevalence of HTLV should not increase as a function of age as much as in areas with great potential for sexual transmission.

The similar age-dependent rises in HTLV and syphilis infections (figure 11.1) support the idea that HTLV age prevalence is a useful indicator of potential for sexual transmission. But HTLV has advantages over syphilis as such an indicator. Unlike bacterial infections, HTLV infections cannot be cured;

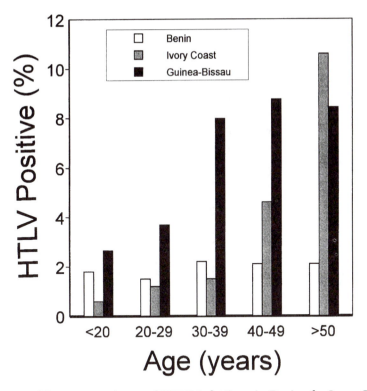

Figure 11.2. The age prevalence of HTLV infections in Benin, the Ivory Coast and Guinea-Bissau (data from Nauclér et al. 1992; Verdier et al. 1994).

hence the cumulative prevalence of HTLV will not be directly altered by temporal or geographic variations in antibiotic usage. Also, because HTLV-I is transmitted slowly, often requiring years of sexual contact, its prevalence should tend to dampen year-to-year variations in sexual behavior much like a running average.

In general, HTLV-I age prevalences tend to agree with sociological data. In the Central African country of Zaire, for example, HTLV-I infections rise rapidly with age (figure 11.1). The age prevalences in the Ivory Coast and Guinea Bissau rise less steeply than those in Zaire (figure 11.2). Overall, the rise in the age prevalences of HTLV infections in West African countries range from moderately less steep than that of Zaire (as in the Ivory Coast) to flat (as in Benin), supporting the idea that the potential for sexual transmission is lower in West Africa than in Central Africa.

As predicted from evolutionary theory, the longitudinal tracking of HIV-2 infections in West Africa indicates that the rate at which HIV-1 infections progress to AIDS is greater than the rate at which HIV-2 infections progress to AIDS. HIV-2 infections in West Africa progress to AIDS slowly relative to HIV-1 infections in East Africa (Kanki et al. 1994; D.W. Mulder et al. 1994; Poulsen et al. 1997). In Senegal, HIV-2 and HIV-1 infections have been tracked since the time of seroconversion. (Prevalence data indicate that HIV-1 was brought into Senegal during the late 1980s by workers returning from other countries; Kane et al. 1993.) In Senegal, HIV-2 infections progressed to AIDS more slowly and were associated with a slower immunological deterioration than HIV-1 infections (Marlink et al. 1994).

These differences between HIV-1 and HIV-2 could represent chance differences in virulence between these two lineages of HIV rather than differences that evolved in response to different opportunities for sexual transmission. This possibility emphasizes the need to assess whether variations in virulence within each of these HIV types are associated with differences in the potential for sexual transmission. Although gathering conclusive evidence will require decades, the evidence documented to date is consistent with an evolutionary effect of the potential for sexual transmission on virulence.

Sociological studies suggest, for example, that the potential for sexual transmission of HIV-2 infections in recent decades has been greater in the Ivory Coast than in Senegal (Ewald 1994a). The age prevalence of HTLV-I infections in the Ivory Coast and Guinea Bissau (figure 11.2) indicate a moderately high potential for sexual transmission. Data on the age prevalence of HTLV-I in Senegal have not been published, but the overall prevalence of HTLV-I in the general population is low, about 1% (S. Mboup and R. Marlink, personal communications). Studies of other regions in West and West-Central Africa indicate that such low percentages are associated with a lack of a significant increase in HTLV prevalence with increasing age, as illustrated by the data from Benin (figure 11.2; Delaporte et al. 1989). The HTLV-I data (figure 11.2) therefore lead to the prediction that the HIV-2 infections in Senegal should be more mild than those in Ivory Coast and Guinea Bissau.

Although long-term longitudinal studies of progression to AIDS from the onset of infection have not been completed in these other areas of West Africa, the counts of CD4 cells are consistent with the predicted association between the potential for sexual transmission and virulence. The CD4 cell counts of people who have been monitored in Senegal show no detectable decline if infected with HIV-2, but a substantial decline if infected with HIV-1 (S. Mboup, personal communication). Analogous studies in the Ivory Coast and Guinea Bissau, where HTLV prevalences indicate a greater potential for sexual transmission, are revealing declines in helper T cell counts, albeit less steep declines than among HIV-1 infections (Lisse et al. 1996; A. Greenberg, personal communication). Mortality rates among HIV-2–infected persons were twice that of uninfected individuals in Guinea Bissau (Poulsen et al. 1997). In a longitudinal study in Senegal with a similar number of person-years of observation of HIV-2–infected individuals, no deaths from AIDS had occurred (Kanki 1992). This lack of mortality in Senegal contrasts with the approximately six deaths above baseline mortality for uninfected individuals in the Guinea Bissau study (Poulsen et al. 1997).

These comparisons of HIV infections from Africa draw attention to the need for additional studies that directly address the inherent virulences of HIV transmitted in populations with different potentials for transmission. Only a small proportion of the viral lineages in different geographic areas have been assessed, and those that have been assessed need to be examined more thoroughly with regard to alternative hypotheses (e.g., hypotheses that ascribe a role to variations in viral dosage, simultaneous infection with other sexually transmitted pathogens, host resistance, or genetic heterogeneity of HIVs within hosts). Molecular phylogenies (i.e., evolutionary trees based on divergences in sequences of macromolecules such as DNA) can also be used to separate HIVs that have been cycling in a study area for different amounts of time, as discussed below in the context of HIV infections in Japan.

Future Evolution of HIV

Although explanation of present trends is scientifically useful, the most practical aspect of the evolutionary approach involves identification of interventions that can control the evolution of HIV virulence. If the inherent virulences of HIVs are dependent, in an evolutionary sense, on the potential for sexual transmission, then interventions that reduce this potential should have a long-term evolutionary effect in addition to their short-term epidemiological effects, i.e., in addition to reducing the spread of HIV infection, they should reduce the harmfulness per infection. In contrast, in areas where the potential for sexual transmission is high, the virulence of HIV is expected to remain relatively high.

This evolutionary framework generates predictions about future changes in the virulence of HIV-1 in specific countries and its ability to spread in its highly virulent form within countries. I predicted in 1993, for example, that the lineages of HIV-1 that had been spreading rapidly through Thailand during the late 1980s and early 1990s (Clements 1992; Lenihan 1992; Nelson et al. 1994) would be particularly virulent (Ewald 1994a). Although data need to be gathered over the next decade to fully evaluate this prediction, the published data are supportive: the decline in CD4 cell counts among Thai patients infected with the subtype that has cycled extensively in Thailand (i.e., subtype E) appears to be more rapid than among Thai patients infected with the subtype that has cycled less in Thailand (i.e., subtype B) (Weniger et al. 1994).

Studies of people infected with HIV-1 for more than a decade without deterioration of the immune system indicate that the mildness of their infections is sometimes attributable to inherently mild viruses (Learmont 1992; Cao et al. 1995; Michael et al. 1995b). Mild variants of HIV-1 therefore seem to be establishing viable infections and circulating in human populations. As would be expected from consideration of HIV's high mutation rate, the mild variants appear to have been generated by a variety of molecular mechanisms (Cao et al. 1995, Huang et al. 1995). The challenge is to identify and invest in interventions that reduce the potential for sexual transmission (e.g., promotion of condom use and educational programs to reduce rates of sexual partner change) and thereby favor such mild variants over the more virulent variants in the population. These mild strains should then inhibit the virulent variants in the manner of a live vaccine.

Natural infections with mild strains differ from prophylactic live vaccines with regard to thresholds of ethical acceptability. Because prophylactic vaccines are given to uninfected people, the ethically acceptable thresholds for safety and effectiveness are much more formidable, particularly for live HIV vaccines (as discussed in section "Vaccination against HIV"). Interventions that reduce the potential for transmission capitalize on the effectiveness of live vaccines at protecting against highly virulent strains, while eliminating the ethical problems of vaccine-induced disease. Even if interventions that reduce the potential for sexual transmission did not cause the evolutionary reductions in virulence suggested above, they would still be ethically acceptable because they reduce the frequency of infection. Any evolutionary reductions in virulence simply add to the overall benefit of the interventions. In comparison with prophylactic live vaccines, there is no safety threshold that must be met by these naturally occurring infections, because unlike live vaccines, naturally occurring infections do not result from active inoculation of uninfected individuals.

The worldwide spread of HIV has inadvertently set up regional "experiments" that will eventually show whether such behavioral interventions can force HIV to evolve toward a more benign state. One experiment is occurring in Japan, where the subtype E of HIV-1 was introduced several years ago from southeast Asia. Japan has a low potential for sexual transmission. Condoms are the primary method of birth control; the birth control pill is rarely used (Trager 1982). Accordingly, variations in the age prevalence of HTLV (an indicator of the potential for sexual transmission) in Japan are largely attributable to a birth cohort effect; infections within an age group increase little as that group grows older, and the majority of HTLV infections of women are acquired from their mothers rather than sexually (Ueda et al. 1989, 1993; Morofujihirata et al. 1993; Take et al. 1993). The rising age prevalence of HTLV infections among Japanese sex workers and their clients (Nakashima et al. 1995) indicates, however, that the Japanese HTLVs can be spread sexually when the potential for sexual transmission is high. All of these considerations lead to the conclusion that the potential for sexual transmission in Japan is very low. If low potential for sexual transmission favors evolution toward mildness, the Japanese type E viruses should evolve toward a milder state over the next few decades.

Molecular phylogenies can yield precise tests of these predictions by distinguishing those viruses that have been recently introduced into a region from those that have been circulating in the region and have thus been exposed to that region's selective pressures for a longer period of time. This application of molecular phylogenies to predict effects of sexual transmission on HIV virulence in Japan is illustrated schematically in figure 11.3.

One of the difficulties inherent in tests of these predictions is the quantification of virulence. The long time interval between seroconversion and the

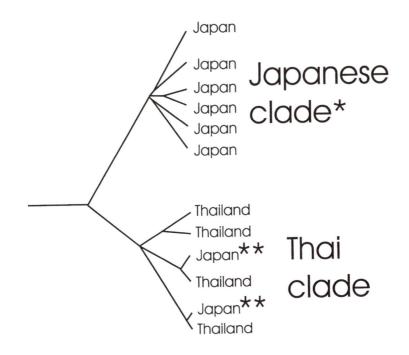

Predicted relative severity: ＊＜＊＊

Figure 11.3. The use of molecular phylogenies for testing predicted effects of the potential for sexual transmission on HIV virulence. The figure represents a hypothetical phylogentic tree for type E HIVs that have been isolated from Japan and Thailand. The country of isolation is given immediately to the right of the branch tip. The Japan branch that is internal to the "Thai clade" therefore represents a type E virus that was isolated in Japan, but descended from a lineage that entered Japan more recently than the viruses that cluster in the "Japanese clade." More generally, this diagram illustrates how phylogenetic analyses generated from nucleotide sequences will allow separation of the HIVs according to their exposure to the local potential for sexual transmission between and within ethnic groups. If such a phylogeny is constructed about two decades after introduction of the type E viruses in Japan, the viruses in the upper half should be more mild than the Japan virus in the lower half because isolates in the upper half would have been circulating for a longer amount of time in Japan (which has a low potential for sexual transmission) and should therefore have become more benign than the Japan virus that is situated in the Thai clade.

onset of AIDS makes this interval years out of date when it is used as an indicator of virulence. The steadily improving understanding of HIV pathogenesis should generate markers of progression that will shorten this lag time between the effects of changes in the potential for sexual transmission on virulence and the measurement of such effects. One of the most promising discoveries is the association between the progression to AIDS and the concentration of HIV RNA in plasma at seroconversion (Mellors et al. 1995). So long as a measurable amount of the variation in HIV RNA concentration is attributable to genetic variation among HIV, this kind of virulence assay could be used to assess the current status of HIV virulence. If, for example, the subtypes E lineages of HIV-1 that have entered Japan from Southeast Asia evolve reduced virulence within a decade of transmission in Japan, their reduced virulence should be detectable through such assays around the year 2005 by measuring RNA densities at the time of seroconversion. Because HIV RNA density, CD4 counts, and other indicators independently improve predictions of progression, we can expect composite indicators of progression that will allow us to test these ideas with increasing accuracy as infections develop after seroconversion. An important refinement will be to account for the contributions to progression from variations in host resistance, HIV inoculum size, and inherent virulences of different strains of HIV. These sources of variation can be assessed by studies of chains of transmission involving rapid versus slow progression (e.g., Learmont et al. 1992), by using molecular phylogenies to separate genetically distinct HIV variants, and by identifying associations between HIV phenotype, genotype, and virulence (e.g., as has been proposed for syncytium-inducing forms).

Invasion of HIV-1 in Areas with HIV-2

The preceding arguments also lead to predictions about the tendency for HIV-1 to supplant HIV-2 in areas where HIV-2 has been endemic. If the potential for sexual transmission remains low in such areas, HIV-1 should not displace HIV-2 because displacement would indicate that HIV-1 is better adapted for transmission than HIV-2 and hence that the attributes of HIV-2, including its virulence, do not reflect the potential for sexual transmission (Ewald 1994a). If, however, the potential for sexual transmission has increased in HIV-2 endemic areas in recent decades, the preceding arguments predict that HIV-1 would be better able than HIV-2 to exploit the situation, and thus HIV-1 would predominate over HIV-2. The data emerging from West Africa support these expectations.

In the Ivory Coast, where sociological studies and the age prevalence of HTLV-1 infections indicate a moderately high potential for sexual transmission (figure 11.2; Sassan-Morokro et al. 1994), HIV-1 has rapidly predominated over HIV-2 (Odehouri et al. 1989; Gershy-Damet et al. 1991). In Senegal, however, where these indicators suggest a low potential for sexual transmission, HIV-1 has not yet overwhelmed HIV-2, but rather has risen to about the same prevalence as HIV-2 (P. Kanki, personal communication). Similarly, on the basis of the lack of an increase in HTLV infection with age in Benin (figure 11.2), this theoretical framework predicts that HIV-1 will not overwhelm HIV-2 in Benin. Accordingly, such a shift has not yet occurred (Adjovi et al. 1994). As expected from the preceding analysis, however, HIV-1 has predominated over HIV-2 among sex workers in Senegal (Kanki et al. 1994; P. Kanki, personal communication). The theoretical framework presented above predicts that HIV-1 will not predominate over HIV-2 in the general population of Senegal and Benin as it has in the Ivory Coast, so long as the potential for sexual transmission in Senegal and Benin does not increase.

The theoretical framework also allows predictions about the simultaneous spread of HIV-2 and HIV-1 into new areas. The strong increase in age prevalence of HTLV in Gabon (LeHesran et al. 1994; Verdier et al. 1994) indicates a relatively high potential for sexual transmission. In accordance with the arguments presented above, HIV-1 has penetrated Gabon more pervasively than HIV-2.

Some Disagreements

The general argument about the relationship between the potential for sexual transmission and virulence has been criticized on the basis of the influence of the timing of transmission during an epidemic on the success of variants with different virulences (Levin 1996). The logic of this criticism is based on population modeling, which indicates that transmission during the early acute phases can account for most of the infections that have been generated during the AIDS pandemic (Jacquez et al. 1994). The conclusion drawn is that transmission early during the epidemic rather than the potential for sexual transmission explains HIV virulence (Levin 1996; Levin et al. 1996).

This argument, however, is a special case of the arguments I have presented rather than an alternative to them. Early during an epidemic the potential for sexual transmission will be higher than when the outbreak enters its endemic phase because the probability of sexual contact with an uninfected individual will be greater when the proportion of uninfected individuals in the population is greater. HIV variants favored by natural selection during the early phases of the epidemic should therefore be more virulent than those favored later during the endemic phase. The key issue is why the viruses that spread early during the pandemic were not of higher or lower virulence. According to the argument I presented, the HIV-1 that spread in East and Central Africa had a relatively high virulence because the potential for sexual transmission was relatively high there. When I suggest that the virulence of HIV will decline in response to a decline in rates of unprotected sexual contact, it is implicit that this decline in virulence is relative to the virulence that would have otherwise occurred (i.e., as a result of the reduction in the potential for sexual transmission due to the reduction in the proportion of the population that is infected; cf. Levin 1966: 96).

Predicted associations between virulence and the potential for sexual transmission can be made between populations at the same phase of an epidemic or between a population with a high potential for sexual transmission during the beginning of the epidemic phase with another that has low potential for sexual transmission due to low rates of partner change and a more endemic phase of the disease. The comparison of East and Central Africa with West Africa probably conformed to the latter situation a decade or two ago, but now is approaching the former situation. The comparisons to date do not allow for the these two components of potential for sexual transmission to be distinguished. Rather, these comparisons represent a first step in assessing whether the hypothesis has merit (Ewald 1994a).

The central point of the trade-off argument that I have presented is that a population with a high potential for sexual transmission should favor HIVs that are more virulent than those favored in populations with a lower potential for sexual transmission that has just passed through the same stage of any epidemic or endemic cycle. Thus, the trade-off argument predicts that an HIV that entered Japan during the early stages of HIV transmission there would not evolve as high a virulence as HIV that entered Thailand during the early stages of HIV transmission in that country. The more testable prediction, however, is that the HIVs that entered Japan will evolve a lower level of virulence after the prevalence has stabilized than will the HIVs that entered Thailand, where the potential for sexual transmission has been relatively high (Ewald 1994a).

Whether or not most of the infections during the current HIV pandemic have been spread during the early acute phase of infections is not germane to the trade-off argument. The more relevant issue is what would have happened had the potential for sexual transmission been higher or lower. Imagine, for example, that sexual partner change always occurred once per year. In that context, no transmission would occur during the acute phase (which lasts no more than a few months), even if the epidemic were at its initial stages. Though the probabilities of transmission per contact are low during the asymptomatic phase and the probability of contact is low during the chronic, symptomatic phase, all transmission would have to occur during these two phases. The empirical evidence shows that transmission does occur from individuals during the post-acute asymptomatic phase (e.g., de Vincenzi et al. 1992). Even though the probability of transmission per contact may be low during this asymptomatic phase, the long period of time over which sexual contacts can occur therefore may still yield substantial transmission.

Other Sexually Transmitted Viruses

Human Papillomaviruses

The general trade-off between the potential for sexual transmission and virulence should also apply to other sexually transmitted viruses. The details of the trade-offs will depend on the particular characteristics of the viruses such as target cell tropisms, reproductive mechanisms, and mutation proneness. Human papillomavirus (HPV), for example, is one of the most common sexually transmitted pathogens and the major risk factor for cervical cancer in a broad range of geographic areas (Fisher 1994; Bosch et al. 1995). The various HPV serotypes, however, are associated with cancer to different degrees (zur Hausen 1991; Fisher 1994; Jamison et al. 1995).

Being a DNA virus, HPV has a relatively low mutation rate and therefore a low potential for keeping ahead of the immune responses through mutation rate. HPV has a lower exposure to the immune system, however, because it infects epithelial cells and can replicate by capitalizing on the division of its host cells.

HPV serotypes vary greatly in their tendencies to trigger cancer. Serotypes 16 and 18 are particularly oncogenic. HPV produces two proteins, called E6 and E7, that can interfere with the cell's mechanism for shutting down cellular replication. In serotype 16, E6 blocks the cellular tumor-suppressing protein p53 directly, whereas E7 appears to block p53's anticancer effects indirectly by influencing p53's effects on another tumor-suppressor protein, pRB (zur Hausen 1991; Slebos et al. 1995). Because p53 arrests cellular replication, blocking p53 allows cellular replication to proceed. E6 and E7 therefore enhance possibilities of massive increases in the number of infected cells. Accordingly, HPV densities increase with increasing degrees of abnormal tissue growth (Fisher 1994). Among HPVs associated with low risks of cancer, E6 and E7 genes do not generate immortalization of cell culture (zur Hausen 1991), indicating that differences in the effects of these genes are at least partly responsible for the differences in virulence between high-risk and low-risk HPVs.

The effects of E6 and E7 from high-risk HPVs on p53, together with other virally altered responses of infected cells, nudge infected cells toward cancer (zur Hausen 1991; Woodworth et al. 1995), but if the potential for sexual transmission is sufficiently great (because of high rates of partner change or low rates of levels of protection during sex), the fitness-enhancing effects of increases in viral production may outweigh the fitness suppressing effects of host death through cancer. If increased partner change rates favor evolutionary increases in HPV virulence, women who have higher numbers of sexual partners should have a disproportionately increased risk of acquiring the more oncogenic HPV serotypes relative to the benign HPV serotypes. Recently published data from Brazil confirm this trend (Franco et al. 1995). Women who had higher numbers of sexual partners had significantly increased chances of acquiring the oncogenic HPVs, but not benign HPVs. Data from Colorado, where oncogenic serotypes predominate to a similar degree, show a similar increase in infection with increasing numbers of sexual partners, but this association was not analyzed according to serotype (Jamison et al. 1995). The data from Brazil indicate that reduction in the numbers of sexual partners would cause an evolutionary decrease in HPV virulence; such reductions should shift women from being more liable to acquire oncogenic than benign HPVs to having about equal probabilities of acquiring oncogenic and benign HPVs. The representation of mild HPVs should increase; that is, HPV should evolve reduced virulence.

Virulence-antigen vaccines could also contribute to reduced virulence of HPV. The key task is to identify the protein variants that contribute to HPV's oncogenicity. The E6 and E7 proteins found in serotype 16 HPVs are prime candidates because of their oncogenicity and immunogenicity. Portions of these proteins trigger specific cytotoxic cellular defenses that can control the outgrowth of cervical tumor cells that express the E7 protein (Ressing et al. 1995). A slightly modified portion of the E6 protein from an oncogenic HPV serotype can trigger strong responses from cytotoxic T lymphocytes that kill cells that were transfected with the E6 gene (Lipford et al. 1995).

The more similar the benign and oncogenic variants are in antigens other than the virulence antigens, the stronger the cross-protection by mild variants against oncogenic variants. Variation in oncogenic potential within serotypes may provide the basis for such protection. Evidence for such variation has recently been documented for serotype 18 (Hecht et al. 1995).

Human T-cell Lymphotropic Viruses

HTLV has characteristics intermediate between HIV and HPV. Like HIV, HTLV is a retrovirus and therefore has the potential for mutation proneness, but like HPV it reproduces largely without producing virions that are released into the intercellular medium. Presumably, this reduced reliance on release of free virions results from the need to remain viable in the host for long periods of time, a critical requirement for maintenance of the virus through vertical transmission. In Japan, for example, most HTLV-I infections are acquired perinatally from mothers (Ewald 1994b). These infections typically must remain viable for decades before being transmitted vertically into the next generation or sexually. HTLV-I thus behaves much like a DNA virus. Instead of relying on production of free virions, the integrated DNA copy of the RNA genome causes increased proliferation of infected cells, much as HPV does. HTLV, however, triggers this proliferation by stimulating production of interleukin 2, which causes cells to divide, and the receptor for interleukin 2, which causes the infected cell to divide in response to its own elevated interleukin 2 production (zur Hausen 1991).

This effect, along with other alterations of cellular control of transcription (Yoshida 1994) elevates cellular multiplication, increasing the probability of cancer. As with HIV and HPV, the harmfulness of HTLV-I infections is associated with the potential for sexual transmission: the average age at which cancers occur is about 60 in Japan, where transmission is mostly vertical, and about 40–45 in Jamaica, where transmission is largely sexual (Ewald 1994b). The same difference occurs among Japanese and Caribbean residents of the United States (Levine et al. 1994), which presumably are harboring Japanese and Caribbean HTLVs. Unlike the HPV and HIV studies, investigations have not yet assessed differences in inherent virulence of the Japanese and Caribbean forms of HTLV-I, nor those for HTLVs from any other regions. But like the situation with HIV and HPV, evolutionary considerations make predictions about what will be found when such measurements are made. HTLV-I forms that have been endemic in Senegal and Benin, for example, should be more mild than the HTLV-I forms endemic in Zaire because both sociological data and the age prevalences of HTLV (figure 11.1) indicate that the potential for sexual transmission is much lower in Senegal than in Zaire. Similarly, the frequency of mild HPV serotypes should be higher in Senegal than in Zaire, and higher in Japan than in the Brazil or the United States. The HTLV-I isolates in Japan fall into two distinct groups, a variant that tends to occur within Japan and a more cosmopolitan variant (Vidal et al. 1994). As the histories of transmission of these two groups becomes better understood, comparisons of their pathology with each other and with other closely related HTLV-I forms could yield additional information.

These considerations suggest a characteristic of evolutionary epidemiology particularly relevant to health science policy. In each of the three sexually transmitted viruses discussed thus far, the observed associations between virulence and sexual transmission are as predicted from evolutionary considerations. If additional studies such as those suggested in this chapter confirm these effects, investment in interventions that reduce the potential for sexual transmission should have a much greater effect on reducing the morbidity and mortality of all these pathogens simultaneously. The evidence for HPV, HIV, and HTLV supports this conclusion across the spectrum of viruses from high to low mutation rate. Reductions in the potential for sexual transmission may therefore have effects across the spectrum of sexually transmitted viruses.

Hepatitis and Epstein-Barr Viruses

Whenever sexually transmitted viruses can benefit from pushing their interactions with host cells toward cancerous growth or from rapidly growing inside of cells, we can expect changes in the potential for sexual transmission to alter virulence. These theoretical considerations should also apply, for example, to hepatitis B virus, which like HTLV has evolved an intermediate position in the spectrum of mutation proneness. Unlike HTLV, the hepatitis B virus has arrived at this intermediate position from the opposite end: HTLV is an RNA virus that reproduces largely through its DNA form, whereas hepatitis B is a DNA virus that has evolved an RNA intermediate, apparently to deal with its high level of exposure to the immune system. The flexibility of viral reproductive strategies offers promise for general theories of viral virulence because it implies that phylogenetic constraints as great as those imposed by the viruses' genetic material (RNA versus DNA) can be overcome. Considering the damage caused by sexually transmitted viruses (e.g., a half million cases of cervical cancer annually, and a similar number of AIDS cases; Fisher 1994), even small reductions in the potential for sexual transmission could have major health benefits.

Although these arguments have dealt with viruses that are transmitted largely by genital contact, similar trade-offs should occur among viruses that are transmitted by intimate oral/oral sexual contact. The agent of infectious mononucleosis, Epstein-Barr virus (EBV), for example, is strongly linked to

Burkitt's lymphoma and nasopharyngeal cancers and may be contributing to Hodgkin's lymphoma and some forms of breast cancer (zur Hausen 1991; Iezzoni et al. 1995; Labrecque et al. 1995; Luqmani & Shousa 1995). The considerations mentioned in this chapter suggest that studies should investigate whether the inherent virulence of EBV is relatively high in populations in which the potential for intimate oral contact is relatively high. Globally, the cancers associated with HTLV, EBV, HPV, and hepatitis B viruses account for approximately 15% of all cancers (zur Hausen 1991). Adding this total to the rising numbers of AIDS cases, it is easy to see that the potential health benefits associated with evolutionary reductions in viral virulence are enormous. As additional sexually transmitted viruses and their links to chronic diseases like cancer are discovered, our recognition of these benefits may continue to increase.

The Past and the Prospects

During the 1960s and early 1970s the general sense among health scientists was that, in economically prosperous countries, infectious diseases were essentially conquered (Burnet and White 1972). In 1967, for example, the Surgeon General of the United States, William Stewart, declared that the United States could "close the book on infectious disease" (Pennisi 1996). Since that time the book has not only been left open but enlarged by the recognition of the vast, open-ended challenge of controlling infectious diseases. The standard solutions of the twentieth century did not provided the expected mastery over infectious diseases largely because of evolutionary processes. Pathogens have been evolving resistance to antibiotics faster than the rate of generation of new classes of antibiotics. The evolutionary versatility of pathogens like HIV have also made vaccine development difficult; once such vaccines are developed, the same evolutionary versatility will likely drain their effectiveness, just as the influenza virus' potential for generating variation has diminished the effectiveness of each new influenza vaccine.

Considering these contributions of evolutionary processes to our current problems with infectious diseases, it seems prudent to consider the ways in which we can harness evolutionary processes so that they help contribute more toward the solution than to the problem. This chapter has suggested a three-pronged approach to evolutionarily mold HIV and other sexually transmitted pathogens into milder forms: use of antivirals that handicap a virus' ability to cause damage, use of vaccines that preferentially suppress the more harmful variants, and interventions that alter the potential for sexual transmission. The relative reliance on each strategy will depend on the specifics of a given situation. Most of the HIV-infected people in poor countries, for example, will not be able to afford the cost of combination antiviral treatment, which typically amounts to more than $10,000 per year. For such countries, interventions to reduce the potential for sexual transmission will undoubtedly be relatively more important. With improved knowledge about molecular mechanisms and epidemiological effects of alternative interventions, the feasibility of evolutionary strategies and the appropriate mix in different places should become apparent.

Acknowledgments
This study was supported by a grant from Leonard X. Bosack and Bette M. Kruger Charitable Foundation, a George E. Burch Fellowship in Theoretic Medicine and Affiliated Sciences awarded by the Smithsonian Institution, and an Amherst College Faculty Research Award. D.S. Burke made helpful comments on the manuscript. I thank A. Greenberg, P. Kanki, R. Marlink, and S. Mboup for informing me about their unpublished results.

References

Adjovi, C., Josse, R., Foundohou, J., Helynck, B., and Davo, N. 1994 HIV seroprevalence rates in Benin. *AIDS* 8:1021–1022.

Almond, N., Kent, K., Cranage, M., Rud, E., Clarke, B., and Stott, E. J. 1995 Protection by attenuated simian immunodeficiency virus in macaques against challenge with virus-infected cells. *Lancet* 345:1342–1344.

Baba, T. W., Jeong, Y. S., Penninck, D., Bronson, R., Greene, M. F., and Ruprecht, R. M. 1995 Pathogenicity of live, attenuated SIV after mucosal infection of neonatal macaques. *Science* 267:1820–1825.

Ball, G.H. 1943. Parasitism and evolution. *American Naturalist* 77:345–64.

Bloom, B. R. 1996 A perspective on AIDS vaccines. *Science* 272:1888–1890.

Bosch, F. X., Manos, M. M., Munoz, N., Sherman, M., Jansen, A. M., Peto, J., Schiffman, M. H., Moreno, V., Kurman, R., Shah, K. V., Alihonou, E., Bayo, S., Mokhtar, H. C., Chicareon, S., Daudt, A., Delosrios, E., Ghadirian, P., Kitinya, J. N., Koulibaly, M., Ngelangel, C., Tintore, L. M. P., Riosdalenz, J. L., Sarjadi, Schneider, A., Tafur, L., Teyssie, A. R., Rolon, P. A., Torroella, M., Tapia, A. V., Wabinga, H. R., Zatonski, W., Sylla, B., Vizcaino, P., Magnin, D., Kaldor, J., Greer, C., and Wheeler, C. 1995 Prevalence of human papillomavirus in cervical cancer: a worldwide perspective. *Journal of the National Cancer Institute* 87:796–802.

Boucher, C. A. B. 1992 Clinical significance of zidovudine-resistant human immunodeficiency viruses. *Research in Virolology* 143:134–136.

Boucher, C. A. B., Lange, J. M. A., Miedema, F. F., Weverling, G. J., Koot, M., Mulder, J. W., Goudsmit, J., Kellam, P., Larder, B. A., and Tersmette, M. 1992 HIV-1 biological phenotype and the development of zidovudine resistance in relation to disease progression in asymptomatic individuals during treatment. *AIDS* 6:1259–1264.

Burke, D. S. 1993 Vaccine therapy for HIV—a historical review of the treatment of infectious diseases by active specific immunization with microbe-derived antigens. *Vaccine* 11:883–891.

Burnet, F. M., and White, D. O. 1972 *Natural History of Infectious Disease*, 4th ed., Cambridge: Cambridge University Press.

Cao, Y. Z., Qin, L. M., Zhang, L. Q.,
Safrit, J., and Ho, D. D. 1995 Virologic and immunologic characterization of long-term survivors of human
immunodeficiency virus type 1 infection. *New England Journal of
Medicine* 332:201–208.

Cheng-Mayer, C., Seto, D., Tateno, M.,
and Levy, J. A. 1988 Biologic features of HIV-1 that correlate with
virulence in the host. *Science* 240:80–
82.

Clements, A. 1992 Thailand stifles
AIDS campaign. *British Medical
Journal* 304:1264.

Clinical trials of zidovudine in HIV
infection. 1989 *Lancet* 334:483–
484.

Cohen, O. J., Pantaleo, G., Holodniy,
M., Schnitman, S., Niu, M., Graziosi,
C., Pavlakis, G. N., Lalezari, J., Bartlett, J. A., Steigbigel, R. T., Cohn, J.,
Novak, R., Mcmahon, D., and Fauci,
A. S. 1995 Decreased human immunodeficiency virus type 1 plasma vi-
remia during antiretroviral therapy
reflects downregulation of viral replication in lymphoid tissue. *Proceedings of the National Academy of Sciences USA* 92:6017–6021.

Condra, J. H., Schleif, W. A., Blahy, O.
M., Gabryelski, L. J., Graham, D. J.,
Quintero, J. C., Rhodes, A., Robbins,
H. L., Roth, E., Shivaprakash, M., Ti-
tus, D., Yang, T., Teppler, H.,
Squires, K. E., Deutsch, P. J., and
Emini, E. A. 1995 In vivo emergence
of HIV-1 variants resistant to multi-
ple protease inhibitors. *Nature* 374:
569–571.

Condra, J. H., Schleif, W. A., Blahy,
O. M., Danovich, R. M., Gabryelski,
L. J., Graham, D. J., Quintero, J. C.,
Rhodes, A., Robbins, H. L., Roth, E.,
Shivaprakash, M., Titus, D., Yang,
T., Zhang, J., Chodakewitz, J. A.,
Deutsch, P. J., Holder, D. J., and Em-
ini, E. A. 1996 In vivo evolution of
HIV-1 resistance to the protease in-
hibitor indinavir sulfate [Abstract].

In *Fourth Workshop on Viral Resistance,* Annapolis, MD, September 10–
12, 1995, p. 43. Annapolis, MD:
Walter Reed Army Institute.

Connor, R. I., and Ho, D. D. 1994 Human immunodeficiency virus type 1
variants with increased replicative
capacity develop during the asymptomatic stage before disease progression. *Journal of Virology* 68:4400–
4408.

Corey, L., and Fleming, T. R. 1992
Treatment of HIV infection—pro-
gress in perspective. *New England
Journal of Medicine* 326:484–485.

Croteau, G., Doyon, L., Thibeault, D.,
McKercher, G., Pilote, L., and La-
marre, D. 1997 Impaired fitness of
human immunodeficiency virus type
1 variants with high-level resistance
to protease inhibitors. *Journal of Virology* 71:1089–1096.

Daly, C. C., Maggwa, N., Mati, J. K.,
Solomon, M., Mbugua, S., Tukei, P.
M., and Hunter, D. J. 1994 Risk factors for gonorrhoea, syphilis, and tri-
chomonas infections among women
attending family planning clinics in
Nairobi, Kenya. *Genitourinary Medicine* 70:155–161.

de Vincenzi, I., Ancelle-Park, R. A.,
Brunet, J. B., Costigliola, P., Ricchi,
E., Chiodo, F., Roumeliotou, A., Pa-
paevengelou, G., Coutinho, R. A.,
Vanhaastrecht, H. J. A., Brettle, R.,
Robertson, R., Kraus, M., Heckmann,
W., Saracco, A., Johnson, A. M.,
Vandenbruaene, M., Goeman, J., Cardoso, J., Sobel, A., Gonzalez-Lahoz,
J., Andres-Medina, R., Casabona, J.,
and Tor, J. 1992 Comparison of female to male and male to female
transmission of HIV in 563 stable
couples. *British Medical Journal* 304:
809–813.

Delaporte, E., Peeters, M., Duran, J. P.
et al. 1989 Seroepidemiological survey of HTLV1 infection among randomized populations of Western
Central African countries. *Journal of*

Acquired Immune Deficiency Syndromes 2:410–413.

DeWolf, F., Lange, J. M. A., Goudsmit, J., Cload, P., DeGans, J., Schellekens, P. T. A., Coutinho, R. A., Fiddian, A. P., and Van Der Noordaa, J. 1988 Effect of zidovudine on serum human immunodeficiency virus antigen levels in symptom-free subjects. *Lancet* 331:373–376.

Egger, M., Neaton, J. D., Phillips, A. N., and Smith, G. D. 1994 Concorde trial of immediate versus deferred zidovudine. *Lancet* 343:1355.

Erice, A., Mayers, D. L., Strike, D. G., Sannerud, K. J., McCutchan, F. E., Henry, K., and Balfour, H. H. 1993 Brief report—primary infection with zidovudine-resistant human immunodeficiency virus type-1. *New England Journal of Medicine* 328:1163–1165.

Ewald, P.W. 1991 Transmission modes and the evolution of virulence, with special reference to cholera, influenza and AIDS. *Human Nature* 2:1–30.

Ewald, P. W. 1994a *Evolution of Infectious Disease.* New York: Oxford University Press.

Ewald, P. W. 1994b Evolution of mutation rate and virulence among human retroviruses. *Philosophical Transactions of the Royal Society of London* B 346:333–343.

Ewald, P. W. 1996 Vaccines as evolutionary tools: the virulence antigen strategy. In S. H. E. Kaufman, ed., *Current Concepts in Vaccine Development.* Berlin: de Gruyter; pp. 1–25.

Fackelmann, K. A. 1992 No survival bonus from early AZT. *Science News* 141:100.

Fahey, J. L., Taylor, J. M. G., Korns, E., and Nishanian, P. 1986 Diagnostic and prognostic factors in AIDS. *Mount Sinai Journal of Medicine* 53:657–663.

Farzadegan, H., Henrard, D.R., Klee-
berger, C.A., Schrager, L., Kirby, A.J., Saah, A.J., Rinaldo, C.R., OGorman, M., Detels, R., Taylor, E., Phair, J.P., and Margolick, J.B. 1996 Virologic and serologic markers of rapid progression to AIDS after HIV-1 seroconversion. *Journal of Acquired Immune Deficiency Syndromes and Human Retrovirology* 13:448–455.

Fischl, M. A., Richman, D. D., Grieco, M. H., Gottlieb, M. S., Volberding, P. A., Laskin, O. L., Leedom, J. M., Groopman, J. E., Mildvan, D., Schooley, R. T., Jackson, G. G., Durack, D. T., King, D., and AZT Collaborative Working Group. 1987 The efficacy of azidothymidine (AZT) in the treatment of patients with AIDS and AIDS-related complex. *New England Journal of Medicine* 317:185–191.

Fischl, M. A., Richman, D. D., Hansen, N., Collier, A. C., Carey, J. T., Para, M. F., Hardy, D. W., Dolin, R., Powderly, W. G., Allan, J. D., Wong, B., Merigan, T. C., McAuliffe, V. J., Hyslop, N. E., Rhame, F. S., Bahour, H. H., Spector, S. A., Volberding, P., Pettinelli, C., Anderson, J., and AIDS Clinical Trials Group. 1990 The safety and efficacy of zidovudine (AZT) in the treatment of subjects with mildly symptomatic human immunodeficiency virus type 1 (HIV) infection: a double-blind, placebo-controlled trial. *Annals of Internal Medicine.* 112:727–737.

Fisher, S. G. 1994 Epidemiology: a tool for the study of human papillomavirus-related carcinogenesis. *Intervirology* 37:215–225.

Fitzgibbon, J. E., Gaur, S., Frenkel, L. D., Laraque, F., Edlin, B. R., and Dubin, D. T. 1993 Transmission from one child to another of human immunodeficiency virus type-1 with a zidovudine- resistance mutation. *New England Journal of Medicine* 329:1835–1841.

Fitzgibbon, J. E., Howell, R. M., Haber-zettl, C. A., Sperber, S. J., Gocke, D. J., and Dubin, D. T. 1992 Human immunodeficiency virus type 1 *pol* gene mutations which cause decreased susceptibility to 2',3'-dideoxycytidine. *Antimicrobial Agents and Chemotherapy* 36:153–157.

FDA. 1991 *FDA Antiviral Drugs Advisory Committee Reviews New AZT Data*. Paper T91–6. Washington, DC: Food and Drug Administration.

Franco, E. L., Villa, L. L., Ruiz, A., and Costa, M. C. 1995 Transmission of cervical human papillomavirus infection by sexual activity: differences between low and high oncogenic risk types. *Journal of Infectious Diseases* 172:756–763.

Friedland, G. H. 1990 Early treatment for HIV. *New England Journal of Medicine* 322:1000–1002.

Gao, Q., Gu, Z. X., Parniak, M. A., Li, X. G., and Wainberg, M. A. 1992a *In vitro* selection of variants of human immunodeficiency virus type-1 resistant to 3'-azido-3'-deoxythymidine and 2',3'-dideoxyinosine. *Journal of Virology* 66:12–19.

Gao, Q., Parniak, M. A., Wainberg, M. A., and Gu, Z. X. 1992b Generation of nucleoside-resistant variants of HIV-1 by in vitro selection in the presence of AZT or DDI but not by combinations. *Leukemia* 6:S192–S195.

Gazzard, B. G. 1992 When should asymptomatic patients with HIV infection be treated with zidovudine. *British Medical Journal* 304:456–457.

Gershy-Damet, G. M., Koffi, K., Soro, B., Coulibaly, A., Koffi, D., Sangare, V., Josseran, R., Guelain, J., Aoussi, E., and Odehouri, K. 1991 Seroepidemiological survey of HIV-1 and HIV-2 infections in the five regions of Ivory Coast. *AIDS* 5:462–463.

Graham, N. M. H., Zeger, S. L., Park, L. P., Phair, J. P., Detels, R., Vermund, S. H., Ho, M., Saah, A. J., and the Multicenter AIDS Cohort Study. 1991 Effect of zidovudine and *Pneumocystis carinii* pneumonia prophylaxis on progression of HIV-1 infection to AIDS. *Lancet* 338:265–269.

Gu, Z. X., Fletcher, R. S., Arts, E. J., Wainberg, M. A., and Parniak, M. A. 1994 The k65r mutant reverse transcriptase of HIV-1 cross-resistant to 2',3'-dideoxycytidine, 2',3'-dideoxy-3'-thiacytidine, and 2',3'-dideoxyinosine shows reduced sensitivity to specific dideoxynucleoside triphosphate inhibitors in vitro. *Journal of Biological Chemistry* 269:28118–28122.

Hamilton, J. D., Hartigan, P. M., Simberkoff, M. S., Day, P. L., Diamond, G. R., Dickinson, G. M., Drusano, G. L., Egorin, M. J., George, W. L., Gordin, F. M., Hawkes, C. A., Jensen, P. C., Klimas, N. G., Labriola, A. M., Lahart, C. J., O'Brien, W. A., Oster, C. N., Weinhold, K. J., Wray, N. P., Zolla-Pazner, S. B., and the Veterans Affairs Cooperative Study Group on AIDS Treatment. 1992 A controlled trial of early versus late treatment with zidovudine in symptomatic human immunodeficiency virus infection. *New England Journal of Medicine* 326:437–443.

Hecht, J. L., Kadlish, A. S., Jiang, G., and Burk, R. D. 1995 Genetic characterization of the human papillomavirus (HPV) 18 E2 gene in clinical specimens suggests the presence of a subtype with decreased oncogenic potential. *International Journal of Cancer* 60:369–376.

Hecht, M. F., Jewett, J., and Bateman, W. B. 1990 Management of HIV infection in the seropositive, asymptomatic patient. In K. K. Holmes, L. Corey, A. C. Collier, and H. H. Handsfield, eds., *AIDS Dx/Rx*. New York: McGraw-Hill; pp. 92–106.

Heckler, M. 1984. The press conference of secretary Margaret Heckler,
Monday, April 23, 1984. Washington, DC: U.S. Department of Health
and Human Services.

Ho, D. D. 1995 Time to hit HIV, early
and hard. *New England Journal of
Medicine* 333:450–451.

Huang, Y. X., Zhang, L. Q., and Ho, D.
D. 1995 Characterization of nef sequences in long-term survivors of
human immunodeficiency virus type
1 infection. *Journal of Virology* 69:93–
100.

Iezzoni, J. C., Gaffey, M. J., and Weiss,
L. M. 1995 The role of Epstein-Barr
virus in lymphoepithelioma-like car-
cinomas. *American Journal of Clinical Pathology* 103:308–315.

Jacquez, J. A., Koopman, J. S., Simon,
C. P., and Longini, I. M. 1994 Role of
the primary infection in epidemics
of HIV infection in gay cohorts. *Journal of Acquired Immune Deficiency Syndromes* 7:1169–1184.

Jamison, J. H., Kaplan, D. W., Ham-
man, R., Eagar, R., Beach, R., and
Douglas, J. M. 1995 Spectrum of
genital human papillomavirus infection in a female adolescent population. *Sexually Transmitted Diseases*
22:236–243.

Japour, A. J., Chatis, P. A., Eigenrauch,
H. A., and Crumpacker, C. S. 1991
Detection of human immunodeficiency virus type 1 clinical isolates
with reduced sensitivity to zidovudine and dideoxyinosine by RNA-
RNA hybridization. *Proceedings of
the National Academy of Sciences
USA* 88:3092–3096.

Kane, F., Alary, M., Ndoye, I., Coll, A.
M., Mboup, S., Gueye, A., Kanki, P.
J., and Joly, J. R. 1993 Temporary ex-
patriation is related to HIV-1 infection in rural Senegal. *AIDS* 7:1261–
1265.

Kanki, P. J. 1992 Virologic and biologic
features of HIV-2. In G. P. Wormser,
ed., *AIDS and Other Manifestations
of HIV Infection*, 2nd ed. New York:
Raven Press; pp. 85–93.

Kanki, P. J., Travers, K. U., Mboup, S.,
Hsieh, C. C., Marlink, R. G., Gueyendiaye, A., Siby, T., Thior, I., Hernan-
dezavila, M., Sankale, J. L., Ndoye,
I., and Essex, M. E. 1994 Slower heterosexual spread of HIV-2 than HIV-
1. *Lancet* 343:943–946.

Kellam, P., C.A.B. Boucher, and B.A.
Larder. 1992. Fifth mutation in human immunodeficiency virus type 1
reverse transcriptase contributes to
the development of high-level resistance to zidovudine. *Proceedings of
the National Academy of Sciences of
the United States of America* 89:
1934–8.

Kess, S., Bresolin, L., and Henning, J.
1991 *HIV Early Care*. AMA physician guidelines. Chicago: American
Medical Association.

Koch, M. A., Volberding, P. A., Lagakos, S. W., Booth, D. K., Pettinelli,
C., and Myers, M. W. 1992 Toxic effects of zidovudine in asymptomatic
human immunodeficiency virus-
infected individuals with CD4+ cell
counts of 0.50x10(9)/l or less: detailed and updated results from protocol–019 of the AIDS clinical trials
group. *Archives of Internal Medicine*
152:2286–2292.

Labrecque, L. G., Barnes, D. M., Fenti-
man, I. S., and Griffin, B. E. 1995
Epstein-Barr virus in epithelial cell
tumors: a breast cancer study.
Cancer Research 55:39–45.

Larder, B. A., Darby, G., and Richman,
D. D. 1989 HIV with reduced sensitivity to zidovudine (AZT) isolated
during prolonged therapy. *Science*
243:1731–1734.

Larder, B. A., and Kemp, S. D. 1989
Multiple mutations in HIV-1 reverse
transcriptase confer high-level resistance to zidovudine (AZT). *Science*
246:1155–1158.

Learmont, J., Tindall, B., Evans, L.,
Cunningham, A., Cunningham, P.,

Wells, J., Penny, R., Kaldor, J., and Cooper, D. A. 1992 Long-term symptomless HIV-1 infection in recipients of blood products from a single donor. *Lancet* 340:863–867.

LeHesran, J. Y., Delaporte, E., Gaudebout, C., Trebuck, A., Schrijvers, D., Josse, R., Peeters, M., Cheringou, H., Dupont, A., and Larouze, B. 1994 Demographic factors associated with HTLV-1 infection in a Gabonese community. *International Journal of Epidemiology* 23:812–817.

Lenihan, F. 1992 Thailand tries again to tackle AIDS. *British Medical Journal* 305:1385.

Levin, B. R. 1996 Evolution and maintenance of virulence in microparasites. *Emerging Infectious Diseases* 2: 93–102.

Levin, B. R., Bull, J. J., and Stewart, F. M. 1996 The intrinsic rate of increase of HIV/AIDS. *Mathematical Biosciences* 132:69–96.

Levine, P. H., Manns, A., Jaffe, E. S., Colclough, G., Cavallaro, A., Reddy, G., and Blattner, W. A. 1994 The effect of ethnic differences on the pattern of HTLV-I-associated T-cell leukemia/ lymphoma (HATL) in the United States. *International Journal of Cancer* 56:177–181.

Levy, J. A. 1989 Human immunodeficiency viruses and the pathogenesis of AIDS. *Journal of the American Medical Association* 261:2997–3006.

Lipford, G. B., Bauer, S., Wagner, H., and Heeg, K. 1995 Peptide engineering allows cytotoxic T-cell vaccination against human papilloma virus tumour antigen, E6. *Immunology* 84: 298–303.

Lisse, I.M., Poulsen, A.G., Aaby, P., Knudsen, K., and Dias, F. 1996 Serial CD4 and CD8 T-lymphocyte counts and associated mortality in an HIV-2-infected population in Guinea-Bissau. *Journal of Acquired Immune Deficiency Syndromes and Human Retrovirology* 13:355–362

Luqmani, Y. A., and Shousha, S. 1995 Presence of Epstein-Barr virus in breast carcinoma. *International Journal of Oncology* 6:899–903.

Marlink, R., Kanki, P., Thior, I., Travers, K., Eisen, G., Siby, T., Traore, I., Hsieh, C. C., Dia, M. C., Gueye, E., Hellinger, J., Gueyendiaye, A., Sankale, J. L., Ndoye, I., Mboup, S., and Essex, M. 1994 Reduced rate of disease development after HIV-2 infection as compared to HIV-1. *Science* 265:1587–1590.

Masquelier, B, E. Lemoigne, I. Pellegrin, D. Douard, B. Sandler, and H.J.A. Fleury. 1993. Primary infection with zidovudine-resistant HIV. *New England Journal of Medicine* 329:1123–24.

McNearney, T., Hornickova, Z., Templeton, A., Birdwell, A., Arens, M., Markham, R., Saah, A., and Ratner, L. 1995 Nef LTR sequence variation from sequentially derived human immunodeficiency virus type 1 isolates. *Virology* 208:388–398.

Mellors, J. W., Kingsley, L. A., Rinaldo, C. R., Todd, J. A., Hoo, B. S., Kokka, R. P., and Gupta, P. 1995 Quantitation of HIV-1 RNA in plasma predicts outcome after seroconversion. *Annals of Internal Medicine* 122:573–579.

Michael, N. L., Chang, G., Darcy, L. A., Ehrenberg, P. K., Mariani, R., Busch, M. P., Birx, D. L., and Schwartz, D. H. 1995b Defective accessory genes in a human immunodeficiency virus type 1-infected long-term survivor lacking recoverable virus. *Journal of Virology* 69:4228–4236.

Michael, N. L., Mo, T., Merzouki, A., Oshaughnessy, M., Oster, C., Burke, D. S., Redfield, R. R., Birx, D. L., and Cassol, S. A. 1995a Human immunodeficiency virus type 1 cellular RNA load and splicing patterns predict disease progression in a longitudinally studied cohort. *Journal of Virology* 69:1868–1877.

Mitsuya, H., Weinhold, K. J., Furman, F. A., St. Clair, M. H., Lehrman, S. N., Gallo, R. C., Bolognesi, D., Barry, D. W., and Broder, S. 1985 Inhibition of the in vitro infectivity and cytopathic effect of human T-lymphotropic virus type III/ lymphadenopathy associated virus (HTLV-III/LAV) by 2′,3′-dideoxynucleosides. *Proceedings of the National Academy of Sciences USA* 82:7096–7100.

Mohri, H., Singh, M. K., Ching, W. T. W., and Ho, D. D. 1993 Quantitation of zidovudine-resistant human immunodeficiency virus type 1 in the blood of treated and untreated patients. *Proceedings of the National Academy of Sciences USA* 90:25–29.

Moore, R. D., Hidalgo, J., Sugland, B. W., and Chaisson, R. E. 1991 Zidovudine and the natural history of the acquired immunodeficiency syndrome. *New England Journal of Medicine* 324:1412–1416.

Morofujihirata, M., Kajiyama, W., Nakashima, K., Noguchi, A., Hayashi, J., and Kashiwagi, S. 1993 Prevalence of antibody to human T-cell lymphotropic virus type-I in Okinawa, Japan, after an interval of nine years. *American Journal of Epidemiology* 137:43–48.

Muckenthaler, M., Gunkel, N., Levantis, P., Broadhurst, K., Goh, B., Colvin, B., Forster, G., Jackson, G. G., and Oxford, J. S. 1992 Sequence analysis of an HIV-1 isolate which displays unusually high-level AZT resistance in vitro. *Journal of Medical Virology* 36:79–83.

Mulder, D.W., Nunn, A.J., Wagner, H U., Kamali, A., Kengeya-Kayondo, J.F. 1994 Two-year HIV-1 associated mortality in a Ugandan rural population. *Lancet* 343:1021–1023.

Mulder, J. W., Cooper, D. A., Mathiesen, L., Sandström, E., Clumeck, J., Gatell, J. M., French, M., Donovan, B., Gray, F., Yeo, J. M., and Lange, J.

M. A. 1994 Zidovudine twice daily in asymptomatic subjects with HIV infection and a high risk of progerssion to AIDS: a randomized, double-bline placebo-controlled study. *AIDS* 8:313–321.

Najera, I., Richman, D. D., Olivares, I., Rojas, J. M., Peinado, M. A., Perucho, M., Najera, R., and Lopez Galindez, C. 1994 Natural occurrence of drug resistance mutations in the reverse transcriptase of human immunodeficiency virus type 1 isolates. *AIDS Research and Human Retroviruses* 10:1479–1488.

Nakashima, K., Kashiwagi, S., Kajiyama, W., Hirata, M., Hayashi, J., Noguchi, A., Urabe, K., Minami, K., and Maeda, Y. 1995 Sexual transmission of human T-lymphotropic virus type I among female prostitutes and among patients with sexually transmitted diseases in Fukuoka, Kyushu, Japan. *American Journal of Epidemiology* 141:305–311.

National Institute of Allergy and Infectious Diseases. 1990 State-of-the-art conference on azidothymidine therapy for early HIV infection. *American Journal of Medicine* 89:335–344.

Nauclér, A., Andersson, S., Albino, P., Paolo Da Silva, A., Cndreasson, P-Å., Biberfeld, G. 1992 Association between HTLV-1 and HIV-2 infections in Bissau, Guinea-Bissau. *AIDS* 6:510–511.

Nelson, K. E., Suriyanon, V., Taylor, E., Wongchak, T., Kingkeow, C., Srirak, N., Lertsrimongkol, C., Cheewawat, W., and Celentano, D. 1994 The incidence of HIV-1 infections in village populations of northern Thailand. *AIDS* 8:951–955.

Nowak, M. A., Anderson, R. M., McLean, A. R., Wolfs, T. F. W., Goudsmit, J., and May, R. M. 1991 Antigenic diversity thresholds and the development of AIDS. *Science* 254:963–969.

Nunberg, J. H., Schleif, W. A., Boots, E. J., Obrien, J. A., Quintero, J. C., Hoffman, J. M., Emini, E. A., and Goldman, M. E. 1991 Viral resistance to human immunodeficiency virus type-1-specific pyridinone reverse transcriptase inhibitors. *Journal of Virology* 65:4887–4892.

Odehouri, K., De Cock, K. M., Krebs, J. W., Moreau, J., Rayfield, M., McCormick, J. B., Schochetman, G., Bretton, R., Bretton, G., Ouattara, D., Heroin, P., Kanga J-M, Beda, B., Niamkey, E., Kadio, A., Gariepe, E., and Heyward, W. L. 1989 HIV-1 and HIV-2 infection associated with AIDS in Abidjan, Côte d'Ivoire. *AIDS* 3:509–12.

Pennisi, E. 1996 U.S. beefs up CDC's capabilities. *Science* 272:1413.

Pison, G., Le Guenno, B., Lagarde, E., Enel, C., and Seck, C. 1993 Seasonal migration: a risk factor for HIV infection in rural Senegal. *Journal of Acquired Immune Deficiency Syndromes* 6:196–200.

Poulsen, A.G., P. Aaby, O. Larsen, H. Jensen, A. Naucler, I.M. Lisse, C.B. Christiansen, F. Dias, and M. Melbye. 1997. 9-year HIV-2-associated mortality in an urban community in Bissau, west Aftrica. *Lancet* 349:911–14.

Ressing, M. E., Sette, A., Brandt, R. M. P., Ruppert, J., Wentworth, P. A., Hartman, M., Oseroff, C., Grey, H. M., Melief, C. J. M., and Kast, W. M. 1995 Human CTL epitopes encoded by human papillomavirus type 16 E6 and E7 identified through in vivo and in vitro immunogenicity studies of HLA-a*0201-binding peptides. *Journal of Immunology* 154:5934–5943.

Richman, D. D. 1991 Antiviral therapy of HIV infection. *Annual Review of Medicine* 42:69–90.

Richman, D. D. 1992 HIV drug resistance. *AIDS Research and Human Retroviruses* 8:1065–1071.

Richman, D. D. 1993 HIV drug resistance. *Annual Review of Pharmacology and Toxicology* 33:149–164.

Richman, D. D. 1996 HIV therapeutics. *Science* 272:1886–1888.

Ruedy, J., Schechter, M., and Montaner, J. S. G. 1990 Zidovudine for early human immunodeficiency virus (HIV) infection: who, when, and how? *Annals of Internal Medicine* 112:721–723.

Rutherford, G. W. 1994 Long term survival in HIV-1 infection. *British Medical Journal* 309:283–284.

Sassan-Morokro, M., Coulibaly, I. M., Coulibaly, D., Sidibé, K., Ackah, A., Abouya, L., Digbeu, H., DeCock, K. M., Greenberg, A. E. 1994 Alarming levels of prostitute contact, STD, and condom neglect among HIV-1, HIV-2, dually reactive, and HIV-negative men with TB in Abidjan, Côte d'Ivoire [Abstract 104C]. In: *Tenth International Conference on AIDS*, Yokohama, Japan, August 7–12, 1994, p. 33.

Schellekens, P. T. A., Tersmette, M., Roos, M. T. L., Keet, R. P., Dewolf, F., Coutinho, R. A., and Miedema, F. 1992 Biphasic rate of CD4+ cell count decline during progression to AIDS correlates with HIV-1 phenotype. *AIDS* 6:665–669.

Schmitz, S. H., Scheding, S, Voliotis, D, Rasokat, H, Diehl, V., Schrappe, M. 1994 Side effects of AZT prophylaxis after occupational exposure to HIV-infected blood. *Annals of Hematology* 69:135–138.

Seligmann, M., Warrell, D. A., Aboulker, J. P., Carbon, C., Darbyshire, J. H., Dormont, J., Eschwege, E., Girling, D. J., James, D. R., Levy, J. P., Peto, T. E. A., Schwarz, D., Stone, A. B., Weller I V D, Withnall, R., Gelmon, K., Lafon, E., Swart, A. M., Aber, V. R., Babiker, A. G., Lhoro, S., Nunn, A. J., and Vray, M. 1994 Concorde: MRC/ANRS randomise double-blind controlled trial of immediate and deferred zidovudine in symptom-free HIV infection. *Lancet* 343:871–881.

Shafer, R. W., Iversen, A. K. N., Winters, M. A., Aguiniga, E., Katzenstein, D. A., and Merigan, T. C. 1995 Drug resistance and heterogeneous long-term virologic responses of human immunodeficiency virus type 1-infected subjects to zidovudine and didanosine combination therapy. *Journal of Infectious Diseases* 172:70–78.

Shafer, R. W., Kozal, M. J., Winters, M. A., Iversen, A. K. N., Katzenstein, D. A., Ragni, M. V., Meyer, W. A., Gupta, P., Rasheed, S., Coombs, R., Katzman, M., Fiscus, S., and Merigan, T. C. 1994 Combination therapy with zidovudine and didanosine selects for drug-resistant human immunodeficiency virus type 1 strains with unique patterns of pol gene mutations. *Journal of Infectious Diseases* 169:722–729.

Shirasaka, T., M.F. Kavlick, T. Ueno, W.Y. Gao, E. Kojima, M.L. Alcaide, S. Chokekijchai, B.M. Roy, E. Arnold, R. Yarchoan, *and others*. 1995. Emergence of human immunodeficiency virus type 1 variants with resistance to multiple dideoxynucleosides in patients receiving therapy with dideoxynucleosides. *Proceedings of the National Academy of Sciences of the United States of America* 92:2398–402.

Shirasaka, T., R. Yarchoan, M.C. Obrien, R.N. Husson, B.D. Anderson, E. Kojima, T. Shimada, S. Broder, and H. Mitsuya. 1993. Changes in drug sensitivity of human immunodeficiency virus type-1 during therapy with azidothymidine, dideoxycytidine, and dideoxyinosine—an invitro comparative study. *Proceedings of the National Academy of Sciences of the United States of America* 90: 562–66.

Slebos, R. J. C., Kessis, T. D., Chen, A.

W., Han, S. M., Hedrick, L., and Cho, K. R. 1995 Functional consequences of directed mutations in human papillomavirus E6 proteins: abrogation of p53-mediated cell cycle arrest correlates with p53 binding and degradation in vitro. *Virology* 208:111–120.

Smith, M. S., Koerber, K. L., and Pagano, J. S. 1994 Long-term persistence of zidovudine resistance mutations in plasma isolates of human immunodeficiency virus type 1 of dideoxyinosine-treated patients removed from xidovudine therapy. *Journal of Infectious Diseases* 169: 184–188.

St. Clair, M. H., Martin, J. L., Tudor-Williams, G., Bach, M. C., Vavro, C. L., King, D. M., Kellam, P., Kemp, S. D., and Larder, B. A. 1991 Resistance to ddI and sensitivity to AZT induced by a mutation in HIV-1 reverse transcriptase. *Science* 253: 1557–1559.

Sugiyama, H., Doi, H., Yamaguchi, K., Tsuji, Y., Miyamoto, T., and Hino, S. 1986 Significance of post-natal mother-to-child transmission of human T-lymphotropic virus type-I on the development of adult T-cell leukemia/lymphoma. *Journal of Medical Virology* 20:253–260.

Swanson, C. E., and Cooper, D. A. 1990 Factors influencing outcome of treatment with zidovudine of patients with AIDS in Australia. *AIDS* 4:749–757.

Swanson, C. E., Tindall, B., and Cooper, D. A. 1994 Efficacy of zidovudine treatment in homosexual men with AIDS-related complex: factors influencing development of AIDS, survival and drug intolerance. *AIDS* 8:625–634.

Swart, A. M., Weller, I., and Darbyshire, J. H. 1990 Early HIV infection: to treat or not to treat? *British Medical Journal* 301:825–826.

Tajima, K., and Ito, S. 1990 Prospective studies of HTLV-I and associated diseases in Japan. In W. A. Blattner, ed., *Human Retrovirology: HTL*. New York: Raven Press; pp. 267–279.

Take, H., Umemoto, M., Kusuhara, K., and Kuraya, K. 1993 Transmission routes of HTLV-I: an analysis of 66 families. *Japanese Journal of Cancer Research* 84:1265–1267.

Tersmette, M., Gruters, R. A., de Wolf, F., de Goede, R. E. Y., Lange, J. M. A., Schellekens, P. T. A., Goudsmit, J. A. A. P., Huisman, H. G., and Miedema, F. 1989 Evidence for a role of virulent human immunodeficiency virus (HIV) variants in the pathogenesis of acquired immunodeficiency syndrome: studies on sequential HIV isolates. *Journal of Virology* 63:2118–2125.

Tisdale, M., Myers, R. E., Maschera, B., Parry, N. R., Oliver, N. M., and Blair, E. D. 1995 Cross-resistance analysis of human immunodeficiency virus type 1 variants individually selected for resistance to five different protease inhibitors. *Antimicrobial Agents and Chemotherapy* 39:1704–1710.

Trager, J. 1982 *Letters from Sachiko*. New York: Atheneum.

Ueda, K., Kusuhara, K., Tokugawa, K., Miyazaki, C., Hoshida, C., Tokumura, K., Sonoda, S., and Takahashi, K. 1989 Cohort effect on HTLV-I seroprevalence in southern Japan. *Lancet* 334:979.

Ueda, K., Kusuhara, K., Tokugawa, K., Miyazaki, C., Okada, K., Maeda, Y., Shiraki, H., and Fukada, K. 1993 Mother-to-child transmission of human T-lymphotropic virus type I (HTLV-I): an extended follow-up study on children between 18 and 22–24 years old in Okinawa, Japan. *International Journal of Cancer* 53:597–600.

Verdier, M., Bonis, J., and Denis, F. A., 1994 The prevalence and incidence of HTLVs in Africa. In M. Essex, S. Mboup, P. J. Kanki, and M. R. Kalenga, eds., *AIDS in Africa*. New York: Raven Press, pp. 173–193.

Vidal, A. U., Gessain, A., Yoshida, M., Mahieux, R., Nishioka, K., Tekaia, F., Rosen, L., and DeThe, G. 1994 Molecular epidemiology of HTLV type I in Japan: evidence for two distinct ancestral lineages with a particular geographical distribution. *AIDS Research and Human Retroviruses* 10:1557–1566.

Volberding, P. A., Lagakos, S. W., Koch, M. A., Pettinelli, C., Myers, M. W., Booth, D. K., Balfour, H. H., Reichman, R. C., Bartlett, J. A., Hirsch, M. S., Murphy, R. L., Hardy, W. D., Soeiro, R., Fischl, M. A., Bartlett, J. G., Merigan, T. C., Hyslop, N. E., Richman, D. D., Valentine, F. T., Corey, L., and the AIDS Clinical Trials Group of the National Institute of Allergy and Infectious Diseases. 1990 Zidovudine in asymptomatic human immunodeficiency virus infection: a controlled trial in persons with fewer than 500 CD4-positive cells per cubic millimeter. *New England Journal of Medicine* 322:941–949.

Volberding, P. A., Lagakos, S. W., Grimes, J M, Stein, D.S., Rooney, J, Meng, T-Z, Fischl, M. A., Collier, A C., Phair, J P, Hirsch, M. S, Hardy, W. D., Balfour, H. H, and Reichman, R. C. 1995 A comparison of immediate with deferred zidovudine therapy for asymptomatic HIV-infected adults with CD4 cells counts of 500 or more per cubic millimeter. *New England Journal of Medicine* 333:401–407.

Wahlberg, J., Fiore, J., Angarano, G., Uhlén, M., and Albert, J. 1994 Apparent selection against transmssion of zidovudine-resistant human immunodeficiency virus type 1 variants. *Journal of Infectious Diseases* 169:611–614.

Wainberg, M. A., Beaulieu, R., Tsou-kas, C., and Thomas, R. 1993 Detection of zidovudine-resistant variants of HIV-1 in genital fluids. *AIDS* 7: 433–434.

Wainberg, M.A., Hsu, M., Gu, Z.X., Borkow, G., and Parniak, M.A. 1996b. Effectiveness of 3TC in HIV clinical trials may be due in part to the M184V substitution in 3TC-resistant HIV-1 reverse transcriptase. *AIDS* (suppl.) 10:S3–S10

Wainberg, M.A., Drosopoulos, W.C., Salomon, H., Hsu, M., Borkow, G., Parniak, M.A., Gu, Z.X., Song, Q.B., Manne, J, Islam, S, Castriota, G, & Prasad, V.R. 1996a. Enhanced fidelity of 3TC-selected mutant HIV-1 reverse transcriptase. *Science* 271: 1282–1285.

Wainberg, M. A., Salomon, H., Gu, Z. X., Montaner, J. S. G., Cooley, T. P., Mccaffrey, R., Ruedy, J., Hirst, H. M., Cammack, N., Cameron, J., and Nicholson, W. 1995 Development of HIV-1 resistance to (−)2'-deoxy-3'-thiacytidine in patients with AIDS or advanced AIDS-related complex. *AIDS* 9:351–357.

Weniger, B. G., Tansuphaswadikul, S., Young, N. L., Pau, C. P., Lohsom-boon, P., Yindeeyoungyeon, W., Limpakarnjanarat, K. 1994 Differences in immune function among patients infected with distinct Thailand HIV-1 strains [Abstract O12C]. In: *Tenth International Conference on AIDS*, Yokohama, Japan, August 7–12, 1994, p. 10.

Woodworth, C. D., Mcmullin, E., Iglesias, M., and Plowman, G. D. 1995 Interleukin 1 alpha and tumor necrosis factor alpha stimulate autocrine amphiregulin expression and proliferation of human papillomavirus-immortalized and carcinoma- derived cervical epithelial cells. *Proceedings of the National Academy of Sciences USA* 92:2840–2844.

Yoshida, M. 1994 Mechanism of transcriptional activation of viral and cellular genes by oncogenic protein of HTLV-I. *Leukemia* (suppl.) 8:s51–s53.

zur Hausen, H. 1991 Viruses in human cancers. *Science* 254:1167–1173.

12

PALEOLITHIC NUTRITION REVISITED

S. BOYD EATON

S.B. EATON III

MELVIN J. KONNER

The nutritional needs of today's humans arose through a multimillion-year evolutionary process, during nearly all of which genetic change reflected the life circumstances of our ancestral species (Eaton & Konner, 1985). But, since the appearance of agriculture 10,000 years ago and especially since the Industrial Revolution, genetic adaptation has been unable to keep pace with cultural progress (Neel, 1994; Rendel, 1970). Natural selection has produced only minor alterations during the past 10,000 years, so we remain nearly identical to our late Paleolithic ancestors (Tooby & Cosmides, 1990), and, accordingly, their nutritional pattern has continuing relevance. The preagricultural diet might be considered a possible paradigm or standard for contemporary human nutrition (Burkitt & Eaton, 1989; Eaton & Konner, 1985; O'Dea & Sinclair, 1983).

Nutrition science is based on epidemiological, physiological, and animal investigations; the ultimate arbiter in this field can only be experimental laboratory and clinical research. Still, these approaches might be complemented by insights from an evolutionary perspective. Over a decade ago (Eaton & Konner, 1985), we proposed a model based on analyzing the nutritional properties of wild game and uncultivated vegetable foods, evaluating archeological remains, and studying the subsistence of recent foragers.[1] This approach (for method see Eaton, 1992; Eaton & Konner, 1985) allows estimation of an "average" preagricultural diet analogous to an "average" American diet—the patterns of Greenland Inuit and Australian Aborigines being as different as those of American vegans and fast-food addicts. Since 1985 our insights have been updated: first, our database has been much expanded. The 1985 analyses were derived from a total of 69 game and wild vegetable items; now we have at least some data on 321. Second, estimates presented in 1985 (Eaton & Konner, 1985) and subsequently amended (Eaton, 1992; Eaton et al., 1988) can be compared with those generated from the current data set, permitting assessment of consistency. Third, the original essay presented estimates on only 4 micronutrients (sodium, potassium, calcium, and ascorbic acid); now, we can make retrojections for many more. Fourth, we now compare and contrast current dietary recommendations with our estimate of ancestral human nutrition. This exercise reveals both gratifying parallels and potentially instructive points of disagreement.

Energy Requirements

That Paleolithic humans were as tall as are members of current affluent societies has only recently been appreciated (Roberts et al., 1994; Walker, 1993), but the necessity for vigorous physical activity engendered by a nomadic hunting and gathering lifestyle has long been apparent. Skeletal remains show that our ancestors typically developed lean body mass considerably in excess of that common today (Ruff et al., 1993). The physical demands of life during the agricultural period were also strenuous (Heini et al., 1995); it took the Industrial Revolution to dissociate productivity from human caloric expenditure. For example, the introduction of mechanized farming in Japan reduced average daily work expenditure by more than 50% (Shimamoto et al., 1989), while, in Britain, proliferation of labor-saving devices between 1956 and 1990 reduced caloric expenditure (in excess of basal metabolic needs) by an astonishing 65% (Ministry of Agriculture, Fisheries and Foods, 1995).

The height, robusticity, and unavoidable physicality of preagricultural humans mandated a caloric intake greater than that of most twentieth-century westerners. Twentieth-century hunter-gatherers are nonetheless lean, with skinfold thicknesses only half those of age-matched North Americans (Eaton et al., 1988). Therefore, our remote ancestors must have existed within an environment characterized by both greater caloric output and greater caloric intake than is now the rule.

Micronutrients and Phytochemicals

Fruits, roots, legumes, nuts, and other noncereals provided 65–70% of the average forager subsistence base (Eaton & Konner, 1985). These foods were generally consumed within hours of being gathered, typically with minimal or no processing (Schroeder, 1971) and often uncooked. Such foods and wild game are characterized by high average content of vitamins and minerals relative to their available energy (table 12.1).

Governmental organizations traditionally recommend levels of essential nutrient intake adequate to meet known metabolic needs (Food and Nutrition Board, 1989). However, assuming a 65:35 plant:animal subsistence pattern (that considered most likely by anthropologists; [Lee, 1968]) and a 3000 kcal/day diet, it seems reasonable that our preagricultural ancestors might have had

Table 12.1 Nutrient values of hunter-gatherer foods

	Vegetable foods (n = 236)			Animal foods (n = 85)		
	n	Mean	Range[a]	n	Mean	SD
Vitamins (mg/100 g)						
Riboflavin	89	0.168	0.001–1.14	26	0.399	0.246
Folate	11	0.0180	0.0028–0.0618	3	0.00567	0.00170
Thiamine	101	0.115	0–0.94	28	0.215	0.197
Ascorbate	123	33.0	0–414	18	4.79[b]	5.43
Carotene	51	0.328	0–6.55	—	—	—
Retinol equivalents	—	54.6	0–1090	—	—	—
Vitamin A	59	1.08	0–8.41	6	0.461	0.368
Retinol equivalents	—	180	0–1400	—	76.8[c]	61.4
Vitamin E	24	1.93	0.007–9.08	—	—	—
Minerals (mg/100 g)						
Iron	167	2.90	0.1–31	22	4.15	2.77
Zinc	91	1.12	0.1–9.5	11	2.67	0.860
Calcium	181	103	1–650	28	22.7[c]	30.9
Sodium	139	13.5	0–352	16	59	23.6
Potassium	112	448	5.1–1665	16	317	43.3
Fiber (g/100 g)	132	6.15	0–44.9	—	—	—
Energy (kJ/100 g)	184	456	16.7–2557	44	527	196
kcal/100g	—	(109)	(4–563) —	(126)	(46.8)	

[a] Range necessary because weighting precludes SD.
[b] See values for seabirds in Mann et al. (1962).
[c] Vitamin A and calcium in animal foods from organ meat, skin, small bones, insects, shellfish, and marrow.

an intake of most vitamins and minerals much in excess of currently recommended dietary allowances, either in absolute terms (table 12.2) or relative to energy intake (table 12.3). For some nutrients (e.g., folic acid) the intake retrojected for Paleolithic humans would have reached theoretically beneficial levels now thought attainable only through use of supplements (Daly et al., 1995).

Paleoanthropological data can be compared with estimates of optimal intake derived from conventional approaches. For example, current recommendations for optimal vitamin C ingestion range from 6mg/day to 750 mg/day (Levine et al., 1995). For a 3000 kcal diet, Paleolithic humans were likely to have averaged about 600 mg of vitamin C—within the spectrum of contemporary estimates, but toward the higher end. In addition, paleonutritional evidence may provide perspective when results of more conventional investigations conflict. Epidemiological studies on antioxidants currently seem at odds (Alpha-tocopherol, Beta-carotene Prevention Study Group, 1994; Sies & Krinski, 1995), but the evidence that ancestral humans consumed more tocopherol and carotene than do current humans appears clearcut (see tables 12.2 and 12.3).

The content of non-nutrient phytochemicals in most wild plant foods is unknown, but it is plausible that their concentrations may be relatively high, like those of micronutrients. The physiological role of these substances (protease inhibitors, organic isothiocyanates, organosulfur compounds, plant phenols, and flavanoids) has not been established, but their possible function as biological response modifiers and/or chemopreventive compounds attracts continuing research attention (Flagg et al., 1994; Hertog et al., 1993; Watanabe et al., 1991; Zhang et al., 1992).

Electrolytes

Typical adult Americans consume nearly 4000 mg of sodium each day (Food and Nutrition Board, 1989), of which 75% is added to food during processing (James et al., 1987); only about 10% is intrinsic to the basic food items. Potassium intake averages 2500–3400 mg/day (Food and Nutrition Board, 1989) so Americans, like nearly all people living today, consume more sodium than potassium. Humans are the only free-living, nonmarine mammals to do so. In the United States, recommended dietary allowances for sodium are 500–2400 mg/day and for potassium 2000–3400 mg/day (Food and Nutrition Board, 1989). These recommendations are roughly intermediate between current and ancestral human experience.

Improved data on the sodium and potassium content of uncultivated plant foods and wild game (table 12.1) allow earlier calculations (Eaton & Konner, 1985) to be updated. The refined estimate of Paleolithic intake contrasts strikingly with our current pattern. Preagricultural humans are calculated to have consumed only 768 mg of sodium while obtaining 10,500 mg of potassium each day. The potential implications of electrolyte intake at the Paleolithic level are apparent in data from the Intersalt Study (Intersalt Cooperative Re-

Table 12.2 Estimated daily Paleolithic intake of selected nutrients compared to recommended and current levels

	Paleolithic intake[a]	U.S. RDA[b]	Current U.S. intake[b]
Vitamins (mg/day)			
Riboflavin	6.49	1.3–1.7	1.34–2.08
Folate	0.357	0.18–0.2	0.149–0.205
Thiamin	3.91	1.1–1.5	1.08–1.75
Ascorbate	604	60	77–109
Carotene	5.56	—	2.05–2.57
Retinol equivalents	927		342–429
Vitamin A	17.2	4.80–6.00	7.02–8.48
Retinol equivalents	2870	1170–429	800–1000
Vitamin E	32.8	8–10	7–10
Minerals (mg/day)			
Iron	87.4	10–15	10–11
Zinc	43.4	12–15	10–15
Calcium	1956	800–1200	750
Sodium	768	500–2400	4000
Potassium	10500	3500	2500
Fiber (g/day)	104	20–30	10–20
Energy (kJ/day)	12558	7326–10465	9209–12139
kcal/day	3000	2200–2900	1750–2500

[a]Based on 913 g meat and 1697 g vegetable food/day = 12558 kJ (3000 kcal). See
Eaton (1992) for method.
[b]RDA, recommended daily allowance; Food and Nutrition Board (1989).

Table 12.3 Dietary micronutrient intake relative to energy intake at estimated Paleolithic and current levels

	mg/4189 kJ (1000 kcal)		Ratio Paleolithic: current
	Paleolithic[a]	Current[b]	
Vitamins			
Riboflavin	2.16	0.6	3.60
Folate	0.119	0.08	1.49
Thiamin	1.30	0.51	2.55
Ascorbate	201	24	8.38
Carotene	1.85	1.09	1.70
Retinol equivalents	309	182	—
Vitamin A	5.74	2.12	2.71
Retinol equivalents	957	353	—
Vitamin E	10.9	3.5	3.11
Minerals			
Iron	28.5	4.9	5.82
Calcium	653	392	1.67
Zinc	14.5	5.3	2.74
Sodium	256	1882	0.136
Potassium	3500	1177	2.97

[a]From table 12.2.
[b]Food and Nutrition Board (1989).

search Group, 1988), whose subjects included Yanamamo and Xingo Amerindians, and Asaro from New Guinea. These groups are rudimentary horticulturists, but they are relatively isolated, little acculturated, and their food, like that of hunter-gatherers, is free of added salt. Their dietary sodium: potassium ratio (0.13) is comparable to that estimated for preagricultural humans (0.07) and contrasts with that for current Americans (~1.35).

The Intersalt investigators found that, on average, Yanamamo, Xingo, and Asaro populations have low blood pressure (102/62, considered low by current American standards), no blood pressure increase with age, and minimal (0.6%) prevalence of hypertension (Carvallo et al., 1989). These findings are consistent with those for other previously studied forager and horticulturist groups (Eaton et al., 1988). In contrast, for 48 Intersalt study groups who did have access to salt, sodium intake averaged 3818 mg/day, but potassium averaged only 2106 mg/day. Median blood pressure for these populations was 119/74 and tended to rise with age; their prevalence of hypertension (blood pressure >140/90) varied from 5.9% to 33.5% (Intersalt Cooperative Research Group, 1988).

The editorial accompanying Intersalt's initial publication was subtitled "Salt has only small importance in hypertension," (Swales, 1988), but a follow-up article focusing on Intersalt's remote study groups concluded that "a minimum intake of salt is required to produce a high frequency of hypertension in populations" (Mancilha Carvalho et al., 1989, p. 244). That minimum appears to exceed the level consumed during the Paleolithic.

Carbohydrate

The typical carbohydrate intake of ancestral humans was similar in magnitude (45–50% of daily energy) to that in currently affluent nations, but there was a marked qualitative difference. Under most circumstances during the late Paleolithic, the great majority of carbohydrate was derived from vegetables and fruit, little was derived from cereal grains, and none from refined flours (Eaton & Konner, 1985), a diet extending the multimillion year experience of primates generally (Milton, 1993). Current recommendations that individuals consume 55% or more of their energy as carbohydrate are not far from human evolutionary experience, but the different makeup of the carbohydrate involved has important implications. Only 23% of American carbohydrate consumption is derived from fruit or vegetable sources (Committee on Diet and Health, 1989) and for Europeans the proportion is lower still (James et al., 1988). The corollary is that preagricultural humans consumed roughly three times the vegetables and fruit typical Westerners do today. Their intake would have equaled or exceeded that of current vegans whose consumption of vegetables, roots, fruit, and berries is 2.6 times that of matched omnivores and whose antioxidant vitamin intake is 247–313% greater (Rauma et al., 1995).

Much current carbohydrate intake is in the form of sugars and sweeteners; in the mid-1980s, American per capita consumption in these categories exceeded 54.6 kg (120 lbs) annually (Committee on Diet and Health, 1989). Such products, together with foods made from highly refined grain flours, provide "empty calories" (i.e., food energy without essential amino acids, essential fatty acids, micronutrients, and perhaps phytochemicals). Energy sources of this type are much less available to hunter-gatherers; wild honey is a seasonal delicacy and varies considerably in availability. The relatively caries-free nature of Paleolithic dental remains (Larsen, 1997) suggests that, over the course of their lives, honey accounted for far less than the 18% contribution that sugars and sweeteners now make to the daily energy intake of Westerners (Committee on Diet and Health, 1989). Not only are sugars and highly refined flour products devoid of nutrients other than energy, they are also low in bulk so that they can be eaten quickly and occupy only a small proportion of gastric capacity. The bulky carbohydrate sources that fueled human evolution had to be eaten more slowly and usually produced more gastric distention for a given caloric load (Duncan et al., 1983).

Hunter-gatherers use many species of fruits and vegetables, often more than 100, as sources of their subsistence. In this respect, inhabitants of affluent industrialized nations, with ready access to an even wider variety of produce, are far better off than traditional agriculturists whose choices were often markedly constrained. Foraging forestalled famine except under the most adverse climatic conditions and, in addition, provided a varied abundance of micronutrients and phytochemicals. Epidemiological investigations have consistently demonstrated an exceptionally strong association between cancer prevention and the consumption of fruits and vegetables (Block et al., 1992; World Cancer Research Fund and American Institute for Cancer Research, 1997). It is tempting to speculate that this association reflects the benefits of the intrinsic micronutrient/phytochemical load of fruits and vegetables. Micronutrients and phytochemicals are much less prominent intrinsic components of sugars and highly refined flour products. It is potentially significant that a similar cancer preventive relationship has not yet been proven to exist for grain products (Block et al., 1992 World Cancer Research Fund and American Institute for Cancer Research, 1997).

Fat

There is near unanimous current opinion that saturated fat should account for less than 10% of each day's dietary energy, perhaps 7–8% (Grundy, 1994). Similar consensus exists that high intakes of cholesterol (>500 mg/day) are to be avoided, an intake of 300 mg/day or less being widely advocated. There is also substantial agreement that high total-fat diets (>40% of dietary energy) are unhealthy; however, there is dispute as to whether the target for fat consumption should be low (10–20% of energy) or moderate (in the 30–35% range) (Grundy, 1994).

Saturated fatty acids are calculated to have provided about 6% of the average total energy intake for Paleolithic humans (Eaton, 1992). Forager cholesterol intake is estimated at 480 mg/day, unavoidable when game makes up a third of the nutrition base, and fat intake is projected at 20–25% total energy, intermediate between the frequently lauded traditional (c. 1960) Japanese (11%) and Mediterranean (37%) patterns (Willett, 1994).

Cholesterol-Raising Fatty Acids

Saturated fats, particularly C_{14} myristic and C_{16} palmitic acid (but not C_{18} stearic acid) are thought to be the most important dietary factors related to coronary heart disease (Grundy, 1994). At present, the chief sources of such fats in affluent nations are meat, dairy products, and tropical oils. Even though they are predominantly monounsaturated, *trans* fatty acids[2] also raise serum cholesterol levels much like C_{14} myristic and C_{16} palmitic acids do. Each of these sources was much less or not at all a factor influencing lipid metabolism in the late Paleolithic.

Game has less fat overall than does modern commercial meat (4.2 g fat/100 g meat versus 20.0 g fat/100 g meat) (Eaton, 1992), while its proportion of C_{14} and C_{16} fatty acids is also lower (0.99 g C_{14} and C_{16}/100 g fat versus 5.64 g C_{14} and C_{16}/100 g fat). For these reasons, game has much less tendency to raise serum cholesterol levels than does meat from today's supermarkets (O'Dea et al., 1990; Sinclair et al., 1987). Because there were no domesticated animals during the Paleolithic, adults and older children then had no dairy foods what-soever. And while tropical plant species, including coconuts and palm nuts, were presumably important regional dietary resources, their entire edible portion (including beneficial nutrients) was consumed, not just their oil. Unlike today, these foods were then available only in areas to which they were indigenous. Some *trans* fatty acids occur naturally in milk, but the overwhelming majority result from commercial hydrogenation (which makes fats more saturated). They effectively raise serum cholesterol levels, but, unlike C_{14} and C_{16} fatty acids, they reduce the health-promoting HDL-cholesterol fraction (Judd et al., 1994). *Trans* fatty acids would have made no contribution to adult Stone Age diets. After subtracting non–cholesterol-raising C_{18} stearic acid and adding *trans* fatty acids, each about 3%, the overall contribution of cholesterol-raising fatty acids to the current American diet is about 13–14% of caloric consumption, far above both recommendations (7–8%) and Paleolithic experience (perhaps 5% if C_{18} stearic acid is subtracted).

Dietary Cholesterol and Serum Cholesterol

Despite high dietary cholesterol intake, foragers studied in this century have manifested very low serum cholesterol levels, averaging around 3.2 mmol/L (125 mg/dL) (Eaton et al., 1988), within the range found for free-living non-human primates (2.3–3.5 mmol/L; 90–135 mg/dL) (Eaton, 1992). This similarity of serum cholesterol values may indicate existence of a natural primate pattern and thus counter contentions that low serum cholesterol levels are irrelevant or even harmful to human health (Dalen & Dalton, 1996; Jacobs et al., 1992; Kritchevsky & Kritchevsky, 1992). Furthermore, the low serum cholesterol of hunter-gatherers, despite their high cholesterol intake, adds credence to observations that the effect of dietary cholesterol on serum levels is mitigated as the ratio between polyunsaturated and saturated fat (P:S) rises (Schonfeld et al., 1982). The P:S ratio for hunter-gatherers is 1.4; for Americans it is 0.4. It is estimated that an increase in dietary cholesterol of 200 mg/day should elevate serum cholesterol about 0.2 mmol/l (8 mg/dl) (Grundy, 1994). Forager intake, at 480 mg/day, is nearly 200 mg/day greater than the most common recommendation (<300 mg/day). Apparently, the high P:S ratio, low saturated fat content, and low total fat intake of hunter-gatherer diets more than offsets the adverse effects of their high cholesterol intake (O'Dea et al., 1990; Sinclair et al., 1987).

Polyunsaturated Fatty Acids

In affluent nations, (ω)-6 polyunsaturated fatty acid (PUFA) intake is now roughly 11 times that of ω-3 PUFA (Adam, 1989; Hunter, 1990), but the ratio in forager diets is more nearly equal, ranging from 4:1 to 1:1 (Sinclair & O'Dea, 1993). At the other extreme are the Greenland Inuit, for whom ω-6:ω-3 ratios of 1:40 have been calculated; however, their extensive exploitation of marine resources is at variance with the currently accepted paleoanthropological view of human savanna origins. The terrestrial food chain provides both linoleic acid (LA; 18:2, ω-6) and alpha-linolenic acid (ALA; 18:3, ω-3) from plant sources (Simopoulos, 1991) as well as their desaturation/elongation products, arachidonic acid (AA; an ω-6 PUFA), and eicosapentaenoic and docosahexaenoic acids (EPA and DHA; ω-3 PUFAs) from animal tissue (Naughton et al., 1986). The prominence of game in typical hunter-gatherer diets produces high levels of AA, EPA, and DHA in their plasma lipids relative to those found in Westerners (Sinclair & O'Dea, 1993).

Arachidonic acid is the metabolic precursor of powerful eicosanoids such as thromboxane and the four-series leukotrienes. The ω-3 PUFAs, especially EPA and DHA, apparently modulate eicosanoid biosynthesis from AA so that proaggregatory and vasoconstrictive eiconsanoid formation is reduced. For this effect, it appears that the ratio of ω-6 to ω-3 PUFA in the diet, rather than the absolute amount of ω-3 PUFA, is the decisive factor (Boudreau et al., 1991). In addition, ω-6 PUFA may act as a promoter in carcinogenesis, a property seemingly absent for ω-3 PUFA. Based on these considerations, some investigators have advocated a return to the dietary ω-6:ω-3 ratio of ancestral humans (Simopoulos, 1991; Sinclair & O'Dea, 1993).

Protein

Retrojected protein intake for Paleolithic humans, typically about 30% of daily energy, is hard to reconcile with the 12% currently recommended for Americans (Committee on Diet and Health, 1989). The recommended daily allowance (RDA) is actually a range, 0.8–1.6 g/kg/day, which contrasts with 2.5–3.5 g/kg/day for foragers. Observed protein intake for other primates, such as chimpanzees, gorillas, baboons, and howler monkeys, is also higher than that advocated by human nutritionists and ranges from 1.6 to 5.9 g/kg/day in the wild (Casimir, 1975; Coelho et al., 1976; Hladik, 1977; Whiten et al., 1991). Furthermore, veterinary recommendations for higher primates in captivity also substantially exceed the RDA for Americans. (Panel on Nonhuman Primate Nutrition, 1978). It would be paradoxical if humans, who, during evolution, added hunting and scavenging skills to their higher primate heritage, should now somehow be harmed as a result of protein intake habitually tolerated or even required by their near relatives.

Protein and Disease

Epidemiological studies have linked high protein intake with cancer (especially of the breast and colon), but such evidence is inconsistent, as is the effect of varying protein intake on spontaneous tumor incidence in animal experiments (Committee on Diet and Health, 1989). Studies comparing different countries also show that diets high in meat have a strong positive correlation with atherosclerotic coronary artery disease, but such diets are now linked with high intakes of total and saturated fat, which probably accounts for a large part of the association (Committee on Diet and Health, 1989). In addition, high-meat diets in industrialized countries tend to have restricted levels of plant foods, especially fruits and vegetables, which may increase susceptibility to both cancer and atherosclerosis. Conversely, the high animal protein intake of preagricultural humans generally occurred within a nutritional context of low fat and high fruit and vegetable consumption. In such circumstances a high protein diet may elevate plasma HDL-cholesterol while lowering total cholesterol and triglyceride levels, thereby reducing cardiovascular risk (Wolfe & Giovannetti, 1991).

High intake of purified, isolated protein increases urinary excretion of calcium, chiefly in experimental settings where phosphorus intake is held constant (Committee on Diet and Health, 1989). But high protein natural diets have increased phosphorus as well (because of the high phosphorous content in meat), and there is little evidence that such diets increase the risk of osteoporosis (Committee on Diet and Health, 1989; Hunt et al., 1995; but see Abelow et al., 1992). Paleolithic humans developed high peak (or maximum) bone mass, probably reflecting their habitual levels of physical activity together with their ample calcium intake. Although there are few older skeletons to evaluate, those available suggest preagriculturalists experienced less bone loss with age than did subsequent low-protein-diet agriculturists (Eaton & Nelson, 1991).

High protein diets should worsen the course of chronic renal failure (Brenner et al., 1982); however, prominent underlying causes of renal disease, such as diabetes and hypertension, are rare among foragers (Eaton et al., 1988). Health evaluations including autopsy studies have shown no increased frequency of renal disease among partially acculturated Alaskan and Greenland Inuit whose protein intake approximates that postulated for preagricultural humans (Arthaud, 1970; Kronmann & Green, 1980; Mann et al., 1962). Because of its largely vegetable nature, the high protein intake of nonhuman primates presumably has less anabolic effect than would a comparable level of animal protein for humans, but the resulting nitrogen load for the kidneys would be equivalent.

Although increased protein intake may not be actively harmful, experiments by exercise physiologists make it hard to understand why humans might require dietary protein at the 2.5–3.5 g/kg/day level, at least in regard to nitrogen balance and lean body mass maintenance (Young, 1986). Perhaps protein intake at the levels characteristic of human evolutionary experience has physiological consequences apart from those directly affecting nitrogen metabolism. The ratio of protein to carbohydrate in human diets affects the relative amounts of insulin and glucagon secreted after a meal (Rossetti et al., 1989; Westphal et al., 1990); in addition, by influencing desaturase activity, protein may modulate eicosanoid biosynthesis (Brenner, 1981; Sears, 1993). In any event, the nutritional habits of preagricultural humans were apparently contrary to recent recommendations that carbohydrate, not protein, intake should be increased to compensate for dietary fat reduction (Committee on Diet and Health, 1989).

Fiber

Analysis of vegetable foods consumed by recently studied hunter-gatherers (tables 12.1 and 12.2) and evaluation of archaic native American coprolith remains suggest preagricultural fiber intake exceeded 100 g/day (Eaton, 1990). Rural Chinese consume up to 77 g of fiber per day (Campbell & Chen, 1994), and estimates of 60–120 g/day have been made for rural Africans (Burkitt, 1983). Chimpanzees and other higher primates obtain up to 200 g of fiber from each day's food (Milton, 1993). In contrast, fiber intake for adult Americans is generally less than 20 g/day and current recommendations range from 20 to 30 g/day. (Butrum et al., 1988).

Because the fiber consumed by Paleolithic humans came primarily from fruits, roots, legumes, nuts, and other noncereal vegetable sources, its content of phytic acid would have been less than that of the fiber consumed now in industrialized nations, which comes largely from grain (Eaton, 1990). For the same reason, the proportion of soluble, fermentable fiber relative to insoluble, nonfermentable fiber was likely to have been higher for preagricultural humans than for current residents of affluent nations. Reservations about increasing the fiber content of Western diets revolve around potential adverse effects on micronutrient absorption, especially of minerals, due to binding by fiber (Committee on Diet and Health, 1989). There is, however, little evidence that diets containing up to 50 g/day have a negative effect on absorption (Committee on Diet and Health, 1989), even when the fiber is predominantly wheat with its high phytic acid content. The bony remains of preagricultural humans suggest that they absorbed minerals adequately, even though their fiber intake exceeded that so far studied by nutritionists. The high proportion of soluble fiber in Paleolithic diets should have favorably affected lipid metabolism (Kritchevsky, 1994; Rimm et al., 1996).

Discussion

In some respects, the nutritional experience of humans during evolution, (i.e., in "the environment of evolutionary adaptedness"; Tooby & Cosmides, 1990, pp. 386) parallels and supports existing dietary recommendations. For example, advice to decrease saturated fat intake below 10% of daily energy (interpreted to mean that cholesterol-raising fatty acids should comprise no more than 7–8% of each day's calories) nearly matches the 5% inake level retrojected for Paleolithic humans. American RDAs advocate increased energy expenditure through physical activity rather than voluntary reduction in energy intake as a preferred way of maintaining health and desirable body weight. This advice fits perfectly with our evolutionary past, which was characterized by obligatory exercise and relatively high caloric intake.

In other cases, the Paleolithic experience can serve as a reference standard when the recommendations of nutritionists are at variance. Many nutritional advisory bodies suggest that fat content not exceed 30% of caloric intake. Still, supporters of the "Mediterranean" dietary paradigm (Grundy, 1994; Willett, 1994) debate with advocates of a traditional "East Asian" (Japanese, Chinese) approach (Campbell & Chen, 1994; Ornish et al., 1990). The latter emphasizes very low fat intake, 10–15% of total energy, and increased carbohydrate intake—up to 70% of daily calories. In contrast, the former allows total fat in the 35–40% range, but stresses substitution of monounsaturated fat (e.g., olive oil) for saturated fat. Critics of the East Asian diet point out its tendency to lower HDL-cholesterol while raising triglyceride levels in the serum (Grundy, 1994). Furthermore, high carbohydrate diets may adversely affect control in diabetic patients (Garg et al., 1994). The Mediterranean diet has been faulted because its high fat content may promote obesity, because high fat diets may elevate serum insulin levels (Feskens et al., 1994), and because breast and colon cancer incidence consistently shows strong positive correlations with dietary fat intake when different countries are compared (Carroll, 1994). The retrojected preagricultural diet would have provided neither high carbohydrate nor high fat content. Intermediate between the Mediterranean and East Asian patterns, it might achieve the benefits of both while avoiding their drawbacks.

The Paleolithic diet was nutrient-rich; could it have provided dangerous amounts of vitamins and minerals? A review (Levine et al., 1995) advises against vitamin C intake in excess of 500 mg/day, about 100 mg less than the estimate for Paleolithic humans, chiefly because of concern about producing oxalate stones. However, the relationship between ascorbate intake and formation of such stones is disputed (Diplock, 1995). Iron intake would have been high, but still below the minimum toxic dose (table 12.4). Vitamin A consumption would have been well within traditionally accepted limits, but close to the 10,000 IU/day level, above which teratogenic effects have been identified (Rothman et al., 1995). In each case the effects of high nutrient intake within a Paleolithic nutritional-developmental-experiential framework might differ from those of the same nutrient level within the typical affluent Western biobehavioral setting.

Two other features of Paleolithic nutrition that differ from contemporary nutritional theory are its content of fiber and of protein, both much above current recommendations. Regarding these nutrients, it must be reemphasized that foragers subsisted almost exclusively on game and wild plant foods; there were no domesticated animals (for dairy foods), and significant use of cereal grains began only about 15,000 years ago—late in evolutionary terms. Given the nutritional properties of game (high in protein relative to energy) and uncultivated plants (high in fiber relative to energy), a preagricultral diet would necessarily have had high protein, high fiber, or both. If the average plant: animal subsistence ratio was 65:35, a 30% protein, 100 g fiber diet emerges. This near inescapable result seems strange to us today because we are used to diets containing large amounts of "empty" calories from sugar, highly refined flour, and nonessential fat. It is actually diets of this sort which are novel in evolutionary and comparative zoological terms.

Table 12.4 Paleolithic micronutrient intake and currently estimated minimum daily toxic doses

	Paleolithic intake[a]	Minimum toxic dose[b]
Vitamins (mg)		
Riboflavin	6.49	1000
Folate	0.357	400
Thiamin	3.91	300
Ascorbate	604	1000–5000
Carotene	927	—
Vitamin A (IU)	9570	25,000–50,000
Vitamin E	32.8	1200
Minerals (mg)		
Iron	85.4	100
Zinc	43.4	500
Calcium	1960	12,000
Sodium	768	—
Potassium	10500	—

[a]From table 12.2.
[b]Food and Nutrition Board (1989:518).

Proponents of evolutionary (or Darwinian) medicine emphasize its heuristic potential and its capacity to suggest new avenues for conventional research. An example is the need for better nutritional analyses of wild game and vegetable foods; analyses available to date are nonstandardized and incomplete. For example, content of vitamin A and carotenoids is often commingled and there are no data (known to us) that would allow distinction of beta-carotene from other carotenoids, even when total carotenoid content is provided. Similarly, information on ω-6/ω-3 partition is uncommon, and data on PUFA other than LA and ALA are even less available. Any reconstruction of Paleolithic nutrition is utterly dependent on such data, so a project to determine the nutritional makeup of foods available to foragers, especially in Africa, Europe, and Asia (since the other continents were inhabited "late"), would be extremely valuable. Even more urgent is the need to study remaining and recently acculturated hunter-gatherer subsistence patterns. Some data on this subject are available, but populations from which we can directly verify and expand the database are vanishing; this irreplaceable human resource needs to be utilized optimally while it still exists.

The numerous inconsistencies between current recommendations and retrojected Paleolithic nutrition suggest obvious investigative possibilities. As advocated by Campbell (Campbell, 1994; Campbell & Chen, 1994), such research might best be conducted in a comprehensive program that studies many nutrients simultaneously and which takes nutritional adaptation into account. For example, high fiber diets initiated by adults could be poorly tolerated, but when begun in childhood might be accommodated without difficulty and perhaps demonstrate heretofore unappreciated plasticity in human gut development. A dietary protein level of 30% total energy may be harmful when the diet also includes excessive sodium and insufficient potassium, especially if individuals engage in nutritional and exercise patterns promoting obesity. When the individuals involved are lean, normotensive, and nondiabetic, abundant dietary protein may be beneficial. Similarly, an extremely high fiber intake may offset potentially adverse consequences of elevated dietary micronutrient levels (e.g., zinc, iron). The most meaningful research designed to reconcile current nutritional recommendations with the nutrition that shaped our metabolic needs during evolutionary experience will probably involve comprehensive integration of multiple dietary variables and exercise activities in studies that begin early in life and proceed through development into adulthood.

Notes

1. We use the terms foragers and hunter-gatherers interchangeably.

2. *Trans* fatty acids are those with hydrogen molecules on opposite sides of the carbon–carbon double bond, as opposed to *cis* fatty acids (predominate in nature), which have hydrogen molecules on the same side of the double bond.

References

Abelow B.J., Holford T.R., and Insogna K.L. (1992). Cross-cultural association between dietary animal protein and hip fracture: a hypothesis. *Calcified Tissues International* 50:14–18.

Adam O. (1989). Linoleic and linolenic acid intake. In C. Galli and A.P. Simopoulos, eds., *Dietary w-3 and ω-6 Fatty Acids. Biological Effects and Nutritional Essentiality*, pp.33–42. Plenum, New York.

The Alpha-tocopherol, Beta-carotene Prevention Study Group (1994). The effect of vitamin E and beta carotene on the incidence of lung cancer and other cancers in male smokers. *New England Journal of Medicine* 330: 1029–35.

Ames B.N., Gold L.S., and Willett W.C. (1995). The causes and prevention of cancer. *Proceedings of the National Academy of Science USA* 92:5258–65.

Arthaud J.B. (1970). Cause of death in 339 Alaskan natives as determined by autopsy. *Archives of Pathology* 90:433–38.

Block G., Patterson B., and Subar A. (1992). Fruit, vegetables, and cancer prevention: a review of the epidemiological evidence. *Nutrition and Cancer* 18:1–29.

Boudreau M.D., Charmugan P.S., Hart S.B., Lee S.H., and Hwang D.J. (1991). Lack of dose response by dietary ω-3 fatty acids at a constant ratio of ω-3 to n-6 fatty acids in suppressing eicosanoid biosynthesis from arachidonic acid. *American Journal of Clinical Nutrition* 54:111–17.

Brenner B.M., Meyer T.W., and Hostetter T.H. (1982). Dietary protein intake and the progressive nature of kidney disease. *New England Journal of Medicine* 307:652–59.

Brenner R.R. (1981). Nutritional and hormonal factors influencing desaturation of essential fatty acids. *Progress in Lipid Research* 20:41–47.

Burkitt D. (1983). *Don't Forget Fiber in Your Diet*, p. 32. Martin Dunitz, Singapore.

Burkitt D.P., and Eaton S.B. (1989). Putting the wrong fuel in the tank. *Nutrition* 5:189–91.

Butrum R.R., Clifford C.K., and Lanza E. (1988). NCI dietary guidelines. *American Journal of Clinical Nutrition* 48:888–95.

Campbell T.C. (1994). The dietary causes of degenerative diseases: nutrients vs. foods. In N.J. Temple and D.P. Burkitt, eds., *Western Diseases. Their Dietary Prevention and Reversibility*, pp. 119–52. Humana Press, Totowa, NJ.

Campbell T.C., and Chen J. (1994). Diet and chronic degenerative diseases: perspectives from China. *American Journal of Clinical Nutrition* 59:1153S-61S.

Carroll K.K. (1994). Lipids and cancer. In K.K. Carroll and D. Kritchevsky, eds., *Nutrition and Disease Update. Heart Disease*, pp. 235–96. American Oil Chemistry Society Press, Champaign, IL.

Casimir M.J. (1975). Feeding ecology and nutrition of an eastern gorilla group in the Mt. Kahuzi region (Republique du Zaire). *Folia Primatologica* 24:84–136.

Coelho A.M., Bramblett C.A., Quick L.B., and Bramblett S.S. (1976). Resource availability and population density in primates: a sociobioenergetic analysis of the energy budgets of Guatemalan howler and spider monkeys. *Primates* 17:63–80.

Committee on Diet and Health, National Research Council (1989). *Diet and Health*. National Academy Press, Washington, DC.

Dalen J.E., and Dalton W.S. (1996). Does lowering cholesterol cause cancer? *Journal of the American Medical Association* 275:67–69.

Daly L.E., Kirke P.N., Molloy A., Weir D.G., and Scott J.M. (1995). Folate levels and neural tube defects. Implications for prevention. *Journal of the American Medical Association* 274:1698–1702.

Diplock A.T. (1995). Safety of antioxidant vitamins and carotene. *American Journal of Clinical Nutrition* 62 (suppl):1510S–16S.

Duncan K.H., Bacon J.A., and Weinsier R.L. (1983). The effects of high and low energy density diets on satiety, energy intake, and eating time of obese and non-obese subjects. *American Journal of Clinical Nutrition* 37: 763–67.

Eaton S.B. (1990). Fibre intake in prehistoric times. In A.R. Leed, ed., *Dietary Fibre Perspectives. Reviews and Bibliography* vol. 2, pp. 27–40. John Libbey, London.

Eaton S.B. (1992). Humans, lipids and evolution. *Lipids* 27:814–20.

Eaton S.B., and Konner M. (1985). Paleolithic nutrition. A consideration of its nature and current implications. *New England Journal of Medicine* 312:283–89.

Eaton S.B., Konner M., and Shostak M. (1988). Stone agers in the fast lane: chronic degenerative diseases in evolutionary perspective. *American Journal of Medicine* 84:739–49.

Eaton S.B., and Nelson D.A. (1991). Calcium in evolutionary perspective. *American Journal of Clinical Nutrition* 54:281S–87S.

Feskens E.J.M., Loeber J.G., and Kromhout D. (1994). Diet and physical activity as determinants of hyperinsulinemia: the Zutphen elderly study. *American Journal of Epidemiology* 140:350–60.

Flagg E.W., Coates R.J., Jones D.P, Byers T.E., Greenberg R.S., Gridley G., McLaughlin J.K., Blot W.J., Haber M., Preston-Martin S., Schoenberg

J.B., Austen D.F., and Fraumeni J.F. (1994). Dietary glutathione intake and the risk of oral and pharyngeal cancer. *American Journal of Epidemiology* 139:453–65.

Food and Nutrition Board, National Research Council (1989). *Recommended Dietary Allowances*, 10th ed. National Academy Press, Washington, DC.

Garg A., Bantle J.P., Henry R.R., Coulston, A.M., Griver K.A., Raatz S.K., Brinkley L., Chen Y.-D.I., Grundy S.M., Huet B.A., and Reaven G.M. (1994). Effects of varying carbohydrate content of diet in patients with non-insulin dependent diabetes mellitus. *Journal of the American Medical Association* 271:1421–28.

Grundy S.M. (1994). Lipids and cardiovascular disease. In D. Kritchevsky and K.K. Carroll, eds., *Nutrition and Disease Update. Heart Disease*, pp. 211–79. American Oil Chemistry Society Press, Champaign IL.

Heini A.F., Minghelli G., Diaz E., Prentice A.M., and Schutz Y. (1995). Free-living energy expenditure assessed by two different methods in lean rural Gambian farmers. [abstract] *American Journal of Clinical Nutrition* 61:893.

Hertog, M.G.L., Feskens E.J.M., Hollman P.C.H., Katan M.D., and Kromhout D. (1993). Dietary antioxidant flavonoids and risk of coronary heart disease: the Zutphen elderly study. *Lancet* 342:1007–11.

Hladik C.M. (1977). Chimpanzees of Gabon and chimpanzees of Gombe: some comparative data on the diet. In T.H. Clutton-Brock, ed., *Primate Ecology: Studies of Feeding and Ranging Behavior in Lemurs, Monkeys and Apes*, pp. 481–501. Academic Press, London.

Hunt J.R., Gallagher S.K., Johnson L.K., and Lykken G.I. (1995). High-versus low-meat diets: effects on zinc absorption, iron status, and calcium, copper, iron, magnesium, manganese, nitrogen, phosphorus, and zinc balance in post menopausal women. *American Journal of Clinical Nutrition* 62:621–32.

Hunter J.E. (1990). n-3 Fatty acids in vegetable oils. *American Journal of Clinical Nutrition* 51:809–14

Intersalt Cooperative Research Group (1988). Intersalt: an international study of electrolyte excretion and blood pressure. *British Medical Journal* 297:319–28.

Jacobs D., Blackburn H., Higgins M., Reed D., Iso H., McMillan G., Neaton J., Nelson J., Potter J., Rifkind B., Rossouw J., Shekelle R., and Yusuf S. (1992). Report of the conference on low blood cholesterol: mortality associations. *Circulation* 86:1046–60.

James W.P.T., Ferro-Luzzi A., Isaksson B., and Szostak W.B. (1988). *Healthy Nutrition. Preventing Nutrition-related Diseases in Europe*. World Health Organization, Denmark.

James W.P.T., Ralph A., and Sanchez-Castillo C.P. (1987). The dominance of salt in manufactured food in the sodium intake of affluent societies. *Lancet* i:426–29.

Judd J.T., Clevidence B.A., Muesing R.A., Wittes J., Sunkin M.E. and Podczasy J.J. (1994). Dietary trans fatty acids: effects on plasma lipids and lipoproteins of healthy men and women. *American Journal of Clinical Nutrition* 59:861–68.

Kritchevsky D. (1994). Dietary fiber and cardiovascular disease. In D. Kritchevsky and K.K. Carroll, eds., *Nutrition and Disease Update. Heart Disease*, pp. 189–210. American Oil Chemistry Society Press, Champaign, IL.

Kritchevsky F.B., and Kritchevsky D. (1992). Serum cholesterol and cancer risks: An epidemiologic perspective. *Annual Review of Nutrition* 12: 391–416.

Kronmann N., and Green A. (1980). Epidemiological studies in the Upernavik district, Greenland. Incidence of some chronic diseases 1950–74. *Acta Medica Scandinavia* 208:401–6.

Larsen C.S. (1997). *Bioarchaeology: Interpreting Behavior from the Human Skeleton*, pp. 67–72. Cambridge University Press, Cambridge.

Lee R.B. (1968). What hunters do for a living, or, how to make out on scarce resources. In R.B. Lee and I. DeVore, eds., *Man the Hunter*, pp. 30–48. Aldine, Chicago.

Levine M., Dhariwal K.D., Welch R.W., Wang Y., and Park J.B. (1995). Determination of optimal vitamin C requirements in humans. *American Journal of Clinical Nutrition* 62(suppl): 1347S–56S.

Mann G.V., Scott E.M., Hursh L.M., Heller C.A., Youmans J.B., Consolazio C.F., Bridgforth E.B., Russell A.L., and Silverman M. (1962). The health and nutritional status of Alaskan Eskimos. *American Journal of Clinical Nutrition* 11:31–76.

Mancilha Carvalho, J.J.M., Baruzzi R.G., Howard P.F., Poulter N., Alpers M.P., Franco L.J., Marcopito L.F., Spooner V.J., Dyer A.R., Elliott P., Stamler J., and Stamler R. (1989). Blood pressure in four remote populations in the Intersalt Study. *Hypertension* 14:238–46.

Milton K. (1993). Diet and primate evolution. *Scientific American* 269:86–93.

Ministry of Agriculture, Fisheries and Foods (1995). *Household Food Consumption and Expenditure, 1990. With a Study of Trends over the Period 1940–1990.* London: Her Majesty's Stationary Office.

Naughton J.M., O'Dea K., and Sinclair A.J. (1986). Animal foods in traditional Australian Aboriginal diets: polyunsaturated and low in fat. *Lipids* 21:684–90.

Neel J.V. (1994) *Physician to the Gene Pool*, pp. 302, 315. John Wiley, New York.

O'Dea K., and Sinclair A. (1983). The modern western diet—the exception in man's evolution. In K.A. Boundy and G.H. Smith, eds., *Agriculture and Human Evolution*, pp. 56–61. Australian Institute Agricultural Science, Melbourne.

O'Dea K., Traianedes K., Chisholm K., Leyden H., and Sinclair A.J. (1990). Cholesterol-lowering effect of a low-fat diet containing lean beef is reversed by the addition of beef fat. *American Journal of Clinical Nutrition* 52:491–94.

Ornish D., Brown S.E., Scherwitz L.W., Billings J.H., Armstrong W.T., Ports T.A., McLanahan S.M., Kirkeeide R.L., Brand R.J., and Gould K.L. (1990). Can lifestyle changes reverse coronary heart disease? The lifestyle heart trial. *Lancet* 336:129–133.

Panel on Nonhuman Primate Nutrition (1978). *Nutrient Requirements of Nonhuman Primates*, pp. 36–41. National Academy of Science, Washington, DC.

Rauma A.L., Torronen R., Hanninen O., Verhagen H., and Mykkanen H. (1995). Antioxidant status in long term adherents to a strict uncooked vegan diet. *American Journal of Clinical Nutrition* 62:1221–27.

Rendel J.M. (1970). The time scale of genetic change. In S.V. Boyden, ed., *The Impact of Civilization on the Biology of Man*, pp. 27–47. Australian National University Press, Canberra.

Rimm E.B., Ascherio A., Giovannucci E., Spiegelman D., Stampfer M.J., and Willett W.C. (1996). Vegetable, fruit, and cereal fiber intake and risk of coronary heart disease among men. *Journal of the American Medical Association* 275:447–51.

Roberts M.B., Stringer C.B., and Parfitt S.A. (1994). A hominid tibia from

middle Pleistocene sediments at Box-
grove, UK. *Nature* 369:311–13.

Rossetti L., Rothman D.L., DeFranzo
R.A., and Shulman G.I. (1989). Effect
of dietary protein on in vitro insulin
action and liver glycogen repletion.
American Journal of Physiology 257:
E212–19.

Rothman K.J., Moore L.L., Singer M.R.,
Nguyen U.-S.D.T., Manning S., and
Milunsky A. (1995). Teratogenicity
of high vitamin A intake. *New England Journal of Medicine* 333:1369–
73.

Ruff C.B., Trinkhaus E., Walker A.,
and Larsen C.S. (1993). Postcranial
robusticity in Homo. 1: Temporal
trends and mechanical interpretation. *American Journal of Physical
Anthropology* 91:21–53.

Schonfeld G., Patsch W., Rudel L.L.,
Nelson C., Epstein M., and Olson
R.D (1982). Effects of dietary cholesterol and fatty acids on plasma lipoproteins. *Journal of Clinical Investigations* 69:1072–80.

Schroeder H.A. (1971). Losses of vitamins and trace minerals resulting
from processing and preservation of
foods. *American Journal of Clinical
Nutrition* 24:562–73.

Sears B. (1993). Essential fatty acids
and dietary endocrinology: a hypothesis for cardiovascular treatment. *Journal of Advances in Medicine* 6:211–24.

Shimamoto T., Komachi Y., Inada H.,
Doi M., Iso H., Sato S., Kitamura A.,
Iida M., Konishi M., Nakanishi N.,
Terao A., Naito Y., and Kojima S.
(1989). Trends for coronary heart
disease and stroke and their risk factors in Japan. *Circulation* 79:503–15.

Sies H., and Krinski N.I. (1995). The
present status of antioxidant vitamins and carotene. *American Journal of Clinical Nutrition* 62:299S–
300S.

Simopoulos A.P. (1991). Omega-3 fatty
acids in health and disease and in

growth and development. *American
Journal of Clinical Nutrition* 54:438–
63.

Sinclair A., and O'Dea K. (1993). The
significance of arachidonic acid in
hunter-gatherer diets: implications
for the contemporary Western diet.
Journal of Food Lipids 1:143–57.

Sinclair A.J., O'Dea K., Dunstan G., Ire-
land P.D., and Niall M. (1987). Effects on plasma lipids and fatty acid
composition of very low fat diet enriched with fish or kangaroo meat.
Lipids 22:523–29.

Smith P., Prokopec M., Pretty G.
(1988). Dentition of a prehistoric
population from Roonka Flat, South
Australia. *Archives of Oceania* 23:31–
36

Steinmetz K.A., and Potter J.D. (1991).
Vegetables, fruit and cancer. I. Epidemiology. *Cancer Causes Control* 2:
325–57.

Swales J.D. (1988). Salt saga continued. Salt has only small importance
in hypertension. *British Medical
Journal* 297:307–8.

Tooby J. and Cosmides L. (1990). The
past explains the present. Emotional
adaptations and the structure of ancestral environments. *Ethology and
Sociobiology* 11:375–424.

Walker A. (1993). Perspectives on the
Nariokotome discovery. In A. Walker and R. Leakey, eds., *The Nariokotome Homo erectus Skeleton* pp.
411–30. Harvard Univ Press,
Cambridge, MA.

Watanabe T., Kondo K., and Oshi M.
(1991). Induction of in vivo differentiation of mouse erythroleukemia
cells by genistein, an inhibitor of tyrosine protein kinases. *Cancer Research* 51:764–68.

Westphal S.A., Gannon M.C., and
Nuttall F.Q. (1990). Metabolic response to glucose ingested with various amounts of protein. *American
Journal of Clinical Nutrition* 52:267–
72.

Whiten A., Byrne R.W., Barton R.A., Waterman P.G., and Henzi S.P. (1991). Dietary and foraging strategies of baboons. *Philosophical Transactions of the Royal Society (London) Series B* 334:187–97.

Willett W.C. (1994). Diet and health: what should we eat? *Science* 264: 532–37.

Wolfe B.M., and Giovannetti P.M. (1991). Short-term effects of substituting protein for carbohydrate in the diets of moderately hypercholesterolemic human subjects. *Metabolism* 40:338–43.

World Cancer Research Fund and American Institute for Cancer Research. (1997). *Nutrition and the Prevention of Cancer: a Global Perspective*, p. 506. American Institute for Cancer Research, Washington, D.C.

Young V.R. (1986). Protein and amino acid metabolism in relation to physical exercise. In M. Winick, ed., *Nutrition and Exercise*, pp. 9–29. John Wiley, New York.

Zhang Y., Talalay P., Cho C.G., and Posner G.H. (1992). A major inducer of anticarcinogenic protective enzymes from broccoli: isolation and elucidation of structure. *Proceedings of the National Academy of Science USA* 89:2399–2403.

13

HUMAN EVOLUTION, LOW BACK PAIN, AND DUAL-LEVEL CONTROL

ROBERT ANDERSON

The Ubiquity of Back Pain

"Oh, my aching back!" is a cry familiar to everyone, and for good reason. From 18% to 26% of the general population in urban industrial societies will be experiencing low back pain (LBP) at any one moment in time (the point prevalence rate). Each year, as much as 45% of a population may experience an attack of LBP (the annual prevalence rate). The lifetime prevalence of LBP in urban industrial societies ranges from 50% to 80%. That is, at one time or another, from half to three-quarters of all men and women are likely to experience at least one episode of back pain severe enough to incapacitate them for up to a month. It is not clear that lifetime prevalence is lower in small-scale or less technologically developed societies, as some have stated (Fahrni, 1976). The evidence on that point is equivocal (Anderson, 1987). Most sufferers recover, but 2% or 3% of those afflicted in any society do not fully recover. They continue to experience back pain as a chronic and more or less incapacitating affliction. Almost all of the LBP that creates these remarkable prevalence rates is mechanical in origin, as will be explained below (Allan and Waddell, 1989, Anderson, 1992; Frymoyer, 1990; Papageorgiou and Rigby, 1991).

Medical Approaches to the Treatment of Low Back Pain

Mechanical back pain results from wear and tear of the spinal joints and associated soft tissues. More than 95% of those afflicted respond well to conservative, nonsurgical care (White & Anderson, 1991). Because the pain-producing lesions of those who suffer from mechanical back pain are highly diverse, no one form of conservative treatment always works well. For this reason, the variety of treatments is enormous, including the prescription of analgesics and muscle relaxants, anesthetic and steroid injections, chiropractic spinal adjustments, acupuncture, physical therapy in various forms (including exercise, hot or cold packs, electrical stimulation, ultrasound, massage, and mobilization), biofeedback, and so on. However, one of the most impressive of the conservative approaches, known as lumbar stabilization or dynamic stabilization, is to train patients how to brace the spine in a neutral position so that it does not move, no matter what the arms, legs and head do, even if one is walking. This technique as it applies to locomotion will be examined and evaluated from an evolutionary perspective below. The fundamental question addressed is not whether stabilization while walking helps in the treatment of LBP and in preventing recurrence, but rather whether lumbar stabilization is advisable as a learned style of locomotion for healthy people based on the assumption that it provides protection against the risk of being stricken with LBP? Prevention rather than cure is the issue.

The Evolutionary Perspective

In evolutionary medicine it is often found that diseases of our time result from, or are exacerbated by, changed conditions. Bodies adapted to the life circumstances of hunters and foragers prove relatively maladapted to the contemporary world (Eaton et al., 1988; Nesse & Williams, 1994; Trevathan, 1995). Back pain offers a salient example. Contemporary urban-industrial peoples have become very sedentary. Growing children compromise their spinal health by sitting for long hours each day in school (Keegan, 1953). Many graduate into occupations in which they sit or move very little most of the time. Even people in more active occupations tend to become quite sedentary during evening hours and weekends (Anderson, 1992). Excessive sitting and poor sitting posture have been implicated as contributors to low back pain (lumbago) and pain radiating into the leg (sciatica) that is so common in complex societies (Bruggeman et al., 1993; Fahrni, 1976).

In addition to unhealthy habits of static posture (standing, sitting, and sleeping), the ways in which people lift and carry or walk and run are also implicated in pain of spinal origin (Farfan, 1973; Howorth, 1946; White & Panjabi, 1978). Each of these behaviors impinges on the spine in unique ways. Because static postures, lifting movements, and habits of locomotion constitute three substantially different ways in which the spine can be injured, and because the cervical, thoracic, and lumbar spines respond differently to stress and injury, I orient my analysis primarily to the relationship between bipedal locomotion (walking) and low back pain that is mechanical in nature (as distinct from disease processes such as rheumatoid arthritis or ankylosing spondylitis).

From a methodological point of view, the problem of LBP is couched here in terms of the paradigm of dual-level control. The concept of dual-level control offers a way to frame the question of what an evolutionary approach can tell us about the relationships between the style, manner, or habitus of walking (the locomotion culture system), the biomechanics and style of walking (the individual body system), and the pain or pathology of lumbago and sciatica (the neurophysiologic system).

When Did Back Pain Evolve as a Human Problem?

It would appear that when relatively sedentary urban societies emerged, beginning around 5000 years ago, many people began to walk less and to sit or squat more. Within the last century, this trend in lifestyle peaked, culminating in the "couch potatoes" of our time with their frequent complaints of back pain. However, the evolutionary record suggests a far more complicated scenario. It can be argued that a high prevalence of LBP characterized humans from when they first made the shift to bipedal stance and walking more than 4 million years ago. The hypertrophic lipping of a lumbar vertebra of Lucy (*Australopithecus afarensis*), an early upright hominid, suggests that if she was not experiencing LBP during her short life, she might have had she lived longer. Lucy lived more than 3 million years ago (Johanson & Coppens, 1976; Johanson & Edey, 1982). Paleopathologists consistently document degenerative changes in prehistoric spines, including bony spur formations around intervertebral disks (Kennedy, 1984; Merbs, 1983; Ortner & Putschar, 1981).

Experimental evidence adds further credence to the idea that LBP affected our early ancestors. Cassidy and his associates (Cassidy et al., 1988) created 21 bipedal rats by means of forelimb amputation. These experimental rats were reared with a control group of 19 unoperated, quadrupedal rats. The bipedal rats became quite proficient at walking upright. However, in contrast to the control group, the lumbar vertebrae of the bipedal rats developed measurable wedging, neural canal narrowing, and, in one-fourth of the animals, large disk herniations in the lumbosacral area. The investigators conclude that these degenerative changes can be attributed to the effects of gravity under the biomechanical circumstances of upright posture and bipedal locomotion. It was estimated that the shearing force acting on the last lumbar intervertebral disk of a bipedal specimen was 2–16 times that occurring in a normal rat (Cassidy et al., 1988). Because the spine of a genetically quadrupedal rat is very different from that of a human being, one cannot merely extrapolate from these findings to spinal degeneration in humans, but the experiment is suggestive.

The logic of this argument was advanced by Krogman many years earlier when he drew attention to changes in the vertebral column inherent in transforming an apelike spine into a human spine adapted to verticality (Krogman, 1951). The ancestral primate spine incorporated a C-shaped curve and the biomechanical principle of an arch resting on two supports (fore- and hindlimbs), the way engineers design a bridge. This conformation of the spine characterizes the fetus and neonate. Several months after birth, as the infant learns to hold its head up, a lordodic curve is introduced into the cervical region in all anthropoids (O'Connor, 1992). Around the end of the first year of life the infant learns to stand, and a second lordodic curve appears in the lumbar spine that is unique to hominids. These changes leave the original C-shape persistent in the remaining thoracic and sacral-coccygeal areas. Krogman referred to the resultant conformation of the spine as an S-curve.

With an S-shaped spine and associated changes in the appendicular skeleton it became possible for humans to stand erect and move bipedally. Many anthropologists have speculated about the benefits conferred by these changes in posture and movement (e.g., Park, 1996). Advantages are thought to include becoming able to pick up and carry objects, being positioned to see better with the head raised higher above the ground, and not the least, acquiring the capability for technological growth permitted by hand–eye coordination directed by larger, more complex brains. Although the benefits were enormous, the upright spine also resulted in identifiable health problems that appeared as adverse side effects.

Krogman (1951) identified these costs of bipedalism as susceptibility to sinus infections, malocclusion of the teeth, impaction of wisdom teeth, susceptibility to neck and mid-back pain, circulatory problems (including hemorrhoids and varicose veins), aching feet due to fallen arches, bunions, and corns, pain and tearing of the perineum experienced by women during parturition, and cranial deformation in newborns, sometimes with hypoxia or nerve damage. Krogman noted that the lumbar spine and sacroiliac joints were biomechanically compromised in ways that could cause lumbago and sciatica.

Dual-Level Control

To evaluate inherent anatomic and physiologic risks for LBP in living populations while concurrently identifying cultural praxis that increases or reduces these risks, I invoke the concept of dual-level control described by Polanyi (1968; see also Anderson, 1996). By this means I hope to evaluate the complex interactions of two simultaneous, interactive, but different processes, biological evolution and cultural evolution.

The concept of dual-level control identifies hierarchies of systems separated by a series of boundary conditions. As Polanyi (1968:1310) explains, "The theory of boundary conditions recognizes the higher levels of life as forming a hierarchy, each level of which relies for its workings on the principles of the levels below it, even while it itself is irreducible to these lower principles." In the section below, I confine the analysis to the boundaries between the locomotion culture, the individual body, and the neurophysiologic systems (figure 13.1).

The Individual Body System

Walking is a biomechanical process that involves coordinated movements of the entire body. The way people walk is constrained by genotype, the product

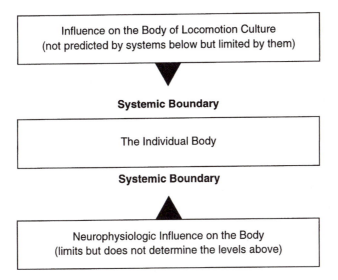

Figure 13.1. The body under dual-level control.

of an evolutionary heritage, but within genetic constraints is free to vary by culture (resulting in the phenotype of walking). The individual body system is taken here as subject to dual-level analysis. That is, the body is examined as subject to constraints from below (the neurophysiologic system) and impositions (controls) from above (the locomotion culture system).

The Neurophysiologic System: Constraints of Homeostasis

Guided by the paradigm of dual-level control, the lumbar spine in motion can be thought of as an organ system that functions to avoid injury and pain by means of homeostatic mechanisms that perpetuate pain-free, efficient bipedalism. That is, in healthy walking, no damage is done to biomechanical structures and no pain signals are transmitted to the central nervous system. Conversely, this homeostasis can be overwhelmed by stress, strain, and trauma, leading to the experience of pain and dysfunction.

Breakdown processes involve deterioration of vertebral joints comprising bone, cartilage, tendons, and ligaments as well as other soft tissues such as muscles, fascia, vasculature, and elements of the nervous system. From a biomechanical perspective, breakdown usually takes place in the three joints (an intervertebral disk and two zygapophyseal joints) that connect each lumbar vertebra to the adjacent vertebra either above or below. The key to understanding how low back pain and spinal nerve root (especially sciatic nerve) inflammation occur is to realize that instability between vertebral segments is often the final common pathway of stress, strain, and trauma.

The main weight-bearing joint is a partially movable, nonsynovial intervertebral disk composed of a fibrous ring (the anulus fibrosus) and a pulpy center (the nucleus pulposus). Under mechanical stress the disk may develop circumferential and radial tears that eventuate in herniation of the nucleus into and through the anulus. This internal disruption of the disk renders it thin and unstable (Cassidy & Wedge, 1988).

The two zygapophyseal joints posterior and lateral to each disk are synovial and thus subject to inflammation (synovitis), capsular fibrosis, and adhesions conducive to hypomobility. However, with ongoing microtrauma in the form of small abnormal movements secondary to disk disruption and loss of disk height, these joints may lose their tightness to become lax, hypermobile, and unstable (Cassidy & Wedge, 1988; Grieve, 1981).

Concomitant changes include muscle pathology, particularly in the multifidus muscles intrinsic to the spinal column, resulting in changes in muscle spindle activity and tension, the stimulation of nociceptive nerves emanating from the area, the release of histamine and leukotriene B-4 by mast cells and neutrophils, the synthesis of pain-producing peptides in the dorsal root ganglia, including substance P and vasoactive intestinal polypeptide (VIP), and vasoconstriction in muscles inducing painful muscle spasms—all contributing further to facet strain, anular tearing, and intervertebral relationships that are unstable (Kirkaldy-Willis, 1988). When the normal integrity of the lumbar spine (homeostasis) is overwhelmed by breakdown processes, the intervertebral complex becomes unstable and pain ensues.

The Locomotion Culture Level: Control by Tradition and Design

The need to preserve the anatomic and physiologic integrity of the neurophysiologic system constrains but does not determine or control what happens above the neurophysiologic boundary in the individual body and locomotion culture systems. The analysis is not positivist or reductionist in the sense that it assumes that human posture and movement are neurophysiologically determined.

Conditions at the neurophysiologic level impose constraints. Human movement is limited by physiologic ranges of motion, muscular strength, and ligamentous elasticity. However, within such constraints, control at higher levels includes learned behaviors deriving from cultural traditions. These include bodily habits learned by unconscious imitation ("osmosis") as modified by the instructions of parents, the role-modeling of peers, the training of dance or martial arts instructors, the coaching of physical education teachers, etc. As a consequence, people in different societies display different styles of walking and body habitus (Bartenieff & Lewis, 1980; Birdwhistell, 1970; Bourdieu, 1984; Desjarlais, 1992; Mauss, 1973). Those suffering from back pain who resort to Western medicine and chiropractic encounter additional mechanisms of control insofar as they are taught posture and movement practices believed to minimize pain, promote healing, and prevent further trauma and deterioration.

One of the most influential techniques for managing LBP is the technique of lumbar stabilization. This technique was developed by physical therapists and orthopedists based on experimentation and clinical experience. It became apparent in their work with LBP patients that microtrauma resulting from the small aberrant movements of unstable spinal joints causes pain. Also, a healed lumbar spine tends to become reinjured because of these small instabilities. Clinical experience (in the absence of clinical trials) encourages the belief that people with LBP will get better if they can be taught to maintain a rigid spine, particularly by learning to brace the lumbar spine and the pelvic girdle. To achieve this kind of bracing requires prescribed exercises to strengthen spinal and abdominal muscles, to acquire stronger leg muscles (permitting the body's center of gravity to be lowered at the knees), and to improve flexibility of the hip joint (achieved by means of stretching exercises). If these coordinated changes can be achieved, the lumbar spine will be protected against harmful micromovements of the spine during walking, twisting, lifting, or carrying. Clinical experience teaches that this kind of change in static and dynamic postures, including the way one walks, permits painful backs to heal and stay well (Saliba & Johnson, 1991).

To achieve this kind of internal, anatomic bracing requires that one learn to move the arms and legs independently of the trunk of the body. The trunk is held rigid as a single unit, with the shoulders and pelvis kept in unshakable alignment with each other. In this way, the spine is braced to avoid lateral bending, twisting, flexion, and extension in the lumbar spine (the latter achieved by bending at the hip joints). In place of twisting, stabilization requires that one learn to turn to the left or the right by rotating the trunk of the body as a single unit, moving the feet instead of the spine. Stabilization includes an additional advantage in that it shifts the practitioner–patient relationship from doctor controlled to patient controlled. The clinician, having made the diagnosis and devised the treatment plan, needs subsequently only to function as a teacher or coach. The person in pain is put in control of his or her own health-care program (White 1991; White & Anderson, 1991).

A Debatable Scenario

The orthopedic and chiropractic literature is replete with descriptions of the practice of stabilization as a recommended approach for the ameliorization or cure of LBP (White & Anderson, 1991). Stabilization is also taught as a preventive measure for those who have recovered from back pain, so that injury does not reoccur (White, 1983). Ignored, however, is a much larger issue: Should people who have never suffered with LBP be taught lumbar stabilization as a preventive measure? Should we all learn to adopt a stabilized spine as our customary style of posture and movement, including walking?

Prevalence data alone would seem to affirm the advisability of such a program. The fact that most people at one time or another will experience LBP supports the assumption that walking with a stabilized spine should be taught to people in good spinal health. Indeed, it is widely noted among health care providers that every adult should regard himself or herself as being at risk for LBP, and it is therefore widely recommended that to reduce that risk everyone should adopt stabilization as a lifelong habit. Because postural habits are difficult to change, and early injury can result in long-term consequences, this widely held belief includes a conviction that even children should be taught to maintain a stabilized spine (Kunse, 1991; White, 1983). But is this feasible?

A reconstruction of cultural evolution and movement styles supports the belief that, indeed, it is feasible for adults to move through the activities of daily life with a solid trunk and, as a correlate, for their children to be brought up to achieve that mode of ambulation. I base this conclusion as an extrapolation from the choreometric research of Lomax (Lomax, 1968, 1976).

Lomax identified postural habits based on films documenting dance and movement in a worldwide sample of 21 cultures. He hypothesized that dances reflect the movement styles of work, noting that the postures of work tend to be the same as those of dance. From a survey of a variety of cultures he concluded that "work and dance movement styles are virtually identical" (Lomax 1968: 225). In other words, his findings are relevant to the activities of daily life as well as to the special circumstances of dancing.

Lomax was especially intrigued with the widespread custom of performing dance and work movements with the trunk of the body consistently held in what Western medicine would call a stabilized spine: "Inspection of more than 200 films from all five continents convinced us that in [technologically] simple cultures most individuals tend to adopt the same body attitude no matter what the activity and in spite of differences of age and sex" (Lomax, 1968: 235). Lomax found that a stabilized spine, which he calls a "single-unit torso," was characteristic of the aboriginal inhabitants of the entire circumpacific region, where it was presumably associated in antiquity with simple stone-age technologies and hunting, gathering, and fishing economies (Lomax, 1968, 1976; Lomax & Paulay, 1984).

As a finding incidental to broader ethnographic interests, Reichel-Dolmatoff described the single-unit biomechanical system as he witnessed it among the Kogi Indians of Colombia, South America, before 1950:

> But the trunk always remains rigidly erect, the chin pressed against the neck, and the shoulders thrown back. This posture is not even changed when walking. Advancing with very short steps, but with the legs somewhat apart and slightly bent at the knees, he walks uphill and down, leaping, jumping, always with hard, terse movements, as if seeking balance for the weight of the trunk, which does not take part in the labor of the legs by any action. (quoted in Devine, 1985: 553)

The claim by Lomax that body habits in dance correlate with body practices in daily life has been illustrated only once in ethnographic research. During two field expeditions in the 1960s, Jablonko and Jablonko filmed the way children and adults move among the Maring of Highland New Guinea. They confirmed that Maring adults tend to maintain a solid trunk, to turn without twisting, and to move their arms and legs with minimal bending of the spine in daily life as well as when dancing. In the film sequence of a man working at clearing a field and planting with a digging stick, they observed, "As he works the man uses his trunk as a solid unit, and moves all the joints of his arms simultaneously" (Jablonko & Jablonko, 1968). Jablonko (1968:123) also notes, "The Maring treatment of the body, as seen in the dance, is to hold the trunk as one unit and leave it unchanged as the arms manipulate weapons or drums. The legs move the whole body through space."

Those who teach stabilization are convinced, as am I from my own clinical experience, that stabilization offers an effective way to rehabilitate those who suffer from LBP. Assumptions based on clinical impressions are not unbiased, however, and need to be tested in clinical trials. The anecdotal accounts of health care providers can be misleading. Unfortunately, no well-designed trials have been carried out, although some have evaluated conservative care programs that include stabilization along with education, exercises, and other methods. Five such trials have been published, but they are of uncertain validity (Lahad et al., 1994). The five together enrolled fewer than 500 subjects, none of whom were followed for more than 2.5 years. Nonetheless, if stabilization consistently produces a large beneficial effect, even these small trials might have demonstrated it. Instead, they provide only minimal support for the use of clinical strategies that include training in the techniques of stabilization.

Research in the non-Western world is equally skimpy. Only one investigation is relevant. It was carried out in New Guinea, where posture and movement would be expected to resemble that of the Maring. Clunie, a physician, examined a convenience sample of 109 adults for the presence of musculoskeletal disorders. Quite unexpectedly, he found that in his sample of 83 men and 26 women of all ages, 75% were experiencing mechanical back pain at the time he saw them (Clunie, 1989; Clunie et al., n.d.).

Taken at face value, this finding would seem to refute the claim that solid trunk body mechanics protect the spine. However, the data have many limitations. The sample was recruited by encouraging individuals above the age of 15 years to come for an examination, especially if they were experiencing back pain. The published and unpublished findings include no information about the movement habits of these Papuans other than to note that they "would take it on themselves to carry vast loads on their backs—which would never cease to amaze" (G. Clunie, personal communication, 1990). The report does not include any information on how the assessments were made or the extent to which mild conditions were included in the total. Even so, it leaves a nagging suspicion that perhaps stabilization does not confer the health benefits we expected to see.

An Alternative Scenario

The Neurophysiologic System: Gracovetsky's Theory of Optimal Control

In applying the principle of dual-level control, we must return to the system of neurophysiologic constraint to see what may be missing in the biomechanical component of the stabilization hypothesis. To this end, let us now incorporate the evolutionary reconstruction of Gracovetsky (1988:4), who points out that, "The structure of the [human] vertebrate spine must have been advantageous in evolution; its description by a wide variety of models is indicative of our failure to grasp the basic principles underlying the need for its very existence."

Gracovetsky offers a model based on a theory of optimal control of the spine system. By optimal control he assumes that, in evolutionary terms, the spine must have evolved to make the most economical use of the energy available within the system. He sees energy efficiency as the single most important criterion for evolutionary success. This requires one to identify, "the conditions under which the normal function of the musculoskeletal system is governed by both the task to be accomplished and the need to do so with the minimum amount of energy" (Gracovetsky, 1988:6). This hypothesis is consistent with that of McHenry (1982:163), who notes that, "The origin of hominid bipedalism may not have involved extraordinary events, but could have arisen as *an energetically efficient mode of terrestrial locomotion* for a small-bodied hominoid moving between arboreal feeding sites" (emphasis added; see also Napier, 1967; Napier & Napier, 1985).

The most basic task to be accomplished, Gracovetsky argues, would be locomotion (walking, jogging, running). Locomotion with the minimum amount of energy cannot be achieved by bracing the trunk and relying primarily upon movement of the legs, as stabilization theory would require. On the contrary, the basic energy of locomotion derives from spinal torsion, converting the spine to what Gracovetsky has termed "the spinal engine." Yet, in experimental work, Farfan has demonstrated that excessive torsion of the lumbar spine is a major cause of disk degeneration and instability (Farfan, 1973).

In normal walking, the spine moves by coupling three kinds of motion: flexion-extension, lateral bending, and axial rotation (torsion). The most basic movement in walking at the level of the lumbar spine is the coupling of flexion-extension with torsion in the presence of lordosis. Gracovetsky builds on this observation to argue that this coupling converts the muscular power of lateral flexors of the spine into an axial torque that moves the pelvis. "This hypothesis implies that the spine drives the pelvis, with the legs following and amplifying the pelvic motion, and not the other way round" (Gracovetsky, 1988:276).

Based on biomechanical evidence, Gracovetsky argues for a model of the spine that is diametrically opposed to that implicit in stabilization theory. Rather than trunk motions as products of movement of the legs, motions that can and should be dampened or eliminated by stabilization, the spinal engine model requires movement of the spine, particularly at the level of the lumbar lordosis, to move the pelvis, and by extension, the legs.

The efficiency of the spinal engine is improved by an energy storage system, so that movement of the pelvis is not solely muscle driven. Kinetic energy initiated by muscle action is stored for the next cycle of movement in the elastic elements of the muscles and spinal ligaments, particularly in the massive lumbodorsal fascia. Hip extensors raise the center of gravity of the trunk. On downward motion, the spine bends laterally, converting the potential energy of lifting into kinetic energy by stretching the posterior ligamentous system. This results in stiffening the intervertebral joints because it has the effect of bracing the facet joints. Bracing of the intervertebral joints permits the large forces of torsion to contribute to locomotion without damaging the disk and facets.

In terms of energy efficiency, Gracovetsky's hypothesis is that the energy of muscle contractions to move the spine and pelvis is not wasted. Most of this energy is stored as elastic potential energy in the lumbodorsal fascia. With every subsequent step in walking, this energy is recovered to rotate the pelvis by means of coupled motion. Elasticity in the facets and disks also stores and releases kinetic energy as the joints rotate and derotate. "From a clinical and rehabilitation point of view, the theory highlights the fundamental importance of axial torsion in the motion of the spine, coupled with a critically timed axial compression" (Gracovetsky, 1988:360).

In illustration of his hypothesis, Gracovetsky was able to demonstrate with the cooperation of a man born without legs that it is possible to walk by striding on the ischial tuberosities alone. The legless man "walked" with a pattern of movment that was quite similar to ordinary walking, relying on bending and torquing of the spine. Gracovetsky (1988:265) reports: "the spinal engine theory explains these facts in a natural way. The legs serve to amplify the motion of the spine; when they are absent, the motion of the trunk becomes more dynamic in order to maintain a reasonable forward velocity, but there is no need to change the basic pattern of motion." In short, to avoid stress and strain in the lumbar spine, with resultant pain and dysfunction, it would seem that a constraint of the biomechanical system is that walking should include (not eliminate) certain kinds of movements in the trunk of the body.

The Cultural System: Walking with a Pelvic Swing

Gracovetsky's theory of locomotion has implications for the maintenance of a healthy spine (see also Bartenieff & Lewis, 1980). It would support a conclusion contrary to that of stabilization theory, namely, that a healthy spine requires people to walk with repetitive movement of the lumbar spine to ensure that it functions at a maximum of energy efficiency. In addition, and consistent with Gracovetsky's theory of optimal control, the health of the spine should benefit from movement of the very joints that stabilization attempts to keep immobile, because nutrient-bearing fluid that is essential for joint and bone health is "pumped" when the joint is exercised. Immobilized joints tend to lose fluid content, putting them at risk for degeneration and instability (Adams & Hutton, 1983; Aegerter & Kirkpatrick, 1968; Kapandji, 1974).

Clinical Implications

At issue is the question of how one should be taught to walk as a child and as an adult being taught protective spinal hygiene. How is one to satisfy the constraints of the neurophysiologic system and thereby minimize the risk of succumbing to mechanical LBP, a risk inherent in the biomechanics of the lumbar spine as it evolved anatomically from a quadrupedal ancestry and later worsened by the evolution of cultural practices of extreme sedentation? On the principle of dual-level control, the analysis is localized at the boundaries on either side of the individual body system and the spine that is central to it.

The neurophysiologic system can be described as a highly complex homeostatic mechanism that preserves and optimizes the anatomic and biochemical integrity of spinal joints. That is, this lower system sets constraints on how the spine can function as an individual body system. However, the lower level does not control higher levels. Within the constraints imposed by the neurophysiologic system, higher levels permit a wide range of variation (i.e., control in the locomotion culture system). As long as patterns of body attitude do not exceed anatomic, physiologic, and biomechanical limits, posture and movement can differ from one society to another (or from one individual to another).

Dual-level control identifies a dynamic, interactive process. Feedback loops from the cultural system can influence the locomotion practices of the body system, which in turn impact on homeostasis in the neurophysiologic system, as when injurious patterns of movement cause degeneration and instability of the spinal joints. Conversely, a spine that has become unhealthy provides feedback across higher boundaries to require that one assert constraints in the cultural system to limit how one stands and moves in the biomechanical system.

In clinical practice, one encounters, as a widespread mechanical cause of pain and dysfunction, disk and facet instabilities at one or several vertebral levels. To permit healing and to prevent injury that can be expected to recur if established patterns of posture and movement are resumed, many clinicians train their patients in stabilization methods. The evidence in support of this approach is mainly anecdotal but includes some weak support from clinical trials. Clearly, this now common practice should be investigated systematically by means of randomized clinical trials.

Untested and unexplored, however, is a seemingly reasonable extension of stabilization to individuals with healthy spines as a protective measure. Because the evolutionary past of all humans contributed spines at risk for instability and pain, and because back pain is prevalent in all societies for which we have data, it would seem logical to advocate that everyone be trained to move in a stabilized manner. However, there is no research evidence to support this concept other than the the the work of Lomax (1976), which suggests that our prehistoric ancestors may have moved with solid trunks.

Further, this recommendation for stabilization flies in the face of the equally impressive finding that throughout Africa, South Asia, and Southeast Asia, gardeners and horticulturalists for millennia have walked (and danced) with marked undulations, rotations, and bendings of the lumbar spine and pelvis (Lomax 1976). Whether these styles of walking are associated with greater prevalences of back pain is not known. The epidemiology of back pain is poorly documented outside of North America and Western Europe.

One other investigation of culture and spinal movement is relevant here. In a study carried out by researchers in the Department of Physical Medicine and Rehabilitation at Istanbul University (Aksoy et al., 1993), 100 healthy Muslim men aged 18–31 who had prayed five times daily for at least 2 years were compared with a control group comparable in every way except that they did not regularly worship. When Muslim men pray, they subject their spines to vigorous lumbar extension and flexion movements, quite contrary to what occurs in the stabilized backs of orthopedics or the single-unit trunks of Lomax. In this research, a LBP history was taken for every subject. The two groups demonstrated no statistically significant difference in the prevalence of LBP. The dramatic movements of Islamic praying seem to have had no effect either in favoring the onset of back pain or in its prevention (Aksoy et al., 1993).Consistent with these findings by Lomax and the researchers from Istanbul, Gracovetsky's theory of optimal control provides a reasonable basis for assuming that healthy walking requires vigorous movement of the lumbar spine, extension and flexion to be sure, but also including associated bending and twisting assumed by stabilization experts to cause spinal degeneration and LBP.

In the locomotion culture system, Gracovetsky's engineering concept of the spinal engine has not been taken by clinicians as an invitation to train their patients to walk with a swinging pelvis. In the experience of clinicians, impaired spines do not heal well unless the movement of unstable joints is minimized or eliminated. However, the logic of the spinal engine hypothesis suggests that the recommendation should not be extended to people with healthy spines, because spines will stay healthy if the way one walks (i.e., control in the locomotion culture system) involves in a swinging gait.

The evidence available at this time therefore supports a policy of recommending that people with apparently healthy spines be encouraged to walk with with ample movement of the pelvis and spine. Walking this way, however, should always be associated with the practice of spinal stabilization when lifting and carrying or when bending and twisting. Also, in static postures of sitting and standing, the maintenance of a lumbar lordosis is recommended, including the use of some form of lumbar support device (White, 1983, 1991). Gracovetsky's spinal engine is not relevant to any of these nonwalking activities. I know of no studies that implicate walking in the etiology of spinal degeneration if the legs and feet are healthy and if proper shoes are worn (Mennell, 1960; White, 1983). Many studies implicate lifting and carrying, stationary bending and twisting, and prolonged sitting or standing in bent-over postures (Kelsey, 1982).

These recommendations are proposed on an interim basis. It is a truism that the most important task of science is not so much to find answers as to identify the questions that should be asked. The question of how healthy people should walk to protect themselves against the onset of LBP requires additional research. In comparative research, investigators should search aggressively for whatever correlations may exist between styles of walking and prevalence rates of LBP. With due regard for ethical issues, experimental programs should be designed to prospectively evaluate effects on the spinal health of healthy subjects when stabilized walking is contrasted with swinging gaits. Such comparative and experimental research will inevitably be multidisciplinary, associating clinical specialists (such as orthopedists, chiropractors, or physical therapists) with field ethnographers and epidemiologists in joint enterprises that no one kind of specialist can effectively carry out alone. Until that research is completed, the findings of evolutionary medicine support the recommendations proposed here.

Acknowledgments
I am grateful to the following for comments, criticisms, corrections, and suggestions relating to earlier drafts of this chapter: Lawrence M. Elson, Serge Gracovetsky, Scott Haldeman, Allison Jablonko, and Arthur H. White.

References

Adams, M. A., and Hutton, W. C. (1983) The effect of posture on the fluid content of lumbar intervertebral discs. *Spine* 8: 665–671.

Aegerter, E., and Kirkpatrick, J. A., Jr. (1968) *Orthopedic Diseases: Physiology, Pathology, Radiology*, 3rd ed. Philadelphia: W. B. Saunders Co.

Aksoy, C., Kozakçioglu, M., Kavuncu, V., Danesghal, S., and Diniz, F. (1993) Effects of Islamic worship on lumbar movements and back pains. In E. Ernst, M. I. V. Jayson, M. H. Pope, and R. W. Porter, eds., *Advances in Idiopathic Low Back Pain*. Vienna: Blackwell-MZV, pp. 241–244.

Allan, D. B., and Waddell, G. (1989) An historical perspective on low backpain and disability. *Acta Ortho-*

paedica Scandinavica, (suppl. 234) 40: 1–23.

Anderson, R. (1987) Investigating back pain: The implications of two village studies in South Asia. In P. Hockings, ed., *Essays in Honor of David G. Mandelbaum.* Berlin: Mouton de Gruyter, pp. 385–396.

Anderson, R. (1992) The back pain of bus drivers: Prevalence in an urban area of California. *Spine* 17: 1481–1488.

Anderson, R. (1996) *Magic, Science and Health: The Aims and Achievements of Medical Anthropology.* Fort Worth, TX: Harcourt Brace.

Bartenieff, I., and Lewis, D. (1980) *Body Movement: Coping with the Environment.* New York: Gordon and Breach Science Publishers.

Birdwhistell, R. L. (1970) *Kinesics and Context: Essays on Body Motion and Communication.* Philadelphia: University of Pennsylvania Press.

Bourdieu, P. (1984) *Distinction: A Social Critique of the Judgement of Taste.* London: Routledge.

Bruggeman, B., Bruggeman, J., and Kooke, H. J. (1993) Rugklachten, natuuren cultuur lordose of kyfose [with English summary]. *Fysio/Therapie 2000* 3: 26–28.

Cassidy, J. D., and Wedge, J. H. (1988) The epidemiology and natural history of low back pain and spinal degeneration. In W. H. Kirkaldy-Willis, ed., *Managing Low Back Pain*, 2nd ed. New York: Churchill Livingstone, pp. 3–14.

Cassidy, J. D., Yong-Hing, K., Kirkaldy-Willis, W. H., and Wilkinson, A. A.(1988) A study of the effects of bipedism and upright posture on the lumbosacral spine and paravertebral muscles of the Wistar rat. *Spine* 13: 301–307.

Clunie, G. (1989) Swelling, spondylitis, and spears. *Annals of the Rheumatic Diseases* 48: 692–695.

Clunie, G., Koki, G., and Prasad, M. L. (n.d.) HLA B27, arthritis and spondylysis in an isolated community in Papua New Guinea. Unpublished ms.

Desjarlais, R. T. (1992) *The Body and Emotion: The Aesthetics of Illness and Healing in the Nepal Himalaya.* Philadelphia: University of Pennsylvania Press.

Devine, J. (1985) The versatility of human locomotion. *American Anthropologist* 87: 550–570.

Eaton, S. B., Shostak, M., and Konner, M. (1988) *The Paleolithic Prescription: A Program of Diet & Exercise and a Design for Living.* New York: Harper & Row.

Fahrni, W. H. (1976) *Backache and Primal Posture.* Vancouver, BC: Musqueam Publishers.

Farfan, H. F. (1973) *Mechanical Disorders of the Low Back.* Philadelphia: Lea & Febiger.

Frymoyer, J. W. (1990) Magnitude of the problem. In J. N. Weinstein and S. W. Wiesel, eds., *The Lumbar Spine.* Philadelphia: W. B. Saunders Company.

Gracovetsky, S. (1988) *The Spinal Engine.* Vienna: Springer-Verlag

Grieve, G. P. (1981) *Common Vertebral Joint Problems.* New York: Churchill Livingstone.

Howorth, B. (1946) Dynamic posture. *Journal of the American Medical Association* 131: 1398–1404.

Jablonko, A. (1968) Dance and daily activities among the Maring people of New Guinea: A cinematographic analysis of body movement style. Ph.D. Dissertation, Columbia University.

Jablonko, A., and Jablonko, M. (1968) *Maring in Motion: A Choreometric Analysis of Movement Among a New Guinea People.* 16 mm film, color, 18 min. Distributor: Pennsylvania State University, University Park.

Johanson, D., and Coppens, Y. (1976) A preliminary anatomical diagnosis of the first Plio-Pleistocene hominid discoveries in the Central Afar, Ethiopia. *American Journal of Physical Anthropology* 45: 217–134.

Johanson, D., and Edey, M. (1982) *Lucy: The Beginnings of Humankind.* New York: Warner Books.

Kapandji, I. A. (1974) *The Physiology of the Joints*, 2nd ed., vol. 3. New York: Churchill Livingstone.

Keegan, J. J. (1953) Alterations of the lumbar curve related to posture and seating. *Journal of Bone and Joint Surgery* 35: 589–609.

Kelsey, J. L. (1982) *Epidemiology of Musculoskeletal Disorders.* New York: Oxford University Press.

Kennedy, K. A. R. (1984) Growth, nutrition, and pathology in changing paleodemographic settings in South Asia. In M. N. Cohen and G. J. Armelagos, eds., *Paleopathology at the Origins of Agriculture.* New York: Academic Press, pp. 169–192.

Kirkaldy-Willis, W. H. (1988) A comprehensive outline of treatment. In W. H. Kirkaldy-Willis, ed., *Managing Low Back Pain*, 2nd ed. New York: Churchill Livingstone, pp. 247–264.

Krogman, W. M. (1951) The scars of human evolution. *Scientific American* 185: 54–57.

Kunse, P. V. (1991) Back school for children. In A. H. White and R. Anderson, eds., *Conservative Care of Low Back Pain.* Baltimore, MD: Williams & Wilkins, pp. 60–61.

Lahad, A., Malter, A. D., Berg, A. O., and Deyo, R. A. (1994) The effectiveness of four interventions for the prevention of low back pain. *Journal of the American Medical Association* 272: 1286–1291.

Lomax, A. (1968) *Folk Song Style and Culture.* Publication no. 88. Washington, DC: American Association for the Advancement of Science.

Lomax, A. (1976) *Dance and Human History: Movement Style and Culture, Number 1.* 16 mm film, color, 40 min. Distributor: University of California Extension Media Center, Berkeley.

Lomax, A., and Paulay, F. (1984) *The Longest Trail.* 16 mm film, color, 58 min. Distributor: University of California Extension Media Center, Berkeley.

McHenry, H. M. (1982) The pattern of human evolution: Studies on bipedalism, mastication, and encephalization. *Annual Review of Anthropology* 11: 151–173.

Mauss, M. (1973 [1934]) Techniques of the body. *Economy and Society* 2: 70–88. [First published in French.]

Mennell, J. McM. 1960 *Back Pain: Diagnosis and Treatment Using Manipulative Techniques.* Boston: Little, Brown and Co.

Merbs, C. F. (1983) *Patterns of Activity-induced Pathology in a Canadian Inuit Population.* Mercury Series, Archaeological Survey of Canada Paper no 119. Ottawa: National Museum of Man.

Napier, J. R. (1967) The antiquity of human walking. *Scientific American* 216: 56–66.

Napier, J. R., and Napier, P. H. (1985) *The Natural History of the Primates.* Cambridge, MA: MIT Press.

Ness, R. M., and Williams, G. C. (1994) *Why We Get Sick: The New Science of Darwinian Medicine.* New York: Vintage Books.

O'Connor, B. L. (1992) The vertebral columns of monkeys, apes, and humans. In A. A. White, III and S. L. Gordon, eds., *American Academy of Orthopaedic Surgeons Symposium on Idiopathic Low Back Pain.* St. Louis, MO: C. V. Mosby, pp. 78–96.

Ortner, D. J., and Putschar W. G. J. (1981) *Identification of Pathological Conditions in Human Skeletal Remains.* Washington, DC: Smithsonian Institution Press.

Park, A.P. (1996) Biological Anthropology, Mountain View, CA: Mayfield Publishing Co.

Papageorgiou, A. C., and Rigby, A. S. (1991) Low back pain. *British Journal of Rheumatology* 30: 208–210.

Polanyi, M. (1968) Life's irreducible structure. *Science* 160: 1308–1312.

Saliba, V. L., and Johnson, G. S. (1991) Lumbar protective mechanism. In A. H. White and R. Anderson, eds., *Conservative Care of Low Back Pain.* Baltimore, MD: Williams & Wilkins, pp. 112–119.

Trevathan, W. (1995) Commentary: Whither our subjects—Whither our species? *Anthropology Newsletter* 36: 26.

White, A. H. (1983) *Back School and Other Conservative Approaches to Low Back Pain.* St. Louis, MO: C. V. Mosby.

White, A. H. (1991) Stabilization of the lumbar spine. In A. H. White and R. Anderson, eds., *Conservative Care of Low Back Pain.* Baltimore, MD: Williams & Wilkins, pp. 106–111.

White, A. H., and Anderson, R. (1991) The challenge of conservative care. In A. H. White and R. Anderson, eds., *Conservative Care of Low Back Pain.* Baltimore, MD: Williams & Wilkins, pp. 427–433.

White, A. A., III, and Panjabi, M. M. (1978) *Clinical Biomechanics of the Spine.* Philadelphia: J. B. Lippincott.

14

WHAT DARWINIAN MEDICINE OFFERS PSYCHIATRY

RANDOLPH NESSE

$\mathcal{S}$ince the late 1970s, psychiatry has accumulated an enormous base of knowledge and remarkably effective new treatments. Despite this new knowledge, however, fundamental disagreements persist about the very nature of psychopathology. While the need for a "bio-psycho-social model" of mental disorders is widely acknowledged, psychiatry remains split into factions, some that emphasize genetic/physiological factors and others that focus on life events and psychological mechanisms. Some emphasize that mental disorders are "diseases like all others" while others decry this "medical model" and insist that many mental disorders can only be understood in their social context.

An evolutionary approach to medicine poses equally thorny conceptual issues. These difficulties include confusion about the distinction between proximate and evolutionary explanations, the need to recognize vulnerabilities instead of diseases as the object of evolutionary explanation, and the difficulties of finding adequate tests for many evolutionary hypotheses. In addition, a wrenching change in perspective is required to look for possible benefits of traits or genes that also cause disease and to begin to ask why all members of a species are susceptible to a disease, instead of why some individuals get sick and others do not. Given difficulties faced by Darwinian medicine, its application to psychiatry might seem to be a case of the blind leading the befuddled. But in fact, psychopathology has been a major early target for systematic and somewhat successful evolutionary approaches to disease. Why? The explanations are partly historical and partly a matter of the obvious need to address some long-standing problems.

While medicine has long had a foundation in the functional understanding provided by physiology, psychiatry has no comparable understanding of normal human behavior. The field of ethology has successfully traversed this path, and psychiatry is now where ethology was in the 1960s: trying to compensate for the lack of a theoretical foundation by striving for scientific rigor in detailed observations and measurements about carefully defined categories. As soon as ethology found its bedrock in the principles of evolutionary biology, it took off in a still-growing burst of scientific advances (Alcock, 1993; Krebs and Davies, 1991). The hope that similar progress may be possible for understanding human behavior explains the early interest of many psychiatrists in an evolutionary approach. In addition to hopes fostered by history, evolutionary approaches are also spurred by the need for a framework that can accommodate the complexity of psychiatric disorders. The human tendency to oversimplify causation has resulted in separate foci on genetic and environmental causes of mental disorder, leaving chasms where there should be studies of interaction effects. Distinguishing carefully between proximate and evolutionary causation and between defenses and defects may help to bridge this gap.

This chapter will not attempt to summarize all the work in evolutionary psychology or even all of evolutionary psychiatry. Excellent overviews of work in evolutionary approaches to human behavior are available (Barkow, 1989; Barkow et al., 1992; Betzig, 1997; Eibl-Eibesfeldt, 1983; Konner, 1983, 1990; Ridley, 1993; Smith and Winterhalder, 1992; Wright, 1994). Work specifically about psychopathology has been summarized in several review articles (McGuire, 1977; McGuire et al. 1992; Nesse, 1984, 1991a), textbook chapters (Gardner, 1995; Konner, 1995) and books on the topic (McGuire and Fairbanks, 1977; McGuire and Troisi, 1998; Stevens and Price, 1996; Wenegrat, 1984, 1990).

Instead of reviewing all the available material, this chapter will use the categories that proved appropriate for Darwinian medicine (Nesse and Williams, 1995; Williams and Nesse, 1991) to see how useful they can be in the field of psychiatry. Evolutionary explanations for the vulnerabilities that make us susceptible to disease fall into just a few categories. First, there are novel environmental factors that change faster than our bodies can evolve. Many of these novel factors are aspects of our human-constructed environment, but novelty also emerges constantly from arms races between competing organisms, either pathogens and hosts, predators and prey, or members of the same species. Second, there are design trade-offs that offer an advantage overall, but that leave us vulnerable to disease. Some such trade-offs are genes that cause disease but give a net fitness advantage. Others are traits that remind us that every aspect of the body is a compromise. Third, there are constraints, that is, limits on what natural selection can do because of its stochastic nature and the impossibility of "fresh starts" because every generation must compete and survive. Finally, there are accidents and mishaps that cause disease that is unrelated to the evolutionary process, except insofar as the body can or cannot prevent or repair the damage (see table 14.1).

Of particular importance for psychiatry is another category, not a true cause of susceptibility to disease, but one that is often confused with diseases— namely, defenses. Examples include the capacities for pain, fever, nausea,

Table 14.1 Evolutionary explanations for disease

1. Novelty
   a. From pathogens or competitors
   b. From aspects of the modern environment
2. Trade-offs
   a. Genes with costs as well as benefits
   b. All traits have positive and negative trade-offs
3. Constraints
4. Accidents and mishaps too rare to shape defenses
(5. Defenses that are often confused with diseases.)

vomiting, cough, diarrhea, and fatigue. Aversive emotions like anxiety and sadness almost certainly have similar origins and functions. We turn first to these defenses.

Defenses, Normality, and Emotions

Much that seems abnormal about the functioning of the brain and the body is not abnormal at all. Many manifestations of illness are not defects in the body's mechanisms, but sophisticated adaptations. They are observed only when aroused by cues that indicate a situation where they may be helpful. For instance, pain is aroused by cues that indicate tissue damage. It is not an abnormality itself, but a useful defense. People born without the capacity for pain are usually dead by their early 30s (Melzack, 1973).

Some manifestations of disease, such as seizures, jaundice, coma, and paralysis, arise from defects in the organism. But many other manifestations of disease are defenses. Coughing clears foreign matter from the respiratory tract. Vomiting eliminates toxins from the stomach; diarrhea clears the colon. The capacity for fatigue protects against tissue damage. The low iron levels seen in chronic infection limit the growth of pathogens (Weinberg, 1984). Fever is a particularly subtle and well-regulated defense against infection (Kluger, 1979), but it sometimes goes so high that seizures result. This is a fine example of a cost that is maintained by the trade-off with its associated benefit. An evolutionary approach encourages sharp attention to distinction between manifestations of disease that are defects versus those that are defenses, and it forces us to acknowledge that much suffering can be adaptive.

Most of the problems people bring to psychiatrists involve aversive emotions. Patients complain of sadness, anxiety, jealousy, anger, or boredom. To understand these complaints thoroughly, we need a comprehensive theory of emotions, something that is just now being formulated. Emotions researchers have reached a consensus that the fundamental emotional capacities have been shaped by natural selection because they give a selective advantage (Ekman, 1992; Frank, 1988; Frijda, 1986; Nesse, 1990). Plutchik, in particular, has advanced this line of reasoning to understand psychiatric disorders (Plutchik,

1980; Plutchik and Kellerman, 1989). Aversive emotions like anxiety and sadness are almost certainly examples of defenses (Barlow, 1991; Morris, 1992; Tooby and Cosmides, 1990a,b). Recognizing the utility of aversive emotions has implications for the whole problem of distinguishing normality versus pathology in general, a long-standing conundrum for psychiatry. It is to this large topic that we first turn.

It is easy to imagine how natural selection could have shaped the coherent and consistent patterns of physiology, behavior, cognition, and subjective experience that we recognize as emotions. If our ancestors repeatedly encountered certain situations that posed consistent adaptive challenges, then individuals whose bodies adjusted so they could cope especially well with those challenges would have a selective advantage (Nesse, 1990).

Take, for instance, male sexual jealousy. Daly and Wilson, among others, have argued that this capacity is a mate-guarding tactic which, if absent, is likely to result in a man having fewer children because his wife may sometimes be pregnant with the children of other men and because he will likely invest much parental effort in children fathered by other men (Daly, Wilson, and Weghorst, 1982; Symons, 1979). Although jealousy is an uncomfortable, undesirable state that can give rise to antisocial acts, it nonetheless is likely to increase reproductive success and therefore be maintained by natural selection. Consistent with this hypothesis, sexual jealousy is more intense in males than in females across a wide variety of cultures, despite the enormous cultural differences in the situations that arouse jealousy and the intensity and nature of its expression (Daly et al., 1982). Furthermore, there are sex differences in the cues that arouse jealousy. For men, cues to infidelity are more potent, while for women, losses of relationships and resources are stronger (Buss et al., 1992).

Such knowledge about the functions of normal jealousy is useful in understanding pathological jealousy. Although psychiatrists often interpret pathological jealousy as reflecting repressed and projected sexual desires, an evolutionary approach encourages the clinician to investigate whether the jealous partner has reason to believe that the spouse may well prefer someone else. Thus, during a period of loss of status or resources, or low self-esteem caused by depression, a man is especially likely to begin acting more jealous. This often precipitates a withdrawal of affection by the spouse, which is interpreted as confirmation of suspicions of infidelity, and the escalating feedback spiral can easily end a marriage or even lead to tragic violence.

Compared with jealousy, the utility of anxiety is more readily recognized, but exactly how it increases fitness is not so easy to describe. More than 60 years ago, Cannon (1929) documented the many benefits of the "fight–flight reaction" in the face of serious danger. He noted that everything about fear—from the cooling sweat, to the trembling that indicates tense muscles, to the increased clotting of blood and shortness of breath and rapid-pounding heartbeat—were all useful when one needed to flee or fight. Today we recognize this same pattern of responses as a panic attack. We now know much more about the proximate mechanisms that mediate these responses, but our knowledge base about its functions and evolutionary origins remains quite crude. Even a basic understanding can be useful, however. In the case of panic, it can account for the syndrome of shortness of breath, a wish to flee, pounding heart, sweating, and thinking of nothing but escape. Agoraphobia, fear of leaving the home, is almost always a complication of panic attacks. What is the connection? There are various proximate explanations, but from an evolutionary point of view, if you have just experienced life-threatening danger, it is wise indeed to stay close to home or to go out only with a companion.

Marks and I have argued that the different subtypes of anxiety disorder and their behavioral characteristics demonstrate a remarkable correspondence to the different kinds of danger that humans are likely to have experienced during the course of evolution (Marks and Nesse, 1994). Thus, fear of heights is manifested by freezing. The sight of blood is the only fear that causes fainting, a useful response if you are bleeding (Marks, 1988). A looming predator causes flight. Social threats cause embarrassment and social anxiety. This approach to the subtypes of anxiety obviates many of the difficulties posed by a system of discrete anxiety disorders. Enormous effort has gone into trying to determine whether the anxiety disorders arise from one underlying disorder or if they are variety of distinct diseases. An evolutionary approach suggests that the various manifestations of anxiety have been partially differentiated by natural selection into incompletely differentiated subtypes, each designed to cope with a particular kind of threat. If this is correct, it becomes less urgent to ask, for instance, whether social phobia and generalized anxiety disorder are essentially different or essentially the same. Both are overlapping response patterns to somewhat related dangers.

An evolutionary view also assists in dismantling the false dichotomy of instinct and learning. Research has begun to reveal the specialized learning mechanisms that mediate fear (Marks, 1987; Marks and Tobena, 1990; Mineka et al., 1980; Öhman and Dimberg, 1984). It has long been apparent that people are more likely to be afraid of heights, spiders, or the dark than they are of leaves, butterflies, or sunshine. For that matter, they are more likely to be afraid of a garter snake, than some dangerous things like guns, knives, and greasy hamburgers. In earlier work on preparedness, certain kinds of cues were found to be more readily paired with certain kinds of responses (Garcia and Ervin, 1968; Seligman, 1970). For instance, it is easy to condition a nausea response to a taste cue, and easy to condition a limb withdrawal response to a pain, but difficult to switch the cues and responses around. Fundamental research on learning is based increasingly on its evolutionary origins and functions, with attempts at integration making good progress (Staddon, 1983).

Mood and Its Disorders

An evolutionary explanation for the capacity for high and low mood has been a major, albeit unachieved, goal of evolutionary psychology. So far, even defining the terms remains problematic. The distinctions among grief, sadness, and depression remain fuzzy, and we do not even know if these intuitive concepts reflect functionally distinct subsets of a single experience, separate subcategories, or variations on the same response.

One major school of thought has emphasized the role of status changes in the regulation and functions of mood. British psychiatrist John Price, after recognizing a connection between pecking orders in birds and primate status hierarchies, began work on the behaviors that follow a rise or fall in status (Price, 1967; Price and Sloman, 1987). This work has been expanded and extended (Gardner, 1982; Gilbert et al., 1995; Price and Sloman, 1987), into a theory that emphasizes the role of depression in creating "involuntary yielding" and the benefits of voluntary yielding—that is, giving up a status competition—in relieving depression (Gilbert, 1992).

The epidemiological studies of Brown and colleagues show the profound effects of losses in precipitating depression, and the role of threats in precipitating anxiety disorders (Brown and Harris, 1978; Finlay-Jones and Brown, 1981). In recent work from an evolutionary vantage point, Brown has reanalyzed these data and found that much of the variance in the loss events is related to experiences of humiliation or entrapment (Brown et al., 1995).

Raleigh et al. (1983, 1991) have studied related mechanisms in vervet monkeys. They found that whole-blood serotonin levels are high in the dominant male, but fall precipitously with loss of rank. Monkeys that receive drugs that increase serotonin levels (versus a placebo) reliably take over a group that has no alpha male, while monkeys that receive drugs that block the effects of serotonin are consistently displaced by those receiving a placebo. I know of no other studies that so clearly demonstrate the two-way street between social circumstances and brain chemistry.

Another tradition of work on depression goes back at least to the psychoanalyst Bibring. He observed that depression often arises when a person is pursuing an unattainable life goal, and often remits when that goal is achieved or given up (Bibring, 1953). Gut (1989) has reinterpreted and extended this work in an evolutionary context. She emphasizes the adaptive significance of stopping current activity after a loss, conserving energy, and thinking hard about the current situation and the available alternatives.

I have emphasized similar functions for mood as a regulator of patterns of resource investment among a variety of enterprises with differing patterns of risk and payoff (Nesse, 1990, 1991b). High mood tends to increase fitness in situations of opportunity, whereas low mood tends to stop investment in hopeless endeavors and facilitates consideration of alternative strategies or enterprises. In short, low mood is a coordinated pattern of responses that are useful in unpropitious situations, whether the loss is discrete or continuing. Momentary changes in positive and negative states may help to fine-tune the tactics required for coping with everyday social life.

The relationship between sadness and depression remains problematic. Sadness is almost certainly adaptive, but depression may arise from dysregulated sadness or from an entirely separate mechanism. Also unanswered is the question of why 15% of the population is especially susceptible to anxiety and mood disorders (Kessler et al., 1994). Vulnerability to the severe mood disorders is mediated substantially by genetic factors. Why the responsible genes persist in the gene pool is another unanswered question. In the case of manic-depressive disorder, it has been proposed that increased reproductive success during periods of high mood may counterbalance the periods of low reproductive success during low mood (Wilson, 1992).

Attachment and Relationships

Human emotions usually arise, of course, in the context of relationships. This is not the place to review the complete history of work in this field, but René Spitz's studies of the high death rates in orphanages inspired Harlow (1974) to carry out his well-known studies that showed infant monkeys prefer milk-less terry-cloth "mothers" to wire-mesh "mothers" with a milk source. These studies showed that both behaviorists and the psychoanalysts were wrong in their supposition that attachment arises out of drive satisfaction. Influenced by these studies, and by contact with ethologist Konrad Lorenz, Bowlby (1969) arrived at a view derived from psychoanalysis but informed by ethology. Bowlby's emphasis on the adaptive functions of attachment led to studies by Ainsworth and others (1978) demonstrating strong correlations between parental behavior and children's attachment styles and between a child's attachment pattern early in life and in later years. The strong implication through most of these studies has been that the attachment style is shaped by the mother's behavior early in life. It is now clear, however, with recent studies by Kagan et al. (1987) that these patterns are strongly influenced by genetic factors. Furthermore, the suggestion has been made by that the "abnormal" styles of attachment, specifically anxious attachment and ambivalent attachment, may in fact be adaptive in certain situations (Chisholm, 1996). If the mother is disinclined to invest in a child, ordinary secure attachment may be less beneficial than anxious or avoidant attachment.

In work at the interface between attachment and depression, Engel and Schmale (1972) have argued that the protest phase of primate infant separation alerts the mother, while the despair phase of quiet huddling conceals the infant from predators and conserves energy in much the same way as hibernation. They suggest that the similarities among depression, hibernation, and the despair phase of separation may arise from this mechanism.

Trivers's theory of parent–offspring conflict (Trivers, 1974) has been remarkably underutilized and even unrecognized in psychiatry. While many clinicians continue to imagine that the interests of the mother and her infant are identical, Trivers has shown that there are many situations in which a mother can maximize her reproductive success by investing less in an offspring than is in the offspring's interest. Because offspring share only half of the mother's genes, the fitness of the mother's genes depends on correctly allocating effort between having more offspring of her own or investing in the offspring she already has. The specific conflicts depend on the stage of life and the available alternatives. Although Trivers noted the significance of conflicts during adolescence, he used the weaning conflict as an exemplar. At some stage of development it is to the infant's advantage to continue to nurse, but the mother's genetic interests would be better served by starting to invest in another offspring. Fierce battles arise that testify to the lack of concordance. As Trivers (1985:155) notes wryly, "An offspring cannot fling its mother to the ground and nurse at will. . . . Given this competitive disadvantage, the offspring is expected to employ psychological tactics. It should attempt to *induce* more investment than the parent is selected to give." In particular, Trivers suggests, the child may pretend to be younger and more helpless than it really is to try to deceive the mother into thinking it is still in her interest to provide more help than she would otherwise. This is, of course, a fine description of regression, and may explain not only the phenomenon of regression but also the annoyance we experience when we think we are being manipulated by people who act more helpless than they are.

Insight into the conflicts between offspring and parents has provided a foundation for a reinterpretation of some core psychoanalytic ideas. Slavin (1992) notes that it is in the interest of parents to suppress conflicts between siblings, and it is also in the parents' interests to manipulate their children to cooperate more than is in the children's best (fitness) interests (Slavin, 1992). Slavin (1992) proposes that if this excess cooperation persists into adulthood, the result is neurosis, or worse. Several other authors, notably Badcock (1988) and Rancour-Laferriere (1985), have gone much further and tried to understand the possible adaptive significance of specific psychoanalytic phenomena such as the Oedipus complex and castration anxiety. Daly and Wilson (1992), however, note the ignorance of most psychoanalysts about fundamental evolutionary principles. They have proposed alternative evolutionary explanations for what have been described previously as Oedipal phenomena.

Trivers (1976) and Alexander (1974) have each proposed that a mechanism for limiting self-knowledge might allow individuals to unconsciously, and thus more successfully, pursue strategies of deception in reciprocity exchanges (Lockard and Paulhus, 1988). This explanation must, however, be incomplete, given that we also repress the prods of conscience as well those of the drives. Lloyd and I (Neese and Lloyd, 1992) have attempted to understand the origins of the psychodynamic defenses, and more specifically to understand ways in which a lack of knowledge about one's own motives might give a selective advantage. Much work remains to integrate psychodynamics with a modern approach to evolution, but a start has been made (Badcock, 1988; Stevens, 1982; Wenegrat, 1990).

The principle of kin selection has proved particularly powerful in explaining cooperation and lack of cooperation in the family (Essock-Vitale and Fairbanks, 1979; Lancaster et al., 1987). Perhaps the single most dramatic finding in the field so far has come been made by the behavioral ecologists Daly and Wilson. On considering the phenomenon of fatal child abuse, they wondered how this apparent anomaly could be reconciled with the evolutionary expectation that parents will go to great lengths to protect their offspring. They hypothesized that perhaps the children who were killed were not, in fact, the biological offspring of both parents in the home. When they did the study to test this hypothesis, they found that the rates of child abuse in homes with at least one step-parent were at least 70 times higher compared to homes with both biological parents present (Daly and Wilson, 1981, 1989). This finding could not be accounted for by confounding factors such as alcoholism, poverty, or mental disorders. The discovery of this extraordinarily powerful effect is highlighted, in contrast by decades of studies by child protection specialists that had uncovered factors that are weak by comparison.

Marriage and Other Relationships

Understanding of the origins of conflicts over mates and mating strategies (Hrdy, 1981) has given significant new insights into marriage and its vicissitudes. Buss (1992) has documented, in a study of 37 human cultures, that in choosing a mate, appearance is consistently more important to men, and wealth is consistency more important to women. Given that variance in reproductive success is quite limited in women, but can be large in men, the two sexes can and do pursue different and often conflicting mating strategies (Daly and Wilson, 1983). These distinct strategies help to account for patterns of cooperation and conflict and are proving useful in understanding associated marital pathology (Kerber, 1994).

Based on these principles, Fisher (1992) has summarized data from many societies that shows an average duration of marriage of less than 5 years. She argues that this is sufficient time to provide initial nurturing to human offspring, and that a mechanism ("planned obsolescence of the pair-bond") may have evolved to end marriages after this period of time. Although the data are clear enough, it remains unclear why a specific mechanism would evolve to end pair-bonds when spouses are certainly capable of discerning their individual interests and pursuing them in a more flexible strategy of staying or leaving (Betzig et al., 1988; Hrdy, 1981). Of course, social pressures on mating patterns are profound, so it should be no surprise that an individual's decisions are often severely constrained.

Reciprocity has long been recognized as the engine motivating nonkin relationships, even before Trivers's (1971) seminal paper, but its special significance to psychology results from two major discoveries. The first is that kin selection now explains much that was previously attributed to group selection (Hamilton, 1964; Williams, 1966). Much altruistic behavior benefits genes identical to those of the altruist that are in another (related) individual. With this perspective, the special relationships between blood relatives are expected. Although opportunities for conflict remain, much self-sacrificing behavior is expected as a result of kin selection.

The second discovery is that tit-for-tat is a robust strategy for maximizing the payoff in reciprocity interactions (Axelrod, 1984; Axelrod and Hamilton, 1981). When people trade favors, an exchange that can be seen as the foundation of human social life, there is always the risk that the other will not reciprocate. If the maximum long-term benefit comes from repeated cooperation, but the maximum short-term benefit comes from defecting, a problem of strategy arises. When Axelrod invited scientists to submit programs for how to deal with this dilemma, the winner was "tit-for-tat," that is, cooperate on the first move of the game, and for every subsequent move, do what the other person did last time. The strategy takes advantage of cooperation when that is possible but avoids exploitation.

Cosmides and Tooby (1989) have provided data supporting the existence of specific mental algorithms to detect cheaters, a finding that could have importance for states of paranoia. Others have argued that sociopathy is a specialized evolved strategy for taking advantage of human dependence on reciprocity exchange (Harpending and Sobus, 1987; Mealey, 1995).

If the social emotions—friendship, anger, suspicion, anxiety, and guilt—are viewed as adaptations designed to deal with different outcomes in a reciprocity exchange, then this begins to explain some of the complexity and randomness of the expression these emotions (Frank, 1988; Nesse, 1990) and their variable development and association with various kinds of psychopathology. In particular, specialized emotions of friendship and affection may arise from trustworthy exchanges. When you think that the other may defect, suspicion is aroused, while if the defection actually takes place, anger signals that the defection is unacceptable and must be remedied if the relationship is to continue. Conversely, when we are tempted to defect, anxiety often inhibits our actions after a defection, most of us feel guilty, which induces states of self-punishment that are hard to understand from other perspectives but that make sense as a way to reestablish a reciprocity relationship after a defection (table 14.2).

This chapter has so far addressed various aspects of human behavior that are often thought to be abnormal but in fact can be recognized either as defenses or as behavior patterns that tend to increase reproductive success. What kinds of research projects will be needed to advance work in this area? It is essential to keep separate several questions about such defenses.

First, there is the question of the function of the basic capacity for a defense, such as anxiety or mood. Recast, this question asks why individuals with this capacity have a selective advantage over those who do not. To address the question, one uses the same approaches used by physiologists in ascertaining

Table 14.2 The social emotions and the Prisoner's Dilemma

	Other cooperates	Other defects
Self cooperates	Friendship, trust	*Before:* Suspicion *After:* Anger
Self defects	*Before:* Anxiety *After:* Guilt	Hatred, mistrust, rejection

the functions of an organ: take it out, block its effects, or look at what happens to people who have innately high or low levels. Such studies face special difficulties. People who report few experiences of the negative emotion may lack the capacity, but they also may simply be living fortunate lives. Then, also, there is the expectation that natural selection will set the regulation of inexpensive defenses to a hair-trigger (the "smoke detector principle"), so that many expressions of the defense will be unnecessary. We don't even have solid documentation of the value of fever in many everyday infections, so it may be some time before we have evidence for the value of low mood.

Second is the question of how a defense is regulated. This involves multiple levels, from the psychological to brain chemicals. Interestingly, it appears that we know more about the brain mechanisms than we do about the psychological mechanisms that regulate mood. The range of factors that can explain within-individual variation is daunting: diet, exercise, sunlight, changes in social status, sexual experiences, martial changes, family conflicts, work stress, social support, community structure changes, substance use and abuse, health, sleep patterns, cognitive patterns, media exposure, trauma, opportunity, and myriad other factors. Studies of the same individuals with frequent sampling over days and months may begin to give answers to these questions.

Third is the question of why people differ in their tendencies to certain emotions like low mood and anxiety. Unpacked, this question includes many facets, such as baseline levels of mood and anxiety, overall emotionality, responsiveness to positive events, responsiveness to negative events, duration of responses, cognitive patterns that mediate responses, and responsiveness to social cues. There is now strong evidence that genetic factors have strong effects on baseline mood levels and tendencies to mood disorders, but we do not yet know whether these effects are direct or mediated by genetic effects on cognition or stress responses.

Fourth, there are questions about differences in these reactions between groups. Rates of major depression vary 10-fold between different countries (Weissman et al., 1996). Although this variation may be explained by the above factors, it may also require systematic assessment of cultural and family structures to see how they interact with the regulation mechanisms.

The research agenda to study emotions and other defenses is crowded. Even as it is just getting started, an evolutionary approach helps to clarify the questions and suggest possible answers.

Novelty

Arms Races

In general medicine, evolutionary arms races are most clearly manifest in pathogen–host competitions. We are still susceptible to infection after all these eons of exposure because pathogens evolve much faster than we do and so can outflank nearly any defense our body evolves (Ewald, 1993). Some psychiatric illnesses may turn out to result from infection with agents that have evolved to evade our defenses. In particular, some cases of schizophrenia may result from in utero exposure to viruses (Kendell and Kemp, 1989). Also, there are some recent data suggesting that some cases of obsessive-compulsive disorder may arise after streptococcal-infection–precipitated autoimmune damage to the basal ganglia, including the caudate nucleus which is strongly implicated in the pathophysiology of obsessive-compulsive disorder (Allen et al., 1995). These findings are by no means conclusive, but would tie together many confusing aspects of obsessive-compulsive disorder and would help to explain the pattern of onset, the association with Sydenham's chorea, the pattern of blood antibodies observed in patients and controls, and the genetic predisposition to obsessive-compulsive disorder.

Another kind of arms race has already been alluded to, that between members of the same species. A number of authors have emphasized the possibility that intense human–human competition has shaped the rapid increase in human brain size and intelligence and cognitive complexity (Alexander, 1979; Dawkins, 1976; Humphrey, 1976). The results of such a competition could also explain, as some of the previous authors of works on evolution and psychodynamics have shown, some of the inordinate complexity of human psychology. If our empathy is a tool to help us manipulate others and avoid being manipulated, then layer on layer of complexity may have been laid down by the process of natural selection (Krebs and Dawkins, 1984).

Novel Aspects of the Physical and Social Environment

A large proportion of disease now results from novel aspects of our physical environment. For instance, atherosclerosis, the complications of which will prove fatal for more than a third of us, is due mainly to our evolved preference for fats and the current ready availability of kinds of fats that were less common in the ancestral environment (Eaton et al., 1988). Substance abuse, auto accidents, many infections, cancer, Alzheimer's disease, and other diseases of old age are also experienced almost exclusively in the modern environment. It is enough to make one long for life in the Stone Age. Until, that is, one thinks for a moment about the even greater burden posed by other diseases in the Paleolithic. The same goes for psychiatric disorders. While it is undoubtedly true that some proportion of current mental disorders arise because of our modern environment (Glantz and Pearce, 1989), it is by no means correct that this implies we would be mentally better off in an earlier environment. It is possible that mood and anxiety disorders are less frequent in more traditional cultures, but we do not yet know.

Substance abuse is the most significant mental disorder that is a product of our novel environment. While people in all cultures and times have used drugs, ready availability of a steady supply of potent agents, in conjunction with novel means of administration such as hypodermic needles and crack pipes, has made substance abuse an international plague. An evolutionary view makes one hesitant to seek simplistic solutions (Nesse, 1994). Organisms are designed to repeat behaviors that stimulate reinforcing mechanisms in the brain. When these mechanisms are stimulated directly by drugs instead of by natural experiences, drug-seeking behavior can take over the organism's behavioral control mechanisms (Nesse and Berridge, 1998; Pomerleau, 1997). Although it is not surprising that people vary in their genetic susceptibility to addiction, it is also not surprising that untoward circumstances also make certain people especially vulnerable. The genes that predispose to substance abuse are sometimes called defects, but given that they probably imposed no harm in the Paleolithic, it would be much more appropriate to call them "quirks."

Obesity and eating disorders also arise from our novel environment. In the Paleolithic, high-calorie foods were apparently rarely available without substantial effort, or else natural selection would have shaped more powerful mechanisms to restrict food intake. Given the selective power of famine, it is also unsurprising that many people have a tendency to add more pounds than are necessary or desirable in a modern environment. Severe eating disorders seem to be peculiarly modern phenomena, arising mainly in the past few score years in technological societies (Kurth et al., 1995). Some combination of modern media, food availability, and cultural factors causes many young people to diet, which then leads to behavior appropriate to starvation, namely, seeking out and devouring any available high-calorie food. This loss of control causes increased fear of obesity and emotional upheaval that initiates a vicious cycle of attempts at self-control by dieting, binge eating, followed by guilt and helplessness, more dieting and more binge eating. A small proportion of people who begin this cycle are able to more and more strictly control their intake to the point it is manifest as anorexia nervosa. The absence of menstrual periods associated with anorexia may well be adaptive in a natural environment, where it would be wasteful to become pregnant when food supplies are inadequate (Condit, 1990; Surbey, 1987; Voland and Voland, 1989).

Some aspects of modern life are so ubiquitous that we rarely notice how unusual they are. For instance, spending time is classrooms learning to read is completely novel. It is hard to see what dyslexia and attention deficit disorder would have been like in the Paleolithic. For another example, the simple invention of electric lights seems to have changed people's sleep patterns profoundly. When people enter a sleep lab with lighting that matches that of the sun, they typically make up a cumulative sleep deficit of 15 hours before settling in to a pattern of sleeping approximately nine hours per night with nightly periods of restful wakefulness.

Even conflicts between the sexes may be viewed as a product of novel environments. When it was not possible for one person to amass significant resources, no one man could control more than a few women. When agriculture became organized, however, men promptly used surpluses to control other men to control large number of women in harems (Betzig, 1986). Smuts (1995) has looked at patterns of social organization and suggested that the relative absence of female–female bonds in some cultures makes women far more vulnerable to male aggression.

One of the more worrisome recent epidemiologic findings suggests that an extraordinary epidemic of depression may be upon us. In nine separate epidemiological studies, the rates of depression in young people are far higher than those over a lifetime for older people in the same country (Cross-National Collaborative Group, 1992). Although these studies deserve criticism on methodological grounds, it also seems plausible that changing social patterns and technological changes may indeed be changing rates of depression. If so, we have an even more urgent need to understand the functions and regulation of normal mood.

Genes

In psychiatry, as in the rest of medicine, genes that predispose to common disorders may often also offer benefits that account for their continued frequency, or they may cause disease only in interaction with aspects of the modern environment. Early hopes that we would be able to identify one or two specific genes to account for schizophrenia or manic depressive illness have foundered, but even if multigenic effects are responsible, an evolutionary explanation for their prevalence can be helpful.

For instance, schizophrenia appears to be quite constant at a prevalence of about 1% in diverse human groups. In less modern societies, the course of the disease may be somewhat more benign, but the prevalence is nonetheless similar (Gottesman and Shields, 1982). Given evidence for strong genetic factors that predispose to schizophrenia, one must ask why these genes have persisted given the reproductive disadvantage they cause (Slater et al., 1971). Many authors have suggested possible pleiotropic benefits to the genes that cause schizophrenia. Early, probably erroneous, observations that people with schizophrenia could withstand cold and disease better than other patients were once thought to account for the finding (Jarvik and Chadwick, 1972). More recently, a variety of proposals for alternative explanations have been put forth, ranging from ability to discern the unseemly motives of others (Allen and Sarich, 1988), to sexual selection (Crow, 1993), and circadian regulation (Feierman, 1994). While one of these ideas could eventually prove correct, it seems equally plausible that the relationship is far less direct, such as the possibility that people with a certain antigenic makeup are protected against certain infections but they are vulnerable to autoimmune or other factors that interfere with brain development.

Depression and anxiety are not defects like schizophrenia, but defenses whose regulation has gone awry. Thus, the genetic trade-offs that cause differences in susceptibility to mood and anxiety disorders are directly related to the higher level trade-off of the regulation of the normal emotion. One cannot help but wonder why there is such between-individual variability in regulation of these emotions. The answer may be that different response patterns are optimally adaptive in different environments. This would be a special case of the proposal that differences in personality are adaptations for different social niches (Buss, 1991; Tooby and Cosmides, 1990a)

Alzheimer's disease may be an example of a manifestation of a pleiotropic gene with benefits early in life and a cost later in life (Albin, 1994). The justification for this speculation is the extraordinary frequency of Alzheimer's disease, affecting more than half of the people over the age of 90. So far, however, no one has identified benefits earlier in life for people who carry the tendencies to Alzheimer's disease, although factors associated with apolipoprotein variants could provide an explanation.

Trait Trade-offs

Every trait in the body is a trade-off between multiple benefits and costs. Heavier bones would break less easily, but they would be more expensive to create and maintain. Lower blood pressure or blood glucose would cause less damage to tissues, but only at the cost of decreased ability to respond to situations that require sudden exertion.

Such trade-offs probably account for many aspects of psychiatric illness. As noted above, too little anxiety may be even worse than too much. On the level of life strategy, there are disadvantages to investing too little effort in parenting, and disadvantages to investing too little in mating. The patterns of modern life are quite peculiar along these lines because the environment does not require us to each allocate effort proportionately among the tasks of growth, development, protection, mating, and parenting. Instead, an individual can devote essentially all his or her resources to one area, neglecting all the rest. Furthermore, thanks to caffeine, electric lights, and grocery stores, an individual can, for instance, spend nearly all of his or her waking life simply reading and writing, a thoroughly abnormal pattern of activity that necessitates neglect of other tasks. Studies that relate overall patterns of allocation of life effort to patterns of symptoms have yet to be done.

Historical Constraints

The body offers many illustrations of natural selection's inability to start fresh and create a sensible design instead the jury-rigged bodies we have. Our eyes are inside out, causing blind spots. Our respiratory passages intersect within the esophagus and food can thus choke us. The awkward design of the spine causes most of us pain at one time or another. There are undoubtedly similar examples in our mental mechanisms, but because the connection between the structure and the function is so much more difficult to ascertain and because of the difficulties of carrying out comparative or historical studies, this is a nebulous area. Our patterns of reasoning may, however, provide an important example. We all routinely make substantial errors in logic that reveal underlying biases and rules of thumb (Nisbett and Ross, 1980; Tversky and Kahneman, 1974) Rules of thumb substitute for logic perhaps because the first cognitive mechanisms were based on these crude rules, and there has never been a selective force that could supplant them or make the leap to primary dependence on Bayesian thinking. The mystery is that we have a capacity for logical thought at all. It seems quite possible that this is either an epiphenomenon or a product of sexual selection (Miller, in press). Our emotional reactions may likewise reflect the continuity with our predecessors. Studies of primate cognition may eventually intersect with the studies of human cognitive and social psychology to give us more knowledge about these possibilities.

Conclusions

Given that there is not yet an evolutionary psychology that can provide psychiatry with knowledge comparable to what physiology provides for the rest of medicine, can an evolutionary approach still be useful to psychiatry? As documented in the preceding pages, the answer is yes. The danger now is in taking this preliminary knowledge too far. Several caveats may help. First of all, I would argue that Darwinian psychiatry is not, and should not be, a field of medical practice, but only a basic science that offers us insight and good questions to ask. A Darwinian approach to human nature and human psychopathology is essential to our understanding, but it gives no direct treatment recommendations whatsoever. As is the case with the rest of medicine, clinical recommendations must come from clinical studies. In the area of psychiatry, one must also reinforce the warning that behavior and ethical precepts cannot be derived from biological knowledge. For instance, just because males have more of a tendency than females to abandon their families tells us nothing whatsoever about the moral significance of such acts. Such warnings about the naturalistic fallacy are common in the literature on evolution and psychology, but this is for the good reason that many people, despite warnings, readily draw norms from facts. Furthermore, our understandings of the origins and functional significance of aspects of culture, and the biological underpinnings that make it possible for individuals to absorb culture, remind us that our understanding is rudimentary and is not to be trusted as a guide for action.

  With these caveats in mind, however, an evolutionary view already offers an extraordinarily useful guide for understanding our patients. Even our crude state of knowledge takes us a giant leap from the idiosyncratic theories without biological foundation that have characterized much of the history of psychology. Furthermore, the age-old problem of "normality" makes much more sense in an evolutionary perspective. From a perspective of reproductive success, many emotions, cognitions, and behaviors patterns that have been thought of as pathological can be recognized as adaptive, even if often distressing or socially prohibited. And for some of these, like anxiety, it is clinically useful to explain their functions to patients. When people who experience panic attacks learn that their symptoms are the normal fight–flight response that is going off at the wrong time, this often helps them give up their fears that the symptoms are caused by heart or brain disease.

  These principles are likely to have early application as a foundation for a revised perspective on psychiatric diagnosis. The successive *Diagnostic and*

Statistical Manuals published by the American Psychiatric Association have been essential in allowing researchers to study comparable groups, but they have also been widely criticized because they lack a foundation in theory and they encourage simplistic thinking of psychiatric syndromes as discrete diseases (Kendell, 1984). An evolutionary view, by contrast, requires us to use what we know about the emotions and their regulation and thus to distinguish several categories of mental disorders. First, there are conditions that are painful and perhaps undesirable, but normal and evolutionarily useful, like much jealousy, anxiety and sadness. Second, there are conditions that arise from an initially normal brain that has been exposed to learning or trauma that leads to emotional pathology or maladaptive behavior. Third, there are conditions, like most depression and panic disorder, that arise from dysregulation caused by genetic vulnerability interacting with environmental factors. Fourth, there are conditions that arise from primary abnormalities of brain tissue, like obsessive-compulsive disorder, schizophrenia, and Alzheimer's disease. In contrast to current diagnostic systems that encourage separation of a patient's difficulties into many "co-morbid" conditions, an evolutionary approach fosters exploration of all the factors that interact to account for a person's emotional and behavioral difficulties.

Finally, the greatest current value of a Darwinian perspective on psychiatry is its heuristic utility. Without an evolutionary perspective, Daly and Wilson never would have asked about the frequency of child abuse in families with a step-parent present. Without an evolutionary perspective, it is far harder to even imagine the possible benefits of anxiety, sadness, and grief. Without an evolutionary perspective, it is hard to ask the right questions about the functional significance of various aspects of human nature, adaptive and maladaptive.

References

Ainsworth, M. D., Blehar, M. C., Waters, E., and Wall, S. (1978). *Patterns of Attachment: A Psychological Study of the Strange Situation.* Hillsdale, NJ: Lawrence Erlbaum.

Albin, R. L. (1994). Antagonistic pleiotropy, mutation accumulation, and human genetic disease. In M. R. Rose and C. E. Finch, eds., *Genetics and the Evolution of Aging.* Amsterdam: Kluwer.

Alcock, J. (1993). *Animal Behavior,* 5th ed. Sunderland, MA: Sinauer Associates.

Alexander, R. D. (1974). The evolution of social behavior. *Annual Review of Systematics,* 5, 325–383.

Alexander, R. D. (1979). *Darwinism and Human Affairs.* Seattle: University of Washington Press.

Allen, A. J., Leonard, H. L., and Swedo, S. E. (1995). Case study: A new infection-triggered, autoimmune subtype of pediatric OCD and Tourette's syndrome. *Journal of the American Academy of Child and Adolescent Psychiatry* 34(3), 307–311.

Allen, J. S., and Sarich, V. M. (1988). Schizophrenia in an evolutionary perspective. *Perspectives in Biology and Medicine* 32(1), 132–153.

Axelrod, R. (1984). *The Evolution of Cooperation.* New York: Basic Books, Inc.

Axelrod, R., and Hamilton, W. D. (1981). The evolution of cooperation. *Science,* 211, 1390–1396.

Badcock, C. (1988). *Essential Freud.* Oxford: Basil Blackwell.

Barkow, J. H. (1989). *Darwin, Sex, and Status: Biological Approaches to Mind and Culture.* Toronto: University of Toronto Press.

Barkow, J., Cosmides, L., and Tooby, J., (eds.) (1992). *The Adapted Mind.* New York: Oxford University Press.

Barlow, D. H. (1991). Disorders of emotions: Clarification, elaboration, and future directions. *Psychological Inquiry* 2(1), 97–105.

Betzig, L. L. (1986). *Despotism and Differential Reproduction: A Darwinian View of History.* New York: Aldine Publishing Company.

Betzig, L., ed., (1997). *Human Nature: A Critical Reader.* New York: Oxford University Press.

Betzig, L., Mulder, M.B., and Turke, P. (eds.). (1988). *Human Reproductive Behavior: A Darwinian Perspective.* Cambridge: Cambridge University Press.

Bibring, E. (1953). The mechanisms of depression. In P. Greenacre, ed., *Affective Disorders* (pp. 13–48). New York: International Universities Press.

Bowlby, J. (1969). *Attachment.* New York: Basic.

Brown, G. W., and Harris, T. (1978). *Social Origins of Depression.* New York: The Free Press.

Brown, G. W., Harris, T. O., and Hepworth, C. (1995). Loss, humiliation and entrapment among women developing depression: A patient and non-patient comparison. *Psychological Medicine* 25(1), 7–21.

Buss, D. (1991). Evolutionary personality psychology. *Annual Review of Psychology,* 42, 459–491.

Buss, D. (1992). Mate preference mechanisms: Consequences for partner choice and intersexual competition. In J. Barkow, L. Cosmides, and J. Tooby, eds., *The Adapted Mind.* New York: Oxford, pp. 249–266.

Buss, D. M., Larsen, R. J., Westen, D., and Semmelroth, J. (1992). Sex differences in jealousy: Evolution, physiology, and psychology. *Psychological Science,* 3, 251–253.

Cannon, W. B. (1929). *Bodily Changes in Pain, Hunger, Fear, and Rage. Researches into the Function of Emotional Excitement.* New York: Harper and Row.

Chisholm, J. S. (1996). The evolutionary ecology of human attachment organization. *Human Nature* 7(1), 1–38.

Condit, V. K. (1990). Anorexia nervosa: Levels of causation. *Human Nature* 1(4), 391–414.

Cosmides, L., and Tooby, J. (1989). Evolutionary psychology and the generation of culture, part II: Case study: A computational theory of social exchange. *Ethology and Sociobiology* 10(1–3), 51–98.

Cross-National Collaborative Group. (1992). The changing rate of major depression. Cross-national comparisons. *Journal of the American Medical Association,* 268(21), 3098–105.

Crow T.J. (1993). Sexual selection, Machiavellian intelligence, and the origins of psychosis. *Lancet* 342(8871), 594–598.

Daly, M., and Wilson, M. I. (1981). Abuse and neglect of children in evolutionary perspective. In R. D. Alexander and D. W. Tinkle, eds., *Natural Selection and Social Behavior.* New York: Chiron Press, pp. 405–416.

Daly, M., and Wilson, M. (1983). *Sex, Evolution, and Behavior,* 2nd ed. Boston: Willard Grant Press.

Daly, M., and Wilson, M. (1989). *Homicide.* New York: Aldine.

Daly, M., and Wilson, M. (1992). Is parent-offspring conflict sex-linked? Freudian and Darwinian models. *Journal of Personality* 58(1), 163–189.

Daly, M., Wilson, M. I., and Weghorst, S. J. (1982). Male sexual jealousy. *Ethology and Sociobiology* 3, 11–27.

Dawkins, R. (1976). *The Selfish Gene.* Oxford: Oxford University Press.

Eaton, S. B., Konner, M., and Shostak, M. (1988). Stone agers in the fast lane: chronic degenerative diseases in evolutionary perspective. *The American Journal of Medicine* 84(4), 739–749.

Eibl-Eibesfeldt, I. (1983). *Human Ethology.* New York: Aldine de Gruyter.

Ekman, P. (1992). An argument for basic emotions. *Cognition and Emotion* 6(3/4), 169–200.

Engel, G., and Schmale, A. (1972). Conservation-withdrawal: A primary regulatory process for organismic homeostasis. In R. Porter and J. Night, eds., *Physiology, Emotion, and Psychosomatic Illness*, vol. 8 (pp. 57–85). Amsterdam: CIBA.

Essock-Vitale, S. M., and Fairbanks, L. A. (1979). Sociobiological theories of kin selection and reciprocal altruism and their relevance for psychiatry. *Journal of Nervous and Mental Diseases* 167(1), 23–28.

Ewald, P. (1993). *Evolution of Infectious Disease.* New York: Oxford University Press.

Feierman, J. R. (1994). A testable hypothesis about schizophrenia generated by evolutionary theory. *Ethology and Sociobiology* 15, 263–282.

Finlay-Jones, R., and Brown, G. (1981). Types of stressful life event and the onset of anxiety and depressive disorders. *Psychological Medicine* 10, 445–454.

Fisher, H. E. (1992). *Anatomy of Love.* New York: W.W. Norton.

Frank, R. H. (1988). *Passions Within Reason: The Strategic Role of the Emotions.* New York: W.W. Norton.

Frijda, N. H. (1986). *The Emotions.* Cambridge: Cambridge University Press.

Garcia, J., and Ervin, F. (1968). Gustatory-visceral and telereceptor-cutaneous conditioning: Adaptation in internal and external milieus. *Communications in Behavioral Biology A* 1, 389–414.

Gardner, R. Jr. (1982). Mechanisms in manic-depressive disorder. *Archives of General Psychiatry* 39, 1436–1441.

Gardner R. Jr. (1995). Sociobiology and its applications to psychiatry. In H. I. Kaplan and B. J. Saddock, eds., *Comprehensive Textbook of Psychiatry/VI* (pp. 365–376). Baltimore, MD: Williams and Wilkins.

Gilbert, P. (1992). *Depression: The Evolution of Powerlessness.* New York: Guilford.

Gilbert, P., Price, J., and Allen, S. (1995). Social comparison, social attractiveness and evolution: How might they be related? *New Ideas in Psychology* 13(2), 149–165.

Glantz, K., and Pearce, J. (1989). *Exiles from Eden.* New York: W.W. Norton.

Gottesman, I., and Shields, J. (1982). *Schizophrenia: The Epigenetic Puzzle.* New York: Cambridge University Press.

Gut, E. (1989). *Productive and Unproductive Depression.* New York: Basic Books.

Hamilton, W. D. (1964). The evolution of social behavior I and II. *Journal of Theoretical Biology* 7, 1–52.

Harlow, H. F. (1974). *Learning to Love.* New York: J. Aronson.

Harpending, H. C., and Sobus, J. (1987). Sociopathy as an adaptation. *Ethology and Sociobiology* 8(3S), 63s–72s.

Hrdy, S. B. (1981). *The Woman That Never Evolved.* Cambridge: Harvard University Press.

Humphrey, N. K. (1976). The social function of intellect. In P. Bateson and R. Hinde, eds., *Growing Points in Ethology* (pp. 303–318). London: Cambridge University Press.

Jarvik, L. F., and Chadwick, S. B. (1972). Schizophrenia and survival. In M. Hammer, K. Salzinger, and S. Sutton, eds., *Psychopathology*. New York: John Wiley, pp. 57–74.

Kagan, J., Resnick, R. S., and Snidman, N. (1987). Biological bases of childhood shyness. *Science* 240, 189–208.

Kendell, R. E. (1984). Reflections on psychiatric classification. *Integrative Psychiatry* 2, 43–47W.

Kendell, R. E., and Kemp, I. W. (1989). Maternal influenza in the etiology of schizophrenia. *Archives of General Psychiatry* 46, 878–882.

Kerber, K. (1994). The marital balance of power and the quid pro quo: An evolutionary perspective. *Ethology and Sociobiology* 15(5–6), 283–298.

Kessler, R. C., McGonagle, K. A., Zhao, S., Nelson, C. B., Hughes, M., Eshelman, S., Willchen, H. U., and Kendler, K. S. (1994). Lifetime and 12-month prevalence of DSM-III-R psychiatric disorders in the United States: Results from the national comorbidity survey. *Archives of General Psychiatry* 51, 8–19.

Kluger, M. J., ed. (1979). *Fever, Its Biology, Evolution, and Function*. Princeton, NJ: Princeton University Press.

Konner, M. (1983). *The Tangled Wing: Biological Constraints on the Human Spirit*. New York: Harper Colophon Books.

Konner, M. (1990). *Why the Reckless Survive*. London: Penguin.

Konner, M. (1995). Anthropology and psychiatry. In H. I. Haplan and B. J. Saddock, eds., *Comprehensive Textbook of Psychiatry*, vol. VI (pp. 337–355). Baltimore, MD: Williams and Wilkins.

Krebs, J., and Davies, N. (1991). *Behavioral Ecology: An Evolutionary Approach*, 3rd ed. Oxford: Blackwell.

Krebs, J. R., and Dawkins, R. D. (1984). Animal signals: mind-reading and manipulation. In J. R. Krebs and N. B. Davies, eds., *Behavioral Ecology: An Evolutionary Approach* (pp. 380–402). Sunderland, MA: Sinauer Associates.

Kurth, C., Krahn, D., Nairn, K., and Drewnowski, A. (1995). The severity of dieting and bingeing behaviors in college women: Interview validation of survey data. *Journal of Psychiatric Research* 29(3), 211–225.

Lancaster, J. D., Altmann, J., Rossi, S. A., and Sherrod, L. R., eds. (1987). *Parenting Across the Life Span*. New York: Aldine de Gruyter.

Lockard, J., and Paulhus, D. L., eds. (1988). *Self-deception: An Adaptive Mechanism?* Engelwood Cliffs, NJ: Prentice Hall.

Marks, I. M. (1987). *Fears, Phobias, and Rituals*. New York: Oxford University Press.

Marks, I. M. (1988). Blood-injury phobia: A review. *American Journal of Psychiatry* 145(10), 1207–1213.

Marks, I. M., and Nesse, R. M. (1994). Fear and fitness: An evolutionary analysis of anxiety disorders. *Ethology and Sociobiology* 15(5–6), 247–261.

Marks, I. M., and Tobena, A. (1990). Learning and unlearning fear: A clinical and evolutionary perspective. *Neuroscience and Biobehavioral Reviews* 14, 365–384.

McGuire, M. T. and Fairbanks, L.A. (eds.) (1977). *Ethological Psychiatry: Psychopathology in the Context of Evolutionary Biology*. New York: Grune and Stratton.

McGuire, M., Marks, I., Nesse, R., and Troisi, A. (1992). Evolutionary biology: A basic science for psychiatry.

Acta Psychiatrica Scandinavica 86, 89–96.

McGuire, M. T., and Troisi, A. (1998). *Darwinian Psychiatry*. New York: Oxford University Press.

Mealey, L. (1995). The sociobiology of sociopathy: an integrated evolutionary model. *Behavioral and Brain Sciences*. 18:523–541.

Melzack, R. (1973). *The Puzzle of Pain*. New York: Basic Books.

Miller, G.F. (In press). *Evolution of the Human Brain through Runaway Sexual Selection*. Boston: MIT Press/ Bradford Books.

Mineka, S., Keir, R., and Price, V. (1980). Fear of snakes in wild- and laboratory-reared rhesus monkeys (*Macaca mulatta*). *Animal Learning and Behavior* 8(4), 653–663.

Morris, W. N. (1992). A functional analysis of the role of mood in affective systems. In M. S. Clark, ed., *Emotion*. Newbury Park, CA: Sage.

Nesse, R. M. (1984). An evolutionary perspective on psychiatry. *Comparative Psychiatry*, 25(6), 575–580.

Nesse, R. M. (1990). Evolutionary explanations of emotions. *Human Nature*, 1(3), 261–289.

Nesse, R. M. (1991a). Psychiatry and sociobiology. In M. Maxwell, ed., *The Sociobiological Imagination*. Albany: SUNY Press, pp. 23–40.

Nesse, R. M. (1991b). What is mood for? *Psycholoquy* 2(9.2).

Nesse, R. M. (1994). An evolutionary perspective on substance abuse. *Ethology and Sociobiology* 15(5–6), 339–348.

Nesse, R. M., and Lloyd, A. T. (1992). The evolution of psychodynamic mechanisms. In J. Barkow, L. Cosmides, and J. Tooby, eds., *The Adapted Mind: Evolutionary Psychology and the Generation of Culture* . New York: Oxford University Press, pp. 601–624.

Nesse, R., and Williams, G. (1994). *Why We Get Sick: The New Science of Darwinian Medicine*. New York: Times Books.

Nesse, R.M., and Berridge, K.C. (1997). Psychoactive drug use in evolutionary perspective. *Science*, 278, 63–66.

Nisbett, R., and Ross, L. (1980). *Human Inference: Strategies and Shortcomings of Social Judgment*. Englewood Cliffs, NJ: Prentice-Hall.

Öhman, A., and Dimberg, U. (1984). An evolutionary perspective on human social behavior. In W. M. Waid, ed., *Sociophysiology*. New York: Springer, pp. 47–86.

Plutchik, R. (1980). *Emotion: A Psychoevolutionary Synthesis*. New York: Harper and Row.

Plutchik, R., and Kellerman, H., eds. (1989). *Emotion: Theory, Research, and Experience*. New York: Academic Press.

Pomerleau, C. (1997). Cofactors for smoking and evolutionary psychobiology. *Addiction* 92, 397–408.

Price, J. (1967). The dominance hierarchy and the evolution of mental illness. *Lancet* 2, 243–246.

Price, J. S., and Sloman, L. (1987). Depression as yielding behavior: An animal model based on Schjelderup-Ebbe's pecking order. *Ethology and Sociobiology* 8, 85s–98s.

Raleigh, M., McGuire, M., Brammer, G., Pollack, D., and Yuwiler, A. (1991). Serotonergic mechanisms promote dominance acquisition in adult male vervet monkeys. *Brain Research* 559(2), 181–190.

Raleigh, M. J., McGuire, M. T., Brammer, G. L., and Yuwiler, A. (1983). Social and environmental influences on blood serotonin concentrations in monkeys. *Archives of General Psychiatry* 41, 405–410.

Rancour-Laferriere, D. (1985). *Signs of the Flesh*. New York: Mouton de Gruyter.

Ridley, M. (1993). *The Red Queen.* New York: Macmillan.

Seligman, M. (1970). On the generality of the laws of learning. *Psychological Reviews* 77, 406–418.

Slater, E., Hare, E. H., and Price, J. S. (1971). Marriage and fertility of psychiatric patients compared to national data. *Social Biology* 18 (suppl.), s60–s73.

Slavin, M. (1992). *The Adaptive Design of the Human Psyche : Psychoanalysis, Evolutionary Biology, and the Therapeutic Process.* New York: Guilford Press.

Slavin, M. O. K., Daniel. (1990). Toward a new paradigm for psychoanalysis: An evolutionary biological perspective on the classical-relational dialectic. *Psychoanalytic Psychology*, 7 (suppl.), 5–31.

Slavin, M., and Kriegman, D. (1992). *The Adaptive Design of the Human Psyche: Psychoanalysis, Evolutionary Biology, and the Therapeutic Process.* New York: Guilford Press.

Smith, E. A., and Winterhalder, B., eds. (1992). *Evolutionary Ecology and Human Behavior.* New York: Aldine de Gruyter.

Smuts, B. (1995). The evolutionary origins of patriarchy. *Human Nature* 6(1), 1–32.

Staddon, J. E. R. (1983). *Adaptive Behavior and Learning.* Cambridge: Cambridge University Press.

Stevens, A. (1982). *Archetype: A Natural History of the Self.* London: Routledge.

Stevens, A., and Price, J. (1996). *Evolutionary Psychiatry: A New Beginning.* London: Routledge.

Surbey, M. K. (1987). Anorexia nervosa, amenorrhea, and adaptation. *Ethology and Sociobiology* 8(3), 47–61.

Symons, D. (1979). *The Evolution of Human Sexuality.* New York: Oxford University Press.

Tooby, J., and Cosmides, L. (1990a). On the universality of human nature and the uniqueness of the individual: The role of genetics and adaptation. *Journal of Personality* 58, 17–67.

Tooby, J., and Cosmides, L. (1990b). The past explains the present: Emotional adaptations and the structure of ancestral environments. *Ethology and Sociobiology*, 11(4/5), 375–424.

Trivers, R. L. (1971). The evolution of reciprocal altruism. *Quarterly Review of Biology* 46, 35–57.

Trivers, R. L. (1974). Parent-offspring conflict. *American Zoologist* 14, 249–264.

Trivers, R. L. (1976). Foreword to *The Selfish Gene* by R. Dawkins. New York: Oxford University Press, pp. v–vii.

Trivers, R. (1985). *Social Evolution.* Menlo Park, CA: Benjamin/Cummings.

Tversky, A., and Kahneman, D. (1974). Heuristics and biases. *Science* 185, 1124–1131.

Voland, E., and Voland, R. (1989). Evolutionary biology and psychiatry: The case of anorexia nervosa. *Ethology and Sociobiology* 10, 223–240.

Weinberg, E. D. (1984). Iron withholding: a defense against infection and neoplasia. *Physiology Reviews* 64, 65–102.

Weissman, M. M., Bland, R. C., and Canino, G. J. (1996). Cross-national epidemiology of major depression and bipolar disorder. *Journal of the American Medical Association*, 276, 293–296.

Wenegrat, B. (1984). *Sociobiology and Mental Disorder: A New View.* Menlo Park, CA: Addison-Wesley.

Wenegrat, B. (1990). *Sociobiological Psychiatry: Normal Behavior and Psychopathology.* Lexington, MA: Lexington.

Williams, G. C. (1966). *Adaptation and Natural Selection: A Critique of Some Current Evolutionary Thought.*

Princeton, NJ: Princeton University Press.

Williams, G. W., and Nesse, R. M. (1991). The dawn of Darwinian medicine. *Quarterly Review of Biology* 66(1), 1–22.

Wilson, D. R. (1992). Evolutionary epidemiology. *Acta Biotheretica* 40:187–189.

Wright, R. (1994). *The Moral Animal.* New York: Vintage.

15

EVOLUTION, SUBSTANCE ABUSE, AND ADDICTION

E. O. SMITH

Magnitude of the Problem

Darwinian evolutionary theory made little impact on the study of human behavior for almost a century. The application of Darwinian theory to the behavior of modern humans was attempted in the 1920s–30s, but with little success. The primary weakness of this attempt, loosely referred to as "social Darwinism" was the failure to differentiate between what is and what ought to be. Darwinian theory, for all its strengths and analytical power, does not dictate a preferred course of action, one that is consistent with the general assumptions of progress, improvement, and competition. Darwinian theory simply provides an explanation for what exists at present and does not offer a master plan. As Beckstrom (1993:2) noted, evolutionary science "can act like travel agents. They cannot tell you where to go, but they can give you information about the costs and benefits of various destinations and help you get there once you finalize your decision."

If, in the last two decades, any advances have been made in the application of evolutionary theory to human behavior, science may now be able to offer some possible insights into what most agree is a pressing societal problem: substance abuse and addiction. Substance abuse is a major problem in Western society, implicated in the deaths of half a million Americans annually, with an associated monetary cost that approximates the annual budget for the Department of Defense (Horgan et al. 1993). Figure 15.1 shows the total costs for all types of substance abuse for 1990. In this chapter, I (1) briefly clarify terminology concerning use and abuse of psychoactive substances; (2) provide a brief historical perspective on drug use/abuse; (3) define the characteristics of a Darwinian trait and the extent to which substance abuse may be considered one; (4) review the available data that assess the magnitude and the costs as-

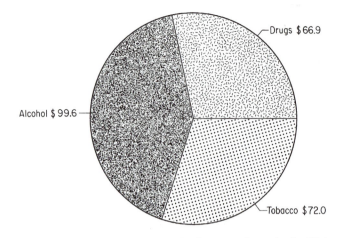

Figure 15.1. The economic costs (in $B) of substance abuse in the United States in 1990. (Redrawn from Horgan et al. 1993: 16.)

sociated with substance abuse; and (5) offer some practical observations that might assist social planners, clinicians, and others in eliminating or at least dealing with the problem from a new perspective.

Terminology

Definitional problems abound in the study of drug use/abuse. Confusion exists over what actually constitutes abuse as differentiated from use, although there seems to be slightly less confusion over what constitutes addiction (see Portenoy & Payne 1992 for a contrary opinion). According to some, if a drug is illegal, then any use is abuse, while others argue that *any use* of legal drugs is not abuse. Some confuse and often equate the use of illegal drugs with the abuse of legal ones. In this paper, I am not concerned with the legality of any particular substance, but I am concerned with a pattern of substance use that has both individual and social costs, substances that are self-administered without medical supervision, and substances that have both psychological and physical withdrawal effects.

It is clear that drug abuse and addiction are far from being a homogenous group of psychiatric problems, and it is also clear that individuals with substance abuse disorders present a heterogeneous collection of clinical symptoms (Bohn & Meyer 1994). Before discussing the evolution of something as complicated as drug addiction and abuse, it is important to present some useful definitions. The American Society of Addiction Medicine has agreed on 32 definitions of terms that are widely used in addiction medicine (Steindler 1994). The result of this adoption is a standardization of the use of terms that have often led to confusion and disagreement.[1]

In general, clinicians and substance abuse professionals distill a few characteristics of central importance in defining drug abuse. Three central features characterize drug abuse/addiction: (1) a psycho/behavioral syndrome which includes (but is not limited to) drug craving, compulsive use, self-administration not under medical supervision, and self-injurious or destructive behavior; (2) physical dependence on the chemical and increasing tolerance to its effects; and (3) physical withdrawal upon cessation of use (Doweiko 1993; Goldstein 1994; Portenoy & Payne 1992; Ray & Ksir 1990). These are general characteristics of drug abuse and are not restricted to any particular psychoactive substance.

Historical Perspective

Preagriculture

Before the rise of agriculture, access to psychoactive substances likely was limited. This is not to say that such use was not important and widespread, but rather that quantities of psychoactive substances available at any one time were limited. Psychoactive substances certainly have a long history of association with religious activities and the achievement of desired altered states of consciousness. Early uses of psychoactive drugs that ameliorated pain, such as analgesics (salicin, morphine) as well as local anesthetics (cocaine), could well have been adopted early in our evolutionary history. The sedative properties of some psychoactive substances might have been used to calm the minds of our ancestors. It may well be that some psychoactive substances initially were used for their therapeutic properties. It is well known that some plants that have hallucinogenic properties often have the side benefit of being toxic to gastrointestinal parasites. Use of hallucinogens could have originated as an effort to control parasitic infection. Diuretics such as caffeine may alter blood pressure, while members of the milkweed family (*Asclepias* spp.) contain cardiac glycosides which can have powerful therapeutic effects (Johns 1990). Plants containing natural stimulants might have provided relief from fatigue as well as elevation of mood, while others may have been used to inhibit anxious responses. Recognition of the importance of psychoactive plants was likely an important part in hominid evolution (Malcolm 1971).

Alcohol

Alcoholic beverages were possibly among the earliest widely consumed psychoactive substances. The earliest record of insobriety is found in Genesis when, after the flood, Noah is reported to have gotten drunk on wine (Gardner 1992). The earliest archaeological evidence of use of alcohol comes from about 6400 BC (Ray & Ksir 1990) and is coincident with the rise of agriculture. The oldest known preserved code of laws, that of the Babylonian king Hammurabi written in 2225 BC, contained regulations for the conduct of business in beer and wine shops, as well as taverns (McKim 1986). Plato established strict rules of conduct for his drinking parties that required one individual, "the master of the feast" to stay sober and determine how much water was to be added to the wine (McKinlay 1951).

It is likely that there are no people on the earth that have not come into contact with alcoholic beverages in some form. Complete abstinence is not widely practiced among modern humans (except for followers of Islam and Alcoholics Anonymous), and there is considerable evidence that consumption of fermented beverages (particularly fruits) has a long history. Many foods that are regularly collected by hunters and gatherers have sufficient sugar content to ferment. In fact, one of the preferred foods of many hunters and gatherers, honey (O'Dea 1991), is perhaps the first food that was fermented and drunk (Crane 1980). Mead, a combination of honey and water, was possibly the first alcoholic beverage, appearing as early as 8000 BC (Ray & Ksir 1990). Indigenous people the world over have developed fermented drinks from a variety of sources: in Siberia red algae was used, North American Indians used maple syrup, Central American Indians produced fermented drinks from agave and cactus, South American Indians used a variety of jungle fruits, and Asians used rice (Siegel 1989).

In the ninth century the Arabs developed a distillation process to increase the alcoholic content of fermented drinks, particularly of wine (Ray & Ksir 1990). Only by distillation can the alcohol content of fermented drinks be elevated above the chemically self-limiting ceiling of approximately 12% (Goldstein 1994). England provides an interesting test case for the potency of distilled beverages and a corresponding case for the late arrival of behavioral problems associated with psychoactive substances. Until the introduction of Dutch gin into England in the early eighteenth century, alcoholism was not seen as a major problem by the government or the public at large (Goldstein 1994). Urban drunkenness, particularly among the poor, became a significant social problem with the availability of alcohol-enhanced drinks, that instead of having an alcohol content of 3–12% for beer and wines, had an alcohol content of as much as 50% or more. Distillation raised the costs for consumption of alcoholic beverages considerably because of the severity of the effects and the costs of consumption; however, it is also likely the benefits to individual consumers were also increased in terms of the quick and sustained levels of intoxication that distilled spirits can deliver. Now, of course, it is well known that public drunkenness is a part of society wherever alcohol is found (Goldstein 1994).

Tobacco

Unlike some other abused psychoactive substances, the origins of tobacco are exclusively in North and South America and are fairly recent (Jarvik & Schneider 1992). One of the first Europeans to cultivate the habit of "drinking" smoke was the Portuguese explorer Rodrego de Jerez who, upon return to Portugal was imprisoned by the court of the Inquisition for this "devilish habit" (Ray & Ksir 1990), only to be released several years later. In spite of its rather chilly reception, smoking would eventually take hold in Europe by the mid-nineteenth century, particularly in France (McKim 1986).

Upon arrival in San Salvador, on October 12, 1492, Christopher Columbus was presented with tobacco leaves. Explorers over the next 50 years found that tobacco use was not restricted to Central America. Jacques Cartier found Indians along the St. Lawrence River smoking pipes in 1535, and in 1495 Amerigo Vespucci found natives chewing tobacco and mixing it with lime (McKim 1986). Tobacco was the most widely cultivated crop in precolonial North America. In the Northwest it was mixed with lime and chewed, while in California it was mixed with *Datura* (a powerful hallucinogen) and drunk. In Eastern North America it was smoked along with sumac (*Rhus* spp.) leaves and the inner bark (phloem) of dogwood (*Cornus* spp.) (Malcolm 1971). In Muslim countries, smoking was forbidden because it was deemed to be an intoxicant and, as such, against the teachings of the Koran. In Russia the czar punished first offenders by slitting their noses and habitual consumers were sent to Siberia or put to death (Argiolas et al. 1986). Both China and Japan instituted equally harsh punishments, but it was not long before most governments realized that there was a considerable amount of money to be made by taxing tobacco, and this tactic was used as a means of discouraging its use (McKim 1986).

Tobacco became a major addictive substance in the last century, largely through a number of technological advances. A new way of curing tobacco was developed that made smoke less irritating to air passages. Mechanization of cigarette production in the 1880s vastly increased production capacity as well as the incentive to increase the market demand. The development of the safety match should not be overlooked as having a powerful influence on the spread of cigarette smoking. New techniques of mass marketing, such as billboards, sponsorship of sporting events, clothing, and apparel have also vastly expanded the appeal of smoking (Goldstein 1994).

Unlike other psychoactive drugs, nicotine from tobacco is almost never administered in its pure form. Nicotine is highly toxic and in its pure form can have lethal consequences. Administration of nicotine in tobacco is a preferred route, for it allows precise control over the concentration of the drug ingested (McKim 1986). Tobacco cigarette smoking is the most common substance abuse disorder in the United States, affecting more than 51 million individuals, and it is the chief avoidable cause of premature deaths in the United States (Jarvik & Schneider 1992). Considerable debate has ensued over the addictive nature of nicotine, but the preponderance of evidence, both animal and human, suggests that it is addictive (Stolerman & Jarvis 1995).

Opium

Opium has a long, distinguished association with *Homo sapiens*; its history goes back more than 6000 years (Sumarians 4000 BC, Egyptians 2000 BC)

(Simon 1992). The pain-reducing properties of opium have been well known for a long time. The extent of the use of opium in early Egyptian and Greek cultures is unclear, but one of the first recorded uses was in Ebers papyrus (ca. 1500 BC), where it was described as being used to prevent the excessive crying of children. Opium was important in Greek medicine, and many believed it was a cure-all. Galen reported that opium cakes and candies were being sold on the streets as a panacea for a wide variety of ailments. Marcus Aurelius, the Roman emperor of this period, was likely addicted to opium and occasionally suffered withdrawal symptoms (Ray & Ksir 1990). In the sixteenth century, there was a marked increase in the interest in opium, largely through the efforts of Paracelsus, who developed a medication called laudanum. Although it is not clear that laudanum contained opium, Paracelsus used opium in a variety of other medications. Dr. Thomas Sydenham, father of clinical medicine, and often referred to as the English Hippocrates, developed a concoction also called laudanum that contained 2 ounces of strained opium, 1 ounce of saffron, and a dram of cinnamon and cloves dissolved in a pint of Canary wine. This is the concoction that Thomas De Quincy took that set the stage for the widespread use of opium in nineteenth century Europe (Ray & Ksir 1990). By the middle of the nineteenth century, there was widespread use of opium as treatment for a variety of clinical conditions including pain, cough, diarrhea, fever, inflammation, delirium tremens, epilepsy, melancholy, mania, asthma, poisoning, diabetes, hemorrhage, skin ulcers, snake bite, rabies, tetanus, spasmodic dysphagia, and constipation (Kramer 1980). Widespread popularity of opiates led to a most unfortunate result: the extensive use of opiates to sedate infants. Because of these practices, an untold number of children became addicted at the hands of their parents.

Morphine was isolated from opium in 1806 by Frederich Sertürner. By 1836, the clinical value of morphine was so clear that Sertürner received the French equivalent of the Nobel Prize. At about the same time in the United States, the Civil War placed a premium on the use of opiates to blunt the pain of battlefield injuries. This widespread use led to an increase in the number of addicts in this country. By the last quarter of the nineteenth century, the nature and extent of opiate dependence was being recognized (Calkins 1871), but there was little understanding of the nature of addiction. Interestingly, while opium was recognized as addictive, heroin (discovered in 1874 and marketed in 1898 by Bayer Laboratories) was thought to be relatively safe (Ray & Ksir 1990).

Use of heroin in the United States has had an interesting and complicated history, which is beyond the scope of this chapter. It is safe to say, however, that the Vietnam War, with the widespread deployment of American troops throughout Southeast Asia, and the ready availability of high-quality, inexpensive heroin did much to introduce heroin to a wide spectrum of the American populous. Today heroin use accounts for slightly more than 25% of the patients in drug treatment facilities (Horgan et al. 1993).

Coca/Cocaine

The history of the use of cocaine has been described elsewhere (Gold & Verebey 1984; Holmstedt & Fredga 1981; Kleber 1988). Briefly, cocaine is a derivative of coca, a shrub native to the eastern Andes, where it has been cultivated for thousands of years. In Bolivia and Peru, millions of Indians chew coca leaves daily. It is estimated that the average native user chews about 60 grams of coca leaves per day. Given that the alkaloid content of a coca leaf is 0.5–0.7%, and taking into account incomplete absorption, it is likely that total dosage is 200–300 mg per 24-hour period (Gawin 1991).[2] In the 1800s, coca became popular in Europe and America in the form of tonics and wines. Coca-Cola began as one of these preparations in 1896; however, by 1906 the coca extract in Coca-Cola was replaced with caffeine. By then, there were 69 imitations of Coca-Cola that contained cocaine (Grinspoon & Bakalar 1976). It was also in the mid-nineteenth century that cocaine was isolated from coca leaves. At first cocaine was used as a local anesthetic; especially for eye operations, but it quickly began to be prescribed for a variety of other ailments. This practice was quickly discontinued when it was recognized that many patients suffered ill effects from cocaine. In the early 1900s laws were passed to control the use of cocaine, but by then a huge black market had developed.

Today cocaine use has declined in the population at large, from a high of 12.2 million users in 1985 to slightly more than 6 million users in 1991; however, the number of heavy users has not significantly changed from 1985 (1985: 647,000; 1991: 625,000). This suggests that heavy use is still a real problem, particularly in urban areas where hard-core users become concentrated and drug-related crime flourishes (Horgan et al. 1993). Today the popularity of "crack" cocaine poses particularly difficult problems. Because of the relatively low price of crack compared with cocaine hydrochloride (powdered form), the rapid "high" (often within 10 seconds) makes crack particularly attractive. However, the short duration of the high (5–15 minutes) relative to that derived for intranasal administration has the net effect of increasing demand (Gold et al. 1992).

Cannabis

The earliest reference to cannabis is found in a pharmacy book written in 2737 BC by the Chinese Emperor Shen Nung, in which he referred to the psychoactive effects of the "Liberator of Sin." Social use of the plant spread to the Moslem world and North Africa by 1000 AD. The use of cannabis has a long history in the Orient and Middle East. Along with exotic spices, coffee, and tea, cannabis was introduced by early explorers to European populations. By the nineteenth century use was widespread. One of the earliest popular accounts of the use of hashish (a potent derivative of marijuana) is found in Alexander Dumas's *Count of Monte Cristo*. The psychoactive properties of cannabis were well known in Europe, and the followers of the Romantic literary tradition, as well as the Impressionist school of art, were searching for a new intellectual experience and are known to have used hashish extensively. The early part of the twentieth century saw increasing concern about marijuana, and in 1937 the U.S. Congress passed the marijuana tax. The net effect was not to directly outlaw marijuana, but following the regulation-by-taxation theme of the Harrison Act of 1914, it taxed the grower, distributor, seller, and the consumer and made marijuana administratively impossible to deal with (Ray & Ksir 1990).

Today marijuana use continues largely unabated, in spite of concerted efforts by law enforcement agencies to reduce supply, grown both domestically and in foreign markets. Data for 1991 report 11% of eighth graders had tried marijuana and 4% said they had smoked it in the last month, while in the overall population 6.6% reported marijuana use in the last month (Horgan et al. 1993). Public opinion varies greatly concerning the dangers of marijuana. During the late 1970s, the U.S. experienced a de facto decriminalization of possession of marijuana, but today the pendulum has swung in the other direction toward a "get tough" policy for offenders (Ray & Ksir 1990). Cannabis poses a problem not only domestically, but on a worldwide basis as well. In Nigeria, estimates range from 20–50% of the male admissions to psychiatric wards are suffering from toxic psychosis from cannabis ingestion. Similar widespread use has also been reported in Uganda (Desjarlais et al. 1995).

Caffeine

Caffeine is the most popular psychoactive agent and is consumed daily by millions of people worldwide. Compared to some other psychoactive substances, caffeine is a recent arrival on the scene. The drug was first isolated in coffee in 1821, although Muslims began using coffee in religious rituals and ceremonies more than a thousand years ago. Coffee came to Europe in the seventeenth century and was soon accepted throughout the continent. Coffee drinking was so popular and widespread by this time that it was labeled an abuse by moralists of the day. In his famous *Kaffee Kantate* (1734), J. S. Bach portrays a father distressed at his daughter's addiction to coffee (Bettmann 1995).[3] Honoré de Balzac was a caffeine addict and required large quantities to work. Balzac wrote in his *Treatise of Modern Stimulants* on the effects that coffee had on him. He observed that a fortnight without coffee caused severe stomach cramps and depression (Lawton 1910). In addition to coffee, there are a variety of sources for caffeine (e.g., tea, cola, guaraná, maté, chocolate). Chocolate has only a small amount of caffeine, but has a considerable amount of theobromine, which has similar behavioral effects to caffeine.

The economics of coffee make it an important product on the worldwide market (second only to oil). The price of coffee dictates consumption patterns, and as the price of coffee goes up, consumption goes down (Ray & Ksir 1990). The per capita consumption of caffeine on a worldwide basis is approximately 70 mg/day, while in the United States this figure exceeds 200 mg/day, and some individuals report an intake of 2–5 grams/day (Greden & Walters 1992).

Caffeine intoxication is recognized by the American Psychiatric Association (American Psychiatric Association 1994) and is characterized by restlessness, nervousness, excitement, insomnia, flushed face, diuresis, and gastrointestinal problems. Coffee intoxication may be induced by consumption of as little as 200 mg of caffeine, but invariably it is induced by consumption of 1 gram or more (Syed 1976). While not posing the immediate social costs that use of other psychoactive substances carry, caffeine can have deleterious consequences for individuals, in fact, at least six deaths have been attributed to an overdose of caffeine. Lethal doses in humans have been estimated at 3–8 grams taken by mouth. Death results from convulsions and respiratory failure (McKim 1986).

Hallucinogens

Along with alcohol, hallucinogens may very well have considerable antiquity as substances widely exploited for their psychoactive properties. Hallucinogens have their origins in naturally occurring plants, and while any drug taken in sufficient quantity can induce hallucinations, hallucinogens induce hallucinations as a property of the drug directly attributable to its mode of action, as opposed to the quantity consumed. Naturally occurring hallucinogens are found in a wide array of plants. For example, they are found in the ergot fungus that infects rye and resembles lysergic acid diethylamide (LSD) in its psychoactive properties. Psilocybin is a serotoninlike substance found in several species of mushrooms (members of the genera *Psilocybe, Conocybe, Paneolis,* and *Stropharia*) native to North America. Lysergic acid is found in the seeds of the morning glory (*Turbina corymbrosa*). Dimethyltryptamine (DMT) can be found in the bark of the members of the genus *Virola*. Mescaline is the psychoactive ingredient in the peyote cactus (*Lephophora williamsii*). Myristicism is a drug that is found in the fruits of members of the genus *Myristica* and in particular *Myristica fragrans* or nutmeg. Other hallucinogens include ibotinic acid, the active ingredient in the *Amanita muscaria* mushroom (McKim 1986).

Hallucinogens are chemically related to natural neurotransmitters and, like all psychoactive drugs, act by disturbing the finely tuned neurochemistry of the brain. Hallucinogens, like other addictive drugs, are self-administered for the purpose of altering mood, emotion, and perception. Native peoples all over the world are known to use hallucinogens ceremonially, on infrequent occasions, or under the strict control of a shaman or religious leaders. This lack of widespread use may reflect a deep-seated concern about the dangerous properties of the drugs, or it may simply be due to reduced availability. Unlike other drugs mentioned here, hallucinogens have low abuse potential. They are not used in an out-of-control fashion as seen with other addictive drugs. Of all the hallucinogens, only phencyclidine might be considered addictive, because it produces dependence and an opiatelike withdrawal (Goldstein 1994).

Evolutionary Characteristics

For a trait to evolve in a population, three conditions must be present: genetic basis, variation in expression, and effect on fitness. It is a central tenet of Darwinian evolutionary theory that a trait could not evolve if it reduced the reproductive success of its carrier. In an evolutionary sense, then, one wonders how the predisposition for drug use, a seemingly maladaptive trait, could persist in the population. If drug use imposes negative fitness costs on its possessor, how could it have evolved? To determine if substance abuse is a phenomenon amenable to evolutionary analysis, let us see how well drug abuse fits the Darwinian model.

Genetic Basis

The first assumption of the evolutionary perspective is that substance abuse (or any potential Darwinian character) has at least a partial underlying genetic basis (Gianoulakis & de Waele 1994; Goodwin 1985; Harford 1992; Karp 1994; Li et al. 1994; Lumeng & Crabb 1994; Svikis et al. 1994; Vesell et al. 1971). The most compelling data supporting this position come from the studies of alcoholism, in particular "twin studies" as well as studies of children of alcoholic parents (Anthenelli & Schuckit 1992; Kendler et al. 1992; 1994). The general logic of twin studies is to assess the relative contributions of genetic and environmental factors for an illness by comparing rates of the illness in monozygotic and dizygotic twins. In the extreme, monozygotic twins should show higher rates than the dizygotic twins if the disorder is heavily genetically influenced. On the other hand, environmentally induced disorders should reveal no differences in rates of the illness between monozygotic and dizygotic twins (Anthenelli & Schuckit 1992). Kaij (1960) found the similarity or concordance rate for alcoholism among dizygotic and monozygotic twins to be vastly different, with monozygotic twins having roughly double the rate for dizygotic twins. Other studies (Vesell et al. 1971) generally support the findings of high heritability of alcoholism, although not all authors agree (Gurling et al. 1984). The preponderance of evidence supports the hypothesis of a genetic basis for alcoholism, but the results of these studies point to the complex nature of the gene–environment interaction (Anthenelli & Schuckit 1992).

Classical adoption studies have also provided researchers with a powerful methodology to study the relative contributions of genetic versus environmental factors for a variety of diseases. A number of studies have shown that offspring raised apart from an alcoholic biological parent had a significantly higher probability of developing alcoholism than those adopted children with nonalcoholic biological parents (Bohman et al. 1981; Gurling et al. 1984).

Recently, Gabel and his colleagues (Gabel et al. 1995) found a significant positive correlation between fathers who were substance abusers (SA) and increased rate of conduct disorders in their sons. The study assessed the relationship between homovanillic acid (HVA), the metabolite of dopamine (DA), an important brain neurotransmitter, and monoamine oxidase (MAO), the enzyme that facilitates the conversion of DA to HVA. Significantly higher MAO activity was found in sons of SA fathers than of non-SA fathers. These findings further support the underlying genetic predisposition toward substance abuse.

At the molecular level, there is some evidence for the genetic basis of substance abuse and, in particular, alcoholism (Bohman et al. 1981; Cotton 1979; Goodwin et al. 1973; Karp 1994; Kendler et al. 1992). One of the most compelling hypotheses about the genetic basis for substance abuse, addictive, impulsive, and compulsive disorders in general (Blum et al. 1996) concerns DA. Blum and his colleagues found at least one reward pathway in the brain that is implicated in a variety of psychiatric disorders. They suggest that an alternative form of the gene for the dopamine D_2 receptor, the A_1 allele, is implicated in altered patterns of DA release in response to challenges from a variety of sources. Although each substance of abuse affects different parts of the neural pathway (serotonin in the hypothalamus, enkephalins in the ventral tegmental area and the nucleus accumbens, and gamma-amino butyric acid (GABA) in the ventral tegmental area and the nucleus accumbens), the results are the same. Dopamine is released in the nucleus accumbens and the hippocampus.

The current model suggests that individuals predisposed toward a variety of addictive, compulsive disorders are often born with an alteration on chromosome 11 that regulates the expression of the gene that codes for the D_2 receptor gene, the A_1 allele. Individuals with the A_1 allele have approximately 30% fewer D_2 receptors than those with the alternative allele (A_2). It is well known that D_2 receptors are responsible for the inhibition of the releasing enzyme adenylate cyclase, suppression of Ca^{2+}, and activation of K^+ currents. The D_2 receptor gene controls the production of the D_2, receptors, therefore possession of the A_1 allele is implicated in a reduced number of D_2 receptors. The precise mechanism of action is unclear, but it appears that the A_1 allele reduces the expression of the D_2 allele relative to carriers of the alternative (A_2) allele. This reduced number of D_2 receptors may translate into lower levels of dopaminergic activity in those parts of the brain implicated in behavior reward. It is possible that A_1 carries may not experience normal rewards associated with dopamine activity, and this may translate into stimulus-seeking behavior or cravings. Alcohol, cocaine, marijuana, nicotine, and theobromine can all increase the level of dopamine produced and can result in a temporary reduction in craving. Blum and his colleagues suggest that affected individuals (A_1 carriers) are likely to attempt to modulate dopamine levels either by consumption of psychoactive substances or by engaging in activities that enhance dopamine production or uptake (Blum et al. 1996).

Variation in Expression

The second assumption requires that there be some phenotypic variation in the characteristic of interest. Again, relevant data come largely from studies of alcoholics, but there are possible similar patterns of variation in the use/

abuse of other drugs. Ethnic differences have been suggested to play a role in the tolerance to alcohol as well as to other drugs (Goedde et al. 1983; Yamashita et al. 1990).[4] Between 30% and 50% of the Asian population lack one of the isoenzyme forms of aldehyde dehydrogenase (ALDH), the major enzyme that degrades the first metabolite of ethanol, acetaldehyde, in the liver. After imbibing alcohol, affected individuals develop heightened blood acetaldehyde levels associated with facial flushing, tachycardia, and a burning sensation in the stomach. Not surprisingly, Asians missing this isoenzyme are less likely to drink heavily and have lower rates of alcoholism (Ewing, et al. 1974). Additionally, Jews demonstrate a low rate of alcoholism, but no underlying genetic factors have been identified. Researchers emphasize possible social practices having the net effects of controlling alcohol consumption: association of alcohol abuse with non-Jews, socialization of children into a culture of moderate drinking, adult primary relationships confined to non drinkers or moderate drinkers, and techniques to avoid excess drinking under social pressure (Glassner & Berg 1980).

On an individual basis, there is considerable difference in the tolerance to psychoactive substances, but it is unclear to what extent these are fundamental genetic differences or differences that are the result of exposure. It seems safe to say that there is modest phenotypic variation in the alcohol metabolism and presumably the metabolism of other psychoactive drugs as well, but certainly ethnic as well as individual differences play a major role in the expression of drug abuse. Figure 15.2 shows the enormous differences in alcohol consumption across a variety of different cultures, ranging from consumption of slightly more than 3 gallons of alcohol per person per year in Luxembourg to less than one-half a quart per person per year in Morocco and Trinidad. American Indians, for example, had no cultural prohibition on alcohol and when they were exposed to frontier alcohol use, were prone to "drink to get drunk." This pattern of drinking and drunkenness was liberated from social rules, and the drunken individual's behavior was tolerated (Westermeyer 1987). Binge or spree drinkers are quite common in Finland; however, these individuals remain sober most of the time and only lapse into drunkenness on occasion (McKim 1986). Aymara and Quechua Indians in the Andean Highlands engage in regular chewing of coca leaves. Periodic chewing (and smoking) of opium by "hill-tribes" in Thailand, Laos, and Burma is also well known and must be seen as a regular part of their culture. These examples are reflective of the enormous variability in drug usage patterns that are in no small part influenced by cultural constraints (Grinspoon & Bakalar 1976).

Fitness Consequences

Assessment of the effects of any particular trait on reproductive success is difficult at best because fitness should be measured in terms of the number of offspring surviving to sexual maturity and subsequently reproducing themselves. For any long-lived species then, it is difficult to accurately assess life-

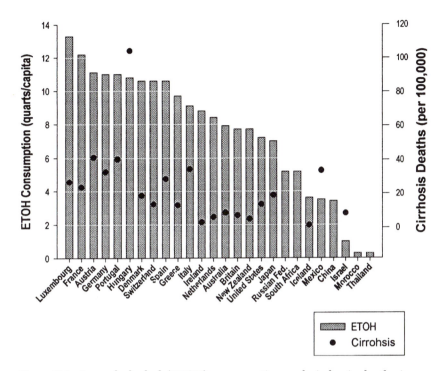

Figure 15.2. Annual alcohol (ETOH) consumption and cirrhosis deaths in selected countries in 1993. (Redrawn from Cronin 1995)

time reproductive success, the only true measure of reproductive differentials
(Brown 1988; Clutton-Brock 1988; Harvey et al. 1986). The only feasible alternative, in most cases, is to measure some correlate of fitness to make any
interindividual comparisons. Data to demonstrate the fitness effects of
substance use and abuse are difficult to quantify, but it can be argued that
there are both positive and negative consequences. The question really
becomes, then, what fitness benefits might have accrued to our ancestors, such
that selection for responsiveness to psychoactive substances could have been
favored? These "ancestral benefits" then may have favored a trait in a historical population that still exists in modern populations, even though its
function may have changed. Such benefits are now liabilities, in modern
fitness terms, but our bodies still function in ways that have considerable
antiquity.

Psychoactive drugs are used for a variety of reasons that ultimately may be
fitness enhancing: (1) sedatives may be used for their sleep-inducing properties; (2) analgesics are used to relieve pain; (3) narcotics can be used to achieve
detachment and euphoria (e.g., opioids acting in the central nervous system
produce analgesia, a decreased sense of apprehension, a sense of tranquillity,
increased self-esteem, and euphoria); (4) stimulants induce feelings of eupho-

ria,[5] an increased sense of energy, enhanced mental acuity, increased sensory awareness, increased self-confidence, and postponement of fatigue; (5) anti-depressants are used to elevate mood and overcome depression; (6) tranquilizers inhibit anxious responses; (7) hallucinogens have been found to break down ego boundaries and heighten perception of sensory stimuli; and (8) alcohol may increase longevity, lower risk of coronary heart disease, and increase levels of high-density lipoproteins (HDL), a negative risk factor for myocardinal infarction. (See Malcolm [1971] for a complete discussion of the potential benefits of drug use.)

The initial use of psychoactive drugs can be traced to a variety of reasons ranging from the direct improvement of existing health conditions to the psychoactive properties of the substance. These initial uses, which could have been fitness-enhancing or at worst adaptively neutral, are now largely overshadowed by the fitness-reducing aspects of substance use and abuse. The key to this hypothesis is the asynchrony of selective forces on the phenotype and the corresponding effects on the genotype. Now there are significant fitness costs to many of these behaviors, but the genotypic response to these pressures is experiencing a time lag (see below).

Perhaps alcohol consumption provides the clearest picture of the costs associated with heavy substance use. Alcohol has been widely found to have a disinhibitory effect on consumers and as such is implicated in the expression of aggressive behavior (Giancola & Zeichner 1995; Laplace et al. 1994; Pihl & Peterson 1995). Although not indicative of the costs in a historical or evolutionary perspective, nearly half of the convicted felons in the United States are alcoholic (Golding 1993; Murdoch et al. 1990) and about half of all police activities in large cities are associated with alcohol-related offenses (Goodwin 1992). Overall, the death rate for alcoholism is approximately 3% of total deaths in the United States (Winick 1992). More than 19,500 deaths were directly attributable to alcohol numbered, while alcohol was indirectly implicated in an additional 88,900 fatalities in 1989 (Horgan et al. 1993). Other adverse consequences of alcohol use include an intimate association with suicide and homicide. In at least 50% of the homicides worldwide, the slayer, the victim, or both had measurable blood alcohol levels. Alcohol is involved in about 75% of the suicides and as many 86% of murderers (Lester 1992; 1995; Rich et al. 1986) in the United States. Cross cultural data have confirmed the association between alcohol and violence. In Papua New Guinea, beer consumption doubled every 4–5 years during the period 1962–1980; this increase was accompanied by a 400% increase in traffic fatalities as well as increases in death and serious injury from blunt trauma, knife, and bullet wounds (Desjarlais et al. 1995). Estimates of the prevalence of alcohol use vary, but it is estimated that 140 million Americans consume alcohol. Of these, 18 million are reported to be alcoholics or alcohol abusers (Nadelmann 1989). The total cost of alcohol abuse in the United States alone is estimated at near $100 billion annually (Horgan et al. 1993) (see Figure 15.3).

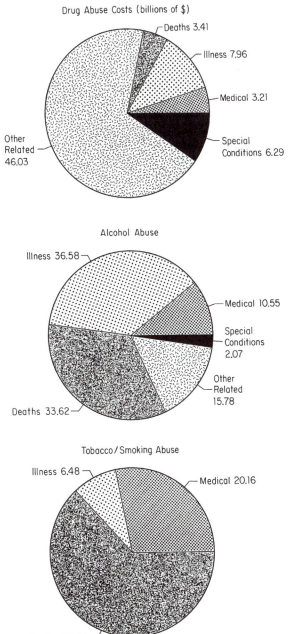

Figure 15.3. Substance abuse costs, by category, in the United States in 1990. (Redrawn from Horgan et al. 1993: 16.)

Perhaps the most unfortunate cost of alcohol consumption is seen in infants of alcoholic mothers. In the overall U.S. population, fetal alcohol syndrome (FAS) affects about 0.4–2.9 births per 1000, but for mothers who are alcoholic the rate climbs to an astonishing 23–29 births per 1000. If all alcohol-related birth defects are counted, the prevalence rate may be as high as several hundred per 1000. It may be that maternal alcohol abuse is the most frequent known environmental cause of mental retardation in the Western world (Ray & Ksir 1990).

In Russia, current estimates suggest that 20–25% of the adult population is alcoholic. There is one female alcoholic to every five male alcoholics. The prevalence varies significantly by occupation; 10% of the workers in the nuclear power industry suffer from alcoholism and 42% of those in the woodworking industry are afflicted. It is further estimated that only one in seven alcoholics seeks professional help in Russia (Matilainen et al. 1994). Alcoholism has reached record proportions among aboriginal groups on Taiwan. Lifetime prevalence rates range from 68.1% to 72.3%. These rates are approximately twice what we see in two other well-studied populations, Peruvian Indians living in Lima (34.8%) (Yamamoto et al. 1993), and Mexican-Americans living in Los Angeles (31.3%) (Karno et al. 1987).

Nicotine addiction provides another useful example. Estimates of the extent of nicotine addiction are difficult to obtain, but it is estimated that 26% of Americans (46 million) smoke, and of these 80% would like to stop and try to do so each year. Only 2–3% of those who try to stop succeed. The economic costs of smoking/tobacco use in 1990 was estimated at $72 billion (Horgan et al. 1993), which is slightly more than the 1990 fiscal budget for the education department ($23.1 billion), the energy department ($12 billion), the justice department ($6.5 billion), and the transportation department ($28.6 billion) combined (see figure 15.3).

Data for other psychoactive drugs are not as detailed, but they also paint a similar picture of the costs of excess consumption (See figure 15.3). For example, even though heroin was introduced into Pakistan only 20 years ago, that country has the highest per capita use of heroin in the world (2.03% of the urban population and 1.36% of the rural population, totaling 1.5 million heroin addicts). Data are not available on the worldwide costs of substance abuse, but total estimated costs of substance abuse in the United States in 1990 exceed $235 billion. This is approximately 23% of the total receipts for the government in 1990. This is an astonishing figure that suggests the benefits of substance abuse must be extraordinary to outweigh the heavy costs. Continuation of the use of substances with such substantial negative effects poses an interesting dilemma. Negative outcomes may occur relatively infrequently, and the possibility of their occurrence may be overwhelmed by the likelihood of pleasurable consequences (Critchlow 1986), but when negative outcomes do occur, the costs are high. The only psychoactive drug that increases negative health outcomes for virtually all users (addicts or not) and for those in the vicinity of the use is tobacco. In these cases, however, the harm is caused by the toxic content of the smoke, not the behavior of the user.

Alternative Hypotheses

In Western society, drug abuse is seen both as a societal problem but more importantly, in many cases, as an individual problem that is to be overcome. Historically, drug abuse has been characterized as a moral or constitutional weakness directly reflecting the character of the individual (Thomason 1938).[6] If, rather than viewing substance abuse as a character flaw or fundamental constitutional weakness, it is viewed as the outcome of complex interactions between biological and social factors, then a new perspective may be developed that is helpful in identifying aspects of the problem that are amenable to clinical intervention. If we consider substance abuse as having underlying biological components, then we are forced to ask how such seemingly maladaptive traits could have evolved. It seems that there are at least several possible hypotheses to explain both the evolution, as well as the maintenance, of this most enigmatic behavior in modern populations. Historically, hypotheses focus on the proximate or neurodevelopmental mechanisms that may contribute to addiction, but rarely have researchers considered the evolutionary basis for the psychiatric phenomenon.[7] I attempt to outline some of the more conspicuous hypotheses below.

Constitutional Weakness—"Pharmacologic Calvinism"

The idea that excessive use of alcohol, or any psychoactive substances for that matter, is a moral problem has a long history in Western thought. King James I (1604) wrote that drunkenness was the root of all sins. Many historical events have signaled the widely held notion that intemperance was among the chief evils of society. Certainly, Prohibition in the United States was an excellent example of the belief in the evil powers of alcohol and drug use. Prohibition was not just a matter of political convention or health concerns, but a complex interplay of these factors with a middle-class, rural, Protestant, evangelical concern that life was being undermined by ethnic groups with different religions, a lower standard of living, and lower standards of morality (Ray & Ksir 1990).

In more recent times, one of the most articulate spokesmen of this opinion was the first "drug czar," the Director of the National Drug Control Policy Center, Dr. William Bennett, who said, "We identify the chief and seminal wrong here as drug use. Drug use, we say is wrong" (Weinraub, 1988:A1). In an editorial the *Wall Street Journal* further echoed these sentiments, "We agree with Drug Czar William Bennett that this [substance abuse] is in no small part a moral question. This nation is suffering a drug epidemic today because of the loosening of societal control in general, and in particular because of the glorification of drugs during the 1960s" (Wall Street Journal 1989:A6).

Those who hold this position see the problem as primarily one of morality and also tend to see enforcement as the key. Bennett also noted "those who use, sell and traffic drugs must be confronted, and must suffer consequences . . . We must build more prisons. There must be more jails. We must have more judges to hear drug cases and more prosecutors to bring them to trial" (Massing 1990:32). The focus of this position is not on the drugs themselves, the behavior of those who consume drugs, the negative health consequences for those that use drugs, or their families, friends, and communities. The focus is on morality.

An evolutionary perspective on drug abuse is distinctly at odds with this hypothesis largely because evolutionary theory makes no assumptions about absolute moral questions. Substance abuse *is* and apparently *has been* for a significant part of human history. Whether it is morally acceptable is *not* the question. It is social problem of growing magnitude and seriousness and in order to control it, we must understand it as completely as possible.

Handicap Hypothesis

Jared Diamond in *The Third Chimpanzee* (1992) applies one of the classic theories of sexual selection to the use and abuse of drugs. Following Zahavi's (1975, 1977, 1991) handicap hypothesis, Diamond (1992) suggests that humans use drugs and engage in other risky behaviors (bungee jumping, hang gliding, sky diving, etc.), particularly in adolescence and early adulthood, as a means of gaining status. Consistent with this view of sexual selection is the observation that males are more likely to engage in this "risk-taking" behavior.[8] The messages of our old and new displays nevertheless remain the same: I'm strong and superior. "Even though I take drugs only once or twice, I must be strong enough to get past the burning, choking sensation of my first puff on a cigarette, or to get past the misery of my first hangover. To do so chronically and remain alive and healthy, I must be superior (so I imagine)" (Diamond 1992: 199).

The handicap hypothesis was developed by Zahavi to explain the existence and maintenance of expensive anatomical accoutrements and behaviors primarily used by males in attracting mates.[9] It is well documented that males engage in drug-taking behavior significantly more than females (Horgan et al. 1993), and by doing so are possibly advertising their fitness to females. While Diamond's argument might make sense for substance abuse by males, it does little to inform questions about females' abuse of psychoactive substances. In general, because of differential parental investment in offspring, males are the sex that must demonstrate their superiority over others of the same sex to secure successful matings. Females, on the other hand, experience disproportionately high costs of producing offspring and are not selected to take risks to demonstrate their fitness.

Cheating and Reproductive Advantage

The cheating and reproductive advantage hypothesis suggests an evolutionary link between alcoholism and antisocial personality disorder (ASPD) (Kofoed 1988; MacMillan & Kofoed 1984). These two disorders co-occur frequently, and there are several hypotheses about their relationship. Seventy percent of men with ASPD have secondary alcohol problems (Anthenelli & Schuckit 1992). The important point is that the disorders are distinct, but they are often found together in the same individual because of selective pressures. Evidence suggests that individuals with both disorders are more successful in sequestering mating opportunities than those with only one. Cheating and deception are viewed as reproductive tactics by many evolutionary biologists (Bond & Robinson 1988; Byrne & Whiten 1992; Kutchinsky 1987; Smith 1987; Welles 1981; Whiten & Byrne 1988). To maximize fitness, individuals will engage in a variety of behaviors that enhance their fitness at the expense of a competitor.

Cheating as a reproductive tactic typically involves males seeking sexual relationships in indiscriminate ways so that they do not invest in the offspring of any particular female, but at the same time they try to inseminate as many females as possible. Females are fooled into believing that these "cheater" males will provide parental investment and consequently allow themselves to be fertilized by these males. According to this hypothesis, individuals with ASPD find themselves in a society that condones the use of alcohol as well as its enhancing effects on the likelihood of sexual activity (albeit at relatively low doses). The confluence of the two disorders provides reproductive opportunities for affected individuals who otherwise might have been precluded from mating. Individuals with the tendency toward positive experiences with alcohol are especially prone to abuse, and when coupled with a predisposition toward ASPD, individuals who carry a genetic predisposition for both conditions are likely to be produced. Phenotypes with both conditions will enjoy enhanced fitness relative to individuals with only one of these conditions. Males, in particular, are predicted to be good at deceiving females, often using alcohol to set the stage for mating attempts. Females must be duped into unreciprocated investment. It is likely that many of these attempts result in failure, but in evolutionary terms, according to this hypothesis, cheating males were sufficiently successful to pass along the genetic characteristic.

Evolutionary By-product

It is possible that the tendency to abuse psychoactive substances is an evolutionary by-product of selection for some other set of characteristics. *Pleitropy* is a phenomenon that occurs when a gene has more than one, apparently independent, phenotypic effect (Hartl 1994). In this case, the propensity to differential responsiveness to psychotropic substances may simply result because of selective pressures for some other character, and one of the multiple effects of selection is the increase in frequency of addictive behaviors. For example, the gene for a highly selected character, enhanced spatial perception, for example, as well as differential psychoactive drug susceptibility, may be pleiotropic effects of the same gene. In that case, the evolutionary costs of such a deleterious trait are balanced against the benefits of the positively selected characteristic. Hence, the maladaptive trait persists in the population in the face of negative consequences because of the compensatory benefits of the selected phenotypic characteristic.

Pleitropy is well known in the evolutionary literature and has been used to explain the evolution of seemingly maladaptive traits in a variety of species. It is possible that the trait in question is subject to directional selection for increase in the mean, but it is genetically negatively correlated with another trait. Individuals who take risks in a variety of types of behaviors may be favored early in life, but those who engage in risky behavior have a shortened life expectancy. This would give the appearance of a positively selected character which has the net effect of lowering fitness (Futuyma 1986).

Phenotypic/Genotypic Asynchrony

If the use and, ultimately, the abuse of psychoactive substances is the outcome of Darwinian evolutionary processes, then what could have been the possible fitness benefits of this behavior? Is it possible that we can treat substance abuse and addiction as an evolved trait like a predisposition to cancer, heart disease, osteoporosis or MS? Do humans, as some have suggested, have a predisposition toward the use of psychoactive substances? Is it possible that like hunger, thirst, and sex, intoxication may be a basic part of the human condition (Siegel 1989)? If this view is correct, then it seems important to analyze the human use of psychoactive substances, not as something dictated entirely by cultural convention and opportunity, but also as a reflection of our evolutionary history.

It is relatively easy to imagine the fitness-enhancing aspects of the use of psychotropic substances in our evolutionary past. From the reduction of stress, improvement in performance, increased sociability, or the simple reinforcing properties of altering psychic state, the use of psychotropic substances could have directly affected fitness. Use of psychotropic substances could have been favored by those who were particularly sensitive to the effects, and in responding they might accrue slight fitness advantages over those less susceptible. This tendency could have been held in check during the course of the evolution of modern humans by the lack of large quantities of highly potent psychoactive drugs. Those substances that were available were all naturally occurring and lacked the concentration of highly refined or synthetically produced substances today. Hence, individual behavior would likely rarely have gotten out of control and become pathological. Today, however, the ready availability and high concentration of psychoactive substances can produce dire consequences.

If as some biological anthropologists have argued (Eaton & Konner 1985; Eaton et al. 1988, 1994), modern humans are basically equipped with the anatomy and physiology of our Paleolithic ancestors, then it may well be that a significant part of the behavior of modern humans directly reflects the evolved behavior of our ancestors. In that case, an evolutionary perspective allows the development of hypotheses about the adaptive significance of a behavior that evolved in a very different ecological setting than the one found in today. Differential responsiveness to psychotropic drugs may have had positive consequences in the past, while imposing heavy costs in the modern setting.

Dopamine Hypothesis

The focus of this chapter is on the evolutionary mechanisms that might have favored drug use in our not-so-distant past. The proximate mechanisms that might be implicated in the maintenance of a genetic predisposition to abuse psychotropic substances are not well developed, but one of the most promising hypotheses concerns the production of a powerful neurotransmitter, dopamine. This hypothesis was originally formulated as a proximate mechanism to explain alcoholism (Blum & Payne, 1991), but has been expanded to include a variety of other obsessive, compulsive disorders (Blum, Cull, Braverman, & Comings, 1996). (See brief discussion above.)

Individuals who exhibit addictive behavior suffer from a neurochemical deficit. Under normal resting conditions a person with this genetic predisposition to drug use cannot achieve feelings of well-being routinely experienced by normal people because not enough dopamine is being released and not enough can bind to the dopamine D_2 receptors in the reward part of the brain. Because of this deficiency to dopamine, a super-sensitivity develops in the nucleus accumbens, the major reward site of the brain. Anything that brings about a release of dopamine, even small amounts of alcohol or other psychotropic drugs, can lead to powerful feelings of well-being. The alcohol- or drug-prone individual experiences a sense of pleasure and marked well-being with the first ingestion of psychoactive substances. The individual is resistant to the adverse effects (loss of motor control, dizziness, and nausea) of the substances. Drugs and/or alcohol temporarily set off the release of dopamine sufficient to mediate the naturally low levels and induce a powerful feeling of well-being. This is precisely the reason that alcoholics consistently report a strong desire to maintain that feeling of euphoria produced by the first few drinks. If, however, the alcohol-prone individual continues to consume alcohol, a number of neurochemical changes can occur, which may include, but are not limited to, a decrease in the number of dopamine recptors (D_2), an increase in the breakdown of dopamine, a decrease of dopamine released at the nucleus accumbens, and a general lowering of neurotransmitter activation at reward sites in the brain. A person drinks more, but the effects decrease and the damage to reward centers increases, intensifying the craving for more alcohol. Again, this explains why the alcoholic will continue to consume alcohol in an effort to regain the euphoria associated with the first few drinks, but is destined, because of the nature of the feedback system, to never be able to experience it. Although the precise neurochemical pathways have not been worked out in detail for other substances, it is likely that similar phenomena are an intrinsic part of most substance abuse (Blum 1989; Blum & Noble 1994; Blum & Payne 1991; Blum et al. 1996; Noble et al. 1994).

Individuals with differential drug responsiveness likely have existed in human populations for many generations, but it is only recently that substances that short-circuit the adapted neural pathways have become widely available and in highly concentrated forms. The rise of agriculture and the domestication of potentially psychoactive plants is a recent phenomenon in human evolution (perhaps no more than 400 or so generations), but the consequences of the human desire to seek pleasure and avoid pain are extraordinarily deeply rooted in our evolutionary past.

Evolution and Addiction: Application of the Theory

If drug abuse/addiction is amenable to an evolutionary analysis, one of the first questions that must be considered is why do humans use psychoactive substances in the first place? It is important to remember that much health damage results from behavior directed toward increasing pleasure or avoiding pain. People use tobacco, alcohol, etc., because these substances make them feel good, get them high, and/or help them relax and forget their problems. In an evolutionary perspective, it makes perfect sense that our neural circuitry, especially in the brain, has undergone strong selective pressure and extensive modification toward this end. Behaviors that increase pleasure or avoid pain have been associated with activities that are essential for survival and reproduction (food, sex, human attachment, rest, athletic proficiency, etc.) and have been the object of intense evolutionary pressure. The fact that individuals find these activities pleasurable generally enhances our Darwinian fitness. If individuals did not find these survival activities pleasurable, then our species would have become extinct a long time ago.

However, as every parent knows, not everything that feels good may be good for us. Natural selection will favor individuals over time who avoid certain potentially harmful substances or experiences. To the extent that avoidance has some genetic basis, natural selection will act on it, but there are a variety of reasons it typically takes a number of generations for selection to act. When a substance or experience is first introduced to the population, the situation is very different—there will be no evolved safeguards to limit an individual's exposure—and this is when selection is most intense. This introduction can occur in two ways: individuals can be exposed to a *completely new* substance or opportunity (e.g., alcohol and tobacco), or individuals can be exposed to *more* of a substance or opportunity that has in the past promoted survival and reproduction, but exposure was in considerably more limited quantities (e.g., salt and fat). This argument may be summarized as the Pleistocene hunter-gatherer model.[10]

These recent changes in the type, as well as extent, of exposure to substances that are harmful to us are a price of the affluence of Western society. These recent changes (e.g., introduction of highly concentrated and readily available psychoactive substances) stimulate old, deeply rooted survival-trait–based neural circuitry that urges us to consume these substances in large quantities in order to reap the pleasure bonanza. The body does not know that the high induced by the consumption of alcohol, smoking of crack cocaine, or injection of heroin is disconnected from the evolutionary antecedent survival behaviors. The body operates with a fairly simple algorithm—maximize pleasure and minimize pain and discomfort. If ingestion of exogenous substances produces emotional responses that mimic those generated from the performance of evolutionarily adaptive behaviors, then it is wholly reasonable to expect that people will continue to engage in those behaviors, ones that generate the maximum pleasure with the minimum effort. Simply put, people want to experience pleasure and avoid pain, drugs short circuit the evolved mechanisms and directly produce pleasure or ameliorate pain and discomfort. Today we no longer see a "goodness of fit" between the performance of fitness-enhancing behaviors and the resulting feelings of pleasure and satisfaction. In a real sense we find ourselves in a rapidly diminishing downward spiral.

Earlier in this chapter I briefly alluded to the magnitude of the problem of substance abuse. Accurate estimates of the numbers of abusers are as elusive as a definition of the phenomenon, but data indicate that overall prevalence rate for substance abuse in public and private psychiatric populations is about one in two. The prevalency rates for addictive disorders varies in the clinical populations: 30% in depressive disorders, 50% in bipolar disorders, 50% in schizophrenic disorders, 80% in antisocial personality disorders, 30% in anxiety disorders, and 25% in phobic disorders (Miller 1994). It is estimated that as much as 20% of the population may be affected.

Figure 15.4 shows the lifetime prevalence of the top 10 major psychiatric disorders in the general population. Combining estimates for alcohol and substance abuse, the prevalence is 19.6%. If these estimates are correct, then one out of five people in the United States have a substance abuse problem (Miller 1994). These data strongly suggest that current approaches to substance abuse treatment have been only marginally effective at best, and if we are going to deal with this most pressing problem we are going to have to develop new and novel ways of looking at the problem.

If the evolutionary model I have suggested is correct, then what can we say to the larger question of substance abuse? Several suggestions come to mind: First, an evolutionary perspective removes substance abuse from the realm of moral judgment. Like cancer, heart disease, and muscular dystrophy, substance abuse arises from biological origins rooted deeply in human evolutionary history. This awareness should begin to undermine the widespread notion that individuals plagued by the disease of addiction lack willpower or lack self-restraint. To discriminate against substance abuse makes about as much sense as discrimination against males with pattern baldness. Second, and closely related to the first, is the establishment of realistic goals for use of psychoactive drugs. Given our evolutionary history, it is unreasonable and unrealistic to aspire to a "zero-intake" society. It is reasonable, however, to expect that we may achieve, not a drug-free society, but one with substantially less drug abuse. By understanding the mechanism of action and the potential basis for widespread use in the face of empirical data about the costs of substance abuse, we may be able to make rational decisions about treatment as well as prevention. Third, serious efforts should be made to reduce the ease of acquisition of legal drugs. This could be accomplished through restriction

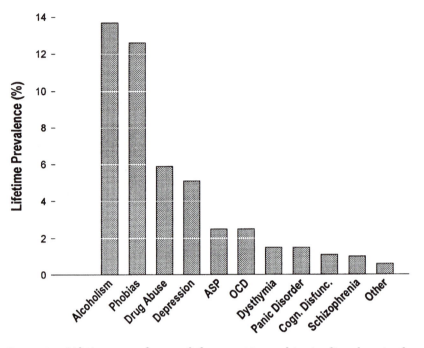

Figure 15.4. Lifetime prevalence of the top 10 psychiatric disorders in the United States (Miller 1994).

of the sale of all alcoholic beverages as well as all tobacco products to state-owned liquor stores. This would make acquisition costs higher, as well as place the sale under tighter control, thus reducing the probability of selling to minors. Fourth, the evolutionary perspective should help focus attention on education and treatment, rather than on punishment and retribution. See Goldstein (1994) for a discussion of several of these ideas. These suggested remedies would, no doubt, be expensive, but relative to the $240 billion per year costs of substance abuse, the current expenditures on treatment costs alone by the federal government, $8.2 billion in 1990 (Rouse 1995), seems insignificant (Johnston 1988). In fact, alcohol and tobacco industries spend slightly more than half the total government expenditure on treatment on advertising annually.[11]

Finally, the dopamine model (Blum et al. 1996) suggests that a profitable area for continued research will be in the development of substances that enhance the production of dopamine or other neurotransmitters without the addictive side effects. Addicts could then maintain a normal neurochemical balance without resorting to destructive behavior. Individuals with a wide array of addictive, impulsive, and compulsive disorders might be able to find relief from potentially lethal situations through modulation of neurotransmitters. Although this is merely speculation at this point, given the advances in other areas of evolutionary biology and neuroscience, it seems to me that continued application of the evolutionary perspective is both critical and is likely to prove very fruitful.

Acknowledgments

Earlier versions of this paper were presented in a symposium "Evolutionary Medicine: New Perspectives and Opportunities" at the Annual Meeting of the American Association for Advancement of Science, Boston, MA, February 1993, and in a symposium "Evolutionary Medicine and Anthropology" at the Annual Meeting, American Anthropological Association, Washington, DC, November, 1993.

Notes

1. Steindler (1994: 2.1–2.2) notes that

> Abuse is the harmful use of a specific psychoactive substance; addiction is a disease process characterized by the continued use of a specific psychoactive substance; addiction is a disease process characterized by the continued use of a specific psychoactive substance despite physical, psychological or social harm; and dependence is either physical dependence, a physiological state of adaptation to a specific psychoactive substance characterized by a withdrawal syndrome during abstinence, which may be relieved in total or in part by the readministration of the substance, psychological dependence, subjective sense of need for a specific psychoactive substance, either for its positive effects or to avoid the negative effects associated with its abstinence or as a category of psychoactive substance use disorder.

2. Compare this to what is considered a large dose in modern populations of 800 mg in single dose (Grinspoon & Bakalar 1976). Information on lethal doses of cocaine are somewhat problematic, but in one experiment (Pickett 1970) notes that the equivalent of an injection of 2 grams in a 150-lb man proved lethal in 50% of the experimental subjects.

3. Bach's affinity for coffee is not definitely known; however, an inventory of his estate revealed that the kitchen was equipped with numerous coffee pots. A particularly impressive one was assessed for 18 thalers (approximately $2000 today) (Bettmann 1995).

4. See Ewing et al. (1974) for a contrary opinion.

5. Athletes have known for a long time that amphetamines enhance performance, but the effect is generally recognized as small. Historically, it has taken about 7 years to decrease the record time for the one mile run by 1%. Imagine the impact a 1% decrease in time could make, particularly at the highest levels of competition (Laties & Weiss 1981; Smith & Beecher 1959). A survey of Olympic records for 68 of the 250 Olympic events revealed that an improvement of 1% in performance would have changed the winner in approximately three-quarters of these events (50/68). In fact, Gemini-Titan astronaut Gordon Cooper was ordered

to take amphetamines before assuming manual control of the reentry of the space vehicle (Ray & Ksir 1990).

6. Many clinicians argue that there is little convincing evidence to suggest that substance abuse is anything more than an environmentally induced social pathology with no underlying basis in biology.

7. Although see Nesse (1992, 1994) for the first application of evolutionary theory to psychiatric disorders in general and substance abuse in particular.

8. Sexual selection (Trivers 1972, 1985) predicts that because males generally have lower investment in offspring, they are predisposed to demonstrate their superiority as mates by engaging in flamboyant, extravagant displays. Females, on the other hand,

due to their greater degree of intrinsic investment in offspring, are favored to be choosy in their selection of mates.

9. Zahavi developed these ideas to counter the argument for "runaway selection" advanced by Fisher (1930).

10. The Pleistocene hunter-gatherer model has been invoked in anthropology to explain a variety of conditions found in modern humans, ranging from our aggressive tendencies, to configuration of sex organs, to social organization, to bipedalism.

11. The tobacco industry is reported to spend $4 billion annually on advertising in the United States (Bristow 1994); the alcohol industry invests $700 million in advertising (Arnesen 1995).

References

American Psychiatric Association. (1994) *Diagnostic and Statistical Manual of Mental Disorders.* Washington, DC: American Psychiatric Association.

Anthenelli, R. M., and Schuckit, M. A. (1992) Genetics. In J. H. Lowinson, P. Ruiz, R. B. Millman, and J. G. Langrod, eds., *Substance Abuse: A Comprehensive Textbook*, pp. 39–50. Baltimore, MD: Williams and Wilkins.

Argiolas, A., Melis, M. R., and Gessa, G. L. (1986) Oxytocin: An extremely potent inducer of penile erection and yawning in male rats. *European Journal of Pharmacology* 130:265–272.

Arnesen, D. W. (1995) Alcohol advertising versus the Constitution: What are the limits of state regulation? *Journal of the Academy of Marketing Science* 23: 66–67.

Beckstrom, J. H. (1993) *Darwinism Applied: Evolutionary Paths to Social Goals.* Westport, CT: Praeger.

Bettmann, O. L. (1995) *Johann Sebastian Bach: As His World Knew Him.* New York: Carol Publishing Group.

Blum, K. (1989) A commentary on neurotransmitter restoration as a common mode of treatment for alcohol, cocaine, and opiate abuse. *Integrative Psychiatry* 6:99–204.

Blum, K., Cull, J. G., Braverman, E. R., and Comings, D. E. (1996) Reward deficiency syndrome. *American Scientist* 84:132–145.

Blum, K., and Noble, E. P. (1994) The sobering d-2 story. *Science* 265:1346–1347.

Blum, K., and Payne, J. E. (1991) *Alcohol and the Addictive Brain: New Hope for Alcoholics from Biogenetic Research.* New York: The Free Press.

Bohman, M., Sigvardsson, S., and Cloninger, C. R. (1981) Maternal inheritance of alcohol abuse: Cross-fostering analysis of adopted women. *Archives of General Psychiatry* 38:965–969.

Bohn, M. J., and Meyer, R. E. (1994) Typologies of addiction. In M. Galanter and H. D. Kleber, eds., *The American Psychiatric Press Textbook of Substance Abuse Treatment*, pp. 11–24. Washington, DC: American Psychiatric Press.

Bond, C. F., and Robinson, M. (1988) The evolution of deception. *Journal of Nonverbal Behavior* 12:295–307.

Bristow, L. R. (1994) Protecting youth from the tobacco industry. *Vital Speeches of the Day* 60:333–336.

Brown, D. (1988) Components of lifetime reproductive success. In T. H. Clutton-Brock, ed., *Reproductive Success*, pp. 439–453. Chicago: The University of Chicago Press.

Byrne, R. W., and Whiten, A. (1992) Cognitive evolution in primates: Evidence from tactical deception. *Man* 27:609–627.

Calkins, A. (1871) *Opium and the Opium Appetite*. Philadelphia: J. B. Lippincott.

Clutton-Brock, T. H. (1988) Reproductive success. In T. H. Clutton-Brock, ed., *Reproductive Success*, pp. 472–520. Chicago: The University of Chicago Press.

Cotton, N. S. (1979) The familial incidence of alcoholism: A review. *Journal of Studies on Alcohol* 40:89–116.

Crane, E. (1980) *A Book of Honey*. Oxford: Oxford University Press.

Critchlow, B. (1986) The powers of John Barleycorn: Beliefs about the effects of alcohol on social behavior. *American Psychologist* 41:751–764.

Cronin, A. (1995) The tipplers and the temperate: Drinking around the world. *New York Times* (January 1): D4

Desjarlais, R., Eisenberg, L., Good, B., and Kleinman, A. (1995) *World Mental Health: Problems and Priorities in Low-Income Countries*. Oxford: Oxford University Press.

Diamond, J. M. (1992) *The Third Chimpanzee: The Evolution and Future of the Human Animal*. New York: Harper Collins.

Doweiko, H. F. (1993) *Concepts of Chemical Dependency*. Pacific Grove, CA:Brooks/Cole.

Eaton, S. B., and Konner, M. (1985) Paleolithic nutrition: A consideration of its nature and current implications. *New England Journal of Medicine* 312:283–289.

Eaton, S. B., Pike, M. C., Short, R. V., Lee, N. C., Trussell, J., Hatcher, R. A., Wood, J. W., Worthman, C. M., Blurton Jones, N. G., Konner, M. J., Hill, K., Bailey, R., and Hurtado, A. M. (1994) Women's reproductive cancers in evolutionary context. *Quarterly Review of Biology* 69: 353–367.

Eaton, S. B., Shostak, M., and Konner, M. (1988) *The Paleolithic Prescription: A Program of Diet and Exercise and a Design for Living*. New York: Harper and Row.

Ewing, J. A., Rouse, B. A., and Pelizzari, E. D. (1974) Alcohol sensitivity and ethnic background. *American Journal of Psychiatry* 131:206–210.

Fisher, R. A. (1930) *The Genetical Theory of Natural Selection*. Oxford: Clarendon Press.

Futuyma, D. J. (1986) *Evolutionary Biology*. Sunderland, MA: Sinauer Associates.

Gabel, S., Stadler, J., Bjorn, J., Shindledecker, R., and Bowden, C. L. (1995) Homovanillic acid and monoamine oxidase in sons of substance-abusing fathers: Relationship to conduct disorder. *Journal of Studies on Alcohol* 56:135–139.

Gardner, E. L. (1992) Brain reward mechanisms. In J. H. Lowinson, P. Ruiz, and J. G. Langrod, eds., *Substance Abuse: A Comprehensive Textbook*, pp. 70–99. Baltimore, Md: Williams and Wilkins.

Gawin, F. H. (1991) Cocaine addiction, psychology, and neurophysiology. *Science* 251:1580–1586.

Giancola, P. R., and Zeichner, A. (1995) Alcohol-related aggression in males and females: Effects of blood alcohol concentration, subjective intoxication, personality, and provocation. *Alcoholism: Clinical and Experimental Research* 19:130–134.

Gianoulakis, C., and de Waele, J.-P. (1994) Genetics of alcoholism: Role of endogenous opioid system. *Metabolic Brain Disease* 9:105–131.

Glassner, B., and Berg, B. (1980) How Jews avoid alcohol problems. *American Sociological Review* 45:647–664.

Goedde, H. W., Agarwal, D. P., Harada, S., Meier-Tackmann, D., Ruofu, D., Bienzle, U., Kroeger, A., and Hussein, L. (1983) Population genetic studies on aldehyde dehydrogenase isoenzyme deficiency and alcohol sensitivity. *American Journal of Human Genetics* 35:769–772.

Gold, M. S., Miller, N. S., and Jonas, J. M. (1992) Cocaine and crack: neurobiology. In J. H. Lowinson, P. Ruiz, R. B. Millman, and J. G. Langrod, eds., *Substance Abuse: A Comprehensive Textbook*, pp. 222–235. Baltimore; MD: Williams and Wilkins.

Gold, M. S., and Verebey, K. (1984) The psychopharmacology of cocaine. *Psychiatric Annals* 14:714–723.

Golding, A. M. B. (1993) Two hundred years of drug abuse. *Journal of the Royal Society of Medicine* 86:282–286.

Goldstein, A. (1994) *Addiction: From Biology to Drug Policy.* New York: W. H. Freeman.

Goodwin, D. W. (1985) Alcoholism and genetics: The sins of the fathers. *Archives of General Psychiatry* 42: 171–174.

Goodwin, D. W. (1992) Alcohol: Clinical aspects. In J. H. Lowinson, P. Ruiz, R. B. Millman, and J. G. Langrod, eds., *Substance Abuse: A Comprehensive Textbook*, pp. 144–151.

Baltimore, MD: Williams and Wilkins.

Goodwin, D. W., Schulsinger, F., Hermansen, L., Guze, S. B., and Winokur, G. (1973) Alcohol problems in adoptees raised apart from alcoholic biologic parents. *Archives of General Psychiatry* 28:238–243.

Greden, J. F., and Walters, A. (1992) Caffeine. In J. H. Lowinson, P. Ruiz, R. B. Millman, and J. G. Langrod, eds., *Substance Abuse: A Comprehensive Textbook*, pp. 357–370. Baltimore, MD: Williams and Wilkins.

Grinspoon, L., and Bakalar, J. B. (1976) *Cocaine: A Drug and Its Social Evolution.* New York: Basic Books.

Gurling, H. M., Oppenheim, B. E., and Murray, R. M. (1984) Depression, criminality and psychopathology associated with alcoholism: Evidence from a twin study. *Acta Geneticae Medicae et Gemellologiae* 33:333–339.

Hartford. T. C. (1992) Family history of alcoholism in the United States: Prevalence and demographic characteristics. *British Journal of Addiction* 87: 931–935.

Hartl, D. L. (1994) *Genetics.* Boston: Jones and Bartlett.

Harvey, P. H., Martin, R. D., and Clutton-Brock, T. H. (1986) Life histories in comparative perspective. In B. B. Smuts, D. L. Cheney, R. M. Seyfarth, R. W. Wrangham, and T. T. Struhsaker, eds., *Primate Societies*, pp. 181–196. Chicago: The University of Chicago Press.

Holmstedt, B., and Fredga, A. (1981) Sundry episodes in the history of coca and cocaine. *Journal of Ethnopharmacology* 3: 113–147.

Horgan, C., Marsden, M. E., and Larson, M. J. (1993) *Substance Abuse: The Nation's Number One Health Problem.* New York: Institute for Health Policy, Brandeis University.

Jarvik, M. E., and Schneider, N. G. (1992) Nicotine. In J. H. Lowinson,

P. Ruiz, R. B. Millman, and J. G. Langrod, eds., *Substance Abuse: A Comprehensive Textbook*, pp. 334–356. Baltimore, MD: Williams and Wilkins.

Johns, T. (1990) *With Bitter Herbs They Shall Eat It: Chemical Ecology and the Origins of Human Diet and Medicine.* Tucson: University of Arizona Press.

Johnston, L. D. (1988) *Reducing Drug Use in America: A Perspective, a Strategy, and Some Promising Approaches.* Lansing: Institute for Social Research, University of Michigan.

Kaij, L. (1960) *Studies on the Etiology and Sequels of Abuse of Alcohol.* Lund, Sweden: University of Lund Press.

Karno, M., Hough, R. L., Burnam, A., Escobar, J. I., Timbers, D. M., Santana, F., and Boyd, J. H. (1987) Lifetime prevalence of specific psychiatric disorders among Mexican-Americans and non-Hispanic whites in Los Angeles. *Archives of General Psychiatry* 44: 697–701.

Karp, R. W. (1994) Genetic studies in alcohol research. *American Journal of Medical Genetics (Neuropsychiatric Genetics)* 54: 304–308.

Kendler, K. S., Heath, A. C., Neale, M. C., Kessler, R. C., and Eaves, L. J. (1992) A population-based twin study of alcoholism in women. *Journal of the American Medical Association* 268: 1877–1882.

Kendler, K. S., Neale, M. C., Heath, A. C., Kessler, R. C., and Eaves, L. J. (1994) A twin-family study of alcoholism in women. *American Journal of Psychiatry* 151: 707–715.

Kleber, H. D. (1988) Cocaine abuse: Historical, epidemiological and psychological perspectives. *Journal of Clinical Psychiatry* 49: 3–6.

Kofoed, L. (1988) Selective dimensions of personality: Psychiatry and socio-

biology in collision. *Perspectives in Biology and Medicine* 31: 228–242.

Kramer, J. C. (1980) The opiates: Two centuries of scientific study. *Journal of Psychedelic Drugs* 12: 89–103.

Kutchinsky, B. (1987) Deception and propaganda. *Social Science and Modern Society* 24(5): 21–24.

Laplace, A. C., Chermack, S. T., and Taylor, S. P. (1994) Effects of alcohol and drinking experience on human physical aggression. *Personality and Social Psychology Bulletin* 20: 439–444.

Laties, V. G., and Weiss, B. (1981) The amphetamine margin in sports. *Federation Proceedings* 40: 2689–2692.

Lawton, F. (1910) *Balzac.* London: Grant Richards Ltd.

Lester, D. (1992) *Why People Kill Themselves.* Springfield, IL: Charles C. Thomas.

Lester, D. (1995) Alcohol availability, alcoholism and suicide and homicide. *American Journal of Drug and Alcohol Abuse* 21: 147–150.

Li, T. K., Lumeng, L., Mcbride, W. J., and Murphy, J. M. (1994) Genetic and neurobiological basis of alcohol-seeking behavior. *Alcohol and Alcoholism* 29: 697–700.

Lumeng, L., and Crabb, D. W. (1994) Genetic aspects and risk factors in alcoholism and alcoholic liver disease. *Gastroenterology* 107: 572–578.

MacMillan, J., and Kofoed, L. (1984) Sociobiology and antisocial personality: An alternative perspective. *Journal of Nervous and Mental Disease* 172: 701–706.

Malcolm, A. I. (1971) *The Pursuit of Intoxication.* Toronto: Alcoholism and Drug Addiction Research Foundation.

Massing, M. (1990) The two William Bennetts. *New York Review of Books*, 37(3): 29–33.

Matilainen, T. K. M., Puska, P., Berg, M.-A. K., Pokusajeva, S., Moisejeva, N., Uhanov, M., and Artemjev, A.

(1994) Health-related behaviors in the Republic of Karelia, Russia, and North Karelia, Finland. *International Journal of Behavioral Medicine* 1: 285–304.

McKim, W. A. (1986) *Drugs and Behavior*. Englewood Cliffs, NJ: Prentice-Hall.

McKinlay, A. P. (1951) Attic temperance. *Quarterly Journal of Studies on Alcohol* 12:61–102.

Miller, N. S. (1994) Prevalence and treatment models for addiction in psychiatric populations. *Psychiatric Annals* 24:399–406.

Murdoch, D., Pihl, R. O., and Ross, D. (1990) Alcohol and crimes of violence: Present issues. *International Journal of Addictions* 25:1065–1081.

Nadelmann, E. A. (1989) Drug prohibition in the United States: Costs, consequences, and alternatives. *Science* 245:939–947.

Nesse, R. M. (1992) Substance abuse and the evolution of emotions. Unpublished manuscript.

Nesse, R. M. (1994) An evolutionary perspective on substance abuse. *Ethology and Sociobiology* 15:339–348.

Noble, E. P., Syndulko, K., Fitch, R. J., Ritchie, T., Bohlman, M. C., Guth, P., Sheridan, P. J., Montgomery, A., Heinzmann, C., Sparkes, R. S., and Blum, K. (1994) D-2 dopamine receptor taqi a alleles in medically ill alcoholic and nonalcoholic patients. *Alcohol and Alcoholism* 29:729–744.

O'Dea, K. (1991) Traditional diet and food preferences of Australian Aboriginal hunter-gatherers. *Philosophical Transactions of the Royal Society of London B* 334:233–241.

Pickett, R. D. (1970) Acute toxicity of heroin, alone and in combination with cocaine or quinine. *British Journal of Pharmacology* 40:145–146.

Pihl, R. O., and Peterson, J. (1995) Drugs and aggression: Correlations, crime and human manipulative studies and some proposed mechanisms. *Journal of Psychiatric Neuroscience* 20:141–149.

Portenoy, R. K., and Payne, R. (1992) Acute and chronic pain. In J. H. Lowinson, P. Ruiz, R. B. Millman, and J. G. Langrod, eds., *Substance Abuse: A Comprehensive Textbook*, pp. 691–721. Baltimore, MD: Williams and Wilkins.

Ray, O., and Ksir, C. (1990) *Drugs, Society, and Human Behavior*. St. Louis: Times Mirror/Mosby College Publishing.

Rich, C. L., Young, D., and Fowler, R. C. (1986) San Diego suicide study. I. Young vs. old subjects. *Archives of General Psychiatry* 43:577–583.

Rouse, B. A. (1995) *Substance Abuse and Mental Health Statistics Sourcebook*. Washington, DC: Superintendent of Documents, U.S. Government Printing Office.

Siegel, R. K. (1989) *Intoxication: Life in Pursuit of Artificial Paradise*. New York: E. P. Dutton.

Simon, E. J. (1992) Opiates: neurobiology. In J. H. Lowinson, P. Ruiz, R. B. Millman, and J. G. Langrod, eds., *Substance Abuse: A Comprehensive Textbook*, pp. 195–204. Baltimore, MD: William and Wilkins.

Smith, E. O. (1987) Deception and evolutionary biology. *Cultural Anthropology* 2:50–64.

Smith, G. M., and Beecher, H. K. (1959) Amphetamine sulfate and athletic performance. *Journal of the American Medical Association* 170: 542–557.

Steindler, E. M. (1994) Addiction terminology. In N. S. Miller, ed., *Principles of Addiction Medicine*, pp. 1–4. Chevy Chase, MD: American Society of Addiction Medicine.

Stolerman, I. P., and Jarvis, M. J.

(1995) The scientific case that nicotine is addictive. *Psychopharmacology* 117:2–10.

Svikis, D. S., Velez, M. L., and Pickens, R. W. (1994) Genetic aspects of alcohol use and alcoholism in women. *Alcohol Health and Research World* 18:192–196.

Syed, I. B. (1976) The effects of caffeine. *Journal of the American Pharmaceutical Association* 10:568–572.

Thomason, G. (1938) *Science Speaks to Young Men: On Liquor, Tobacco, Narcotics and Marijuana.* Mountain View, CA: Pacific Press.

Trivers, R. L. (1972) Parental investment and sexual selection. In B. G. Campbell, ed., *Sexual Selection and the Descent of Man*, pp. 136–179. Chicago: Aldine.

Trivers, R. L. (1985) *Social Evolution.* Menlo Park, CA: Benjamin Cummings.

Vesell, E. S., Page, J. G., and Passancanti, G. T. (1971) Genetic and environmental factors affecting ethanol metabolism in man. *Clinical Pharmacology and Therapeutics* 12:192–201.

Wall Street Journal (1989) The devil you know. *Wall Street Journal* (December 29):A6.

Weinraub, B. (1989) President offers strategy for U.S. on drug control. *New York Times* (September 6):A1 & B7.

Welles, J. (1981) The sociobiology of self-deception. *Human Ethology Newsletter* 3:14–19.

Westermeyer, J. (1987) Cultural patterns of drug and alcohol use: An analysis host and agent in the cultural environment. *United Nations Bulletin on Narcotics* 39:11–17.

Westermeyer, J. (1992) Cultural perspectives: native Americans, Asians, and new immigrants. In J. H. Lowinson, P. Ruiz, R. B. Millman, and J. G. Langrod, eds., *Substance Abuse: A Comprehensive Textbook*, pp. 890–896. Baltimore, MD: Williams and Wilkins.

Whiten, A., and Byrne, R. (1988) Tactical deception in primates. *Behavioral and Brain Sciences* 11:233–273.

Winick, C. (1992) Epidemiology of alcohol and drug abuse. In J. H. Lowinson, P. Ruiz, R. B. Millman, and J. G. Langrod, eds., *Substance Abuse: A Comprehensive Textbook*, pp. 15–29. Baltimore, MD: Williams and Wilkins.

Yamamoto, J., Silva, J. A., Sasao, T., Wang, C., and Nguyen, L. (1993) Alcoholism in Peru. *American Journal of Psychiatry* 150:1059–1062.

Yamashita, I., Ohmori, T., Koyama, T., Mori, H., Boyadjive, S., Kielholz, P., Gastpar, M., Moussaoui, D., Bouzekraoui, M., and Sethi, B. B. (1990) Biological study of alcohol dependence syndrome with reference to ethnic difference: A report of a WHO collaborative study. *Japanese Journal of Psychiatry & Neurology* 44:79–84.

Zahavi, A. (1975) Mate selection: A selection for a handicap. *Journal of Theoretical Biology* 53:205–214.

Zahavi, A. (1977) The cost of honesty: further remarks on the handicap principle. *Journal of Theoretical Biology* 67: 603–605.

Zahavi, A. (1991) On the definition of sexual selection, Fisher's model, and the evolution of waste and of signals in general. *Animal Behaviour* 42:501–503.

16

MENOPAUSE IN EVOLUTIONARY PERSPECTIVE

LYNNETTE E. LEIDY

$\mathbf{M}$ enopause, the cessation of menses, is an event experienced by all human females who live beyond 60 years of age (Diczfalusy 1986). From a life-history perspective, emphasizing life span stages, it is the universal aspect of menopause that draws the attention of evolutionary biologists, physical anthropologists, and physicians. Although other mammals experience reproductive senescence (Comfort 1979; Packer et al. 1998), and individual primates have demonstrated endocrinologic and histologic changes similar to the human menopause (Gould et al. 1981; Graham et al. 1979; Hodgen et al. 1977; Lapin et al. 1979; Walker 1995), no other species of primates has yet demonstrated a universally experienced, permanent cessation of reproductive cyclicity followed by such a prolonged period of postreproductive life (Lancaster and Lancaster 1983; Pavelka and Fedigan 1991).

Although menopause is experienced by individual women as a process that may span years (Kaufert 1988), this chapter will treat menopause as a one-time life-history event, marking the transition from reproductive to postreproductive life. Menopause as a life-history event will be considered in three ways. First, it will be considered as a beneficial trait, either selected for directly, and therefore adaptive, or "exaptive" (Gould and Vrba 1982) in the sense that it occurred for another reason but is now understood to be beneficial. Second, menopause will be considered as a neutral trait that was uncovered by an increased maximum life span potential (MLP) during human evolution. Maximum life span potential represents "the length of life possible for a given species under optimum nutritional and minimal environmental hazard conditions" (Cutler 1978:463). Third, menopause will be considered as a deleterious trait in the context of the more recent increase in average life span. Menopause, from this perspective, is an epidemiologic risk factor for chronic diseases. This third perspective is currently emphasized by biomedicine.

Differences between the maximum and average life span are important to recognize when considering the origins and consequences of menopause. The maximum, or MLP, for *Homo sapiens* has exceeded 110 years (Finch 1990). This potential life span evolved thousands of years before the present. The average human life span (or life expectancy), on the other hand, is a population-specific trait that continues to increase as a result of cultural changes, such as improvements in the distribution of food and health resources. When a population's average life span lengthens, more women experience menopause (Lancaster and King 1992).

The life-history perspective emphasizes the invariate nature of menopause to better contrast human menopause with the reproductive senescence of other primates (Pavelka and Fedigan 1991). This chapter will differ from a strict life-history perspective in that variation in the timing of menopause will be treated as an important consideration. There are several reasons to emphasize variation. First, natural selection acts on phenotypic variation in a trait. For example, some adaptive scenarios postulate that the cessation of menses originally occurred at an earlier age (premature ovarian failure, see below) and gradually shifted to a later age (menopause) as the MLP lengthened (Donaldson 1994; Peccei 1995). In addition, variation in age at menopause continues, both within and between contemporary populations.[1]

Variation in the timing of menopause is also important when considering menopause to be a neutral trait. Cross-species evidence suggests that 50 years may be the limit of mammalian ovarian function and, as will be discussed, diminishing ovarian function regulates the onset of menopause. A third reason to emphasize variation in age at menopause is that early or late ages at menopause are epidemiologic risk factors for different diseases (Eaton et al. 1994). Specifically, late age at menopause is a risk factor for breast cancer (Kelsey et al. 1993); early age at menopause is a risk factor for bone loss (Lindquist et al. 1983).

Cross-sectional, retrospective, and longitudinal studies place the median age of menopause at 50–51 years in well-nourished, industrialized societies (Gosden 1985; Gray 1976; Greene 1984). More women experience menopause before 50 years than after (Treloar 1974). The minimum age at menopause (40 years) is an arbitrary cutoff between a clinically normal menopause and premature ovarian failure (POF), which can occur as early as 29 or 30 years of age. The same factors related to POF, such as galactosemia (Cramer et al. 1989), autoimmune disfunction (Leidy 1990), and genetics (Gosden 1985; Mattison et al. 1984) are probably involved in the clinically normal cessation of menses.

The Medical Treatment of Menopause

A great deal of thought has already been given to the evolutionary significance of menopause by biologists, anthropologists, and physicians. It is argued, for instance, that menopause was adaptive in the past due to the selective advantage of grandmothering, but that menopause is currently a deleterious trait due to the lengthened average life span. Such ideas are applied clinically to support the use of hormone replacement therapy as a widespread treatment for postmenopausal women (Austad 1994; Donaldson 1994; Naftolin et al. 1994) and in creating rationale for assisted reproduction for women in their 60s (Schiff and Utian 1994).

Menopause has a history of treatment as a disease event that precedes the discovery of estrogen in the early 1930s (Tepperman and Tepperman 1987). During the nineteenth century, for example, the "change of life" was considered to be a "critical period" that determined the future mental and physical health of all women (Currier 1897; Tilt 1857). By the nineteenth century, hot flashes had long been attributed to the menopause transition, or, more correctly, to the climacteric—the transition from the reproductive to the postreproductive phase of life.

Since the discovery of estrogen, menopause has been defined as a deficiency disease similar to diabetes or hypothyroidism (Bell 1987; McCrea 1983). Synthetic estrogen replacement became widely available in the 1960s, prescribed so that women could remain "feminine forever" (Wilson 1966). During the 1970s, the use of estrogen declined dramatically due to the scare of endometrial cancer (Kennedy et al. 1985).

When treatment protocol shifted to the use of estrogen combined with progestin, hormone replacement therapy (HRT) gained popularity in the 1980s. Progestin demonstrated protection against endometrial cancer, and hormone replacement was hailed (for a second time) as a rejuvenator (Nachtigall and Heilman 1986). By 1992, all women were urged to consider HRT (American College of Physicians 1992).

While menopause has been treated as both a deficiency disease and an endocrinopathy (disease of the endocrine glands) (Utian 1987), menopause is now more often considered a risk factor for osteoporosis and cardiovascular disease rather than a disease in and of itself (Worcester and Whatley 1992). Estrogen continues to be portrayed as a rejuvenating agent (Brody 1995), but the strongest argument for the use of HRT on a broad scale comes from cost–benefit analyses that demonstrate reduced social costs specific to osteoporosis and cardiovascular disease among women taking the medication (Daly et al. 1994; Mishell 1989; Notelovitz 1989; Utian 1978). This emphasis on social costs and benefits has most profoundly supported the medicalization of menopause as an event requiring clinical intervention.

The Physiology of Menopause

The generally recommended use of HRT, from the perimenopausal period until the loss of tissue receptor sensitivity (Ettinger and Grady 1993), raises the question of why menopause, with its corresponding drop in estrogen levels, occurs at about the halfway point of the MLP. Whether menopause is understood to be a deficiency disease or a risk factor for chronic disease, the same question is raised. Why do human females, with an MLP of approximately 120 years, experience menopause somewhere between the ages of 40 and 60? Before considering evolutionary explanations for the onset of menopause, the biology of menopause needs to be briefly reviewed.

The timing of menopause is determined by two factors: the number of eggs formed in the female ovary during the fetal period of development, and the rate of loss of those same eggs across the life span through the processes of ovulation and degenerative atresia (Gosden 1985; Leidy 1994; Richardson et al. 1987).

The gametes (eggs) formed in the ovary, like all cells in the human body, originate from the fertilized egg. That initial cell forms not only the embryo, but also the yolk sac needed during the first weeks of gestation. The primordial germ cells (in the embryo before sex differentiation) develop from the yolk sac and migrate, in ameboid fashion, to the ovary by the fifth or sixth week of gestation (Moore 1988). Note that, almost from the beginning of development, germ cells demonstrate a separate life history from somatic cells. This addresses the question of how ovarian function, which ceases by the age of 50, can differ so dramatically from the prolonged function of the heart, liver, kidneys, and lungs (Diamond 1996; Hill and Hurtado 1991). Germ cells have a different life history. They go on to produce oogonia, transitional diploid cells produced by mitosis and capable of repeated mitotic divisions, from which oocytes arise through the first prophase of meiosis. Oocytes, surrounded by follicles within the ovaries, complete meiosis after ovulation to become viable haploid ova or eggs (Baker 1986; Crisp 1992).

In human females approximately 7 million oogonia proliferate by the fifth month of fetal development. This number declines to 2 million oocytes in the human ovaries at birth (Baker 1986); 400,000 oocytes are present in the human ovary by the onset of puberty (Byskov 1978). After puberty, ovulation begins; however, monthly ovulation will only account for the disappearance of, at maximum, 400 or so oocytes. Thousands more simply degenerate until comparatively few remain (Nelson and Felicio 1985; Novak 1970; Richardson et al. 1987; Thomford et al. 1987). At this point the monthly menstrual cycles, initiated at puberty, cease.

The degeneration of oocytes occurs through follicular atresia, a process that involves the shrinking of either the oocyte itself or its surrounding follicle. The process of follicular atresia, which occurs before birth and continues throughout the life span, remains incompletely understood (Crisp 1992; Gougeon 1996; Guraya 1985).

Both the peak number of oogonia formed by mitosis during fetal development and the rate of follicular atresia are species specific (Byskov 1978; Sadlier 1973). Cross-species comparisons suggest that the peak number of oogonia formed is roughly correlated with the length of the life span (Finch and Gosden 1986). By way of brief example, the peak number of oogonia stored by the rat is 71,000, in contrast to the peak number of 2,700,000 in the cow (Perry 1971). The maximum stored during the fifth month of gestation in fetal rhesus monkeys is 3,518,000, compared with the almost 7 million estimated during the fifth month of gestation among humans (Baker 1986). That the number of oogonia appears to be species specific suggests genetically determined parameters. The process of atresia also appears to be under a degree of genetic control, perhaps as a type of apoptosis, or programmed cell death (Hughes and Gorospe 1991; Tilly et al. 1992).

The production of millions of oocytes by human females when only a maximum of 400 are used in ovulatory cycles may be viewed as an evolutionary carryover from ancestors that engaged in external fertilization such as fish (Martin 1985). Most fish, amphibians, and reptiles continue oogenesis (germ cell mitosis) throughout adult life (Sadlier 1973). Birds and mammals, however, cease oogenesis just before or just after birth (Guraya 1985; Peters and McNatty 1980).[2] In the human ovary, ova formed at the beginning of life wait, halted in meiotic prophase for up to 50 years (Peters and McNatty 1980).

The following biological points should be kept in mind throughout the ensuing discussion of human menopause. (1) Germ cells that develop into fertilizable oocytes separate early in fetal development from somatic cells. (2) The peak number of oogonia produced by mitosis before or immediately after birth is species specific and roughly correlated with the life span. (3) The oocytes formed during fetal development wait in the ovary for years until claimed by ovulation or atresia. (4) The rate of atresia appears to be preprogrammed, although subject to changes in the ovarian environment (Crisp 1992). And (5) the timing of menopause is determined by the number of oogonia formed coupled with the rate of atresia.

Menstruation ceases when there are no longer ovarian follicles in numbers sufficient to produce estrogen levels capable of maintaining the ovarian–pituitary–hypothalamic hormonal loop. There are different sources and forms of estrogen. For example, an extragonadal estrone (from fat, muscle, and other cells) continues to be produced by women beyond menopause (Metcalf et al. 1981; Rannevik et al. 1986). However, the peri- and postmenopausal periods are characterized by low levels of estradiol, the most physiologically active estrogen. Low estradiol levels are associated with symptoms attributed to menopause, including hot flashes, vaginal dryness, memory changes, and later chronic changes in bone density and cholesterol profiles (Utian 1987; Whitehead et al. 1993).

Evolutionary Perspectives

A long postreproductive life, as experienced by human females, is an anomaly within the animal kingdom. Because postreproductive life is an anomaly, and because estrogen receptor tissues continue to function in response to estrone, some argue that it is medically sensible to "correct" falling estradiol levels in human females to mimic the higher estrogen levels of youth (Austad 1994; Donaldson 1994; Naftolin et al. 1994). While the hope of Darwinian medicine has been to shed new light on the treatment of clinical conditions (Williams and Nesse 1991), thus far an evolutionary perspective of menopause has consistently supported the use of HRT. One reason for this is the ethnocentric nature of the current discussion of Darwinian medicine (Nesse and Williams 1994; Oliwenstein 1995). Another reason is the bias toward emphasizing the positive effects of HRT in the biomedical literature on menopause (Worcester and Whatley 1992).

Chronic conditions associated with declining estrogen levels need to be individually examined across cultural and environmental contexts in particular populations to be certain that these chronic conditions belong in a species-level discussion of menopause. Cross-cultural studies suggest that menopause is not the most important risk factor for osteoporotic fractures (Beall 1987; Martin et al. 1993) or heart disease (Crews and Gerber 1994). For example, Martin et al. (1993) demonstrated that postmenopausal Mexican Mayan women have estradiol levels at or below the values expected for women in the United States. In addition, these women experience bone demineralization as expected in conjunction with low estrogen levels; however, the women do not have a high incidence of osteoporotic fractures. The authors suggest that dietary factors, activity patterns, and/or cultural values may explain the lack of osteoporotic fractures (Martin et al. 1993). Cross-cultural differences indicate that an evolutionary history of menopause may not, and perhaps should not, necessarily illuminate our understanding of osteoporosis or heart disease.

Another point to consider when applying a Darwinian perspective to the study of menopause is that while menopause is a visible marker of permanent infertility, variation in age at menopause has little effect on the length of the reproductive span, or on realized family sizes (Gage et al. 1989; Wood 1994). Instead, variation in age at menopause, by definition, reflects variation in the cessation of menstrual cycles. When menopause is examined from an evolutionary perspective, it is most often used as a proxy for the end of fertility. More accurately, menopause is a direct measure of the point after which estrogen-receptor tissues are no longer exposed to high, cyclic levels of estrogens. For some tissues (breast) this may be beneficial; for other tissues (bone) this may be deleterious. Physical anthropologists Pavelka and Fedigan (1991) challenge evolutionary theorists to take into account the adaptive value of the hormonal changes of menopause. By attending to the positive and negative hormonal consequences of menopause, HRT is less clearly justified as a species-wide treatment for postmenopausal women.

Menopause as an Adaptation

There are four ways in which menopause and postreproductive life can be construed to be adaptive. This chapter will examine, first, how menopause ensures that mothers are young enough to survive pregnancy, parturition, and the infancy of their offspring. Relatedly, menopause curtails the ovulation of old, possibly abnormal, oocytes. Second, advocates of the popular "grandmother hypothesis" speculate that "females can make a greater contribution to the population gene pool by investing in kin (particularly grandchildren) than they could by producing their own offspring" (summarized by Hill and Hurtado, 1991:321).

The third scenario is a variation on the second, except that, instead of suggesting that menopause was selected for the purpose of grandmothering, it is suggested that menopause is a by-product of the adaptive process of atresia, coupled with an extended MLP.

Finally, menopause is considered to be adaptive within contemporary, industrialized society. Constant ovulatory cycles, without breaks for multiple pregnancies and lactation, is a recent human condition (Eaton et al. 1994; Harrell 1977). Various lines of evidence suggest that estrogen exposure over the reproductive span is positively related to risk of breast cancer (Kelsey et al. 1993). Menopause marks the end of breast tissue (and other estrogenically sensitive tissue) exposure to high, cyclic levels of estrogen. Therefore, menopause lowers risk of breast cancer which can strike well before the end of the fertile period and hence directly reduce reproductive success.

Aging Mothers, Aging Eggs

For the first time, biomedicine has directly acted to lengthen, rather than shorten (through sterilization and hysterectomies), the species-specific length of the reproductive period by assisting postmenopausal women to carry pregnancies to term (Antinori et al. 1993; Sauer et al. 1992). With the help of donated eggs, laboratory technology, and exogenous progesterone, postmenopausal women can now maintain a pregnancy and deliver an infant (Horn 1994). Interestingly, physicians are called on to rationalize the benefits of making reproduction possible for women who "may not be in a position to supply support to their children" (Schiff and Utian 1994:179–180). Some physicians have reacted by arguing that there are no similar taboos attached to men who become fathers in their 70s (Paulson and Sauer 1994). This discrepancy is directly related to the evolutionary argument presented below for why a female, rather than a male, menopause evolved.

Menopause is the visible marker of an infertility experienced by all women. In contrast, male reproductive senescence is "subtle, variable, and probably of less functional significance" (Wood 1994:487). Some men demonstrate sustained androgen and sperm production well into advanced age (Nieschlag and Michel 1986). Why, then, would a female menopause evolve? It has been suggested that female menopause is an adaptation related to differential parental investment (Gaulin 1980), specifically in terms of lactation and protection of altricial (helpless) young. This implies that maternal death, more than paternal death, threatens the survival of the youngest offspring, and that "menopause ensures that children are born to mothers likely to live long enough to rear them" (Pollard 1994:214). Human women cease fertility early enough to allow the last child to remain dependent, perhaps for 15 to 16 years (Lancaster and Lancaster 1983). More conservatively, Hill and Hurtado (1991) use data from the Ache foragers of eastern Paraguay to show that offspring survivorship is low when mothers die in the first 5 years of the child's life. (However, they also demonstrate a very low probability of a woman dying within the first five years after a birth until she is very old.)

A slight variation in the aging mother hypothesis focuses on the stresses of pregnancy. Pregnancy makes numerous physiological demands on the mother and, in the context of nutritional stress, older mothers invest more in their offspring. The "*proportion* of available maternal reserves diverted to the fetus will increase with maternal age" (Peacock 1991:375, original emphasis). Even as the proportion of reserves increases, older mothers are increasingly at risk for negative birth outcomes, including low birth weight and prematurity (Aldous and Edmonson 1993; Fretts 1995). At a certain point, it may be adaptive to cease reproducing and invest in offspring already born.

Shifting the focus from maternal viability to that of the offspring, menopause also ensures that defective ova are not fertilized (Martin 1985; O'Rourke and Ellison 1993; Pollard 1994). As already discussed, human oocytes wait in meiotic prophase for 15–50 years before ovulation. Across species, the period of waiting is related to an age-related increased risk of chromosomal abnormalities (Kohn 1978). The increase in risk of fetal loss with maternal age "is attributable primarily to an increased incidence of chromosomal aberrations in the eggs of older women" (see also Wood 1994:266; Kline and Levin 1992). Again, rather than risk a compromised pregnancy or birth, it may be adaptive to invest in offspring already born.

The Grandmother Hypothesis

Menopause as an adaptation to aging mothers and aging eggs can account for the cessation of fertility, but not for the extreme length of the human female's postreproductive life. Many authors have pointed to the latter to argue that menopause and postreproductive aging were selected for by the evolutionary benefits gained through grandparenting. During human evolution, as offspring dependency increased and more adult care was required, grandmothers gained more, in terms of genetic fitness, from investing in their grandchildren than they gained from continuing to produce children of their own (Alexander 1974; Dawkins 1976; Hamilton 1966; Williams 1957).

Many adaptationist vignettes portray menopause as an advantageous trait within the context of extended families (Donaldson 1984, 1994; Evans 1981; Gaulin 1980; Hawkes et al. 1997; Kirkwood 1985; Sherman 1998). For example, menopause may have been a reproductive compromise for women facing "dry season starvation" during "year-round terrestrial life on the open biseasonal savannah" (Donaldson 1994:212, 215). Alternatively, menopause prevented competition between generations by ensuring that women refrained from having children once their daughters were of reproductive age (Cavalli-Sforza 1983). The difficulty lies in testing the grandmother hypothesis with human data (Hames 1984; Hill and Hurtado 1991; Mayer 1982; Peccei 1995; Rogers 1993; Turke 1988). Hill and Hurtado (1991), for example, construct various demographic models to estimate the expected benefits and costs of reproductive senescence. They conclude that reproductive senescence would not be favored by natural selection because the grandmother effect is small, and most women have few living adult offspring to help.

From a different angle, Peccei (1995) proposed that perhaps an original correspondence between follicular function and the somatic life span, as seen in chimpanzees (Graham 1979), changed during the course of human evolution. As infants and children became increasingly dependent, women who experienced premature ovarian failure experienced greater lifetime reproductive success by investing in extant offspring. Premature ovarian failure, a Mendelian trait (Mattison et al. 1984), would have provided phenotypic variation on which selection could act through differential fertility. Peccei (1995) goes on to suggest that the evolution of menopause was "a dynamic process"; the age of onset gradually increased to keep pace both with the dependence of hominid offspring, as well as increasing life spans.

Grandmotherhood as an Exaptation

One of the limits of adaptation theory lies in the broad use of the term "adaptive." As Gould and Vrba (1982) demonstrate with an example borrowed from Darwin, some traits are considered adaptive for a particular role (e.g., skull sutures assist mammalian parturition); however, cross-species comparisons show that those features were not designed for that purpose (e.g., birds and reptiles also have skull sutures—to allow for brain growth). Gould and Vrba call such features exaptations. Exaptations are the "useful" traits that "evolved for other usages (or for no function at all), and later 'coopted' for their current role" (Gould and Vrba 1982:6). To differentiate between an exaptation and an adaptation the historical pathway of a trait's origin must be reconstructed. Species that share the same trait must be identified and the evolutionary relationships among those species described.

The above is not unrelated to the phenomenon of pleiotropy (Hamilton 1966; Williams 1957). Pleiotropy refers to the multiple phenotypic effect of genes; again, for example, cranial sutures both ease parturition and allow for brain growth. Genes that regulate follicular atresia, the lifelong depletion of ovarian follicles, have several effects. One effect in humans is the menopause. Another suggested effect is the initiation of reproductive cyclicity at puberty via waves of developing and atretic follicles that stimulate the hypothalamic–pituitary–ovarian hormonal rhythms (Vihko and Apter 1981).

Looking across species, another function of the reduction of ova through atresia is the use of progesterone from atretic follicles to maintain pregnancies. The Canadian porcupine, for example, forms large numbers of corpora lutea at estrus by luteinizing atretic follicles (Perry 1971). More examples are needed, but it may eventually be arguable that atretic follicles are adaptive for the production of hormones needed to maintain pregnancies among mammals with whom we share distant, common ancestry.

If menopause appeared during human evolution as an effect of the selection for an earlier function of atresia among mammals, reptiles, and birds, then, to use Gould and Vrba's (1982) terminology, atresia for hormonal regulation is an adaptation, and atresia as the mechanism that causes menopause and postreproductive grandmotherhood is an exaptation. This does not discount the hypotheses that menopause is adaptive (in the broad meaning of the word) in terms of aging mothers/aging eggs or the grandmother hypothesis; however, the concept of exaptation suggests that the appearance of menopause is a by-product of an earlier selection—for the process of atresia.

Menopause as Protection from Estrogen-dependent Cancers

The life history of contemporary women in industrialized societies is unique within human evolution. Women no longer experience many cycles of pregnancy and lactation (Harrel 1977), but rather are exposed to repeated ovulatory hormonal cycles that involve periodic elevations in levels of estrogen (Eaton et al. 1994, this volume). While menopause may have been adaptive under past conditions by curtailing fertility so that women were free to invest in the well-being of children and grandchildren, more recent conditions draw attention to the curtailment of hormonal cycles.

Risk of breast cancer increases with early ages at menarche, late ages at menopause, delayed pregnancy, and nulliparity (Bulbrook 1991; Kelsey et al. 1993). Consistent with these observations, Henderson et al. (1985) and Pike et al. (1993) postulate an elevated risk of breast cancer with increased exposure to menstrual cycles across the life span. Without the cessation of ovulatory cycles at menopause, women, particularly in societies characterized by low rates of parity, would continue to be exposed to cyclic high levels of estrogen if they continued to menstruate into the seventh or eighth decade of life. More specifically, their estrogen-receptive target tissues would continue to be exposed to cyclic high levels of estrogen.

In the context of the parity patterns of contemporary Western societies, the cessation of menstrual cycles is adaptive. Although the dropping estradiol levels associated with menopause are frequently characterized as deleterious, as a risk factor for osteoporosis and cardiovascular disease (see below), that same drop in estradiol protects against risk of breast and endometrial cancers. The adaptive value of menopause could be measured in terms of lifetime reproductive success because breast cancers can occur at relatively young ages.

Menopause as a Neutral Trait

Outside of adaptationist scenarios, the emergence of postreproductive life and the timing of menopause may be explained by the dissociation of ovarian function from the somatic life span, a heterochronic change. Heterochrony refers to changes in the relative time of appearance, or rate of development, of features already present in ancestors (Gould 1977; McKinney and McNamara 1991). Heterochronic changes alter the timing of appearance, growth, and/or shape of discrete traits within an organism.

In most vertebrate species the upper limits of the life span correlate with the function of the gonads (Comfort 1979). Humans, short-finned pilot whales (*Globicephala macrorhynchus*), possibly other whales, and possibly African elephants (Finch 1990) are exceptional in their display of a species-universal postreproductive period. The chimpanzee, with an MLP of 50 years, does not outlive its reproductive function (Graham 1979). The ability of humans, whales, and possibly elephants to live well beyond a reproductive span supports the hypothesis that an approximately 50-year limit exists on oocyte viability among species that finish oogenesis during the fetal period (Pavelka and Fedigan 1991).

It is possible that female hominids were originally fertile to the end of the life span, similar to present-day chimpanzees, but then the human MLP extended, perhaps through the progressive prolongation of life stages and in relation to enlarged brain size (Bogin and Smith 1996; Cutler 1976; Gould 1977; Lovejoy 1981; Sacher 1978; Watts 1986). The life span more than doubled, from the 50-year MLP for chimpanzees to >110 years for humans; however, ovarian characteristics such as oocyte number and rate of atresia did not change to the same extent. Menopause, the cessation of menses, was "uncovered" by the extension of the human life span. In this scenario the appearance of menopause is architectural (Gould and Lewontin 1979). The body cells and germ cells finally demonstrate that they have had separate life histories all along, from the first 5 weeks of gestation.

When in hominid evolution did menopause first appear? Fossil evidence suggests that even the more recent bipedal hominids did not live very long (e.g., Trinkaus and Thompson 1987). What we cannot tell from the reconstruction of hominid life tables is whether the shortness of life resulted from reaching the MLP, or whether the short life represented the average life span possible.

Indirect evidence offers a glimpse of an answer. Both cranial capacity and body mass measurements used to reconstruct life span patterns in fossil hominids estimate an MLP of more than 50 years in *Homo erectus* (Smith 1991; Bogin and Smith 1996). Menopause is, potentially, a very old trait. However, MLP does not necessarily reflect life expectancy, so although *H. erectus* possessed a lengthy life span potential, they did not leave evidence of the long average life span (life expectancy) necessary for the experience of menopause. In other words, as is true for ourselves, menopause was possible for *H. erectus* only if individuals lived long enough.

Menopause as a Deleterious Trait

Evolutionary perspectives allow the consideration of menopause as an adaptive or neutral trait (Austad 1994; Pavelka and Fedigan 1991; Sherman 1998). To consider menopause as a deleterious trait requires a historical, nonevolutionary perspective. Menopause as a risk factor for the chronic conditions of osteoporosis (Nuti and Martini 1993; Pouilles et al. 1993) and cardiovascular disease (Stevenson et al. 1993) is a historical demographic, not an evolutionary, problem (Lancaster and King 1992) because demography demonstrates historical changes in population structure, notably how many women live beyond the age of menopause.

While an evolutionary perspective can be used to estimate when the maximum hominid life span exceeded the age of 50, a historical demographic perspective shows just how recently significant percentages of modern humans began to live beyond 50. "The negative aspects of ovarian failure in midlife have only become significant since *life expectation* passed the age of menopause and allowed the emergence of a maladaptive low-oestrogen state" (Faddy et al. 1992:1342, emphasis added). As more and more women live beyond 50 years of age (because an evolved MLP made it possible to do so), these women experience both menopause and more chronic diseases (Diczfalusky 1986).

However, menopause is but one risk factor for osteoporosis and cardiovascular disease. Cross-cultural research indicates that among the Maya declining estrogen levels do not correspond with increasing bone fracture rates (Martin et al. 1993). Other explanations for persistently low fracture rates include high levels of calcium in the diet, exposure to sunshine, stature (a shorter distance to fall), body weight (more padding), and heavy activity patterns. Cross-cultural research among Greenland and Alaskan Eskimos and African Maasai also points to variation in risk of cardiovascular disease (Crews and Gerber 1994). Chronic diseases now associated with menopause require their own evolutionary investigation, particularly because these diseases are not universal but are influenced by variation in nutrition, activity levels, and environmental stressors. The prolongation of the MLP coupled with the recent extension of the average life span allowed for the expression of degenerative diseases. However, the causes of these diseases are multifactorial (Crews and Gerber, this volume).

Clinical Implications

Menopause can be understood to be evolutionarily adaptive or neutral or, from a historical perspective, deleterious. HRT *is* therapeutic for women with debilitating hot flashes or at high risk for heart disease or osteoporosis (Daly et al. 1994; Kiel et al. 1987; The Writing Group 1995). However, HRT is not an appropriate therapy for menopause per se. Menopause cannot be viewed as a deficiency disease in the same sense as, for example, diabetes (McCrea 1983). Diabetes is not a human universal, whereas menopause is experienced by all women who live to 60 years of age. Osteoporosis and heart disease, like diabetes, are diseases, not universals. Osteoporosis and heart disease have their own life histories that need to be considered, separate from the life history of ovarian follicles and menopause.

Pavelka and Fedigan (1991) challenge researchers to consider the adaptive value of the hormonal changes of the climacteric transition. The lowered estrogen levels that are characteristic of postmenopausal women shrink fibroid tumors (Bernstein et al. 1992) and lower the risk of estrogen-promoted endometrial and breast cancers. Twenty years ago, the first round of estrogen replacement therapy stalled because of women's increased risk of endometrial cancer (Kennedy et al. 1985). Currently, progestin in conjunction with HRT protects against estrogen-promoted endometrial cancer (Voigt et al. 1991; Grady et al. 1995); however, the risk of breast cancer from HRT continues to foster popular and medical worries (Colditz et al. 1995). Despite evidence of increased breast cancer risk with prolonged use of HRT (Steinberg et al. 1991), the consideration of HRT by almost all postmenopausal women continues to be encouraged (American College of Physicians 1992) for the prevention of osteoporotic fractures and cardiovascular disease.

The etiologies of osteoporosis and cardiovascular disease are multifactorial and, to a large extent, related to contemporary Western diet and activity patterns (Crews and Gerber, this volume). Therefore, from an evolutionary perspective, an emphasis on healthful changes in diet and exercise is just as appropriate as HRT for the prevention of osteoporosis and cardiovascular disease (Austad 1994). Nonetheless, the medical emphasis remains on HRT (Utian and Schiff 1994) because menopause continues to be characterized as a risk factor for chronic disease. Alternative therapies for hot flashes (Freedman et al. 1995; Wyon et al. 1995), osteoporosis, and heart disease that would not increase the risk of breast cancer need to be encouraged.

Physicians who assist pregnancy in women after a clinically normal menopause (defined as occurring from ages 40 to 60) disregard the possibility that menopause evolved as an adaptive means to curtail reproduction before the onset of chronic disease (Pollard 1994) and before the effects of maternal depletion (Peacock 1991). Likewise, the treatment of the postmenopausal state with estrogen replacement (in the form of HRT) defies the adaptive advantage that menopause provides to women in Western cultures who, because of low rates of parity, have been repeatedly exposed to cyclic, high levels of estrogen and run a higher risk of breast cancer.

Prospects for the Future

More cross-cultural study is needed to determine the universal versus culture-dependent aspects of hot flashes (Lock 1993), osteoporosis (Martin et al. 1993), and cardiovascular disease. One of the most interesting areas for future research concerns the hormonal effects of breast-feeding on the regulation of the menopausal symptoms (Austad 1994; Lancaster and Lancaster 1983). Lancaster and King (1992:12) hypothesize that "unpleasant symptoms caused by abrupt fluctuations in circulating hormones may have been masked or regulated by the continuous presence of prolactin and oxytocin." Further research is needed.

In addition, Naftolin (1994) proposes that the changes of menopause be considered in relation to the female body's adaptive responses to the puer-

perium, the postpartum period, which is also a hypogonadal phase of reproductive life. During this period, women experience vasomotor instability,
sleep disturbances, and bone and fat mobilization. Naftolin (1994) suggests
these changes are adaptive during the postpartum period in that vasomotor
flushing warms the newborn and calcium and lipids are mobilized for milk
production. He concludes that adverse effects of "the reappearance of these
responses to hypoestrogenism during the climacteric is an artifact of the new-
found longevity of women" (Naftolin 1994:225).

The challenge remains for evolutionary biologists, anthropologists, and
physicians to incorporate an evolutionary perspective that considers the adaptive value of the hormonal changes of menopause and reduces the medicalization of menopause as a risk factor for chronic disease.

Acknowledgments
I thank Laurie Godfrey, Jean Nienkamp, Brooke Thomas, Wenda Trevathan,
and three anonymous reviewers for helpful discussions and suggestions.

Notes

1. Mean age at menopause among Filipina women living in Manila is 48 (Ramos-Jalbuena 1994); among Indian South African women the mean is 47.3 (du Toit 1990); in a Massachusetts population the median is 51.3 (McKinlay et al. 1992). For more details at menopause, see Gosden (1985), Gray (1976), Greene (1984), Leidy (1991), and Wood (1994).

2. Some prosimians with "nests" of germ cells in their ovaries appear to be the exception to the rule (Baker 1986).

References

Aldous, M.B., and Edmonson, M.B. (1993) Maternal age at first child-birth and risk of low birth weight and preterm delivery in Washington State. *Journal of the American Medical Association* 270:2574–2577.
Alexander, R.D. (1974) The evolution of social behavior. *Annual Review of Ecology and Systematics* 5:325–383.
American College of Physicians. (1992) Guidelines for counseling postmenopausal women about preventive hormone therapy. *Annals of Internal Medicine* 117:1048–1051.
Antinori, S., Versaci C., Hossein Gholami G., Panci C., and Caffa, B. (1993) Oocyte donation in menopausal women. *Human Reproduction* 8:1487–1490.
Austad, S.N. (1994) Menopause: an evolutionary perspective. *Experimental Gerontology* 29:255–263.
Baker, T.G. (1986) Gametogenesis. In: Dukelow, W.R. and Erwin, J., eds., *Comparative Primate Biology, vol. 3. Reproduction and Development*, pp.195–213. New York: Alan R. Liss.
Beall, C.M. (1987) Nutrition and variation in biological anthropology. In: Johnson, F.E., ed., *Nutritional Anthropology*, pp.197–221. New York: Alan R. Liss.

Bell, S.E. (1987) Changing ideas: the medicalization of menopause. *Social Science and Medicine* 24:535–542.

Bernstein, S.J., McGlynn, E.A., Kamberg, C.J., Chassin, M.R., Goldberg, G.A., Siu, A.L., and Brook, R.H. (1992) *Hysterectomy: A Literature Review and Ratings of Appropriateness*. Santa Monica, California: RAND.

Bogin, B., and Smith, B.H. (1996) Evolution of the human life cycle. *American Journal of Human Biology* 8:703–716.

Brody, J.E. (1995) Restoring ebbing hormones may slow aging. *The New York Times*, July 18, C1–3.

Bulbrook, R.D. (1991) Hormones and breast cancer. *Oxford Review of Reproductive Biology* 13:175–202.

Byskov, A.G. (1978) Follicular atresia. In: Jones, R. E., ed., *The Vertebrate Ovary*, pp. 533–562. New York: Plenum.

Cavalli-Sforza, L.L. (1983) The transition to agriculture and some of its consequences. In: Ortner, D.J., ed., *How Humans Adapt: A Biocultural Odyssey*, pp. 103–126. Washington, DC: Smithsonian Institute Press.

Colditz, G.A., Hankinson, S.E., Hunter, D.J., Willett, W.C., Manson, J.E., Stampfer, M.J., Hennekens, C., Rosner, B., and Speizer, F.E. (1995) The use of estrogens and progestins and the risk of breast cancer in postmenopausal women. *New England Journal of Medicine* 332:1589–1593.

Comfort, A. (1979) *The Biology of Senescence*. New York: Elsevier.

Cramer, D.W., Harlow, B.L., Barbieri, R., and Ng, W.G. (1989) Galactose-1-phosphate usridyl transferase activity associated with age at menopause and reproductive history. *Fertility and Sterility* 51:609–615.

Crews, D.E., and Gerber, L.M. (1994) Chronic degenerative diseases and aging. In: Crews D.E., and Garruto

R.M., eds., *Biological Anthropology and Aging: Perspectives on Human Variation over the Life Span*, pp. 154–181. New York: Oxford University Press.

Crisp, T.M. (1992) Organization of the ovarian follicle and events in its biology: oogenesis, ovulation or atresia. *Mutation Research* 296:89–106.

Currier, A.F. (1897) *The Menopause: A Consideration of the Phenomena which Occur to Women at the Close of the Child-Bearing Period, with Incidental Allusions to their Relationship to Menstruation. Also a Particular Consideration of the Premature (Especially the Artificial) Menopause*. New York: Appleton & Co.

Cutler, R.G. (1976) Evolution of longevity in primates. *Journal of Human Evolution* 5:169–202.

Cutler, R.G. (1978) Alterations with age in the informational storage and flow systems of the mammalian cell. In Bergsma, D., and Harrison, D., eds., *Genetic Effects on Aging*, pp. 463–498. New York: Alan R. Liss.

Daly E., Vessey, M.P., Barlow, D., Gray, A., McPherson, K., and Roche, M. (1994) Hormone replacement therapy in a risk-benefit perspective. In: Berg G., and Hammar, M., eds., *The Modern Management of the Menopause: A Perspective for the 21st Century*, pp. 473–497. New York: Parthenon.

Dawkins, R. (1976) *The Selfish Gene* New York: Oxford University Press.

Diamond, J. (1996) Why women change. *Discover* 17:130–137.

Diczfalusy, E. (1986) Menopause, developing countries and the 21st century. *Acta Obstetrica Gynecologica Scandinavia* (suppl.) 134:45–57.

Donaldson, J. F. (1984) Did the experience of the African savanna bring about the human menopause? *Central African Journal of Medicine* 30:198–201.

Donaldson, J.F. (1994) How did the human menopause arise? *Menopause* 1:211–221.

du Toit, B.M. (1990) *Aging and Menopause among Indian South African Women*. Albany: State University of New York Press.

Eaton S.B., Pike, M.C., Short, R.V., Lee, N.C., Trussell, J., Hatcher, R.A., Wood, J.W., Worthman, C.M., Blurton Jones, N.G., Konner, M.J., Hill, K.R., Bailey, R., and Hurtado, A.M. (1994) Women's reproductive cancers in evolutionary context. *Quarterly Review of Biology* 69:353–366.

Ettinger B., and Grady D. (1993) The waning effect of post-menopausal estrogen therapy on osteoporosis. *New England Journal of Medicine* 329:1192–1193.

Evans, J.G. (1981) The biology of human ageing. *Recent Advances in Medicine*. 18:17–38.

Faddy, M.J., Gosden, R.G., Gougeon, A., Richardson, S.J., and Nelson J.F. (1992) Accelerated disappearance of ovarian follicles in mid-life: implications for forecasting menopause. *Human Reproduction* 7:1342–1346.

Finch, C.E. (1990) *Longevity, Senescence, and the Genome*. Chicago: University of Chicago Press.

Finch, C., and Gosden, R. (1986) Animal models for the human menopause. In: L. Mastroianni and C.A. Paulsen, eds., *Aging, Reproduction, and the Climacteric*, pp. 3–34. New York: Plenum Press.

Freedman, R.R., Woodward, S., Brown, B., Javaid, J.I., and Pandey, G.N. (1995) Biochemical and thermoregulatory effects of behavioral treatment for menopausal hot flashes. *Menopause* 2:211–218.

Fretts, R.C., Schmittdiel, J., McLean, F.H., Usher R.H., and Goldman, M.B. (1995) Increased maternal age and the risk of fetal death. *New England Journal of Medicine* 333:953–957.

Gage, T.B., McCullough, J.M., Weitz, C.A., Dutt, J.S., and Abelson, A. (1989) Demographic studies and human population biology. In: M.A. Little and J.D. Haas, eds., *Human Population Biology: A Transdisciplinary Science*, pp.45–65. New York: Oxford University Press.

Gaulin, S.J.C. (1980) Sexual dimorphism in the human post-reproductive life-span: Possible causes. *Journal of Human Evolution* 9:227–232.

Gosden, R.G. (1985) *Biology of Menopause: The Causes and Consequences of Ovarian Ageing*. New York: Academic Press.

Gougeon A. (1996) Regulation of ovarian follicular development in primates: facts and hypotheses. *Endocrine Reviews* 17:121–155.

Gould, K., Flint, M. and Graham C. (1981) Chimpanzee reproductive senescence: a possible model for evolution of the menopause. *Maturitas* 3: 157–166.

Gould, S.J. (1977) *Ontogeny and Phylogeny*. Cambridge, MA: Harvard University Press.

Gould S.J., and Lewontin R.C. (1979) The spandrels of San Marco and the Panglossian paradigm: a critique of the adaptationist programme. *Proceedings of the Royal Society of London, B* 205:581–598.

Gould, S.J., and Vrba, E.S. (1982) Exaptation—a missing term in the science of form. *Paleobiology* 8:4–15.

Grady, D., Gebretsadik, T., Kerlikowske, K., Ernster, V., and Petitti, D. (1995) Hormone replacement therapy and endometrial cancer risk: a meta-analysis. *Obstetrics and Gynecology* 85:304–313.

Graham, C.E. (1979) Reproductive function in aged female chimpanzees. *American Journal of Physical Anthropology* 50:291–300.

Graham, C., Kling, R., and Steiner R. (1979) Reproductive senescence in female nonhuman primates. In: Bowden, D., ed., *Aging in Nonhuman Primates*, pp. 183–202. New York:Van Nostrand Reinhold.

Gray, R.H. (1976) The menopause— Epidemiological and demographic considerations. In: Beard, R.J., ed., *The Menopause: A Guide to Current Research and Practice*, pp. 25–40. Baltimore, MD: University Park Press.

Greene, J.G. (1984) *The Social and Psychological Origins of the Climacteric Syndrome.* Brookfield, VT: Gower Publishing.

Guraya, S.S. (1985) *Biology of Ovarian Follicles in Mammals.* New York: Springer-Verlag.

Hames, R. (1984) On the definition and measure of inclusive fitness and the evolution of menopause. *Human Ecology* 12:87–91.

Hamilton, W.D. (1966) The moulding of senescence by natural selection. *Journal of Theoretical Biology* 12:12–45.

Harrel, B.B. (1977) Lactation and menstruation in cultural perspective. *American Anthropologist* 83:796–823.

Hawkes, K., O'Connell, J.F., Blurton Jones, N.G. (1997) Hadza women's time allocation, offspring provisioning, and the evolution of long postmenopausal life spans. *Current Anthropology* 38(4):551–577.

Henderson B.E., Ross, R.K., Judd, H.L., Krailo, M.D., and Pike M.C. (1985) Do regular ovulatory cycles increase breast cancer risk? *Cancer* 56:1206–1208.

Hill, K., and Hurtado, A.M. (1991) The evolution of premature reproductive senescence and menopause in human females: an evaluation of the 'grandmother hypothesis.' *Human Nature* 2:313–350.

Hodgen, G., Goodman, A., O'Connor, A., and Johnson D. (1977) Menopause in rhesus monkeys: model for

study of disorders in the human climacteric. *American Journal of Obstetrics and Gynecology* 127: 581–584.

Horn, D.G. (1994) Outside in: the spaces of medicine and cultural critique. Paper presented at the meetings of the American Anthropological Association, Atlanta, GA, November 30–December 4.

Hughes R.M., and Gorospe W.C. (1991) Bio-chemical identification of apoptosis (programmed cell death) in granulosa cells: evidence for a potential mechanism underlying follicular atresia. *Endocrinology* 12:2415–2422.

Kaufert, P. (1988) Menopause as process or event: the creation of definitions in biomedicine. In Lock M., and Gordon, D.R., eds., *Biomedicine Examined*, pp. 331–349. New York: Kluwer Academic.

Kelsey, J.L., Gammon, M.D., and John, E.M. (1993) Reproductive factors and breast cancer. *Epidemiologic Reviews* 15:36–47.

Kennedy, D.L., Baum, C., and Forbes, M.B. (1985) Noncontraceptive estrogens and progestins: use patterns over time. *Obstetrics and Gynecology* 65:441–446.

Kiel, D.P., Felson, D.T., Anderson, J.J., Wilson, P.W.F., and Moskowitz, M.A. (1987) Hip fracture and the use of estrogens in postmenopausal women: the Framingham Study. *New England Journal of Medicine* 317:1169–1174.

Kirkwood, T.B.L. (1985) Comparative and evolutionary aspects of longevity. In: Finch, C.E., and Schneider, E.L., eds., *Handbook of the Biology of Aging*, pp. 27–44. New York: Van Nostrand Reinhold.

Kline, J., and Levin, B. (1992) Trisomy and age at menopause: predicted associations given a link with rate of oocyte atresia. *Paediatric and Perinatal Epidemiology* 6:225–239.

Kohn, R.R. (1978) *Principles of Mammalian Aging*. Englewood Cliffs, NJ: Prentice-Hall.

Lancaster, J.B., and King, B.J. (1992) An evolutionary perspective on menopause. In: Kerns, V., and Brown, J.K., eds., *In Her Prime: New Views of Middle-Aged Women*, pp. 7–15. Urbana: University of Illinois Press.

Lancaster, J.B., and Lancaster, C.S. (1983) The parental investment: the hominid adaptation. In: Ortner, D.J., ed., *How Humans Adapt: A Biocultural Odyssey*. Washington, DC: Smithsonian Institution Press.

Lapin, B.A., Krilova, R.I., Cherkovich, G., and Asanov, S. (1979) Observations from Sukhumi. In: Bowden, D., ed., *Aging in Nonhuman Primates*, pp. 183–202. New York: Van Nostrand Reinhold.

Leidy, L.E. (1990) Early age at menopause among left-handed women. *Obstetrics and Gynecology* 76:1111–1114.

Leidy, L.E. (1991) The timing of menopause in biological and sociocultural context: a lifespan approach. PhD thesis. State University of New York at Albany.

Leidy, L.E. (1994) Biological aspects of menopause: across the lifespan. *Annual Review of Anthropology* 23:231–253.

Lindquist, O., Bengtsson, C., Hansson, T., and Jonsson, R. (1983) Changes in bone mineral content of the axial skeleton in relation to aging and the menopause. *Scandinavian Journal of Clinical and Laboratory Investigation* 43:333–338.

Lock, M. (1993) *Encounters with Aging: Mythologies of Menopause in Japan and North America*. Berkeley: University of California Press.

Lovejoy, C.O. (1981) The origin of man. *Science* 211:341–350.

Martin, C.R. (1985) *Endocrine Physiology*. New York: Oxford University Press.

Martin, M.C., Block, J.E., Sanchez, S.D., Arnaud, C.D., and Beyene, Y. (1993) Menopause without symptoms: the endocrinology of menopause among rural Mayan Indians. *American Journal of Obstetrics and Gynecology* 168:1839–1845.

Mattison, D.R., Evans, M.I., Schwimmer, W., and White, B. (1984) Familial premature ovarian failure. *American Journal of Human Genetics* 36:1341–1348.

Mayer, P.J. (1982) Evolutionary advantage of the menopause. *Human Ecology* 10:477–494.

McCrea, F.B. (1983) The politics of menopause: The 'discovery' of a deficiency disease. *Social Problems* 31:111–123.

McKinlay, S.M., Brambilla, D.J., and Posner, J.G. (1992) The normal menopause transition. *American Journal of Human Biology* 4:37–46.

McKinney, M.L., and McNamara, K.J. (1991) *Heterochrony: The Evolution of Ontogeny*, New York: Plenum Press.

Mishell, D.R. (1989) Estrogen replacement therapy: An overview. *Obstetrics and Gynecology* 161:1825–1827.

Metcalf, M.G., Donald, R.A., and Livesey J.H. (1981) Pituitary-ovarian function in normal women during the menopausal transition. *Clinical Endocrinology* 14:245–255.

Moore, K.L. (1988) *The Essentials of Human Embryology*. Philadelphia: Decker.

Nachtigall, L., and Heilman, J.R. (1986) *Estrogen: The Facts Can Change Your Life*. Los Angeles: Price Stern Sloan.

Naftolin, F., Whitten, P., and Keefe, D. (1994) An evolutionary perspective on the climacteric and menopause. *Menopause* 1:223–225.

Nelson, J.F. and Felicio, L.S. (1985) Reproductive aging in the female: an etiological perspective. *Reviews of*

Biological Research on Aging 2:251–314.

Nesse, R.M., and Williams G.C. (1994) *Why We Get Sick: The New Science of Darwinian Medicine*. New York: Times Books.

Nieschlag, E., and Michel, E. (1986) Reproductive functions in grandfathers. In pp. 59–71. Mastroianni, L., and Paulsen, C.A., eds., *Aging, Reproduction, and the Climacteric*, New York: Plenum Press.

Notelovitz, M. (1989) Hormonal therapy in climacteric women: Compliance and its socioeconomic impact. *Public Health Reports Supplement* 140:70–75.

Novak, E.R. (1970) Ovulation after fifty. *Obstetrics and Gynecology* 36:903–910.

Nuti, R., and Martini, G. (1993) Effects of age and menopause on bone density of entire skeleton in healthy and osteoporotic women. *Osteoporosis International* 3:59–65.

Oliwenstein, L. (1995) Dr. Darwin. *Discover* 16:110–117.

O'Rourke, M.T., and Ellison, P.T. (1993) Menopause and ovarian senescence in human females. *American Journal of Physical Anthropology* Suppl. 16:154.

Packer, C., Tatar, M., Collins, A. (1998) Reproductive cessation in female mammals. *Nature* 392:807–811.

Paulson, R.J., and Sauer, M.V. (1994) Oocyte donation to women of advanced reproductive age: 'How old is too old?' *Human Reproduction* 9:571–572.

Pavelka, M.S., and Fedigan, L.M. (1991) Menopause: a comparative life history perspective. *Yearbook of Physical Anthropology* 34:13–38.

Peacock, N. (1991) An evolutionary perspective on the patterning of maternal investment in pregnancy. *Human Nature* 2:351–385.

Peccei, J.S. (1995) A hypothesis for the origin and evolution of menopause. *Maturitas* 21:83–89.

Perry, J.J. (1971) *The Ovarian Cycle of Mammals*. Edinburgh: Oliver and Boyd.

Peters, H., and McNatty, K.P. (1980) *The Ovary*. New York: Granada Publishing.

Pike, M.C., Spicer, D.V., Dahmoush, L., and Press, M.F. (1993) Estrogens, progestogens, normal breast cell proliferation, and breast cancer risk. *Epidemiology Reviews* 15:17–35.

Pollard, I. (1994) *A Guide to Reproduction: Social Issues and Human Concerns*. Cambridge: Cambridge University Press.

Pouilles J.M., Tremollieres, F., and Ribot, C. (1993) The effects of menopause on longitudinal bone loss from the spine. *Calcified Tissue International* 52:340–343.

Ramoso-Jalbuena J. (1994) Climacteric Filipino women: a preliminary survey in the Philippines. *Maturitas* 19:183–190.

Rannevik G., Carlstrom, K., Jeppsson, S., Bjerre, B., and Svanberg, L. (1986) A prospective long-term study in women from premenopause to postmenopause: changing profiles of onadotrophins, oestrogens, and androgens. *Maturitas* 8:297–307.

Richardson, S.J., Senikas, V., and Nelson, J.F. (1987) Follicular depletion during the menopausal transition: evidence for accelerated loss and ultimate exhaustion. *Journal of Clinical Endocrinology and Metabolism* 65:1231–1237.

Rogers, A.R. (1993) Why menopause? *Evolutionary Ecology* 7:406–420.

Sacher, G.A. (1978) Evolution of longevity and survival characteristics in mammals. In: Schneider, E.L., ed., *The Genetics of Aging*, pp.151–168. New York: Plenum Press.

Sadleir, R.M. (1973) *The Reproduction of Vertebrates*. New York: Academic Press.

Sauer, M.V., Paulson, R.J., and Lobo, R.A. (1992)Reversing the natural decline in human fertility: an extended clinical trial of oocyte donation to women of advanced reproductive age. *Journal of the American Medical Association* 268:1275–1279.

Schiff, E., and Utian W. (1994) Editorial: The genesis of menopause. *Menopause* 1:179–180.

Sherman, P.W. (1998) The evolution of menopause. *Nature* 392:759–761.

Smith, B.H. (1991) Dental development and the evolution of life history in Hominidae. *American Journal of Physical Anthropology* 86:157–174.

Steinberg, K.K., Thacker, S.B., Smith, S.J., Stroup, D.F., Zack, M.M., Flanders, W.D., and Berkelman, R.L. (1991) A meta-analysis of the effect of estrogen replacement therapy on the risk of breast cancer. *Journal of the American Medical Association* 265:1985–1990.

Stevenson, J.C., Crook, D., and Godsland, I.F. (1993) Influence of age and menopause on serum lipids and lipoproteins in healthy women. *Atherosclerosis* 98:83–90.

Tepperman, J., and Tepperman, H.M. (1987) *Metabolic and Endocrine Physiology*. Chicago: Year Book Medical Publishers.

Thomford, P.J, Jelovsek, F.R., and Mattison, D.R. (1987) Effect of oocyte number and rate of atresia on the age of menopause. *Reproductive Toxicology* 1:41–51.

Tilly, J.L., Kowalski, K.I., Schomberg, D.W., and Hsueh, A.J.W. (1992) Apoptosis in atretic ovarian follicles is associated with selective decreases in mRNA transcripts for gonadotropin receptors and cytochrome P450 aromatase. *Endocrinology* 131:1670–1676.

Tilt, E.J. (1857) *The Change of Life in Health and Disease. A Practical Treatise on the Nervous and Other Affections Incidental to Women at the Decline of Life*. London: Churchill.

Treloar, A.E. (1974) Menarche, menopause, and intervening fecundability. *Human Biology* 46:89–107.

Trinkaus, E., and Thompson, D.D. (1987) Femoral diaphyseal histomorphometric age determinations for the Shanidar 3,4,5, and 6 Neandertals and Neandertal longevity. *American Journal of Physical Anthropology* 72:123–129.

Turke, P. W. (1988) Helpers at the nest: childcare networks on Ifaluk. In: Betzig, L., Mulder, M.B., and Turke, P., eds., *Human Reproductive Behavior: A Darwinian Perspective*, pp. 173–188. New York: Cambridge University Press.

Utian, W.H. (1978) Application of cost-effectiveness analysis to post-menopausal estrogen therapy. *Frontiers of Hormone Research* 5:26–39.

Utian, W.H. (1987) Overview on menopause. *American Journal of Obstetrics and Gynecology* 156:1280–1283.

Utian, W.H., and Schiff, I. (1994) NAMS-Gallup survey on women's knowledge, information sources, and attitudes to menopause and hormone replacement therapy. *Menopause* 1:39–48.

Vihko, R., and Apter, D. (1981) Endocrine maturation in the course of female puberty. In: Pike, M.C., Siiteri, P.K., and Welsh, C.W. *Hormones and Breast Cancer*, pp. 57–69. Cold Spring Harbor, NY: Cold Spring Harbor Laboratory.

Voigt, L.F., Weiss, N.S., Chu, J., Daling, J.R., McKnight, B., and van Belle, G. (1991) Progestogen supplementation of exogenous oestrogens and risk of endometrial cancer. *Lancet* 338:274–277.

Walker, M.L. (1995) Menopause in female rhesus monkeys. *American Journal of Primatology* 35:59–71.

Watts, E.S. (1986) Evolution of the human growth curve. In: Falkner, F., and Tanner, J.M., eds., *Human Growth: A Comprehensive Treatise,* vol.1, pp. 153–166. New York: Plenum Press.

Whitehead, M.I., Whitcroft, S.I.J. and Hillard, T.C.(1993) *An Atlas of the Menopause.* New York: Parthenon.

Williams, G.C. (1957) Pleiotropy, natural selection, and the evolution of senescence. *Evolution,* 11:398–411.

Williams, G.C., and Nesse, R.M. (1991) The dawn of Darwinian medicine. *Quarterly Review of Biology* 66:1–22.

Wilson, R.A. (1966) *Feminine Forever.* New York: Evans.

Wood, J.W. (1994) *Dynamics of Human Reproduction: Biology, Biometry, Demography* New York: Aldine De Gruyter.

Worcester, N., and Whatley, M.H. (1992) The selling of HRT: playing on the fear factor. *Feminist Review* 41:1–26.

The Writing Group for the PEPI Trial. (1995) Effects of estrogen or estrogen/progestin regimens on heart disease risk factors in postmenopausal women. The Postmenopausal Estrogen/Progestin Interventions (PEPI) Trial. *Journal of the American Medical Association* 273:199–208.

Wyon, Y., Lindgren, R., Lundeberg, T., Hammar, M. (1995) Effects of acupuncture on climacteric vasomotor symptoms, quality of life, and urinary excretion of neuropeptides among postmenopausal women. *Menopause* 2:3–12.

17

BREAST CANCER IN EVOLUTIONARY CONTEXT

S. BOYD EATON
S. BOYD EATON III

An evolutionary perspective on health and disease affords new insight into the development and operation of human psychology, sheds light on common symptoms like anemia, fever, and morning sickness, and provides new understanding about the relationship between humans and infectious organisms. The Darwinian view also emphasizes that there is now discordance between genetically determined human biology and the biobehavioral circumstances of our lives. In short, our genes and our lives are out of step.

The existing human gene pool has been selected through eons during which our prehuman and early human predecessors interacted with slowly changing environmental factors that, in turn, induced gradual genetic evolution. Humans today are about 1.7% different, genetically, from our chimpanzee relatives (Sibley, 1990). Because our respective ancestors diverged, in the evolutionary sense, approximately 5 million years ago, evolutionary geneticists can roughly calibrate a rate for the evolution of nuclear DNA (Sibley, 1990). This allows speculation that humans shopping in London or watching a tennis match in Sydney are only about 0.003% different, genetically, from Late Paleolithic Stone Agers of 10,000 years ago who gathered roots and berries in pristine meadows and whose competitive instincts were honed by hunting mammoths.

This assertion, that we remain genetic Stone Agers despite the veneer of our civilized ways, is hard to prove. Still, the converse, that we have changed substantially from the physiology of our remote ancestors, is even more difficult to support. Evolutionary geneticists believe that Africans, Asians, and Europeans are "distant" from each other, genetically, by 40,000 years or more (Nei, 1987). Admittedly, there are superficial differences between these population groups, but they are vastly outweighed by biomedical similarities: for all overeating tends to promote obesity, aerobic exercise enhances endurance, pregnancy lasts nine months, etc.

Often, epidemiologists have shown differences in disease incidence that initially appeared to indicate differential susceptibility to varying medical disorders (Trowell, 1980). But in instance after instance, studies of migrants or changing incidence rates over time have subsequently demonstrated that the original incidence patterns were the result of differences in lifestyle factors, rather than underlying biological variation (Eaton et al., 1988). For example, breast cancer rates for ethnically Chinese women in Boston are identical to those for Caucasians, but they are 35% lower in Hawaii and almost 80% lower in China (Trichopoulos et al., 1984). If Africans, Europeans, and Asians are biomedically almost identical to each other despite 40,000 years of genetic "distance," then the differences between ourselves and an individual ancestor of 10,000 years ago must be negligible.[1]

Our genetic nature is related to health and disease through a simple expression:

$$\text{Genetic Makeup} + \text{Lifestyle} \sim \text{Health Status}$$

If, for the purposes of discussion, we accept that the genetic factor in this expression has been essentially constant, then the lifestyle component becomes the variable. The changes in our lives have been staggering: the Neolithic "revolution" of 10,000 years ago altered human diet fundamentally, introduced settled rather than nomadic living, fostered epidemic infectious disease, and began the population expansion which continues at an accelerated pace in the present. Lifestyle changes resulting from industrialization and the Information Age have been more explosive still: it is now possible to exist for weeks at a time eating synthetic foods, never stepping out of doors, and without the need for any physical exertion whatsoever.

We remain biologically adapted for a type of existence that was common in the remote past, in the "environment of evolutionary adaptedness" (Tooby and Cosmides, 1990), but we live in the comparatively affluent present. This is the essence of genetic–lifestyle discordance, a mismatch between our genes and our lives, which has a broadly adverse influence on human health. This discordance affects killers such as heart attack, stroke, diabetes, and high blood pressure and also promotes less deadly, but still costly, uncomfortable, and even disabling conditions like dental caries, osteoporosis, plantar fasciitis, and deafness (Nesse and Williams, 1994). Malignancies of the lung, colon, and prostate are affected as well (Eaton et al., 1988), and women's cancers—of the ovary, endometrium, and, especially, the breast—are particularly good examples of how the mismatch between our genes and our lives promotes disease (Eaton et al., 1994).

Reproductive Contrasts: Ten Thousand Years' Perspective

If one accepts that human biology is still adapted for the lifeways of our remote ancestors, then the reproductive experience of Late Paleolithic women remains the paradigm for women today, the pattern for which the human female is genetically constituted. However, it is highly likely that the experience of women then was far different from that typical of today's population, especially that of well-educated, career-oriented individuals who defer childbearing because of professional or other considerations. No one knows with absolute certainty the precise reproductive patterns that prevailed 10,000 years ago, but because women then were hunter-gatherers, it is more likely that their reproductive experience resembled that of recently studied forager women than that they anticipated the experiences of affluent Westerners. People still living in this way constitute a vanishing human resource that is our best window into the remote past when all humans lived by gathering and hunting. There are important differences of course; the !Kung San of Botswana, the Australian Aborigines, and the Agta of Luzon all live or lived in economically marginal areas unattractive to agriculturists and almost certainly less bountiful than the typical human inhabitants of 10,000 years ago. Still, the similarities must outweigh the differences, and it seems defensible to assume that such people are reasonable surrogates for our remote ancestors.

For forager women, menarche is later and first birth earlier than for typical Americans (table 17.1), so that the menarche–first birth interval is substantially longer for the latter. Early menarche is typical in all the world's industrialized nations now, but it is nevertheless a relatively new phenomenon. As recently as the nineteenth century, menarche in European nations occurred at about the same age as is now observed among forager women (Marshall and Tanner 1986; Tanner, 1973). In rural China it occurred at 17.0 years as late as the 1940s (Key et al., 1990).

Earlier age at menarche is commonly attributed to "better" nutrition, but might as well (or more accurately) reflect "overnutrition": infants and children may now receive excessive caloric intake relative to their caloric output. The

Table 17.1 Reproductive contrasts

Reproductive variable	Gatherer-hunters[a]	Americans
Age at menarche	16.1	12.5
Age at first birth	19.5	26
Menarche–first birth interval (years)	3.4	13.5
Duration of lactation per birth (years)	2.9	0.25
Lifetime duration of lactation (years)	17.1	0.4
Completed family size[b]	5	1.8
Age at menopause	47	50.5
Total lifetime ovulations[c]	160	450

Modified from Eaton et al. (1994) and Forrest (1993).

[a]!Kung, Ache, Agta, Traditional Inuit, Australian Aborigines, Hadza, Hiwi, Efe.
[b]Mean number of live births in women who survive to age 50.
[c]Calculations in Eaton et al. (1994).

energy an organism consumes is used to maintain basal metabolism, to fuel physical exertion, to support growth, and to permit development/maintenance of reproductive function; any excess is stored as adipose tissue. Forager women necessarily breast-fed, but in affluent Western nations infants are generally bottle-fed, often exclusively (Ross Laboratories, 1991). Formula-fed babies are fatter than breast-fed babies because of increased total energy intake; the latter appear to self-regulate energy intake at a lower level, resulting in less body-fat accumulation (Dewey et al., 1993). After infancy, current nutritional and physical activity patterns have resulted in increasing rates of childhood obesity (Gazzaniga and Burns, 1993). Conversely, high levels of physical activity in childhood (the rule for hunter-gatherers, but decreasingly common in affluent nations) are associated with delayed menarche (Merzenich et al., 1993; Cumming et al., 1994). During the 150 years since industrialization, height has increased, while age at puberty has decreased. These parameters are asymptotically approaching maxima and minima, respectively (Sokoloff and Villaflor, 1982; van Wieringen, 1986), but the prevalence of obesity continues to escalate (Kuczmarski et al., 1994). These temporally related observations suggest energy intake relative to expenditure now exceeds the optimum.

A higher proportion of total dietary energy from fat during childhood is associated with accelerated sexual maturation and earlier menarche (Gazzaniga and Burns, 1993; Merzenich et al., 1993). This effect is probably one component of the complex relationship between dietary fat and breast cancer. There is a clear-cut epidemiological association between national levels of fat intake and breast cancer rates (Carroll, 1980), but there is little or no correlation between individual fat intake and breast cancer rates in adult women (Howe, 1992). These observations suggest that the important dietary events relative to breast cancer may occur during fetal development, infancy, and childhood (Micozzi, 1995).

Fat-cell number is influenced by both maternal (McLeod et al., 1972; Udal et al., 1978) and infant (Winick and Noble, 1967; Knittle and Hirsch, 1968; Kramer, 1981; Lewis et al., 1986) nutrition; after infancy, the interaction between physical activity and food intake affects adipocyte formation (Oscai et al., 1974; Gazzaniga and Burns, 1993). It is tempting to speculate that overnutrition during pregnancy and childhood might increase the total number of breast epithelial and/or stem cells in a manner analogous to its effect on adipocytes. If so, the number of potential sites for carcinogenic mutation could be substantially expanded, an effect which could, to some degree, account for the observed relationship between national fat intake and breast cancer mortality.

The latest mean age at puberty for a reproducing society is probably around 18.4 years, in New Guinea, whereas the earliest is 12.4 in Hong Kong (Eveleth and Tanner, 1990). Thus, at 16.1 years (table 17.1), average menarcheal age for hunter-gatherers is much nearer the midpoint, 15.4 years, of the observed human range. This seems intuitively appropriate, whereas the current American mean of 12.5 years seems skewed far toward the youthful extreme. The exceptional human ability to transmit acquired knowledge from generation to generation has developed in conjunction with progressively delayed sexual maturation: we have the latest puberty of any mammal (Short, 1984). The increasing complexity of modern life suggests that still longer delays rather than earlier menarche would be advantageous, because women need more time to acquire the societal skills now necessary for child-rearing. However, since the nineteenth century, decreasing age at menarche has reversed the evolutionary trend towards delayed maturation at a time when occupational, educational, and social forces have led to deferred reproduction. If earlier puberty were accompanied by earlier first pregnancy, the implications for women's cancers would be less ominous; as it is, the resulting prolongation of nubility heightens the breast's susceptibility to carcinogenesis (Eaton et al., 1994).

The length of the menarche–first birth interval is inversely related to breast cancer incidence (Eaton et al., 1994). Breast development at puberty is associated with lobular development and differentiation, but the duct system remains immature. Terminal ductal lobular units (similar to Russo's lobule type I) are the most common individual element until the first full-term pregnancy, during which considerable further maturation and lobular differentiation occurs. In the latter part of the initial full-term pregnancy, type I lobules become transformed into type II and, especially, the fully mature type III lobules. For example, only 9% of total lobules in parous women aged 29–33 are type I, whereas 72% are type I in nulliparous women of the same age. The terminal ductal lobular units (i.e., type I lobules) are critically important because their cell proliferation rates are 20 times those of type III lobules, and it is in the former structures that invasive breast cancer carcinoma originates (Russo et al., 1992). These observations emphasize the importance of the menarche–first birth interval, now four times longer for Americans than for hunter-gatherers.

Calculation of total number of lifetime ovulations and of age at menopause are both subject to methodological difficulties. Precise calculations are impossible, but a forager woman clearly has far fewer ovulations during her lifetime than do her Western counterparts. Part of this difference results from later menarche, part is due to a greater number of pregnancies, and part results from different nursing patterns. Traditional !Kung San mothers, like other foragers, nurse on demand during the day and several more times at night throughout the child's first 2 years. Nursing of this nature suppresses ovulation—as much as 30 months for traditional Australian Aborigines. Age at menopause is another contributing factor. Estimating age for older foragers is difficult, but an endocrine marker, persistently elevated luteinizing hormone, has been used to gauge menopausal status among the Gainj (rudimentary horticulturists in New Guinea) and the Hadza (foragers in Tanzania). Both groups appear to begin menopause at about 47 years. For Americans the average is 50.5 years (Eaton et al., 1994).

Carcinogenesis

The development of cancer is currently viewed as a multistep process involving successive mutational changes within a single cell lineage—most commonly somatic cells. The cell within which the original mutation ("induction") occurs gives rise to a clone of successor cells, in one of which a second mutation takes place. This cell, now carrying two potentially carcinogenic mutations, in turn divides repeatedly until, in one of its descendants, yet a further mutation is produced. This process, sometimes termed "promotion" continues until, finally, a cell with mutational damage sufficient to achieve malignancy is produced. The whole process, from initiation through promotion to malignancy, may involve up to six or seven mutations and take decades for completion. The mutations must affect specific classes of genes, particularly protooncogenes and tumor-suppressor genes (Bishop, 1991).

Protooncogenes suffering mutational damage can be transformed into oncogenes, which tend to accelerate cell proliferation. Only one of the cell's pair of homologous chromosomes need be affected to achieve carcinogenic potential (i.e., the heterozygous state is oncogenic). Conversely, when only one of a cell's two tumor-suppressor alleles is inactivated, the effect is inconsequential because the other allele continues to function effectively. A further event such as deletion of the remaining normal allele or chromosomal translocation resulting in inactivation of the normal allele must occur for expression of the carcinogenic influence (Bishop, 1991; Weinberg, 1996).

Each cell carries multiple different protooncogenes and tumor-suppressor genes: more than 100 have already been identified (Levine, 1995). The physiological function of nonmutated protooncogenes and tumor-suppressor genes is only now being elucidated. At least some of the former are essential for progression of cells through the normal division cycle. The protein products of these protooncogenes show remarkable conservation over the evolutionary time scale: the same functional entities exist in yeast, worms, flies, and humans. They appear to activate (by phosphorylation)[2] critical targets at "go-no go" points in the cell cycle. Overlying this protooncogenic control of cell division is a series of "check point" controls that stop progression of the cell cycle if something has gone wrong. The tumor-suppressor genes apparently regulate this aspect of normal cellular operation. These special genes and their protein products have played a pivotal role in cellular duplication across 2 billion years of evolution (Levine, 1995).

There are two major types of carcinogens: genotoxic and nongenotoxic. The former act by directly damaging DNA. Radiation, some synthetic chemicals (e.g., dyes, pesticides), the much-maligned free radicals, and a host of naturally occurring chemical substances fall into this category. In contrast, nongenotoxic factors are not directly harmful to DNA, but create their effect chiefly through increasing rates of cell division. More rapid cell division raises the likelihood of both spontaneous and induced mutations. Furthermore, it increases the risk of chromosomal translocations and deletions which can reduce a defective tumor suppressor allele to homozygosity. And, finally, a more rapid cell turnover rate decreases the time available for DNA repair. Both external and internal stimuli can enhance cellular mitotic activity: drugs, other chemicals, infectious agents, physical or mechanical trauma, and other sources of chronic irritation can all be effective in this regard. For reproductive cell tumors, including breast cancer, hormones are critically important in controlling cell proliferation rates (Trichopoulos et al., 1996).

New Reproductive Patterns and Increasing Breast Cancer Incidence

For breast cancer, rapid ductal cell turnover enhances and fosters the course of carcinogenesis. This is the mechanism which, to a large extent, explains how changes in female reproductive experience have adversely affected breast cancer incidence. Our "new" reproductive pattern in effect increases the average cell turnover rate in breast duct epithelial and glandular cells, creating a dynamic that favors carcinogenic mutation. For example, breast glandular cell proliferation is minimal during intense nursing (Newcomb et al., 1994; Kelsey and John, 1994), a state which, in total, lasted perhaps 30 times as long for Stone Agers (and of most women until relatively recently) as for American women. Conversely, the period between menarche and first full-term pregnancy, a time during which cell turnover in the breast is at maximum, now averages 13.5 years; for preagriculturalists the average is 3.4 years. There has been a fourfold increase in this "nubility interval," during which susceptibility to carcinogenic mutation is at its peak (table 17.1).

A further, at least theoretical, cause of increased breast cancer incidence involves pregnancy interruption. The potential effects of pregnancy on breast cancer risk are biphasic. During the first trimester, high levels of estrogen and progesterone lead to rapid cell turnover rates within the breast, accordingly enhancing promotion of any initiated cell clones and increasing the likelihood of further carcinogenic mutations. During the last half of a pregnancy (especially the first pregnancy), however, there is extensive differentiation of stem cells in concert with lobular maturation, changes that greatly reduce subsequent cell turnover rates and decrease susceptibility to both carcinogenic mutation and ongoing clonal promotion. The net result of these factors in a normal, noninterrupted pregnancy is to decrease long-term breast cancer risk while transiently increasing breast cancer incidence for a few years following delivery (Lambe et al., 1994). Termination of pregnancy before the sixth month would therefore be expected to increase overall breast cancer risk (especially prior to the first full-term pregnancy) because the early effect of rapid cell proliferation is not counterbalanced by the stem cell differentiation and lobular maturation that normally occur toward the end of the term. Numerous epidemiological investigations pertaining to this hypothesis have been performed, but there are formidable methodological difficulties (to say nothing of the political implications), and overall results are as yet inconclusive (Somerville, 1994).

Relative Breast Cancer Risks

There is no way to determine breast cancer incidence among hunter-gatherers, although the observations of medical anthropologists suggest that neoplastic disease is generally uncommon (Truswell and Hansen, 1976; Hildes and Schaefer, 1984). Analysis of skeletal remains from the Late Paleolithic up until about 1500 AD reveals little evidence of cancer, even in circumstances where it should be identifiable paleopathologically if present (Micozzi, 1991). Still, there are no hard data on breast cancer rates among forager women. This deficiency may be partially addressed by using an epidemiological model devised by Pike and his colleagues (Pike et al., 1983; Pike, 1987) which integrates pertinent reproductive variables to allow cross-cultural prediction of breast cancer incidence. The concept is analogous to anticipating the reduction in coronary heart disease mortality that might result from changes in risk factors for atherosclerosis (e.g., Tsevat, 1991).

Pike's model suggests that the relative cumulative risk of breast cancer, to age 60,[3] for hunter-gatherer women should be approximately one-hundredth that for American women (Eaton et al., 1994). A 100-fold difference sounds extreme, but in 1975 the Japanese rate was one-fourth and the West African only one-twelfth that of North Americans (Parkin et al., 1984). The model predicts the rate for forager women to be about one-eighth that of West Africans.[4] Viewed in this light, an overall 100-fold difference seems credible. In any case, the difference in breast cancer rates between recent foragers (surrogates for ancestral humans) and ourselves must be great. If the model underestimates forager risk by 50%, the difference between Late Paleolithic and current American breast cancer rates would still be impressive: 1 in 9 for us, 1 in 450 for hunter-gatherer women.

Evolutionary Considerations and Breast Cancer Prevention

For many chronic diseases (osteoporosis, type II diabetes, hypertension, coronary heart disease, lung cancer, and others), the concept of discordance between genes and lifestyle suggests relatively straightforward preventive strategies. In each case, readopting the essential elements of preagricultural life, in terms of nutrition, physical exercise, and other behaviors, should markedly reduce disease incidence. Some of the necessary lifestyle modifications might entail practical difficulties: reducing sodium intake to less than a gram a day, adopting a 20% fat diet, consuming more nearly equal proportions of the ω-6 and ω-3 essential fatty acids, eliminating tobacco usage, and making time for adequate exercise (both strength and endurance training) (Eaton et al., 1988). But for each, the basic rationale is clearcut and, for most, there is emerging consensus among health professionals regarding preventive value.

Reordering the reproductive experiences of women in affluent nations to mirror the Paleolithic paradigm should produce analogous beneficial effects on the incidence of women's cancers, including carcinoma of the breast. However, while it might be possible to successfully promote nursing similar to that practiced by forager women, the other components of preagricultural reproduction, especially earlier first birth (at age 19) and, most of all, greater parity (six children per woman) would have disastrous socioeconomic and demographic consequences (e.g., overpopulation). Such a reversal of recent reproductive experience is therefore both unlikely and undesirable, personally and societally. Faced with this dilemma, a second-order solution seems more practicable. Modern pharmacology could be employed to recreate an ancestral hormonal milieu without early and frequent childbirth.

The preventive effects of later menarche, earlier first birth, multiple pregnancies, prolonged intense nursing, and earlier menopause arise from associated hormonal perturbations, which in turn affect cell turnover rates in reproductive tissues. The reproductive experiences of women in affluent nations increase breast cancer risk because they alter hormonal levels in a way which over a lifetime tends to accelerate the proliferation of epithelial cells in the breast ducts and lobules (Russo et al., 1982). It is this effect that enhances cellular susceptibility to carcinogenic mutations and, accordingly, increases likelihood of breast cancer as much as 100-fold. While most women would prefer not to have six children beginning at age 19, those faced with an especially high risk of breast cancer, as determined by family history or genetic testing, might well chose a prophylactic approach based on exogenous hormonal manipulation with the aim of duplicating the ancestral endocrine state.

Endocrine therapy can delay menarche; it would be a simple matter to use existing contraceptive technology for this purpose. A 5-year Norplant gestagen implanted at age 11 would postpone menarche to age 16, the forager mean. Simple oral administration of progesterone would have a similar effect, but would require more patient compliance. Epidemiological data suggest a 20% reduction in breast cancer risk for each year of pubertal delay, even without any additional preventive measures (Henderson et al., 1988).

Experiments on rats have shown that, after menarche, high doses of estrogen and progesterone, producing serum levels that match those during pregnancy, can induce breast duct epithelial/lobular maturation with stem cell differentiation similar to the changes observed after a woman's first full-term pregnancy (Russo et al., 1982; Grubbs et al., 1985). This transformation has been achieved without harming subsequent reproductive success or lactation capability. The treatment program in effect induces a pseudopregnancy, and its efficacy in reducing breast cancer risk is impressive. Animal studies have demonstrated an 88% reduction in total tumor formation. Of rats given a carcinogen (N-nitroso-N-methylurea) without prior pseudopregnancy treatment, 61% developed breast cancers, most more than one, while of those animals who were first treated to induce pseudopregnancy and then given a similar carcinogen, only 13% developed cancers—a statistically significant result (Grubbs et al., 1985).

The serum estrogen levels of forager women are significantly lower than those of women living in affluent nations (table 17.2). Pike and his associates

Table 17.2 Comparative serum estrogen levels

Nonpregnant cycling women	Serum estradiol (pg/ml)[a]
!Kung San (Botswana)	112
Chinese (Shanghai)	136
Americans (Los Angeles)	164

Data from Konner and Worthman, (1980); van der Walt et al.
(1978); Bernstein et al. (1990).

[a] Average of follicular and luteal phases; pg $\times$ 3.671 = pmol.

have developed a unique approach to contraception which, in effect, reduces serum estrogen levels to those found among hunter-gatherers (Pike and Spicer, 1983). This method involves oral administration of a gonadotropin-releasing hormone agonist (GnRHA) which totally, but reversibly, inhibits ovarian production of both estrogen and progesterone. In addition, a low dose of estrogen is added back to guard against osteoporosis and other ill effects of hypoestrogenism; the dose administered produces serum levels similar to those found among foragers. Finally, progesterone is given at intervals, as much as 3 or even 6 months apart, to slough the endometrium and induce menstruation (Henderson et al., 1993). This approach is currently undergoing clinical trials. It has been estimated that 5 years of usage by a woman during her reproductive years would reduce lifelong breast cancer risk by 31% and that 15 years of usage would cut risk by 70% (Henderson et al., 1993).

Interventional Endocrinology and "Natural" Breast Cancer Prevention

The use of oral or implanted medications to alter the endocrine status of American or other affluent women seems an intuitively radical, "unnatural" approach to reducing breast cancer risk. As such, any advocacy of this technique is certain to be highly contentious. However, those opposing this suggestion must come to grips with uncomfortable facts. The American Cancer Society has tracked cancer mortality rates since 1930: the most recently published report (Boring et al., 1995) presents data through 1990—60 years of disease experience. These 60 years have witnessed introduction of chemotherapy, high-dose radiation therapy, and lumpectomy—all major innovations in breast cancer treatment. In addition, mammography was introduced during this period and has been greatly improved in the past two decades. These developments notwithstanding, the mortality rate from breast cancer in 1990 was moderately greater than in 1930. This observation validates Burkitt's (1993) axiom that improvements in diagnosis and therapy alone, no matter how effective, can never lower disease incidence.

It is premature to categorically advocate interventional endocrinology for prevention of breast cancer, but two recommendations do seem appropriate. The first is for increased research about the benefits and side effects of menarcheal delay, pseudopregnancy, and low-estrogen oral contraception. This could be conducted on nonhuman primates and, in the case of contraceptives, on women at especially high risk for development of breast cancer—as is, in fact, already underway. The second recommendation concerns the need for discussion about the desirability of such measures. While these interventions seem foreign and somehow repugnant, they really differ little from the concept of oral contraception, which, while highly controversial 35 years ago, has now become widely accepted.

Breast cancer prevention through interventional endocrinology is consistent with discordance theory: reinstitution of the microanatomical and hormonal milieu for which human physiology has been designed through evolution. It is, in fact, our current lifestyle and its effects on reproductive experience that is unnatural. In recent decades our new lifeways have led to increases in breast cancer incidence that have outweighed improvements in both diagnosis and treatment to the point where one woman in nine may now expect to become a victim of breast cancer during her lifetime. As it becomes increasingly possible to identify women at especially high risk through sophisticated genetic testing, articles about cancer counseling have begun to mention prophylactic mastectomy as an alternative (Bilimoria and Morrow, 1995; Hoskins et al., 1995). If this is so, then surely increased research and social discourse about the merits and drawbacks of interventional endocrinology are entirely in order. An understanding of ancient evolutionary experience may allow us to refocus our thinking and to generate a new, powerful paradigm for disease prevention.

Notes

1. Single gene mutations are an exception, but the most numerically important chronic degenerative diseases (e.g., coronary arteriosclerosis, cancer, diabetes) are polygenic.

2. Phosphorylation is biochemical activation through donation of a high-energy phosphate group (e.g. from ATP).

3. About 9% of forager women attain this age (Howell, 1979).

4. Data primarily from Ibadan (Nigeria) and Dakar (Senegal), urban if not affluent centers (Parkin et al., 1984).

References

Bernstein, L., J.-M. Yuan, R.K. Ross, M.C. Pike, R. Lobo, F. Stanczyk, Y.-T. Gao, and B.E. Henderson (1990) Serum hormone levels in premenopausal Chinese women in Shanghai and white women in Los Angeles: Results from two breast cancer case-control studies. *Cancer Causes Control* 1:51–58.

Bilimoria, M.M., and M. Morrow

(1995) The woman at increased risk for breast cancer: Evaluation and management strategies. *Ca- A Cancer Journal for Clinicians* 45:263–78.

Bishop, J.M. (1991) Molecular themes in oncogenesis. *Cell* 64: 235–248.

Boring, C.C., T.S. Squires, and T. Tong (1995) Cancer statistics, 1995. *Ca- A Cancer Journal for Clinicians* 45:19–38.

Burkitt, D.P. (1993) The Bower science lecture. Philadelphia, January 15.

Carroll, K.K. (1980) Lipids and carcinogenesis. *Journal of Environmental Pathology and Toxicology* 3:253–71.

Cumming, D.C., G.D. Wheeler, and V.J. Harper. (1994) Physical activity, nutrition, and reproduction. *Annals of the New York Academy of Science* 709:55–76.

Dewey, K.G., M.J. Heinig, L.A. Nommsen, J.M. Peerson, and B. Lonner dal (1993) Breast-fed infants are leaner than formula-fed infants at 1 y of age: The DARLING study. *American Journal of Clinical Nutrition* 57:140–45.

Eaton, S.B., M. Konner, and M. Shostak (1988) Stone agers in the fast lane: chronic degenerative diseases in evolutionary perspective. *American Journal of Medicine* 84:739–49.

Eaton, S.B., M.C. Pike, R.V. Short, N.C. Lee, J. Trussell, R.A. Hatcher, J.W. Wood, C.M. Worthman, N.C. Blurton Jones, M.J. Konner, K.R. Hill, R. Bailey, and A. M. Hurtado (1994) Women's reproductive cancers in evolutionary context. *Quarterly Review of Biology* 69:353–67.

Eveleth, P.B., and J.M. Tanner, eds. (1990) *Worldwide Variation in Human Growth,* 2nd ed. Cambridge: Cambridge University Press.

Forrest, J.D. (1993) Timing of reproductive life stages. *Obstetrics and Gynecology* 82:105–11.

Gazzaniga, J.M., and T.L. Burns (1993) Relationship between diet composition and body fatness, with adjustment for resting energy expenditure and physical activity, in preadolescent children. *American Journal of Clinical Nutrition* 58:21–28.

Grubbs, C.J., D.R. Farnell, D.L. Hill, and K.C. McDonough (1985) Chemoprevention of N-nitroso-N-methylurea-induced mammary cancer by pretreatment with 17-estradiol and progesterone. *Journal of the National Cancer Institute* 74:927–31.

Henderson, B.E., R.K. Ross, and L. Bernstein (1988) Estrogens as a cause of human cancer: The Richard and Hinda Rosenthal Foundation Award Lecture. *Cancer Research* 48: 246–53.

Henderson, B.E., R.K. Ross, and M.C. Pike (1993) Hormonal chemoprevention of cancer in women. *Science* 259:633–38.

Hildes, J.A., and O. Schaefer (1984) The changing picture of neoplastic disease in the western and central Canadian Arctic (1950–1980). *Canadian Medical Association Journal* 130:25–52.

Hoskins, K.F., J.E. Stopfer, K.A. Calzone, S.D. Merajver, T.R. Rebbeck, J.E. Garber, and B.L. Weber (1995) Assessment and counseling for women with a family history of breast cancer. *Journal of the American Medical Association* 273:577–85.

Howe, G.R. (1992) High fat diets and breast cancer risk. The epidemiological evidence. *Journal of the American Medical Association* 268:2080–81.

Howell, N. (1979) *Demography of the Dobe !Kung.* New York: Academic Press.

Kelsey, J.L., and E.M. John (1994) Lactation and risk of breast cancer. *New England Journal of Medicine* 330: 136–37.

Key, T.J.A., J. Chen, D.Y. Wang, M.C. Pike, and J. Boreham (1990) Sex hormones in women in rural China and

Britain. *British Journal Cancer* 62: 631–36.

Knittle, J.L., and J. Hirsch (1968) Effect of early nutrition on the development of rat epididymal fat pads: cellularity and metabolism. *Journal of Clinical Investigation* 47:2091–98.

Konner, M., and C. Worthman (1980) Nursing frequency, gonadal function, and birth spacing among !Kung hunter-gatherers. *Science* 207:788–91.

Kramer, M.S. (1981) Do breast feeding and delayed introduction to solid food protect against subsequent obesity? *Journal of Pediatrics* 98:883.

Kuczmarski, R.J., K.M. Flegal, S.M. Campbell, and C.L. Johnson (1994) Increasing prevalence of overweight among U.S. adults. The natural health and nutrition examination surveys, 1960 to 1991. *Journal of the American Medical Association* 272: 205–11.

Lambe, M., C.-C. Hsieh, D. Trichopoulos, A. Ekbom, M. Pavia, and H.-O. Adami (1994) Transient increase in the risk of breast cancer after giving birth. *New England Journal of Medicine* 331:5–9.

Levine, A.J. (1995) The genetic origins of neoplasia. *Journal of the American Medical Association* 273:592–93.

Lewis, D.S., H.A. Bertrand, C.A. McMahon, H.C. McGill, K.D. Carey, and E.J. Masoro (1986) Preweaning food intake influences the adiposity of young adult baboons. *Journal of Clinical Investigation* 78:899–905.

Marshall, W.A., and J.M. Tanner (1986) Puberty. In F. Falkner and J.M. Tanner, eds., *Human Growth*, 2nd ed., vol. 2. New York: Plenum Press, pp. 171–209.

McLeod, K.I., R.B. Goldrick, and H.M. Whyte (1972) The effect of maternal malnutrition on the progeny in the rat. *Australian Journal of Experimental Biology and Medical Science* 50:435–46.

Merzenich, H., H. Boeing, and J. Wahrendorf (1993) Dietary fat and sports activity as determinants for age at menarche. *American Journal of Epidemiology* 138:217–24.

Micozzi, M.S. (1991) Disease in antiquity: The case of cancer. *Archives of Pathology and Laboratory Medicine* 115:838–44.

Micozzi, M.S. (1995) Breast cancer, reproductive biology, and breastfeeding. In P. Stuart-Macadam and K.A. Dettwyler, eds., *Breastfeeding: Biocultural Perspectives*. New York: Aldine and de Gruyter, pp. 347–384.

Nei, M. (1987) *Molecular Evolutionary Genetics*. New York: Columbia University Press.

Nesse, R.M., and G.C. Williams (1994) *Why We Get Sick. The New Science of Darwinian Medicine*. New York: Times Books.

Newcomb, P.A., B.E. Storer, M.P. Longnecker, R. Mittendorf, E.R. Greenberg, R.W. Clapp, K.P. Burke, W.C. Willett, and B. MacMahon (1994) Lactation and a reduced risk of premenopausal breast cancer. *New England Journal of Medicine* 330:81–87.

Ocsai, L.B., S.P. Babirak, F.B. Dubach, J.A. McGarr, and C.N. Spirakis (1974) Exercise or food restriction: Effect on adipose tissue cellularity. *American Journal of Physiology* 227: 901–904.

Parkin, D.M., J. Stjernsward, and C.S. Muir (1984) Estimates of the worldwide frequency of twelve major cancers. *Bulletin of the World Health Organization* 62:163–82.

Pike, M.C. (1987) Age-related factors in cancers of the breast, ovary, and endometrium. *Journal of Chronic Diseases* 40 (suppl. 2):595–695.

Pike, M.C., M.D. Krailo, B.E. Henderson, J.T. Cassagrande, and D.G. Hoel (1983) 'Hormonal' risk factors, 'breast tissue age,' and the age-incidence of breast cancer. *Nature* 303:767–70.

Pike, M.C., and D.V. Spicer (1983) Oral contraceptives and cancer. In D. Shoupe and F. Hasseltine, eds., *Contraception*. New York: Springer Verlag, pp. 67–84.

Ross Laboratories Mother's Surveys (1991) *Recent Trends in Breast Feeding*. Columbus, OH: Ross Laboratories.

Russo, J., L.K. Tay, and I.M. Russo (1982) Differentiation of the mammary gland and susceptibility to carcinogenesis. *Breast Cancer Research and Treatment* 2:5–73.

Russo, J., R. Rivera, and I.M. Russo (1992) Influence of age and parity on the development of the human breast. *Breast Cancer Research and Treatment* 23:211–18.

Short, R.V. (1984) Species differences in reproductive mechanisms. In C.R. Austin and R.V. Short, eds., *Reproduction in Mammals*. Book 4: Reproductive Fitness. Cambridge: Cambridge University Press; pp. 24–61.

Sibley, C.G. (1990) DNA hybridization evidence of hominoid phylogeny: A reanalysis of the data. *Journal of Molecular Evolution* 30:202–36.

Sokoloff, K.L., and G.C. Villaflor (1982) The early achievement of modern stature in America. *Social Science and History* 6:453–81.

Somerville, S.W. (1994) Does abortion increase the risk of breast cancer? *Journal of the Medical Association of Georgia* 83:209–10.

Tanner, J.M. (1973) Trends toward earlier menarche in London, Oslo, Copenhagen, the Netherlands, and Hungary. *Nature* 243:95–96.

Tooby, J., and L. Cosmides. (1990) The past explains the present. Emotional adaptations and the structure of ancestral environments. *Ethology and Sociobiology* 11:375–424.

Trichopoulos, D., S. Yen, J. Brown, P. Cole, and B. MacMahon. (1984) The effect of Westernization on urine estrogens, frequency of ovulation, and breast cancer risk. A study of Ethnic Chinese women in the Orient and the USA. *Cancer* 53:187–92.

Trichopoulos, D., F.P. Li, and D.J. Hunter (1996) What causes cancer? *Scientific American* 275(3):80–87.

Trowell, H.C. (1980) From normotension to hypertension in Kenyans and Ugandans 1928–78. *East African Medical Journal* 57:167–73.

Truswell, A.S., and D.L. Hansen (1976) Medical research among the !Kung. In R.B. Lee and I. Devore, eds., *Kalahari Hunter-Gatherers*. Cambridge, MA: Harvard University Press; pp. 166–95.

Tsevat, J., M.C. Weinstein, L.W. Williams, A.N.A. Tosteson, and L. Goldman (1991) Expected gains in life expectancy from various coronary heart disease risk modifications. *Circulation* 83:1194–1201.

Udal, J.G., G.G. Harrison, Y. Vaucher, P.D. Wilson, and G.D. Morrow (1978) Interaction of maternal and neonatal obesity. *Pediatrics* 62:17–21.

Van der Walt, L.A., E.N. Wilmsen, and T. Jenkins (1978) Unusual sex hormone patterns among desert-dwelling hunter-gatherers. *Journal of Clinical Endocrinology and Metabolism* 46:658–63.

Van Wieringen, J.C. (1986) Secular growth changes. In F. Falkner and J.M. Tanner, eds., *Human Growth. A Comprehensive Treatise, 2nd ed., vol. 3. Methodology. Ecological, Genetic, and Nutritional Effects on Growth*. New York: Plenum Press; 307–31.

Weinberg, R.A. (1996) How cancer arises. *Scientific American* 275(3):62–70.

Winick, M., and A. Noble (1967) Cellular response to increased feeding in neonates. *Journal of Nutrition* 91:179–82.

18

EVOLUTIONARY PERSPECTIVES ON CHRONIC DEGENERATIVE DISEASES

LINDA M. GERBER

DOUGLAS E. CREWS

In recent decades, evolutionary perspectives on chronic diseases and aging have become a major aspect of theory formation in human population biology and biological anthropology (Eaton et al. 1988; Armelagos 1991; Crews 1993; Crews and Gerber 1994). Human biologists have long examined physical and health variation between races, types, and ethnic groups (Coon 1965; Marks 1994; Molnar 1994). Application of an evolutionary perspective to problems such as defining races, categorizing human variation, and identifying population affinities provided a basis for going beyond racial typologies, the search for "pure races," and the use of external characteristics (skin color, hair texture, and color) to determine population relations. The molecular revolution has contributed to a greater understanding of how genetic variation is packaged and altered over generations and provides a new tool for understanding population affinities and disease patterns. These advances also have led to an improved understanding of the evolutionary and genetic basis of specific chronic degenerative diseases (CDD) and a general model for the etiology of particular types of CDD. For those interested in the health and well-being of humankind, a basic understanding of evolutionary pressures that have shaped human physiological responses to the environment is a necessity. It is useful to inform practitioners that different races, ethnic groups, or sexes may respond differently to the same lifestyle or pharmaceutical intervention. An evolutionary perspective also aids practitioners in counseling their various patients as to how behavioral and lifestyle factors may influence longevity and health. Last, an evolutionary perspective provides the basis for genetic counseling and explaining the basis of heritable disease to both patients and concerned families.

Evolutionary models and perspectives on chronic diseases have not yet had any great influence on medical practitioners. It is possible that clinicians' focus on practical patient care has diverted attention away from, or possibly even obscured, the practical and applied value of these evolutionary perspectives for patient counseling and care. Medical practitioners may see little practical use in restating research results showing that patterns of CDD incidence are related to complex interactions of genes, environments, and lifestyles. One reason for this view is that genetic propensities are thought by many to be unalterable. Given the lack of substantial results from any ongoing genetic replacement trials, it would appear that we remain a long way from the promise of any practical genetic engineering as is often depicted in science fiction literature and popular science magazines.

Another reason for the deemphasis of evolutionary pressures is that environments commonly are seen as either inescapable or, at the very least, beyond individual control. In addition, lifestyle attributes are individual characteristics that are extremely resistant to modification by medical intervention or public health activities. Although the national reduction in smoking since the 1960s has been one noteworthy success, personal factors such as low physical activity levels, high fat, high-calorie, low-fiber diets, multiple sex partners, smoking, and alcohol consumption are lifestyle choices that many individuals do not wish to change. Such lifestyle-related risk factors are also among the leading public health problems facing humankind in the twenty-first century.

One purpose of this chapter is to provide practitioners with several practical approaches to integrate evolutionary thinking into their clinical practices. This is accomplished by reviewing what is currently known and hypothesized regarding the evolutionary biology of the major lifestyle-related CDDs in general, along with a discussion of the genetic bases and environmental correlates of several specific CDDs. How this knowledge may be useful in patient counseling and care is also examined. Our perspective is populational, evolutionary, and biomedical, with a specific concentration on diseases that most affect middle-aged and older humans in "cosmopolitan" societies (see McGarvey et al. 1989).

In the first section of this chapter, we review progress toward an integrated evolutionary model to explain the current epidemic of CDD in many of the world's societies (Crews and Gerber 1994). Previously, we have labeled such societies "cosmopolitan" (outward looking, partaking of the worldwide culture promoted by expanding communications and technologies and integration of economies on a worldwide scale), to avoid use of such pejorative terms as "modern," "western," "developed," or "first world," which have been used by others to describe similar societies. We follow this with an examination of clinical conditions that may be termed "chronic degenerative diseases" along with a brief review of how genes and risk factors may be related to the etiology of CDDs. Part of this review is focused on how changing environments and lifestyles have contributed to improved life expectancies and how changes in life expectancy have acted to reveal ancient genes that today contribute to the etiology and pathogenesis of CDDs. We then examine how medical practitioners may make use of such evolutionary knowledge in their everyday clinical practice, with particular emphasis on risk factor reduction, pharmacogenetics, and genetic counseling. These sections are followed by an overall discussion that includes directions and areas for continued research and collaboration between health care professionals and evolutionarily oriented biomedical researchers.

An Evolutionary Model for Chronic Degenerative Diseases

The fields of chronic disease epidemiology and geriatrics/gerontology share much common ground. Both areas include study of risk factors, whether genetic, behavioral, or environmental, for longevity and CDD. Phenomena under study by both are longitudinal in nature. Both use a life history perspective to understand incremental change and degeneration. Last, the outcomes of interest—disease processes and aging—are highly individualistic in nature, and require analyses at the individual, familial, and population level. Some have even speculated that aging-related change and change due to CDDs cannot be studied in isolation from one another because they are inextricably intertwined during each individual's unique life history (see Arking 1991). Furthermore, all current theories of senescence are based on decrements and losses of function that increase risk of CDDs and death. Still, there has been little overt interaction between, or synthesis of theories and methods across, these two disciplines.

Recently, an integrative model was proposed to help explain the epidemic of CDDs that is today afflicting middle-aged and older persons (Crews and Gerber 1994). It seems worthwhile to revisit and expand this theoretical model within the context of evolutionary medicine. In the discussion section, we address similar and different models proposed by others (Williams 1957; Neel 1962, 1982; Hamilton 1966; Eaton and Konner 1985; Eaton et al. 1988; Eaton and Nelson 1991; Jackson 1991; Wilson and Grim 1991; Weder and Schork 1994), many of which may be subsumed under a "thrifty/pleiotropic gene model" of CDDs. This model is based on the assumption that evolutionary biology has programmed all organisms to have "thrifty" genetic mechanisms that enhance extraction, retention, and resorption of scarce but necessary metabolic components (e.g., cholesterol, energy, salt). A second assumption is that most genetic loci and alleles have multiple activities and functions in different metabolic and physiological pathways: loci with multiple functions are said to be "pleiotropic" (see Crews and Gerber 1994 for further elaboration of this model).

Over the last 10,000 years, humans have dramatically altered their environmental and ecological circumstances; this change has been particularly rapid during the last few centuries (Fenner 1970; Stini 1971; Harrison 1973; Baker and Baker 1977; Eaton et al. 1988). Numerous alleles that conferred selective advantages during earlier phases of human evolution may today put their carriers at a disadvantage with respect to CDDs and longevity due to associated late-acting ill effects, but may have little impact on the genetic success of their carriers (Williams 1957 Levi and Anderson 1975; Eaton et al. 1988; James et al. 1989; McGarvey et al. 1989; Crews 1990; Crews and James 1991; Crews and Gerber 1994). Thus, a major contributor to the increased incidence and prevalence of CDDs in present-day cosmopolitan societies, in addition to life expectancies of 70 and more years, may be a lack of fit of humankind's "Stone Age genes" with our technological and cosmopolitan lifestyles (see Eaton and Konner 1985; Eaton et al., 1988; James and Baker 1990; Crews 1990; Crews and Gerber 1994).

Alterations in lifestyle have affected both our social and our physical environments (Baker 1984). Included among these alterations are new nutritional patterns and dietary factors (Stini 1971; Eaton and Konner 1985; Eaton and Nelson 1991), increased public health activities (vaccinations, chlorinated water), improved medical technologies (chemotherapy, perinatal surgery), altered exposures to new and old pathogens (Ebola, HIV, multidrug-resistant bacteria tubercle bacillus), increases in exposure to environmental and human-made toxins (Schell 1991), and reductions or changes in physical activity. Such environmental, lifestyle, and demographic changes are being played out against a background of co-adapted gene complexes that show multiple epistatic (i.e., when one gene or genetic locus influences the activity of another gene or locus), multifactorial, and pleiotropic relationships. Our current gene pool is the product of millions of years of evolution in a mobile, primarily plant-eating, but omnivorous, ecological niche with short (or at least shorter than present) life expectancy. The phenotypes who carry these genes are now being exposed to sedentary lifestyles, fat-rich/fiber-poor diets, and an extended life span. Our genes and gene complexes have never before been exposed to the selective pressures imposed by longevity. Until recently, most humans did not live long enough for the possible late-acting effects of pleiotropic genes to be expressed. For example, estimated life expectancies are only 15 years for Australopithecines, 18 years for Neanderthals, 25 years for Neolithic agriculturalists and Classic/Medieval Europeans, 43 years during the nineteenth century, 55 years in the early twentieth century, and have attained 75 years only in contemporary cosmopolitan settings (see Weiss 1981, 1984, 1989; Crews 1990). Thus, our current burden of CDDs must be seen as a product of evolutionary "lag time," the recency of improved life expectancies, and the declining force of natural selection with increasing age (Rose 1993; Crews and Gerber 1994). Before examining practical and clinical applications, it is necessary to understand physiological implications of the "thrifty/pleiotropic" model.

Pleiotropy and "Thrifty" Genes

In 1957, G.C. Williams proposed that pleiotropic genes or "genes that have opposite effects on fitness at different ages, or, more accurately, in different somatic environments" (p. 400), were the biological bases for age-related decline and senescence. In brief, Williams (1957) hypothesized that multiple genetic loci have multiple alleles (today it is assumed that most loci have multiple alleles) and that various alleles are selected against if they predispose to detrimental outcomes prior to reproduction. This prereproductive selection occurs regardless of any beneficial effects the allele might confer at later ages. Therefore, alleles that have been retained in the gene pool are those associated with enhanced early survival, growth, development, and maturation to reproductive age, regardless of any late-acting detrimental effects they may have. Alleles with early-acting ill effects were selected against, regardless of their possible associations with disease resistance or longevity following attainment of maximum reproductive potential. Alleles with beneficial effects early in life, but detrimental effects sometime after the age of maximum reproductive probability are said to exhibit antagonistic pleiotropy. Such alleles could enhance maturation, attainment of maximum reproductive probability, and subsequent reproductive success; however, they also could predispose their carriers to physiological degeneration or loss of function after the point of maximum reproductive output, that is during middle-age and later years.

Subsequently, in 1962, J.V. Neel proposed that "thrifty" genes have been incorporated into the human genome because of their selective advantage over less "thrifty" ones during earlier phases of human evolution. Thrifty genes would predispose carriers to more efficient extraction of scarce resources from their environment (diet). According to Neel, thrifty genes could increase the efficiency of energy extraction and storage from dietary sources, such that in times of plenty more energy was stored by those with this predisposition than in those without. Subsequently, during times of dietary energy shortage, thrifty individuals would have an advantage, using previously stored energy to maintain homeostasis, while those without thrifty genes would be at a disadvantage. Neel (1962, 1982) applied his model only to energy, glucose metabolism, and non–insulin-dependent diabetes mellitus (NIDDM). However, the generalizability of the thrifty alleles model to multiple dietary and metabolic components (e.g., cholesterol, salt, essential fatty acids) is obvious (Crews and Gerber 1994).

Combining thrifty genotypes with antagonistic pleiotropy yields a "thrifty/pleiotropic gene" model (Crews and Gerber 1994). This model provides a theoretical basis that may aid in explaining multiple CDDs and some patterns of senescence in humans. A thrifty/pleiotropic model suggests that some CDDs should occur secondary to excessive accumulations of once scarce but now abundant resources. Such resources likely include calories, fats, protein, cholesterol, and salt, in addition to glucose/energy as described in Neel's original model. The thrifty/pleiotropic model also implies that another group of CDDs should occur secondary to decreased dietary and metabolic availability of what were once plentiful resources. Resources likely to show such an inverse relationship include calcium, fiber, ascorbic acid, and antioxidants. Thrifty genes for extracting or retaining naturally abundant resources would be un-likely to increase under direct selection. However, if physiologically available levels naturally predisposed to a level of toxicity that reduced reproductive success, natural selection would act to reduce extractive and retention mech-anisms. During hominid evolution these compounds likely were abundant in all preagricultural diets. They are, however, less available in today's more cosmopolitan societies subsisting on processed, canned, and artificial food products. Thrifty genotypes, beneficial during youth and maturity, predispose to improved metabolic extraction, endogenous production, or retention/con-servation of once scarce resources. However, their continued efficiency, or thriftiness, during middle-age and postreproductive life leads to detrimental pleiotropic effects that predispose individuals to CDD. For example, many previously abundant resources are now scarce in modern diets due to socio-cultural changes, (e.g., processing of grains, preserving of foods, social desir-ability). These sociocultural processes have contributed to a relative physio-logical scarcity of many once abundant dietary nutrients and have exposed the detrimental consequences of humankind's poorly developed biological ex-traction, conservation, or resorption processes, resulting in conditions such as osteoporosis, bowel pathologies, and cancer.

This thrifty-pleiotropic gene model of CDDs may be applied at two levels. First, specific risk factors for CDDs should occur secondary to the accumula-tion (thrifty) or scarcity/loss (nonthrifty) of specific nutrients, metabolites, stores, or deposits. Second, some CDDs should result from the antagonistic pleiotropy of thrifty genotypes with high selective value during developmental and reproductive years that are associated with debilitation or increases in risk factors for CDD in later decades.

Exploration of the thrifty/pleiotropic model follows directly from Williams' (1957) formulation of pleiotropy and senescence. Most protein variants, whether common enzymatic, structural, transport, or receptor proteins, found in human populations should complete their primary function(s) equally well, whether it be apolipoprotein E variants, the insertion/deletion polymorphism of angiotensin-converting enzyme (ACE), a restriction-fragment-length poly-morphism of atrial naturitic peptide (ANP), or the various forms of collagen. However, any allele at any one of these loci also may have secondary, tertiary, and other pleiotropic effects that are similar to or greatly different from its primary function(s). This is so because individual variants may show second-ary pleiotropic effects that increase risks of related or even seemingly unre-lated diseases. Different alleles may be associated with lower average life ex-pectancies of their carriers, while the different protein products show no variability in completing their primary function. The ACE and apolipoprotein E loci are useful for illustrating such a phenomenon.

ACE comes in two common forms, one with an insertion (I) and one with a deletion (D) of 274 bases; thus, the ACE loci is said to have an insertion/deletion (I/D) polymorphism. In multiple population studies, neither the I nor the D variant is associated consistently and significantly with blood pres-sure (Barley et al., 1991, 1994; Cambien et al., 1992; F.M. Yatsu, personal communication). However, the D variant is associated with coronary heart disease and stroke in some populations and may show age-specific selective pressures in others. The deletion allele is a potent risk factor for myocardial infarction in several European populations (Cambien et al. 1992) and for stroke in African Americans (Yatsu, personal communication). These associations are likely due to some as yet unknown pleiotropic effects or another unknown primary function of ACE (see Barley et al. 1991, 1994).

The apolipoprotein E (Apo E) locus provides a second example. It has three major alleles (Apo **E*2**, **E*3**, **E*4**) that produce three major proteins (apo E2, apo E3, apo E4). Apo E aids in the metabolism, transport, and processing of lipids and LDL-cholesterol. The Apo **E*4** allele is associated with increased risks for coronary heart disease, hypercholesterolemia, and late-onset Alzheimer's disease in U.S. and European samples. In several groups, including the Yanomami, Mayans, Japanese, and Malays from Singapore, the Apo **E*4** allele is associated with lower cholesteral levels, while in most population samples this allele is associated with higher cholesterol levels (reviewed in Crews et al. 1993). Interestingly, apo E4, but not apo E2 or apo E3, shares amino acid sequences with a protein product expressed in brain tissue. In addition, apo E4 does not appear to have as high an affinity for the LDL-receptor protein as do apo E2 and apo E3. These structural differences underlie increased risks for Alzheimer's disease and hyperlipidemia among Apo **E*4** carriers and homozygotes. The Apo **E*4** allele's association with Alzheimer's disease appears to be unrelated to Apo E's primary functions in lipid transport and catabolism. However, Apo **E*4**'s association with hyperlipidemia seems to be directly linked to Apo E's basic functions in lipid metabolism. Thus, Apo **E*4**'s association with a neurodegenerative disease appears to be an untoward pleiotropic effect, while its association with hyperlipidemia appears to be a primary outcome of its basic function.

These examples illustrate one of the difficulties in genetic epidemiology. Genetic variants at a locus commonly do not show any statistically significant association with variation in the particular phenotypic trait that is regarded as the protein's primary function. However, these variants may still be associated with increased risk of other disease outcomes and be associated with decreased longevity in general (see Williams 1957; Hamilton 1966; Crews and Harper 1998). This simply may be one more twist on the concept of pleiotropy as articulated by Williams (1957). ACE may be the most striking example of antagonistic pleiotropy observed to date. Both common allelic variants at the ACE locus, the insertion and deletion alleles, are poorly correlated with blood pressure variation in all ethnic groups examined (Jeunemaitre et al. 1992; Cambien et al. 1992; Barley et al. 1994). However, the ACE deletion allele is strongly associated with stroke in African-Americans (Yatsu, personal communication), myocardial infarction in European samples (Cambien et al. 1992), and appears to be under age-specific selective pressure in American Samoans (Crews and Harper 1998).

Chronic Degenerative Diseases

Definitional Issues

What is a CDD? One definition is that CDDs are conditions that lead to progressive deterioration in one or more clinical, metabolic, or physiological traits that can be measured on a continuous scale (Crews and James 1991). Clinically, diagnoses of chronic degenerative diseases are commonly made when a particular parameter (e.g., blood glucose, serum cholesterol, bone density, blood pressure, T cells, or hematocrit) rise above or fall below a particular cutpoint (see Sing et al. 1985). These critical cut-points often are based on retrospective or prospective data showing that individuals with elevated or depressed values are at greater risk for morbidity and/or mortality (Sing et al. 1985). However, such cut-points often are arbitrary, changing as diagnostic expertise and evaluation improve. This has been particularly true of serum cholesterol, blood glucose, and blood pressure in recent decades (see AHA, 1988).

Desirable serum cholesterol levels for U.S. residents are a particularly good example. In the 1960s and 1970s, when major research on cholesterol levels was first being reported, the U.S. population average cholesterol level was about 240 mg/dl, and this was thought to be a clinically desirable level. Later, as reports showed increased coronary heart disease (CHD) mortality at even lower cholesterol levels, the desirable level dropped to <200 mg/dl in 1993. In addition, more attention is now given to the level of high-density lipoprotein (HDL), with the desirable level being 35 mg/dl or greater (NCEP Panel 1993). The same has happened with blood pressure cut-points for diagnosing hypertension. Previously, cut-points for clinical hypertension were 160 mm Hg for systolic and 95 mm for diastolic hypertension (Rowland and Robert 1982). Today these cut-points are set at 140 mm Hg and 90 mm Hg, respectively (JNC V 1993). Some reports even use both sets of cut-points, and we predict that as population mean blood pressure declines, even lower cut-points will be established.

Some CDDs are more insidious in their onset and, at present, are only diagnosable when their symptoms lead to clinically recognizable pathology (Crews and James 1991; Crews and Gerber 1994). Examples include many neoplasms, slow virus infections, human immunodeficiency viruses, and the autoimmune diseases. These conditions do not necessarily show progressive and continuous decline to some arbitrary critical value, although they often do show progressive deterioration. The majority of noninfectious CDDs result from a complex, but often poorly understood, series of steps that begin at some unknown point in an individual's life and then progress over the remainder of their life span. The most commonly recognized CDDs include the leading causes of death among adults living in cosmopolitan societies of the twentieth century: CHD, neoplastic diseases, cerebrovascular diseases (CVD), neurological syndromes (e.g., Alzheimer's and Huntington's), and NIDDM. In addition, chronic infectious conditions (e.g., Kuru, human immunodeficiency viruses) represent

CDDs, as do osteoporosis, rheumatoid arthritis, multiple sclerosis, amyotrophic lateral sclerosis, and chronic obstructive pulmonary disease. Neither age of onset nor eventual mortality provide necessary or sufficient criteria for determining what conditions are CDDs (Crews and Gerber 1994). Rather, any disease, condition, or syndrome that continues over an extended period of time and is associated with decreasing physiological function over that period may be labeled a CDD.

Risk Factors and Chronic Degenerative Diseases

As noted in the previous section, numerous conditions may be included as CDDs. Of greatest importance today are those that lead morbidity and mortality in cosmopolitan societies and their associated risk factors: hyperlipidemia, hypertension, hyperinsulinemia, obesity, nitrates, tobacco smoke, chemical exposures, fat-rich/fiber-poor diets, alcohol consumption, and overnutrition. Exposure to and development of these associated risk factors find their fullest expression in the radically altered ecological and environmental circumstances that much of humankind experiences today relative to what our ancestors experienced. Such risk factors and their associated causes of death may have their genesis in large part in thrifty/pleiotropic genes and gene complexes that developed and were retained as part of our genome during earlier phases of human existence. To fully appreciate this concept, it is necessary to understand earlier phases of human existence.

Physiologically, humans are the product of more than 65 million years of primate evolution and of about 5 million years of evolution and adaptation to a mobile, bipedal, omnivorous ecological niche with a diet composed of mostly high-fiber, low-fat, and calorically limited plant products—seeds, grains, nuts, fruits, roots, and vegetables, along with insects and protein from carrion and, possibly, prey animals (Cohen 1989). Such diets were low in cholesterol, fats, and salt, contained no refined sugars or grains, and did not expose humans to nitrates or overnutrition. Nor were our ancestors exposed to pollution or cigarette smoke. When our fully modern human form (Cro-Magnon) appeared 35,000–40,000 years ago, life expectancy was less than 20 years, few individuals lived past 45 years, and adaptive mechanisms for the retention and conservation of scarce resources (calories, cholesterol, salt, fats) were likely already fully developed. Conversely, adaptive mechanisms for the retention of plentiful resources (calcium, vitamin C, fiber) or the elimination of nonexisting chemicals (polychlorinated biphenyls, nitrates, n-butyl acetate) were never developed. Today, well-developed genetic-based adaptive mechanisms likely underlie numerous risk factors for CDD, including salt-sensitive hypertension, arteriosclerosis, obesity, hyperlipidemia, and longevity, while underdeveloped genetic mechanisms may underlie such conditions as osteoporosis, some types of cancer, and, possibly, multiple chemical sensitivity syndrome.

Many such risk factors and morbid conditions appear together as syndromes (Crews and Gerber 1994; Crews and James 1991). For example, the risk factors hyperinsulinemia, hypertension, hyperlipidemia, and obesity frequently occur together in the same individual along with diabetes, atherosclerosis, and CHD. In recent years researchers have taken to naming such phenomena, for example, syndrome X or the New World syndrome (DeFronzo and Ferrannini 1991; Ferrannini et al. 1991; Reaven 1988; Weiss et al., 1984). These "syndromes" probably represent multiple genetic loci, including loci coding for proteins with direct, secondary, and multilevel effects on individual phenotypes. Thus, the genome most likely encodes multiple thrifty/pleiotropic proteins that predispose individuals to specific or multiple risk factors or that enhance progression of disease.

Thus, in addition to genes or gene complexes for specific diseases, alleles that increase susceptibility to disease through their pleiotropic influence on risk factors are an additional underlying mechanism generating CDDs. The latter may be viewed as of two types; variants that enhance and variants that decrease long-term survivability. For lack of more precise terms, we may describe variants that increase individual survivability as enhancing "vitality," and those that decrease survivability as increasing "frailty" (see also Vaupel et al. 1979; Manton and Stallard 1984; Weiss 1989). As an example, any variants at a locus or at multiple loci that predispose to low through moderate blood pressure response to stress may be advantageous in cosmopolitan societies, and thereby increase vitality. Those predisposing to a rapid pressor response may increase risk for CHD or CVD, and thereby increase frailty. Variants at the locus (or loci) in question need not have a direct or even secondary effect on blood pressure. Rather, the effect may be indirect through multiple pathways. This phenomenon also may be illustrated using the ACE and Apo E loci.

Recall that, although the ACE I/D genotype does not explain significant variation in blood pressure, the ACE-D allele is associated with higher disease and mortality levels. This likely represents some pleiotropic effect. For one, protein products of both ACE alleles are equally capable of completing their primary function, "conversion of the largely inactive, decapeptide angiotensin I, after it is cleaved by renin, to the active octapeptide angiotensin I" (Erdös and Skidgel 1987: 345). However, ACE functions in a wide variety of additional pathways, cleaving a variety of peptides; for example, ACE inactivates bradykinin by the release of a C-terminal dipeptide (Erdös and Skidgel 1987). Thus, association of the ACE insertion/deletion polymorphism with stroke, myocardial infarction, and longevity in different samples may have little to do with its relationship to angiotensin. Such associations may be secondary to inactivation of bradykinin by ACE, or it may be due to some other pleiotropic influence of ACE on an alternative peptide or pathway.

Apo E*4's association with late-onset Alzheimer's disease provides a second example. The Apo E*4 allele differs from the E*2 and E*3 alleles in its DNA base structure, such that apo E4 includes a sequence of amino acids identical to that of a protein expressed in brain tissue. Due to this sequence similarity, over time, apo E4 molecules become irreversibly cross-linked as part of the neurological tangles characteristic of Alzheimer's disease. Apparently, this problem has little relationship to apo E4's primary functions of lipid transport and interaction with the LDL receptor. Rather, it seems to be a pleiotropic effect of apo E4's protein structure, not shared with either apo E2 or apo E3.

Why discuss molecular aspects of ACE and Apo E? First, most peptides, such as amylin, insulin, pyruvate kinase, collagen, and ANP, likely have multiple functions in a variety of organs and tissues. Primary functions of such proteins often are difficult to identify and need not be related to the first function identified (see Erdös and Skidgel 1987 for a discussion of the multitude of activities currently ascribed to ACE). Such pleiotropic activities of molecules may lead to associations with one or more CDDs. Second, protein polymorphisms often affect physiological and metabolic variation in multiple systems (as do all those listed previously), thereby decreasing overall vitality. Cohort differences in allele frequencies at a particular locus can be used to guage such untoward pleiotropic effects. Among American Samoans, genotypes at the ACE and ANP loci are in Hardy-Weinberg equilibrium (Crews and Harper, 1998). However, in age-specific analyses, younger Samoans (at or below the median age of 41 years), show a higher frequency of ANP heterozygotes than do older Samoans, while older Samoans show a higher frequency of ACE heterozygosity than do younger Samoans (Crews and Harper 1998). Although neither polymorphism is closely related to blood pressure levels, age-related selection could be occurring at both loci. Susceptibility alleles for CDDs may not always produce proteins that show a major difference in completing their primary function. Rather, proteins somewhat removed from direct maintenance of functional capacity and homeostasis in the particular system may be major contributors to CDD processes. For example, insulin and insulin receptor protein variants may have little effect on circulating glucose levels and diabetes because each completes its primary function adequately. Insulin receptor substrate-I, amylin, or glucokinase, proteins not directly responsible for maintenance of glucose levels may be more closely associated with diabetes and circulating glucose levels because of their less direct associations with glucose homeostasis (figure 18.1).

Changing Environments, Lifestyles, and Morbidity

Clinical Practice and Evolutionary Biology

Blood Pressure and Hypertension

Crews and Gerber (1994) suggested that salt-sensitive hypertension in blacks may be an example of a thrifty/pleiotropic genotype. Several questions were identified for future research: (1) Do African Americans show a greater frequency of salt-sensitive hypertension than do European Americans? (2) On a controlled low-salt diet, do African Americans show greater sodium retention than European Americans?

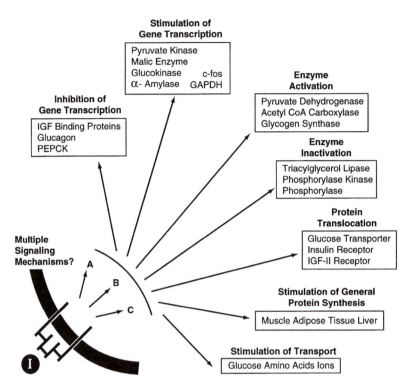

Figure 18.1. Actions of insulin. The binding of insulin (I) to its receptor generates one or more intracellular signals (A, B, C), which stimulate various intracellular processes. (Reprinted with permission from Granner and O'Brien 1992.)

In response to the first question, there is a good deal of evidence to suggest that African Americans with hypertension are more sensitive to changes in dietary sodium chloride than are European Americans (Weinberger et al. 1986; Sullivan and Ratts 1988; JNC V 1993; Pecker 1995). Additionally, there is now evidence that sodium sensitivity can be identified at young ages and that, among young adults, a significantly higher prevalence of sodium sensitivity occurs in blacks than in whites (Falkner and Kushner 1990). Other research has shown that on average, blacks have lower plasma renin levels than whites, a condition that further predisposes individuals to salt sensitivity (Sever et al. 1979; James et al. 1986; Luft et al. 1991; Pecker 1995).

Another line of evidence suggests that African Americans exhibit greater sodium retention than do European Americans. Wassertheil-Smoller et al. (1991) report from the Trial of Antihypertensive Interventions and Management (TAIM) that both diet and drug therapy have different effects on hypertension among blacks and whites. In sodium-restricted groups, both black and white participants given the placebo achieved similar reductions in 24-hour sodium excretion. However, in the groups administered chlorthalidone (a diuretic), blacks showed no reduction in sodium excretion, whereas whites showed highly significant reductions of 33.5 mmol and 41.2 mmol, among men and women, respectively. In another study, Luft et al. (1991) report that blacks are more "sluggish in their natriuretic responses after a saline load compared with white persons"; not only was their excretion of sodium delayed following a load, but blacks also excreted more sodium during the night.

Such variation in blood pressure responses to sodium loading and depletion suggest that recommendations for lifestyle modifications and prescribed drug treatments among hypertensives must consider ethnic variability. For example, the Fifth Report of the Joint National Committee on Detection, Evaluation, and Treatment of High Blood Pressure (JNC V 1993) states that blacks are more responsive to diuretics and calcium antagonists than they are to either beta-blockers or ACE inhibitors. The report goes on to state that diuretics should be the therapy of first choice in blacks. Additionally, the JNC V suggests that lifestyle modifications (e.g., reducing dietary sodium) are particularly important among blacks because of their high prevalence of salt sensitivity. The TAIM study (Langford et al. 1991) and Gillum (1979) also have shown that diuretics are more effective at reducing blood pressure in blacks than in whites, while beta-blockers are more effective in whites than in blacks at reducing diastolic blood pressure. Although a low sodium diet, as recommended by the JNC V, was an efficient dietary intervention for reducing cardiovascular disease risk among whites, such an intervention *increased* the risk for blacks in the TAIM study (Wassertheil-Smoller et al. 1991). Thus, this dietary intervention may not be effective in reducing cardiovascular risk for all ethnic groups. As recently cautioned by Taubes (1995), low sodium diets may represent a case where extrapolation of epidemiological data to clinical practice may not be as clear and unambiguous as one would like.

Cholesterol and Coronary Heart Disease

Efforts to reduce morbidity and mortality from CHD have traditionally focused attention on lowering cholesterol concentrations in persons with coronary disease. The question remained whether lowering cholesterol in persons with elevated levels who were free of coronary disease would result in similar benefits.

A Consensus Conference on Lowering Blood Cholesterol to Prevent Heart Disease was convened in 1984 and concluded that lowering elevated blood cholesterol levels would reduce the rate of CHD (Consensus Conference 1985). The conference also stated that although direct intervention studies had not been conducted in women, there was no reason to propose a separate treatment schedule for women. Based on results of studies in a Japanese population, in which migrants to the United States have increased their cholesterol levels as well as their risk of CHD relative to that of nonmigrants, and in the Finnish, whose cholesterol levels and risk of CHD are greater than U.S. citizens, the panel recommended for men, women, and children 2 years and older a reduction in calories obtained from fat from the current 40% to 30% of total daily calories.

The Second Report of the National Cholesterol Education Program (NCEP) Expert Panel (1993) further highlighted CHD risk status as a guide to type of therapy, suggested that more attention be given to high-density lipoprotein as a CHD risk factor, and increased emphasis on physical activity and weight reduction as components of behavioral therapy for high blood cholesterol (NCEP Panel, 1993). Risk status based on presence of CHD risk factors now includes males ⩾45 years old and females ⩾55 years old or who have experienced premature menopause and are not receiving estrogen replacement therapy.

These consensus statements (Consensus Conference 1985; NCEP Panel 1993) made little reference to either race or ethnicity. Is the prevalence of high cholesterol similar across ethnic groups within the United States? Is hypercholesterolemia a risk factor for CHD in all subpopulations, and should the definition of "high" cholesterol be the same for all? Should people of all ethnicities hope to achieve the same mean levels of lipids? Finally, given that the risk factor status associated with elevated cholesterol levels is similar in all subpopulations (this is by omission, for which we have little evidence), should cholesterol-lowering regimens be the same for all? Should everyone try diet modification first? (Clearly, neither all ethnic groups nor all individuals of any ethnicity who have high mean levels of cholesterol eat high-fat diets).

Trials on lowering cholesterol, either by diet modification or by medication, rarely examine results by ethnicity or race. Furthermore, many of the major trials do not include nonwhite subjects. In fact, neither of the two most recent trials, the Scandinavian Simvastatin Survival Study (4S; 1994, 1995) and the West of Scotland Coronary Prevention Study (Shepherd et al. 1995), provided a demographic breakdown nor an examination of ethnicity or race.

These two trials, however, demonstrated positive results from lowering cholesterol by medication. In the 4S study, lowering total cholesterol was found to decrease both total mortality and mortality due to coronary disease in patients with coronary disease. The West of Scotland Coronary Prevention Study further reported a reduction in nonfatal heart attacks and deaths from coronary disease in men on medication with moderate hypercholesterolemia and no history of myocardial infarction.

Results obtained in earlier studies were mixed. A number of investigators recommended targeting only those individuals most at risk because some evidence suggested that total mortality was not reduced and that the effects of medication and/or cholesterol lowering may even increase noncardiac mortality (Muldoon et al. 1990; Oliver 1992). A recent meta-analysis of six primary prevention trials (on only male participants) found that although the treated groups had fewer deaths due to CHD, they had excess deaths from suicides, homicides, and accidents in both dietary and drug studies (Muldoon et al. 1990). Such consistent results across multiple studies and populations do not bode well for those who insist on cholesterol reduction as the therapy of choice for all patients. Additionally, a recent analysis of controlled trials found the relation between individual cholesterol changes and outcomes to be nonsystematic or absent (Ravnskov, 1992). This study also demonstrated the tendency of reports and reviews to preferentially cite supportive results, leading Ravnskov (1992) to conclude that claims that lowering serum cholesterol reduces either total or CHD mortality are based on the preferential citation of supportive trials. Not only are supportive studies cited more often; they are also published more often (Ravnskov, 1992). Recently, Easterbrook et al. (1991) reported that 44% of studies with null results remained unpublished, whereas only 15% of studies with significant positive results are unpublished. Lack of publication of failure to find expected or significant results weakens the ability of clinicians to provide appropriate counseling and interventions to their patients.

In addition to concerns that all-cause mortality does not decline along with coronary mortality, the association of low total cholesterol levels with noncoronary mortality, especially cancer mortality, has been suggested (D'Agostino et al. 1995). A U-shaped relationship along with excess cancer mortality and increased risk of noncardiovascular, noncancer mortality has been reported in persons with total cholesterol levels below 160 mg/dl (figure 18.2; Jacobs et al. 1992; Neaton et al. 1992; D'Agostino et al. 1995). This excess risk of cancer in those with cholesterol levels below 160 mg/dl may be due to cigarette smoking. In a 32-year follow-up of the Framingham Study, the association of low serum cholesterol with cancer mortality was observed in male smokers but not in nonsmokers (D'Agostino et al. 1995). The Framingham Study included only white males aged 35–69 years at baseline.

Thus, most of our knowledge of the relationship between cholesterol and CHD, along with the risks and benefits of lowering cholesterol to prevent (either primary or secondary) CHD, stem from observations and trials predominantly on white males. Whether genetic predisposition as identified by race/ ethnicity, sex, or genotype influences which cholesterol-lowering regimen to follow or, more to the point, whether cholesterol is an important risk factor for CHD in all groups, has been demonstrated less effectively. Clearly, dietary habits and lifestyle influence to some extent mean cholesterol levels of an individual. Whether different cholesterol extraction, resorption, or conservation mechanisms occur in different populations has not been adequately explored. Available data suggest caution in generalizing that lowering cholesterol by any means reduces CHD and mortality risk in all individuals.

Pharmacogenetics

The area of pharmacogenetics, a term first used by Vogel in 1959 (see Kalow 1990), is the study of genetically determined variations in response to drugs (Stedman, 1976). Because drug metabolism has a significant hereditary component, the interaction of genetic predisposition with environment and social practices has much relevance and applicability to evolutionary medicine.

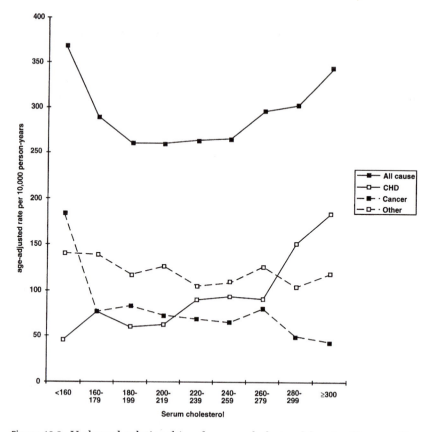

Figure 18.2. U-shaped relationship of serum cholesterol level with mortality. (Reprinted with permission from D'Agostino et al. 1995.)

Ethnic differences in response to different classes of antihypertensive medications have already been addressed (see Freis 1986). Whites are more responsive to beta-adrenergic blocking agents and ACE inhibitors than are blacks. In contrast to their poor response to beta blockers and ACE inhibitors, blacks are more responsive to diuretics (Freis 1986). This may be due to the greater tendency among blacks, compared to whites, for hypertension to be sodium and volume related (Laragh 1973; Freis 1986).

A study conducted among men of Chinese descent found an even more pronounced blood pressure response to propranolol, a beta blocker, than seen among whites. Chinese participants had significantly greater reductions of both heart rate and blood pressure. The plasma propranolol concentration required to produce a 20% reduction in heart rate was significantly lower (37–38%) in Chinese than in the white participants in both supine and upright positions. Chinese subjects also had a 10-fold greater sensitivity to the hypotensive effects of propranolol as compared to white subjects; to achieve a 10%

reduction in mean blood pressure required a concentration of propranolol 10 times higher in white than in Chinese subjects (Zhou et al. 1989).

Primaquine provides another example. This drug, first administered during the course of large-scale cooperative antimalarial research during World War II, was found to cause "primaquine hemolysis" predominantly in black American soldiers stationed in Southeast Asia (Kalow 1991). In some black servicemen, primaquine was toxic and led to anemia due to intravascular hemolysis. A search for the explanation of this phenomenon led to the discovery of glucose-6-phosphate-dehydrogenase (G6PD) deficiency, an X-linked genetic abnormality (McKusick 1983), which is most frequent among descendants of and populations that have been exposed to falciparum malaria. Glucose-6-phosphate-dehydrogenase deficiency confers some protection against childhood malaria, and, as a result, individuals heterozygous for this polymorphism seem to have a selective advantage. As primaquine is the prototype of more than 50 drugs and other substances, G6PD deficiency also predisposes affected individuals (mostly males) to hemolytic crises from exposure to antimalarials, sulfonamides, antipyretics, analgesics, vitamin K analogs, and fava beans (Gilman et al. 1985).

Debrisoquine hydroxylase offers another example of a polymorphic drug-metabolizing enzyme with multiple genetic variants. Frequencies of debrisoquine hydroxylase deficiency vary greatly in different ethnic groups (Eichelbaum and Gross 1990; Kalow 1991). Individuals with this deficiency are cautioned to avoid antiarrhythmic drugs such as perhexiline. Quinidine and neuroleptics also produce debrisoquine hydroxylase inhibition in certain individuals.

Genetic variation in red blood cell pyruvate kinase (PK) is a leading cause of hereditary nonspherocytic hemolytic anemia; PK-deficiency is also the most frequent enzymopathy (disease or condition associated with a defective enzyme) found in the human glycolytic pathway (Baronciani and Beutler 1995). Pyruvate kinase-deficient hemolytic anemia is inherited as an autosomal recessive condition, and more than 59 mutations of the human PK gene have been reported (Rouger et al. 1978; Baronciani and Beutler 1995). Pyruvate kinase deficiency is not commonly viewed as a pharmacogenetic polymorphism in humans because it does not directly affect the metabolism of pharmacological agents. However, a number of drugs and chemicals, including oral contraceptives (Kendell 1977) and aspirin (Glader 1976), have been observed to precipitate episodes of hemolytic anemia in PK-deficient patients. Similarly, in animal models, when the livers of rats are exposed to aspirin, decreased PK activity is observed (Yanagei et al., 1985). In vitro, when PK-deficient erythrocytes are cloned, 2,3-*bis*-phosphoglyceric acid buildup leads to an inhibition of hexose monophosphate shunt activity (Tomoda et al. 1983). This inhibition would be expected to increase susceptibility to any agent that induces oxidative stress, including many of the drugs that lead to hemolytic anemia in G6PD-deficient patients. Wilcox (1996) reported a possible association of PK deficiency with multiple chemical sensitivity syndrome. It appears that many of the chemicals that cause problems for multiple chemical sensitivity patients either are themselves PK inhibitors or may induce PK inhibitors.

Genetic Counseling

With the molecular revolution, genetic counseling has become a growing medical specialty. How patients are counseled depends a great deal on what we know about genetic diseases. Most well known are the single locus phenotypes—phenylketonuria, alkaptonuria, cystic fibrosis, adenosine deaminase deficiency, cystinuria, Tay Sachs, and such. The vast majority of genetic counseling effort is directed toward such single locus conditions. However, given what is known of the genetic and evolutionary bases of CDDs, physicians need to begin counseling their patients regarding their possible genetic risk.

One simple appraisal of coronary risk that has long been in use includes the question, "Did your mother or father die of cardiovascular disease before age 50?" Such simple information, acknowledged by the patient, may do more to influence lifestyle risk factors than all the educational materials currently available. Predispositions to almost all known CDDs are segregating in families and kindreds (Sing et al. 1985; Armelagos 1991; Crews and James 1991). Many lay people have an intuitive grasp of this phenomenon. It is necessary that health educators and clinicians use their patients' knowledge to guide them toward a fuller understanding of their genetic predispositions and susceptibilities to CCDs. The lay public, and patients in particular, must see that "predisposed" and "susceptible" are not synonymous with "preordained," as many believe. Persons with a genetic predisposition to lung cancer may not express this outcome in a nonpolluted and smoke-free environment; those with a genetic susceptibility to NIDDM may not show elevated glucose or obesity in a low-calorie, high-fiber dietary environment.

It is incumbent on health care workers and clinicians to know and understand their patients' ethnic, cultural, and familial backgrounds. Is this patient a member of an ethnic group or kindred at particularly high risk for a condition featuring a strong genetic predisposition (e.g., familial hypercholesterolemia, early-onset Alzheimer's, premature stroke)? Does the patient's ethnic–cultural background expose him or her to risk factors for specific CDDs for which genetic susceptibilities are known (e.g., overnutrition and diabetes among South Pacific Islanders and Native Americans, polycyclic hydrocarbons and amines and lung cancer among many traditional-living peoples, consumption of fermented soy products and digestive cancers among Asians and Asian Americans)? Understanding by clinicians of possible ethnic and cultural risks for specific CDDs with known genetic predispositions, along with provision of clear information on these risks to patients, provides a basis for genetic counseling. It is not necessary to genotype each individual for each specific risk locus; rather, attention to familial health histories along with knowledge of the ethnic and cultural backgrounds of patients will go a long way toward providing health promotion in relationship to the CDDs.

Discussion

This chapter has provided an overview of the evolutionary basis for current patterns of CDDs. A major focus was how and why genetic predispositions toward multiple CDDs may have been retained in humanity's still evolving gene pool. Two specific mechanisms were cited as being influential in producing today's epidemic of CDDs. First is retention of thrifty genes and genotypes in humanity's gene pool over evolutionary time due to enhancement of early development and attainment of reproductive maturity; however, these genotypes may also come with detrimental influences on subsequent longevity and increase the risk of CDDs through their pleiotropic effects. Second is the lack of similar thrifty genes and genotypes for maintaining physiologically adequate levels of resources that previously were abundant in human diets but are now scarce or not included in current diets. Both mechanisms are components of the thrifty/pleiotropic model of CDDs articulated here and elsewhere (Crews and Gerber 1994).

Other aspects of this model have been addressed. Neel (1962, 1982) first developed the thrifty genotype model; Williams (1957) was the first to explain the link between pleiotropy and senescence. Eaton and colleagues (Eaton and Konner 1985; Eaton et al. 1988; Eaton and Nelson 1991) have examined nutrition and CDDs from an evolutionary perspective. Their examinations of Paleolithic nutrition (Eaton and Konner 1985) and changing calcium intakes from prehistoric to modern times (Eaton and Nelson 1991) rely heavily on an evolutionary perspective and easily may be viewed as supportive of the thrifty/pleiotropic model. Calcium, a component of most wild and unprocessed grains and foods, would have been widely available in most prehistoric diets. Today, due to an increased reliance on processed foods and artificial substances (pesticides, cyclamates, preservatives), calcium is now less abundant in diets of cosmopolitan cultures; in fact, calcium in the diet is so rare that many Americans ingest calcium-containing tablets.

Weder and Schork (1994) have argued that essential hypertension in some individuals results from allometric relationships among general somatic growth (particularly height), growth of the kidney, kidney function, and renal homeostasis. They hypothesize that "the renal defect underlying hypertension is a direct consequence of the accelerated growth and development that characterize modern life" (Weder and Schork, 1994:151). Thus, they suggest that the "thrifty genotype" as described by Neel (1962, 1982) for NIDDM does not fit the available data on essential hypertension. Rather, Weder and Schork (1994:154) suggest that "natural selection could lead to the preservation of genes promoting hypertension without invoking the action of novel alleles or genetic disequilibrium." This view is suggestive of the thrifty/pleiotropic model as proposed by Crews and Gerber (1994). Loci that contribute to or limit kidney growth and function may have some, as yet unknown, pleiotropic effect that acts to maintain renal homeostasis by increasing blood pressure. We agree that it is unlikely that any novel alleles to limit blood pressure's allometric increase with body size during growth and development would have been retained during previous phases of human evolution through natural selection, because the high-calorie, rapid growth environment experienced by children in modern cosmopolitan societies has never before been experienced by humankind. In today's cosmopolitan environments, essential hypertension may increase in response to the need for increased renal flow unfettered by specific genetic limiters, thereby resulting in adult hypertension and CHD as an allometric (pleiotropic?) response to rapid and sustained growth. This also may help explain why blood pressure is often related to adult stature. This is the case in the Samoan, African American, Native American, and European American samples we have studied.

The epidemiology of CDDs holds great promise for collaborative investigations among health care professionals and evolutionarily oriented biomedical researchers. The evolutionary model is based on the concept of human variation and natural selection. Risk factors for CDDs are not likely to have the same influence in all ethnic and cultural groups. Salt intake is less of a risk factor for hypertension in European Americans than in African Americans. Interventions to reduce risk factors may need to be applied differentially across populations. In general, diuretics control blood pressure better in African Americans than do beta blockers, whereas dietary interventions to reduce cholesterol and CHD seem to work better among white than nonwhite populations. Each ethnic group or population has to some degree adapted to a different environment. These local adaptive responses may pattern and structure human variation between populations in such a fashion as to make them susceptible to different risk factors and disease processes.

Pharmacogenetics is one area where these patterns of population variation have become most obvious. Population differences in blood pressure reductions in response to various drugs have been detailed here. These hereditary differences make extrapolation of drug trial results across ethnic boundaries extremely hazardous. It cannot be assumed that what works for European American males in a clinical trial will work for non-European Americans, women, children, or the aged. However, this assumption is the basis for much of our knowledge on pharmaceutical intervention on risk factors for CDDs. Dosage and frequency of dosages for most pharmaceuticals are determined for samples of white males, although such drugs also are prescribed for women and children, young and old, and persons of varied ethnic background. As evolutionary biologists, we must examine why more drugs do not show variable activities across ethnic groups. Perhaps they do so in more subtle ways than does primaquine, for instance, leading to differences that are not clinically apparent.

Last, we discussed the need for greater genetic counseling with respect to the CDDs. Multiple loci contribute to risk of CDDs. All of these loci are not likely to be elucidated any time soon. However, frailty with respect to certain CDDs (early-onset Alzheimer's, breast cancer, diabetes, cardiovascular disease) is known to be segregating in particular families and kindreds. Knowledge of such family background is a valuable tool in lifestyle counseling and primary prevention. Clinicians need to include family and lifestyle assessments as part of their initial patient evaluation and to integrate this knowledge with preventive and intervention protocols. Primary prevention will do more to reduce disease burdens and loss of productivity than all the secondary and tertiary interventions currently available. Unfortunately, primary prevention is not a large part of clinical training, leaving clinicians largely on their own in this area. A better understanding of the genetic and evolutionary bases for CDDs should provide additional information clinicians need to evaluate, treat, and counsel their patients regarding the major lifestyle diseases of the twenty-first century.

Chronic degenerative diseases will remain a burden to the human species for the foreseeable future. To better understand the etiology, persistence, and prevalence of CDDs in modern human populations, it is necessary to view today's patients as the products of 5 million years of hominid evolution. During this evolutionary journey, humankind's gene pool was shaped by then-prevailing environmental pressures; pressures that may have little, if any, relationship to the current ecological settings of humans in contemporary cosmopolitan populations. Instead of environmental pressures, humankind's elaboration of culture over the last 10,000 years or so has greatly altered our ecological circumstances.

Today, rather than an expectation of life of only 15 or 18 years, as was characteristic of our earliest ancestors, we can expect to live 75 years and hope for 100. Rather than having high mortality among infants and children aged 5 years and less, today more than 95% of all children born in the United States survive to their fifth birthday. This increased life expectancy has been the result of sociocultural and behavioral changes (public health, hygiene, antibiotics) and development rather than any obvious biological or genetic change in humans. The most rapid increase in life expectancy followed on the heels of the industrial revolution and the development of modern scientific and medical models. These processes have led to life expectancies of 45 years in the nineteenth century, 55 years in the early twentieth century, and more than 75 years in some contemporary societies as we enter the twenty-first century.

As life expectancies have increased over the past two centuries, humankind's innate susceptibilities to various chronic debilitating conditions have been exposed. These innate susceptibilities appear to be the legacy of our evolutionary past and reflect our own evolutionary inertia. Many such susceptibilities may be traced to our phylogenetic background as mobile, omnivorous primates; others likely reflect on our 5 million years of evolution as bipedal hominids, and still others likely reflect our dependency on culture and culturally elaborated social systems for continued survival and reproduction. It is our species' ability to culturally manipulate the environment that has allowed us to improve general survival of infants and children, such that during the twentieth century most individuals in cosmopolitan societies survive to the age of reproduction and many survive sufficiently long to be grandparents and great grandparents.

Today, we must examine our species' evolutionary history to understand patterns of CDDs, their associated risk factors, and population differences in response to both lifestyle and pharmacological interventions. Populations that have over the long term reproduced in areas with an overabundance of sodium chloride in dietary staples may have experienced strong selective pressures against genetically based sodium retention mechanisms; conversely, populations from regions with low availability of sodium chloride likely experienced strong selective pressures against genetically based sodium-eliminating mechanisms. Today, in places like the United States and Europe where salt is a readily available and a highly desired food flavoring, we can predict that the latter would be at greater risk of salt-sensitive hypertension than the former (see Wilson 1986; Jackson 1991). In hindsight, it also is clear why the latter might respond more to a diuretic that in some sense would counteract the effects of salt retention than would the former without innate salt-retaining mechanisms. Similar models also apply to other dietary factors, such as cholesterol, calories, fats, sugars, and alcohol.

An examination of human evolutionary biology suggests that clinicians need to become more familiar with their patients' ethnic, cultural, and family histories to aid in both the prevention of CDDs and in counseling and intervention activities for CDDs. Major CDDs are multifactorial in nature; they result from a complex interplay of genes, environment, and lifestyle, rather than from some single cause such as a virus or bacterium. As we enter the twenty-first century, it is imperative that all individuals concerned about the health and well-being of humankind understand not only our place in nature, but also how nature and our adaptations to it have contributed to our current patterns of health and disease.

References

Arking, R. (1991) *Biology of Aging: Observations and Principles.* Englewood Cliffs, NJ: Prentice Hall.

AHA (American Heart Association) (1988) The Joint National Committee on Detection, Evaluation, and Treatment of High Blood Pressure, The 1988 Report. *Archives of Internal Medicine* 148:1020–38.

Armelagos, G.J. (1991) Human evolution and the evolution of disease. *Ethnicity & Disease* 1:21–25

Baker, P.T. (1984) The adaptive limits of human populations. *Man* (ns) 19: 1–14.

Baker, P.T., and Baker, T.S. (1977) Biological adaptations to urbanization and industrialization: some research strategy considerations. In R.K. Wetherington, ed., *Colloquia in Anthropology*, vol. I, pp. 107–18. Fort Burgwin Research Center, Taos, New Mexico.

Barley, J., Blackwood, A., Carter, N.D., Crews, D.E., Cruickshank, J.K., Jeffery, S., Ogunlesi, A.O., and Sagnella, G.A. (1994) Angiotensin converting enzyme insertion/deletion polymorphism: association with ethnic origin. *Journal of Hypertension* 12:955–957.

Barley, J., Carter, N.D., Cruickshank, J.K., Jeffery, S., Charlett, A., and Webb, D. (1991) Renin and atrial natriuretic peptide restriction fragment length polymorphisms: association with ethnicity and blood pressure. *Journal of Hypertension* 9:993–996.

Baronciani, L., and Beutler, E. (1995) Molecular study of pyruvate kinase deficient patients with hereditary non-spherocytic hemolytic anemia. *Journal of Clinical Investigations* 95:1702–1709.

Cambien, F., Poirier, O., Lecerf, L., Evans, A., Cambou, J.P., Arveiler, D., Luc, G., Bard, J.M., Bara, L., Ricard, S., Tiret, L., Amouyei Falhenc-Gelas, P., and Soubrier, F. (1992) Deletion polymorphism in the gene for angiotensin-converting enzyme is a potent risk factor for myocardial infarction. *Nature* 359:641–644.

Cohen, M.N. (1989) *Health and the Rise of Civilization.* New Haven, CT: Yale University Press.

Coon, C.S. (1965) *The Living Races of Man.* New York: Alfred A. Knopf.

Consensus Conference. (1985) Lowering blood cholesterol to prevent heart disease. *Journal of the American Medical Association* 253:2080–2086.

Crews, D.E. (1993) Biological anthropology and human aging: some current directions in aging research. *Annual Reviews of Anthropology* 22:395–423.

Crews, D. (1990) Anthropological issues in biological gerontology. In R.C. Rubinstein, ed., *Anthropology and Aging: Comprehensive Reviews.* Norwell, MA: Kluwer Academic Publishers, pp. 11–38.

Crews, D.E., and Bindon, J.R. (1990) Biocultural responses to modernization: lessons from Samoa. *16th School of Biological Anthropology,* pp. 103–115.

Crews, D.E., and Gerber, L. (1994) Chronic degenerative diseases and aging. In D.E. Crews and R. Garruto, eds., *Biological Anthropology and Aging,* pp. 154–181. New York: Oxford University Press.

Crews, D.E., and Harper, G.J. (1998)

Renin, ANP, ACE polymorphisms, blood pressure, and age in American Samoans. *American Journal of Human Biology* 10(4):439–449.

Crews, D.E., and James, G.D. (1991) Human evolution and the genetic epidemiology of chronic degenerative diseases. In C.G.N. Mascie-Taylor, G.W. Lasker, eds., *Applications of Biological Anthropology to Human Affairs,* pp. 185–201. New York: Cambridge University Press.

Crews, D.E., Kanboh, M.I., Bindon, J.R., and Ferrell, R.E. (1991) Genetic studies of human apolipoproteins XVIII: population genetics of apolipoprotein polymorphisms in American Samoa. *American Journal of Physical Anthropology* 84:165–170.

D'Agostino, R.B., Belanger, A.J., Kannel, W.B., and Higgins, M. (1995) Role of smoking in the u-shaped relation of cholesterol to mortality in men. The Framingham Study. *American Journal of Epidemiology* 141:822–827.

De Fronzo, RA, Ferrannini, E. (1991) Insulin resistance. A multifaceted syndrome responsible for NIDDM, obesity, hypertension, dyslipidemia, and atherosclerotic cardiovascular disease. *Diabetes Care* 14:173–194.

Easterbrook, P.J., Berlin, J.A., Gopalan, R., and Matthews, D.R. (1991) Publication bias in clinical research. *Lancet* 1:867–872.

Eaton, S.B., and Konner, M.J. (1985) Paleolithic nutrition: a consideration of its nature and current implications. *New England Journal of Medicine* 312:283–289.

Eaton, S.B., Konner, M.J., and Shostak, M. (1988) Stone agers in the fast lane: chronic degenerative diseases in evolutionary perspective. *American Journal of Medicine* 84:739–749.

Eaton, S.B., and Nelson, D.A. (1991) Calcium in evolutionary perspective. *American Journal of Clinical Nutrition* 54 (suppl):2815–2875.

Eichelbaum, M., and Gross, A.S. (1990) The genetic polymorphism of debrisoquine/sparteine metabolism—clinical aspects. *Pharmacology and Therapy* 46:377–394.

Erdös, E.S. and Skidgel, R.A. (1987) The angiotensin I-converting enzyme. *Laboratory Investigation* 56(4): 345–348.

Falkner, B., and Kushner, H. (1990) Effect of chronic sodium loading on cardiovascular response in young blacks and whites. *Hypertension* 15: 36–43.

Fenner F. (1970) The effects of changing social organisation on the infectious diseases of man. In S.V. Boyden, ed. *The Impact of Civilization on the Biology of Man.* Toronto: University of Toronto Press, pp. 48–68.

Ferrannini E., Haffner, S.M., Mitchell, B.D., and Stern, M.P. (1991) Hyperinsulinemia: the key feature of a cardiovascular and metabolic syndrome. *Diabetologia* 34:416–422.

Freis, E.D. (1986) Antihypertensive agents. In W. Kalow, H.W. Goedde, and D.P. Agarwal, eds., *Ethnic Differences in Reactions to Drugs and Xenobiotics*, pp. 313–322. New York: Alan R. Liss.

Gillum, R.F. (1979) Pathophysiology of hypertension in blacks and whites. *Hypertension* 1:468–475.

Gilman, A.G., Goodman, L.S., Rall, T.W., and Murad, F. (1985) *Goodman and Gilman's The Pharmacological Basis of Therapeutics.* New York: Macmillan.

Glader, B.E. (1976) Salicylate-induced injury of pyruvate-kinase-deficient erythrocytes. *New England Journal of Medicine* 294(17):916–918.

Granner, D.K., and O'Brien, R.M. (1992) Molecular physiology and genetics of NIDDM. *Diabetes Care* 15(3):369–395.

Hamilton, W.D. (1966) The molding of senescence by natural selection. *Journal of Theoretical Biology* 12:12–45.

Harrison, G.A. (1973) The effects of modern living. *Journal of Biosocial Science* 5:217–228.

Hinkle, L.E. (1987) Stress and disease: the concept after 50 years. *Social Science and Medicine* 25:561–566.

Jacobs, D., Blackburn, H., Higgins, M., et al. (1992) Report of the conference on low blood cholesterol: mortality associations. *Circulation* 86:1046–1060.

Jackson, F.L. (1991) An evolutionary perspective on salt, hypertension, and human genetic variability. *Hypertension* 17 (suppl.):129–132.

James, G.D., Sealey, J.E., Mueller, F., Alderman, M., Madhavan, S., and Laragh, J.H. (1986) Renin relationship to sex, race and age in a normotensive population. *Journal of Hypertension* 4 (suppl 5):S387–S389.

James, G.D., and Baker, P. (1990) Human population biology and hypertension: evolutionary and ecological aspects of blood pressure. In J.H. Laragh and B.M. Brenner, eds., *Hypertension: Pathophysiology, Diagnosis, and Management*, pp. 137–145. New York: Raven Press.

James, G.D., Crews, D.E., and Pearson, J. (1989) Catecholamine and stress. In M.A. Little and J.D. Haas, eds., *Human Population Biology: A Transdisciplinary Science*, pp. 280–295. New York: Oxford University Press.

Jeunemaitre X., Lifton R.P., Hunt S.C., Williams, R.R., and Lalouel J.M. (1992) Absence of linkage between the angiotensin converting enzyme locus and human essential hypertension. *Nature and Genetics* 1:72–75.

JNC V (1993) The Fifth Report of the Joint National Committee on Detection, Evaluation, and Treatment of High Blood Pressure (JNC V). *Ar-*

chives of Internal Medicine 153:154–183.

Kalow, W. (1990) Pharmacogenetics: past and future. *Life Sciences* 47: 1385–1397.

Kalow, W. (1991) Interethnic variation of drug metabolism. *Trends in Pharmacological Sciences* 12:102–107.

Kendall, A.G. (1977) Red cell pyruvate kinase deficiency: adverse effect of oral contraceptives. *Acta Haematologica* 57(2):116–120.

Langford, H.G., Davis, B.R., Blaufox, M.D., Oberman, A., Wassertheil-Smoller, S., Hawkins, C.M., and Zimbaldi, N. for the TAIM Group (1991) Effect of drug and diet treatment of mild hypertension on diastolic blood pressure. *Hypertension* 17:210–217.

Laragh, J.H. (1973) Vasoconstriction-volume analysis for understanding and treating hypertension. The use of renin and aldosterone profiles. *American Journal of Medicine* 55: 261–274.

Levi, L., and Anderson, L. (1975) *Psychosocial Stress.* New York: Spectrum Publications.

Luft, F.C., Miller, J.Z., Grim, C.E., Fineberg, N.S., Christian, J.C., Daugherty, S.A., and Weinberger, M.H. (1991) Salt sensitivity and resistance of blood pressure. Age and race as factors in physiological responses. *Hypertension* 17 (suppl I):I102–I108.

Manton, K.G., and Stallard, E. (1984) *Recent Trends in Mortality Analysis.* Academic Press: New York.

Marks, J. (1994) *Human Biodiversity: Genes, Race, and History.* New York: Aldine de Gruyter.

McGarvey, S.T., Bindon, J.R., Crews, D.E., and Schendel, D.E. (1989) Modernization and adiposity: causes and consequences. In M.A. Little and J.D. Hass, eds., *Human Population Biology: A Transdisciplinary Science,* pp. 263–279. New York: Oxford University Press.

McKusick, V.A. (1983) *Mendelian Inheritance in Man.* Baltimore, MD: Johns Hopkins University.

Molnar, S. (1994) *Human Variation: Races, Types, and Ethnic Groups.* Englewood Cliffs, NJ: Prentice Hall.

Muldoon, M.F., Manuck, S.B., and Matthews, K.A. (1990) Lowering cholesterol concentrations and mortality: a quantitative review of primary prevention trials. *British Medical Journal* 301:309–314.

NCEP Panel (1993) Expert Panel on Detection and Treatment of High Blood Cholesterol in Adults. Summary of the second report of the National Cholesterol Education Program (NCEP) Expert Panel on Detection, Evaluation, and Treatment of High Blood Cholesterol in Adults (Adult Treatment Panel II). *Journal of the American Medical Association* 269:3015–3023.

Neaton, J.D., Blackburn, H., Jacobs, D., et al. (1992) Serum cholesterol level and mortality findings for men screened in the Multiple Risk Factor Intervention Trial. *Archives of Internal Medicine* 152:1490–1500.

Neel, J.V. (1962) Diabetes mellitus: a "thrifty" genotype rendered detrimental by progress? *American Journal of Human Genetics* 14:353–362.

Neel, J.V. (1982) The thrifty genotype revisited. In J. Köbberling and R. Tattersall, eds., *The Genetics of Diabetes Mellitus,.* London: Academic Press.

Oliver, M.F. (1992) Doubts about preventing coronary heart disease. Multiple interventions in middle aged men may do more harm than good. *British Medical Journal* 304:393–394.

Pecker, M. (1995) Salt sensitivity in hypertensive patients: pathogenesis, identification, and treatment. In J.H. Laragh and B.M. Brenner, eds., *Hypertension: Pathophysiology, Diagnosis, and Management,* 2nd ed., pp. 2543–2555. New York: Raven Press.

Ravnskov, U. (1992) Cholesterol lowering trials in coronary heart disease: frequency of citation and outcome. *British Medical Journal* 305:15–19.

Reaven, G.M. (1988). Banting Lecture. Role of insulin resistance in human disease. *Diabetes* 37:1595–1607.

Rose, M.R. (1991) Evolutionary Biology of Aging. New York: Oxford University Press.

Rouger, H., Bernard, J.F., Torres, M., Schlegel, N., Amar, M., Lopez, M., and Boivin, P. (1978) Five unknown mutations in the LR pyruvate kinase gene associated with severe hereditary nonspherocytic haemolytic anemia in France. *British Journal of Haematology* 92(4):825–830.

Rowland, R., and Roberts, J. (1982) Blood pressure levels and hypertension in persons ages 6–74 years: United States, 1976–1980. In *Vital Health Statistics*, No. 84. Department of Health and Human Services Publication 82–1250. Washington, DC: Government Printing Office.

Scandinavian Simvastatin Survival Study Group. (1994) Randomized trial of cholesterol lowering in 4444 patients with coronary heart disease: the Scandinavian Simvastatin Survival Study Group (4S). *Lancet* 1994; 344:1383–1389.

Scandinavian Simvastatin Survival Study Group. (1995) Baseline serum cholesterol and treatment effect in the Scandinavian Simvastatin Survival Study (4S). *Lancet* 345:1274–1275.

Schell, L.M. (1991) Pollution and human growth: lead, noise, polychlorobiphenyl compounds and toxic wastes. In C.G.N. Mascie-Taylor and G.W. Lasker, eds., *Applications of Biological Anthropology in Human Affairs*. New York: Cambridge University Press pp. 83–116.

Sever, P.S., Peart, W.S., Meade, T.W.,

Davies, I.B., and Gordon, D. (1979) Ethnic differences in blood pressure with observations on noradrenaline and renin. 1. A working population. *Clinical and Experimental Hypertension* 1:733–744.

Shepherd, J., Cobbe, S.M., Ford, I., et al. (1995) Prevention of coronary heart disease with pravastatin in men with hypercholesterolemia. *New England Journal of Medicine* 333:1301–1307.

Sing, C.F., Boerwinkle, E., and Moll, P.P. (1985) Strategies for elucidating the phenotypic and genetic heterogeneity of a chronic disease with a complex etiology. In R. Chakraborty and E.J.E. Szathmary, eds., *Diseases of Complex Etiology in Small Populations*, pp. 39–66. New York: Alan R. Liss.

Stedman, T.L. (1976) *Stedman's Medical Dictionary*, 24th ed, p. 1065. Baltimore, MD: Williams and Wilkins.

Stini, W.A. (1991) The biology of human aging. In C.G.N. Mascie-Taylor and G.W. Lasker, eds. *Applications of Biological Anthropology to Human Affairs*, pp. 207–209. New York: Cambridge University Press.

Sullivan, J.M., and Ratts, T.E. (1988) Sodium sensitivity in human subjects: hemodynamic and hormonal correlates. *Hypertension* 11:717–723.

Taubes, G. (1995) Epidemiology faces its limits. *Science* 269:164–169.

Tomoda, A., Lachant, N.A., Noble, N.A., and Tanaka, K.R. (1983) Inhibition of the pentose phosphate shunt by 2,3-diphosphoglycerate in erythrocyte pyruvate kinase deficiency. *British Journal of Haematology* 54:475–484.

Vaupel, J.W., Manton, K.G., and Stallard, E. (1979) The impact of heterogeneity in individual frailty on the dynamics of mortality. *Demography* 16:439–454.

Wassertheil-Smoller, S., Davis, B.R., Oberman, A., Blaufox, M.D., Langford, H., Wylie-Rosett, J., Hawkins, M., and Zimbaldi, N. (1991) The TAIM Study: sex-race differences in effects of diet and drugs on cardiovascular risk. *Cardiovascular Risk Factors* 1:3–11.

Weder, A.B., and Schork, N.J. (1994) Adaptation, allometry, and hypertension. *Hypertension* 24(2):145–156.

Weinberger, M.H., Miller, J.Z., Luft, F.C., Grim, C.E., and Fineberg, N.S. (1986) Definitions and characteristics of sodium sensitivity and blood pressure resistance. *Hypertension* 8 (suppl II):I127–I134.

Weiss, K.M. (1981) Evolutionary perspectives on human aging. In P.T. Amoss and S. Harrell eds., *Other Ways of Growing Old: Anthropological Perspectives*. Stanford, California: Stanford University Press, pp. 25–58.

Weiss, K.M. (1984) On the number of members of Genus Homo who have ever lived, and some evolutionary interpretations. *Human Biology* 56: 637–649.

Weiss, K.M. (1989) A survey of biodemography. *Journal of Quantitative Anthropology* 1:79–151.

Weiss, K.M. (1989) Are known chronic diseases related to the human lifespan and its evolution? *American Journal of Human Biology* 1:307–319.

Weiss, K.M., Ferrell, R.E., and Hanis, C.L. (1984) A New World "Syndrome" of metabolic diseases with a genetic and evolutionary basis. *Yearbook of Physical Anthropology* 27:153–178.

Wilcox, P.P. (1996) Inherited and acquired enzyme defects in multiple chemical sensitivity. In Salem H, Olajos EJ, Katz, S.A., Rhodes, J.E., Cowan, H.L., and Randall, C.J. (compilers), *Proceedings of the Symposium on Alternatives in the Assessment of Toxicity: Theory and Practice* (May, 1994), Technical Report #ERDEC-SP-033, U.S. Army Edgewood Research, Development, and Engineering Center, pp. 487–503.

Williams, G.C. (1957) Pleiotropy, natural selection, and the evolution of senescence *Evolution* 11:398–411.

Wilson, T.W. (1986) Africa, Afro-Americans, and hypertension: an hypothesis. *Social Science History* 10:489–500.

Yanagi, S., Sakamoto, M., Takahashi, S., Hasuike, A., Konishi, Y., Kumazawa, K., and Nakatano, T. (1985) Enhancement of hepatocarcinogenesis by sorbitan fatty acid ester, a liver pyruvate kinase activity-reducing substance. *Journal of the National Cancer Institute* 75(2):381–384.

Zhou, H-H., Koshakji, R.P., Silberstein, D.J., Wilkinson, G.P., and Wood, A.J.J. (1989) Racial differences in drug response: altered sensitivity to and clearance of propranolol in men of Chinese descent as compared with American whites. *New England Journal of Medicine* 320:565–570.

INDEX